Exploring Autodesk Revit 2017 for Structure

Includes Application of Robot Structural Analysis Professional 2017

(7th Edition)

CADCIM Technologies
525 St. Andrews Drive
Schererville, IN 46375, USA
(www.cadcim.com)

Contributing Author
Sham Tickoo
Professor
Purdue University Northwest
Hammond, Indiana, USA

CADCIM Technologies

Exploring Autodesk Revit 2017 for Structure, 7th Edition
Sham Tickoo

CADCIM Technologies
525 St Andrews Drive
Schererville, Indiana 46375, USA
www.cadcim.com

ISBN 978-1-942689-35-5

www.cadcim.com

DEDICATION

To teachers, who make it possible to disseminate knowledge to enlighten the young and curious minds of our future generations

To students, who are dedicated to learning new technologies and making the world a better place to live in

SPECIAL RECOGNITION

A special thanks to Mr. Denis Cadu and the ADN team of Autodesk Inc. for their valuable support and professional guidance to procure the software for writing this textbook

THANKS

To employees of CADCIM Technologies and Tickoo Institute of Emerging Technologies (TIET) for their valuable help

Online Training Program Offered by CADCIM Technologies

CADCIM Technologies provides effective and affordable virtual online training on various software packages including Computer Aided Design, Manufacturing and Engineering (CAD/CAM/CAE), computer programming languages, animation, architecture, and GIS. The training is delivered 'live' via Internet at any time, any place, and at any pace to individuals as well as the students of colleges, universities, and CAD/CAM/CAE training centers. The main features of this program are:

Training for Students and Companies in a Classroom Setting

Highly experienced instructors and qualified Engineers at CADCIM Technologies conduct the classes under the guidance of Prof. Sham Tickoo of Purdue University Northwest, USA. This team has authored several textbooks that are rated "one of the best" in their categories and are used in various colleges, universities, and training centers in North America, Europe, and in other parts of the world.

Training for Individuals

CADCIM Technologies with its cost effective and time saving initiative strives to deliver the training in the comfort of your home or work place, thereby relieving you from the hassles of traveling to training centers.

Training Offered on Software Packages

CADCIM Technologies provide basic and advanced training on the following software packages:

CAD/CAM/CAE: *CATIA, Pro/ENGINEER Wildfire, SolidWorks, Autodesk Inventor, Solid Edge, NX, AutoCAD, AutoCAD LT, Customizing AutoCAD, AutoCAD Electrical, EdgeCAM, and ANSYS*

Architecture and GIS: *Autodesk Revit (Architecture/Structure/MEP), AutoCAD Civil 3D, AutoCAD Map 3D, Autodesk Navisworks, Bentley Staad.Pro, Oracle Primavera P6, MS Project, and ArcGIS*

Animation and Styling: *Autodesk 3ds Max, 3ds Max Design, Autodesk Maya, Autodesk Alias, Pixologic ZBrush, and CINEMA 4D*

Computer Programming: *C++, VB.NET, Oracle, AJAX, and Java*

For more information, please visit the following link:
http://www.cadcim.com

> **Note**
> If you are a faculty member, you can register by clicking on the following link to access the teaching resources: ***http://www.cadcim.com/Registration.aspx***. The student resources are available at ***http://www.cadcim.com***. We also provide **Live Virtual Online Training** on various software packages. For more information, write us at ***sales@cadcim.com***.

Table of Contents

Chapter 2: Getting Started with a Structural Project

Chapter 3: Setting up a Structural Project

Chapter 4: Structural Columns and Walls

Chapter 5: Foundations, Beams, Floors, and Open Web Joists

Chapter 6: Editing Tools

Chapter 8: Standard Views, Details, and Schedules

Chapter 9: 3D Views, Sheets, Analysis, Reinforcements, and Massing

Chapter 10: Linking Revit Model with Robot Structural Analysis

Preface

Autodesk Revit 2017 for Structure

Autodesk Revit is a Building Information Modeling software developed by Autodesk. The structural tools in Revit are primarily used in structural engineering and they help the users to design structural models using both parametric 3D modeling and 2D drafting elements. The Revit building information model comprises of a physical representation of a building which is fully associated with analytical representation. The building information model in Revit can be used for structural designing, production of drawings, coordination of the project, and also for third party structural analysis application.

The **Exploring Autodesk Revit 2017 for Structure** textbook explains the concepts and principles of Revit Structure through practical examples, tutorials, and exercises. This enables the users to harness the power of BIM with Autodesk Revit 2017 for Structure for their specific use. In this textbook, the author emphasizes on physical modeling, analytical modeling, rebar modeling, and quantity scheduling. Also, this textbook covers the various stages involved in analyzing the model in Robot Structural Analysis software. This textbook is specially meant for professionals and students in structural engineering, civil engineering, and allied fields in the building industry.

In this textbook, special emphasis has been laid on the concepts of structural modeling, which have been explained using relevant graphical examples and illustrations. The accompanying tutorials and exercises, which relate to the real world projects, helps you understand the usage and abilities of the tools available in Autodesk Revit for Structure. Along with the main text, the chapters have been punctuated with tips and notes to make various concepts clear, thereby enabling you to create your own innovative projects.

As you go through this textbook, you will work on tutorials and exercises that can be used to build a complete project. Each of these tutorials and exercises, though complete in themselves, will be a step toward accomplishing the larger projects.

The main features of this textbook are as follows:

- **Project-based Approach**
  The author has adopted a project-based approach and the learn-by-doing theme throughout the textbook. This approach guides the users through the process of creating the designs given in the tutorials.

- **Real-World Designs as Projects**
  The author has used real-world building designs and architectural examples as projects in this textbook so that the users can correlate them to the real-time designs.

- **Tips and Notes**
  Additional information related to various topics is provided to the users in the form of tips and notes.

- **Learning Objectives**
  The first page of every chapter summarizes the topics that are covered in that chapter.

- **Self-Evaluation Test, Review Questions, and Exercises**
  The chapter ends with Self-Evaluation Test so that the users can assess their knowledge of the chapter. The answers to the Self-Evaluation Test are given at the end of the chapter. Also, the Review Questions and Exercises are given at the end of the chapters and they can be used by the Instructors as test questions and exercises.

- **Heavily Illustrated Text**
  The text in this book is heavily illustrated with about 200 line diagrams and screen capture images.

Symbols Used in the Textbook

Note

The author has provided additional information to the users about the topic being discussed in the form of notes.

Tip

Special information and techniques are provided in the form of tips that help in increasing the efficiency of the users.

New

This symbol indicates that the command or tool being discussed is new in this release.

Enhanced

This symbol indicates that the command or tool being discussed has been enhanced in this release.

- Names of the files saved are italicized. Example: *c03_Industrial-Complex_tut2.rvt*
- The methods of invoking a tool/option from the ribbon, Application Menu, or the shortcut keys are given in a shaded box.

Ribbon: Architecture > Build > Column drop-down > Structural Column
Application Menu: New
Shortcut Keys: CTRL+N

Formatting Conventions Used in the Textbook

Please refer to the following list for the formatting conventions used in this textbook.

- Names of tools, buttons, options, browser, palette, panels, and tabs are written in boldface. Example: The **Beam** tool, the **Modify** button, the **Structure** panel, the **Home** tab, the **Properties** palette, **Project Browser**, and so on.
- Names of dialog boxes, drop-downs, drop-down lists, list boxes, areas, edit boxes, check boxes, and radio buttons are written in boldface. Example: The **Options** dialog box, the **Column** drop-down in the **Structure** panel of the **Structure** tab, the **Name** edit box in the **Name** dialog box, the **Chain** check box in the **Options Bar**, and so on.
- Values entered in edit boxes are written in boldface. Example: Enter **4'** in the **Offset** edit box.
- When you select an element or a component, a contextual tab is displayed depending upon the entity selected. In this textbook, this contextual tab is referred to as **Modify | (Elements / Components)**

Ribbon: Modify | (Elements / Components) > Modify > Move
Shortcut Keys: MV

Naming Conventions Used in the Textbook

Tool

If you click on an item in a panel of the ribbon and a command is invoked to create/edit an object or perform some action, then that item is termed as **tool**.

For example:
Beam tool, **Structural Column** tool, **Isolated** tool
Filled Region tool, **Trim/Extend to Corner** tool, **Rotate** tool

If you click on an item in a panel of the ribbon and a dialog box is invoked wherein you can set the properties to create/edit an object, then that item is also termed as **tool**, refer to Figure 1.

For example:
Load Family tool, **Beam** tool, **Wall** tool
Structural Wall tool, **Visibility/Graphics** tool

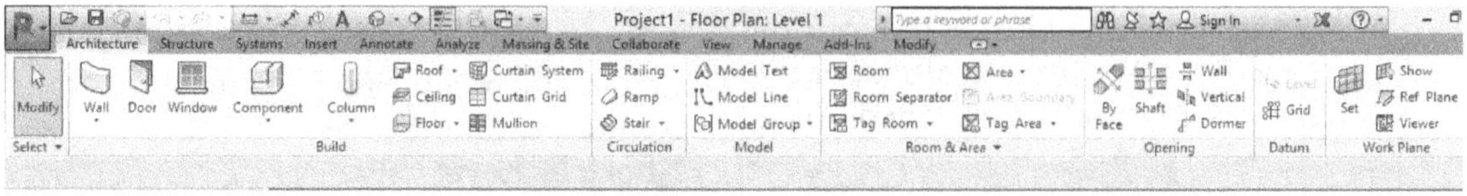

Figure 1 Tools in the ribbon

Button

The item in a dialog box that has a 3d shape like a button is termed as **button**. For example, **OK** button, **Cancel** button, **Apply** button, and so on. If the item in a ribbon is used to exit a tool or a mode, it is also termed as button. For example, **Modify** button, **Finish Edit Mode** button, **Cancel Edit Mode** button, and so on; refer to Figure 2.

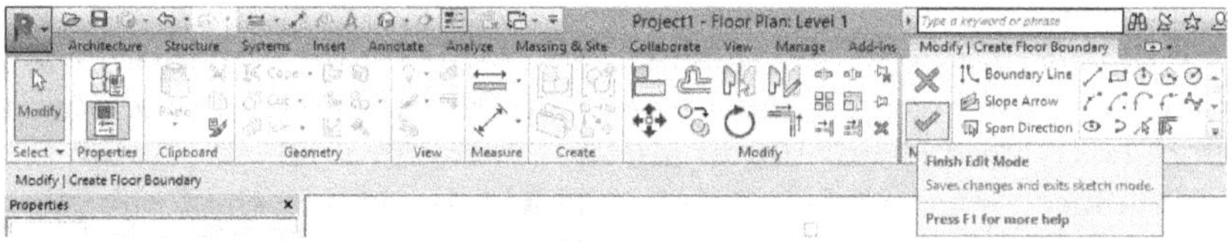

*Figure 2 Choosing the **Finish Edit Mode** button*

Dialog Box

In this textbook, different terms are used for referring to the components of a dialog box. Refer to Figure 3 for the terminology used.

Drop-down

A drop-down is one in which a set of common tools are grouped together for creating an object or performing some action. You can identify a drop-down with a down arrow on it. These drop-downs are given a name based on the tools grouped in them. For example, **Wall** drop-down, **Component** drop-down, **Region** drop-down, and so on; refer to Figure 4.

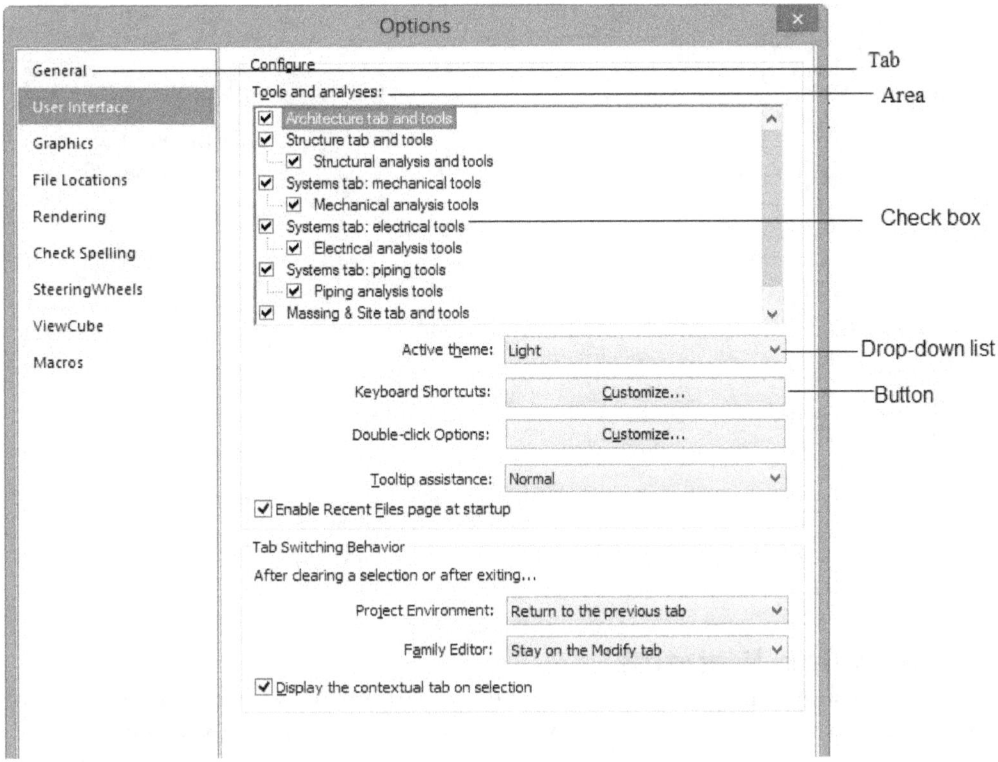

Figure 3 The partial view of components of a dialog box

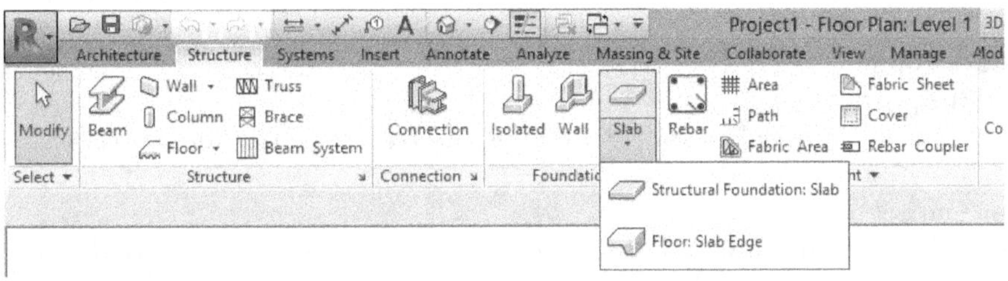

Figure 4 Choosing a tool from the drop-down

Drop-down List

A drop-down list is the one in which a set of options are grouped together. You can set various parameters using these options. You can identify a drop-down list with a down arrow on it. For example, **Type Selector** drop-down list, **Units** drop-down list, and so on; refer to Figure 5.

Options

Options are the items that are available in shortcut menus, drop-down lists, dialog boxes, drop-down lists, and so on. For example, choose the **Zoom In Region** option from the shortcut menu displayed on right-clicking in the drawing area; refer to Figure 6.

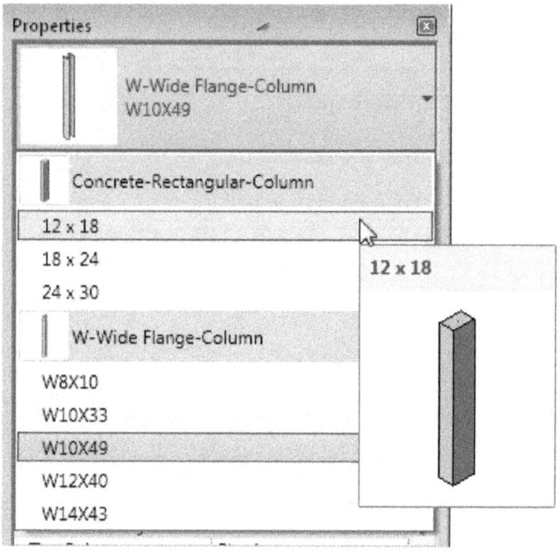

*Figure 5 Selecting an option from the **Type Selector** drop-down list*

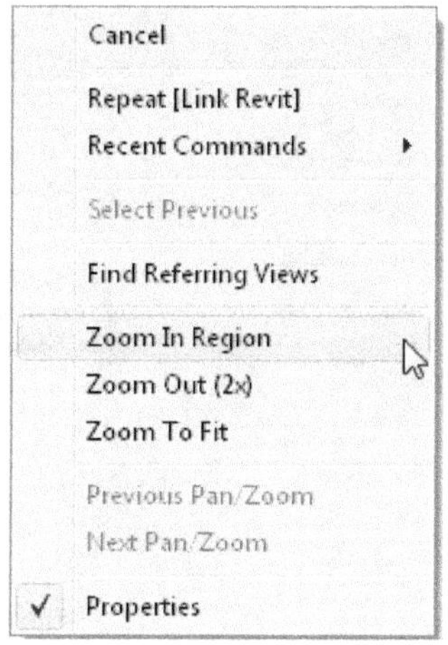

Figure 6 Choosing an option from the shortcut menu

Free Companion Website

It has been our constant endeavor to provide you the best textbooks and services at affordable price. In this endeavor, we have come out with a Free Companion website that will facilitate the process of teaching and learning of Autodesk Revit 2017 for Structure. If you purchase this textbook, you will get access to the files on the Companion website.

The resources available for the faculty and students in this website are as follows:

Faculty Resources

- **Technical Support**
  You can get online technical support by contacting ***techsupport@cadcim.com***.

- **Instructor Guide**
  Solutions to all review questions and exercises in the textbook are provided in this link to help the faculty members test the skills of the students.

- **PowerPoint Presentations**
  The contents of the book are arranged in PowerPoint slides that can be used by the faculty for their lectures.

- **Revit Files**
  The Revit files used in tutorials and exercises are available for free download.

Student Resources

- **Technical Support**
  You can get online technical support by contacting ***techsupport@cadcim.com***.

- **Revit Files**
  The Revit files (*.rvt*) used in tutorials and examples are available for free download.

- **Learning Resources**
  Additional learning resources at *http://revitxperts.blogspot.com* and *http://youtube.com/cadcimtech*.

If you face any problem in accessing these files, please contact the publisher at ***sales@cadcim.com*** or the author at ***stickoo@pnw.edu*** or ***tickoo525@gmail.com***.

Stay Connected

You can now stay connected with us through Facebook and Twitter to get the latest information about our textbooks, videos, and teaching/learning resources. To get such updates, follow us on Facebook (***www.facebook.com/cadcim***) and Twitter (***@cadcimtech***). You can also subscribe to our YouTube channel (***www.youtube.com/cadcimtech***) to get the information about our latest video tutorials.

Chapter 1

Introduction to Autodesk Revit 2017 for Structure

Learning Objectives

After completing this chapter, you will be able to:

- *Understand the basic concepts and principles of Revit 2017*
- *Understand various terms used in Revit*
- *Describe the parametric behavior of Revit*
- *Start the Revit 2017 program*
- *Understand the linking of Analytical Model to Analysis Software*
- *Understand the interface of Revit*
- *Explain the concept of Worksharing using Revit Server*
- *Access the Revit 2017 Help*
- *Work with BIM 360*

INTRODUCTION TO Autodesk Revit for Structure

Welcome to the realm of Autodesk Revit, a powerful software for structural engineering that provides purpose-built tools for structural design, engineering, and analysis.

Autodesk Revit software provides the Engineers and Designers with tools for structural analysis and design. Revit is a BIM software that helps the users in a project to coordinate the documentation of structural design with other disciplines of engineering. Its integrated parametric modeling technology is used to create the information model of a project and to collect and coordinate information across all its representations. In Autodesk Revit, drawing sheets, 2D views, 3D views, and schedules directly represent the same building information model (BIM). In this software, the physical model is associated with an analytical model. As a result, the model created in Revit is ready to be analyzed in a compatible structural analysis software, such as Autodesk Robot Structural Analysis. Using its parametric change engine, you can modify a design at any stage of its creation. The change in the project is automatically made and represented in all its views resulting in the development of better designs along with an improved coordination. The use of Autodesk Revit provides a competitive advantage and a higher profitability to structural engineers and building industry professionals.

Autodesk Revit AS A BUILDING INFORMATION MODELER

The history of computer aided design and documentation dates back to the early 1980s when architects and engineers began using this technology for documenting their projects. Realizing its advantages, information sharing capabilities were developed especially to share data with other consultants. This led to the development of object-based CAD systems in the early 1990s. Before the development of these systems, objects such as structural walls, beams, columns, and slabs were stored as a non-graphical data with the assigned graphics. These systems arranged the information logically but were unable to optimize its usage in a building project. Realizing the advantages of the solid modeling tools, the mechanical and manufacturing industry professionals began using the information modeling CAD technology. This technology enabled them to extract data based on the relationship between model elements.

In 1997, a group of mechanical CAD technologists began working on a new software for the building industry. The Building Information Modeling (BIM) provided an alternative approach to building design, construction, and management. This approach, however, required a suitable technology to implement so as to reap its benefits. In such a situation, the use of parametric technology with the Building Information Modeling approach was envisaged as an ideal combination. They developed a software that was suitable for creating building projects. This led to the development of Autodesk Revit.

Autodesk Revit provides powerful tools for structural design and documentation, in which a digital structural model is created using the parametric elements such as structural walls, beams, columns and so on. All building elements have inherent relationship with one another, which can be tracked, managed, and maintained by the computer.

BASIC CONCEPTS AND PRINCIPLES

Autodesk Revit enables you to envisage and develop a structural model with actual 3D parametric structural elements. It provides a new approach to the structural design and the implementation process. It replicates the way structural engineers conceive the structure of a building. The 2D CAD platforms, such as AutoCAD, mostly use lines to represent the elements, as shown in Figure 1-1. However, in Autodesk Revit, you can create the structural model of a building project using 3D elements such as structural floors, columns, beams, and so on, as shown in Figure 1-2.

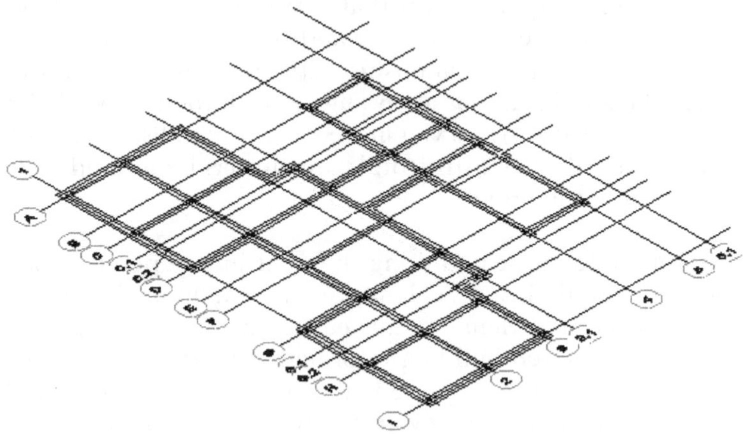

Figure 1-1 *The project created using 2D lines*

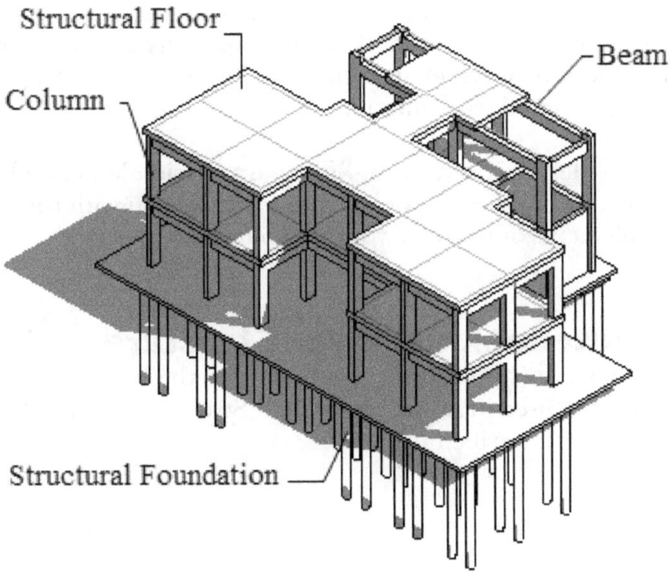

Figure 1-2 *A structural project created using parametric structural elements*

Using these 3D elements, you can visualize the structural project with respect to its scale, volume, and proportions. This enables you to study design alternatives and develop superior quality design solutions. Autodesk Revit automates routine drafting and coordination tasks and assists in reducing errors in documentation. This in turn saves time, improves the speed of documentation, and lowers the cost for users.

Understanding the Parametric Building Modeling Technology

A project in Autodesk Revit is created using the inbuilt parametric building elements. The term 'parametric' refers to the parameters that relate to various building elements. Some relationships are made by Autodesk Revit itself and others by the user. For example, doors, which have an inherent parametric relationship with the structural walls cannot be created without first creating a host wall. A door always moves with the host wall. Similarly, floors too are parametrically linked to walls. When you move structural walls, the structural floor extents are also modified automatically. Each structural element has inbuilt bidirectional associativity with many other elements in the project.

A building information model is created using different interdependent parametric building elements such as structural walls, beams, columns, structural floors, foundations, and so on. As they are bidirectionally associated elements, any change made in one element is automatically adopted by others. The integrated building information model created contains all data for a project. You can then create project presentation views such as structural plans, sections, elevations, and so on for documentation. As you modify the model while working in certain views, Autodesk Revit's parametric change engine automatically updates other views. This capability is, therefore, the underlying concept in Autodesk Revit.

Autodesk Revit's parametric change engine enables you to modify design elements at any stage of the project development. As changes are made immediately and automatically, it saves the time and effort of coordinating them in all other associated views, which, for most projects, is an inevitable part of the design process. Autodesk Revit's capability to coordinate between various aspects of the building design provides immense flexibility in the design and development process along with an error-free documentation.

Autodesk Revit also provides a variety of in-built parametric element libraries that can be selected and used to create a building model. It also provides you with the flexibility to modify the properties of these elements or to create your own parametric elements, based on the project requirement.

Terms Used in Autodesk Revit for Structure

Before using Autodesk Revit, it is important to understand the basic terms used for creating a building model. Various terms used in Autodesk Revit for Structure such as project, level, category, family, type, and instance are described next.

Autodesk Revit Project

A project in Autodesk Revit is similar to an actual structural project. In an actual project, the entire documentation such as drawings, 3D views, specifications, schedules, cost estimates, and so on are inherently linked and read together. Similarly, in Autodesk Revit, a project not only includes the digital 3D building model but also its parametrically associated documentation. Thus, all components such as the building model, its standard views, structural drawings, and schedules combine together to form a complete project. A project file contains all project information such as building elements used in a project, drawing sheets, schedules, cost estimates, 3D views, renderings, and so on. A project file also stores various settings such as environment, lighting, and so on. As data is stored in the same file, so it becomes easier for Autodesk Revit to coordinate the entire database.

Levels in a Building Model

In Autodesk Revit, a building model is divided into various levels. These levels may be understood as infinite horizontal planes that act as hosts for different elements such as roof, floor, ceiling, and so on. The defined levels in a building model can, in most cases, relate to different floor levels, or stories of the building project. Each element that you create belongs to a particular level.

Subdivisions of Elements into Categories and Subcategories

Apart from building elements, Autodesk Revit project also contains other associated elements such as annotations, imported files, links, and so on. These elements have been divided into the following categories:

Model Category : Consists of various structural elements used in creating a building model such as structural walls, structural floors, foundations, beams, braces, and columns.

Annotation Category : Consists of annotations such as dimensions, text notes, tags, symbols, and so on

Datum Category : Consists of datums such as levels, grids, reference planes, and so on

View Category : Consists of interactive project views such as structural floor plans, elevations, sections, 3D views, and renderings

In addition to these four categories, other categories such as **Imported**, **Workset**, **Filter**, and **Revit Categories** can also exist if the project has imported files, enabled worksets, or linked Autodesk Revit projects, respectively.

Families in Autodesk Revit

Another powerful concept in Autodesk Revit is family. A family is described as a set of elements of the same category that can be grouped together based on certain common parameters or characteristics. Elements of the same family may have different properties, but they all have common characteristics. For example, **Concrete-Rectangular-Column** is a concrete column family, but it contains different sizes of columns. Family files have the *.rfa* extension. You can load additional building component families from the libraries provided in Autodesk Revit package.

Families are further divided into certain types. Type or family type, as it is called, is a specific size or style of a family. For example, **Concrete-Rectangular-Column: 12 x 18** is a column type in Imperial system and **M_Concrete-Rectangular-Column: 300 x 450mm** in Metric system. All uses of the same family type in a project have the same properties. Family and family types can also be used to create new families using the **Family Editor**.

Instances are the actual usage of model elements in a building model or annotations in a drawing sheet. A family type created in a new location is identified as an instance of the family type. All instances of the same family type have the same properties. Therefore, when you modify the properties of a family type, the properties of all its instances also get modified. The family categorization of Revit elements is given below:

Model Category	: Column
Family	: **Concrete-Rectangular-Column** in Imperial **M_Concrete-Rectangular-Column** in Metric
Family type	: 12 x 18 (300 x 450mm)
Instance	: Particular usage of a family type

The hierarchy of building elements in Autodesk Revit plays an important role in providing the flexibility and ease of managing a change in a building model. Figure 1-3 shows the hierarchy of categories and families in a typical Autodesk Revit project.

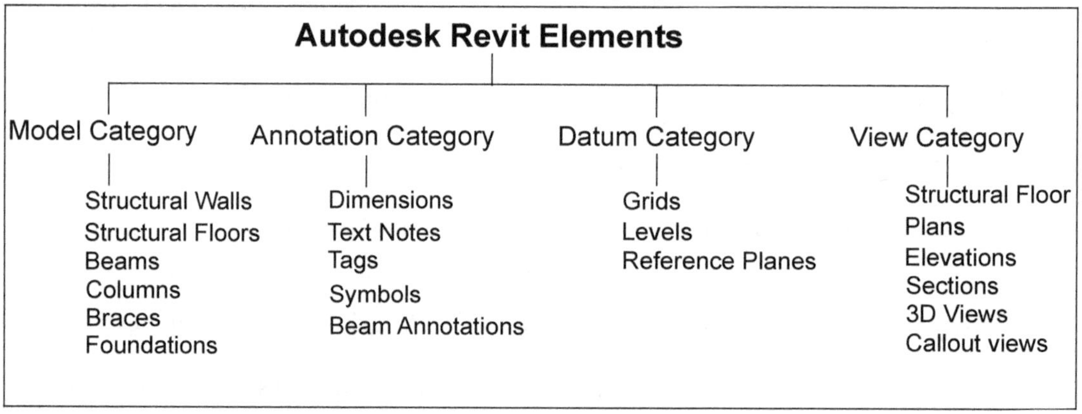

Figure 1-3 *Hierarchy of Autodesk Revit categories and families*

Creating a Structural Model Using Parametric Building Elements

Another classification of categories of elements followed in Autodesk Revit is based on their usage. Autodesk Revit uses five classes of elements: Host, component, annotation, view, and datum. Hosts are the element categories that form the basic structure of a building model and include model elements such as structural walls and floors. Components are the elements that are added to the host elements or act as stand-alone elements such as doors, windows, and foundations. Annotations are the 2D, view-specific elements that add content to the project documentation such as dimensions, tags, text notes, and so on. Views represent various orientations of a building model such as plans, elevations, sections, 3D views, and so on. Datum refers to the reference elements that assist you in creating a building model. The reference elements which include grids, levels, reference planes, and so on.

There is no specific methodology available for creating a building model in Autodesk Revit. It provides you with the flexibility of generating the building geometry based on the project requirement, design complexity, and other factors. However, the following steps describe a general procedure that may be followed for creating an architectural building model using the built-in parametric elements provided in Autodesk Revit.

The first step is to define the levels of the structural model based on the story height of the building and then create grids for inserting columns and foundation at the lowest level. Next, add columns, foundation slab, structural wall, and foundations in that level. You can also link the control height of the structural walls and columns to the levels. Next, create framing members and floors using the defined levels. You can add loads to the model and define load conditions and various analytical settings for the model.

After creating the structural analytical model, you will transfer it to Autodesk Robot Structural Analysis software and analyze the structural entities based on the loads applied to it. After performing the analysis and retrieving the design detail of the model, you will import the structural model from the analysis software into Autodesk Revit. Next, you will create drawing sheets with the desired views for its presentation. You can also add reinforcements to the concrete elements in the structural model. Autodesk Revit also provides tools to create rendered 3D views and walkthroughs. Figure 1-4 shows an example of a building elevation with various structural elements.

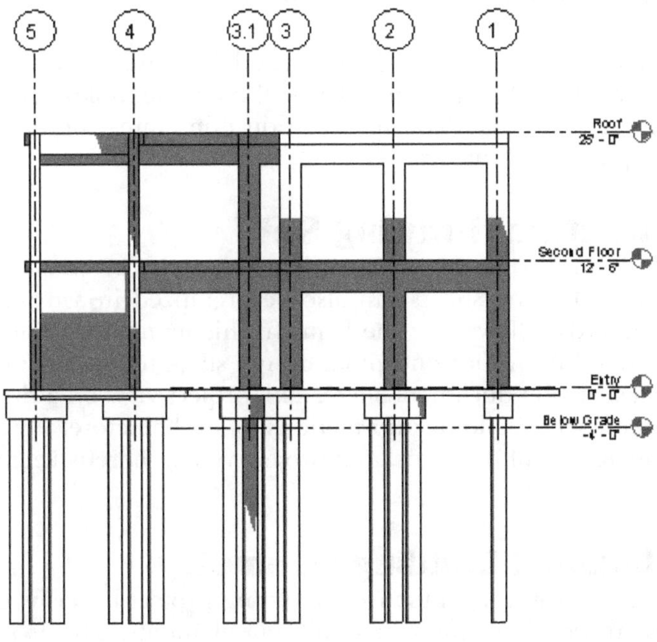

Figure 1-4 *Building section showing building elements and levels*

Visibility/Graphics Overrides, Scale, and Detail Level

Autodesk Revit enables you to control the display and graphic representation of a single element or the element category of various elements in project views by using the visibility and graphics overrides tools. You can select a model category and modify its linetype and detail level. This can also be done for various annotation category elements and imported files. These settings can be done for each project view based on its desired representation. You can also hide an element or an element category in a view using the **Hide in view** and **Isolate** tools. You can override the graphic representation of an element or an element category in any view using the **Visibility/ Graphics** tool.

The scale is an important concept in a Revit project. You can set the scale from the available list of standard scales such as 1/16"=1'0", 1/4"=1'0", 1"=1'0", 1/2"=1'0" (for Imperial) and 1:1, 1:2, 1:3, 1:4, 1:5, 1:10 (for Metric). As you set a scale, Autodesk Revit automatically sets the detail level that is appropriate for it. There are three detail levels provided in an Autodesk Revit project: **Coarse**, **Medium**, and **Fine**. You can also set the detail level manually for each project view. Each detail level has an associated linetype and the detail lines associated with it. The details of annotations such as dimensions, tags, and so on, are also defined by the selected scale.

Extracting Project Information

A single integrated building information is used to create and represent a building project. You can extract project information from a building model and create area schemes, schedule, and cost estimates, and then add them to the project presentation.

Autodesk Revit also enables you to export the extracted database to the industry standard Open Database Connectivity (ODBC) compliant relational database tables. The use of the building information model to extract database information eliminates the error-prone method of measuring building spaces individually.

Creating a Structural Drawing Set

After creating the building model, you can easily arrange the project views by plotting them on the drawing sheets. Drawing sheets can also be organized in a project file based on the established CAD standards followed by the firm. In this manner, the project documentation can easily be transformed from the conceptual design stage to the design development stage and finally to the construction document stage. The project view on a drawing sheet is only a graphical representation of the building information model. Therefore, any modification made in it is immediately updated in all the associated project views, thereby keeping the drawing sets updated.

Creating an Unusual Building Geometry

Autodesk Revit also helps you conceptualize a building project in terms of its volume, shape, and proportions before working with actual building elements. This is possible by using the **Massing** tool, which enables you to create quick 3D models of buildings and conduct volumetric and proportion study on overall masses. It also enables you to visualize and create an unusual building geometry. The same massing model can then be converted into a building model with individual parametric building elements. It provides continuity in the generation of building model right from sketch design to its development.

Flexibility of Creating Special Elements

Autodesk Revit provides a large number of in-built family types of various model elements and annotations. Each parametric element has the associated properties that can be modified based on the project requirement.

Autodesk Revit also enables you to create the elements that are designed specifically for a particular location. The in-built family editor enables you to create new elements using family templates. This provides you with the flexibility of using in-built elements for creating your own elements. For example, using the furniture template, you can create a reception desk that is suitable for a particular location in the design.

Creating Structural Layouts

Autodesk Revit's structural tools enable you to add structural elements to a building model. An extensive in-built library of structural elements has been provided in Autodesk Revit. You can add structural columns, beams, walls, braces, and so on to the project. Thus, structural consultants can also incorporate their elements in the basic architectural building model and check for inconsistency, if any.

Working on Large Projects

In Autodesk Revit, you can work on large projects by linking different building projects together. For a large project that consists of a number of buildings, you can create individual buildings as separate projects and then link all of them into a single base file. The database recognizes the linked projects and includes them in the project representation of the base file.

For example, while working on a large educational institution campus, you can create separate project files for academic building, administration area, gymnasium, cafeteria, computer centre, and so on, and then link them into the base site plan file. In this manner, large projects can be subdivided and worked upon simultaneously.

Working in Large Teams and Coordinating with Consultants

Worksets, in Autodesk Revit, enable the division of the building model into small editable set of elements. The worksets can be assigned to different teams working on the same project and then their work can easily be coordinated in the central file location. The effort required to coordinate, collaborate, and communicate the changes between various worksets is taken care of by computer. Various consultants working on a project can be assigned a workset with a set of editable elements. They can then incorporate their services and modify the associated elements.

For example, a high rise commercial building project can be divided into different worksets with independent teams working on exterior skin, interior walls, building core, toilet details, finishes, and so on. The structural consultants can be assigned the exterior skin and the core workset, in which they can incorporate structural elements. Similarly, the rest of the teams can work independently on different worksets.

STARTING Autodesk Revit 2017

You can start Autodesk Revit by double-clicking on the **Revit 2017** icon on the desktop. Alternatively, you can choose the **Start** button at the lower left corner of the screen (default position) in the taskbar and then choose **All Programs > Autodesk > Revit 2017 > Revit 2017**, as shown in Figure 1-5; the interface will be displayed, as shown in Figure 1-6. Note that this path is used in the Windows 7 operating system.

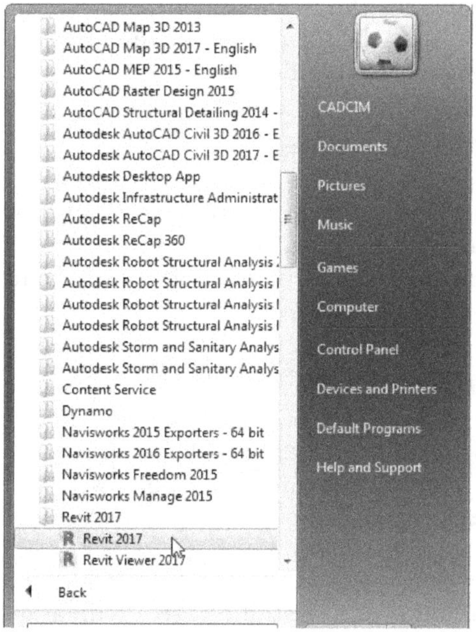

Figure 1-5 Starting Autodesk Revit 2017 using the taskbar

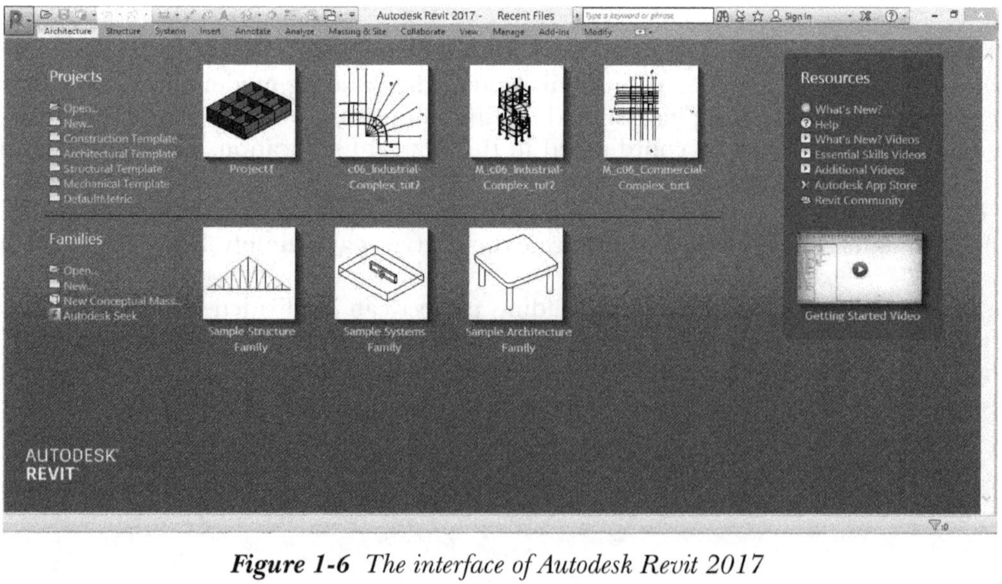

Figure 1-6 The interface of Autodesk Revit 2017

Note
The path for starting Autodesk Revit depends on the operating system being used.

The screen interface has three sections: **Projects**, **Families**, and **Resources**. The options in the **Projects** section are used to open a new or an existing project. In the **Projects** area, you can add some of the frequently used templates such as the the **Construction** template, just below the **New** option. To do so, choose the **Application** button and then choose the **Options** button from the flyout displayed. On doing so, the **Options** dialog box will be displayed. In this dialog box, choose the **File Locations** tab and then choose the **Add Vaue** button; the **Browse for Template File** dialog box will be displayed.In this dialog box, select the required template from the appropriate location and choose the **Open** button; the template will be added under the **Name** column and will also be available under the **Projects** area. Similarly, options in the **Families** section are used to open a new or an existing family. You can also invoke the Conceptual Mass environment from this section to create a conceptual mass model. If you choose the **Autodesk Seek** option from the **Families** section, you will be directed to the *http://seek.autodesk.com/localeTaxBrowse.htm?category=en_us:adsk:revit-arch&locale=en-us&globaldd=globaldropdown.option.b* link and the **Autodesk® Revit Structure Web Library - US Edition** page will open. From this page, you can download various components for your project.

In the **Resources** section, you can choose the **What's New?** option to get information about the new tools and features in Revit 2017. In addition, you can choose the **Help** option from the **Resources** section to get help on various tools. When you choose this option, you will be directed to ***http://help.autodesk.com/view/RVT/2017/ENU/***. Also, the **Autodesk Revit 2017** page with the **Welcome to Revit 2017 Learning** area will be displayed. To access information related to additions and enhancements in Revit 2017 release, you can expand the **What's New** node from the left pane and then click on the **New in Revit 2017** link. On doing so, the **New in Revit 2017** page will be displayed with various links. You learn about various enhancements in Revit 2017 by accessing these links.

The **What's New? Videos** option in the **Resources** section is an enhancement in Revit 2017. When you click on this option, you are directed to the **Autodesk Revit 2017** page that has a list of videos of newly added features in Revit. You can click on a link to view the corresponding video.

You can choose the **Essential Skills Videos** option to view the videos related to basic and advance concepts in Revit 2017. These videos and their associated information help you to learn about the complete software. Moreover, you can choose the **Autodesk App Store** option to access various add-ons that can be used to enhance the productivity of Revit. On choosing this option, the **AUTODESK APP STORE** page will be displayed. In this page, various links are available as add-ons which can be used in Revit applications. In the **Revit Community** option of the **Resources** section, you can access information related to various communities and their contribution in the form of articles, tutorials, and videos.

In the **Projects** section, choose the **Open** option; the **Open** dialog box will be displayed. Browse to the desired location in the dialog box and select the file. Now, choose the **Open** button to open the file.

To open a new project file, choose the **New** option from the **Projects** section. Alternatively, choose **New > Project** from the **Application Menu**; the **New Project** dialog box will be displayed. In this dialog box, make sure that the **Project** radio button is selected, and then choose the **OK** button; a new project file will open and the interface screen is activated.

To create a new project template file, choose the **New** option from the **Projects** section. The **New Project** dialog box will be displayed. In this dialog box, select the base template from the **Template file** drop-down list and make sure that the **Project template** radio button is selected. Choose the **OK** button; a new project template file will open and the interface screen is activated.

USER INTERFACE

In Autodesk Revit, the user interface consists of the ribbon, Drawing Area, **Properties** palette, **Status Bar**, and the **View Control Bar**, as shown in Figure 1-7. In Autodesk Revit, ribbon is an interface from where you can invoke tools. The ribbon, which contains task-based tabs and panels, streamlines the structural workflow and optimizes the project delivery time. In Autodesk Revit, when you select an element in the drawing area, the ribbon displays a contextual tab that comprises of tools corresponding to the selected element. The interface of Autodesk Revit is similar to the interfaces of many other Microsoft Windows-based programs. The main parts in the Revit interface are discussed next.

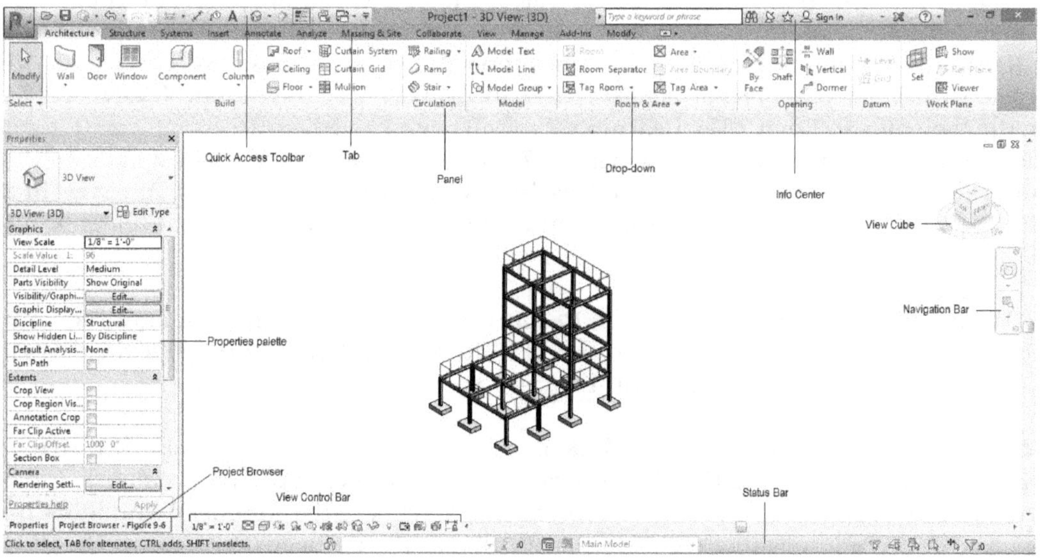

*Figure 1-7 The **Autodesk Revit 2017** user interface*

Title Bar

The Title Bar, docked on the top portion of the user interface, displays the program's logo, name of the current project, and the view opened in the viewing area. **Project 1- Structural Plan: Level 1** is the default project name and view.

Ribbon

The ribbon, as shown in Figure 1-8, is an interface that is used to invoke tools. When you open a file, the ribbon is displayed at the top in the screen. It comprises task-based tabs and panels, refer to Figure 1-8, which provide all the tools necessary for creating a project. The tabs and panels in the ribbon can be customized according to the need of the user. This can be done by moving the panels and changing the view states of the ribbon (the method of changing the ribbon view state is discussed later in this chapter). The ribbon contains buttons, drop-downs, panels, tabs, and tools. These buttons and the tools can be selected from the corresponding panels.

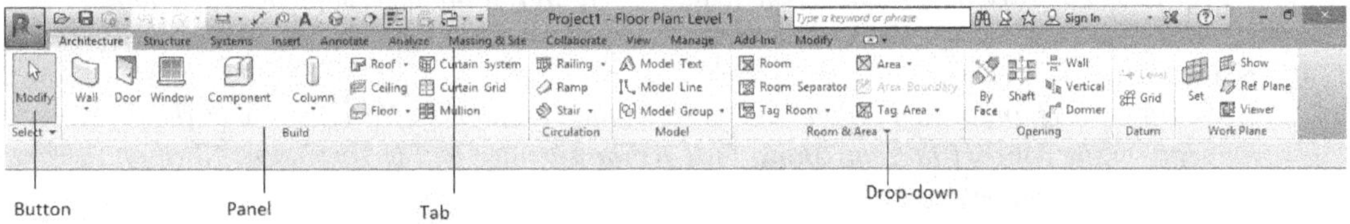

***Figure 1-8** Different components of a ribbon*

Tip

Tooltips appear when you place the cursor over any of the tool icons in the ribbon. The name of the tool appears in a box helping you in identifying the tool icon.

In the ribbon, you can move a panel and place it anywhere on the screen. To do so, press and hold the left mouse button on the panel label in the ribbon, and then drag the panel to a desired place on the screen. Next, use the tools of the moved panel and place the panel back in the ribbon. To do so, place the cursor on the moved panel and choose the **Return Panels to Ribbon** button from the upper right corner of this panel, as shown in Figure 1-9; the panel will return to the ribbon.

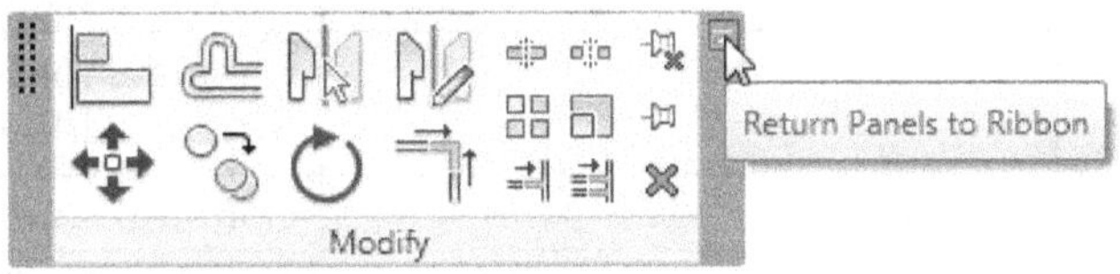

***Figure 1-9** Choosing the **Return Panels to Ribbon** button*

Changing the View States of the Ribbon

The ribbon can be displayed in three view states by selecting any of the following four options: **Minimize to Tabs**, **Minimize to Panel Titles**, **Minimize to Panel Buttons**, and **Cycle through All**. To use these options, place the cursor over the second arrow on the right of the **Modify** tab in the ribbon, refer to Figure 1-10; the arrow will be highlighted. Next, click on the down arrow; a flyout will be displayed, as shown in Figure 1-10. From this flyout, you can choose the **Minimize to Tabs** option to display the only tabs in the ribbon. If you choose the **Minimize to Panel Titles** option, the ribbon will display the titles of the panels along with the tabs. You can choose the **Minimize to Panel Buttons** option to display panels as buttons in the ribbon along with tabs.

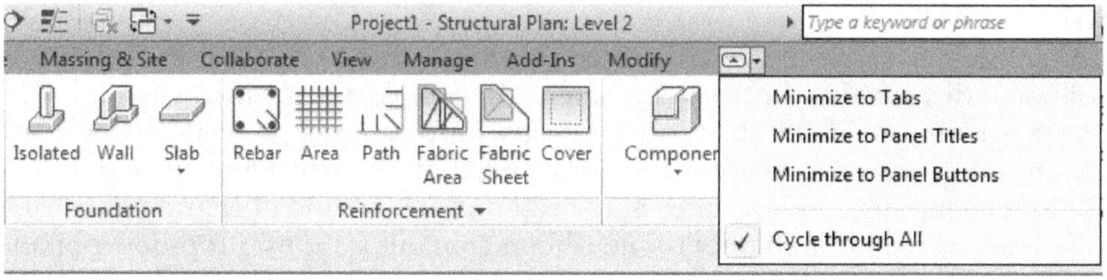

Figure 1-10 *Various options in the flyout for changing the view state of the ribbon*

Note

*If the ribbon is changed to a different view state, then on placing the cursor over the first arrow on the right of the **Modify** tab, the **Show Full Ribbon** tooltip will be displayed. Click on the arrow; the full ribbon will be displayed.*

The table given next describes various tabs in the ribbon and their functions.

Tab	Description
Structure	Contains tools for creating a structural model
Architecture	Contains tools for creating an architectural model
Insert	Contains tools for inserting or managing secondary files such as raster image files and CAD files
Annotate	Contains tools for documenting a building model such as adding texts and dimensions
Analyze	Contains tools for analyzing the structural model
Massing & Site	Contains tools for creating massing and site elements
Collaborate	Contains tools for collaborating the project with other team members (internal and external)
View	Contains tools for managing and modifying the current views, switching views, and so on.
Manage	Contains tools for specifying the project and system parameters and settings
Add-Ins	Contains add ins tools available to subscription customers to use it with the Revit software
Modify	Contains tools for editing elements in the model

Contextual Tabs in the Ribbon

These tabs are displayed when you choose certain tools or select elements. They contain a set of tools or buttons that relate only to a particular tool or element.

For example, when you invoke the **Beam** tool, the **Modify | Place Beam** contextual tab is displayed. This tab has the following panels: **Select**, **Properties**, **Clipboard**, **Geometry**, **Modify**, **View**, **Measure**, **Create**, **Mode**, **Draw**, **Multiple** and **Tag**. The **Select** panel contains the **Modify** tool. The **Properties** panel contains the **Properties** button and the **Type Properties** tool. The **Mode** panel has some necessary tools that are used to load model families or to create the model of a window in a drawing. The other panels, apart from those discussed above, contain the tools that are contextual and are used to edit elements when they are placed in a drawing or selected from a drawing for modification.

Application Frame

The application frame helps you manage projects in Autodesk Revit. It consists of the **Application** button, **Application Menu**, **Quick Access Toolbar**, **InfoCenter**, and **Status Bar**. These options are discussed next.

Application Button

The **Application** button is displayed at the top-left corner of the Autodesk Revit 2017 interface. This button is used to display as well as close the **Application Menu**.

Application Menu

The **Application Menu** contains tools that provide access to many common file actions such as **Open**, **Close**, and **Save**. Click on the down arrow on the **Application** button to display the **Application Menu**, as shown in Figure 1-11. Alternatively, press ALT+F to display tools in the **Application Menu**.

*Figure 1-11 The **Application Menu***

Quick Access Toolbar

The **Quick Access Toolbar**, shown in Figure 1-12, contains the options to undo and redo changes, open and save a file, create a new file, and so on.

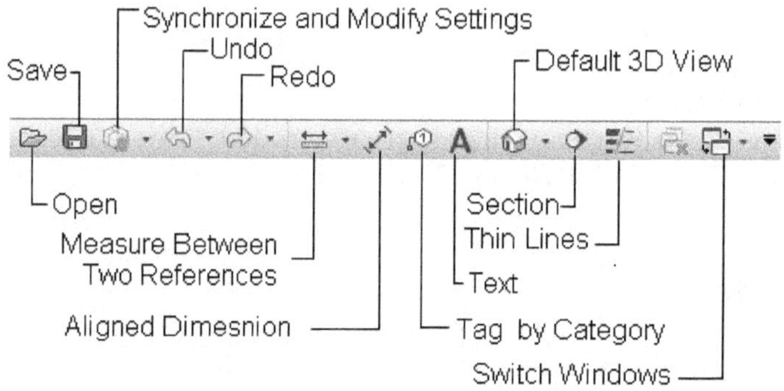

Figure 1-12 The ***Quick Access Toolbar***

By default, the **Quick Access Toolbar** contains the options such as **Open**, **Save**, **Redo**, **Undo**, and so on. You can customize the display of the **Quick Access Toolbar** by adding more tools and removing the unwanted tools. To add a tool or a button from the panel of the ribbon to the **Quick Access Toolbar**, place the cursor over the button; the button will be highlighted. Next, right-click; a flyout will be displayed. Choose **Add to Quick Access Toolbar** from the flyout displayed; the highlighted button will be added to the **Quick Access Toolbar**. The **Quick Access Toolbar** can be customized to reorder the tools displayed in it. To do so, choose the down arrow next to the **Switch Windows** drop-down, refer to Figure 1-12; a flyout will be displayed. Choose the **Customize Quick Access Toolbar** option located at the bottom of the flyout; the **Customize Quick Access Toolbar** dialog box will be displayed. Use various options in this dialog box and choose the **OK** button; the **Customize Quick Access Toolbar** dialog box will close and the tools in the **Quick Access Toolbar** will be reordered.

InfoCenter

You can use **InfoCenter** to search for information related to Revit (Help) to display the **Subscription Center** panel for subscription services and product updates, and to display the **Favorites** panel to access saved topics. Figure 1-13 displays various tools in **InfoCenter**.

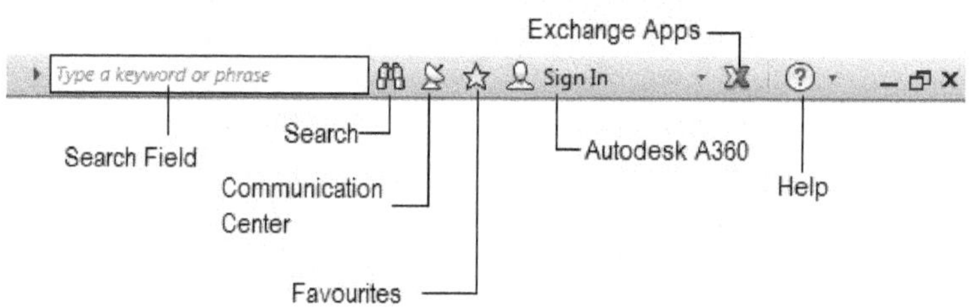

Figure 1-13 The ***InfoCenter***

Status Bar

The **Status Bar** is located at the bottom of the interface screen. When the cursor is placed over an element or a component, the **Status Bar** displays the name of the family and type of the corresponding element or components. It also displays prompts and messages to help you use the selected tools.

View Control Bar

The **View Control Bar** is located at the lower left corner of the drawing window, as shown in Figure 1-14. It can be used to access various view-related tools. The **Scale** button shows the scale of the current view. You can choose this button to display a flyout that contains standard drawing scales. From this flyout, you can then select the scale for the current view. The **Detail Level** button is used to set the detail level of a view. You can select the required detail level as **Coarse**, **Medium**, and **Fine**. Similarly, the **Visual Style** button enables you to set the display style. The options for setting the display style are: **Wireframe, Hidden Line, Shaded, Consistent Colors, Realistic**, and **Raytrace**.

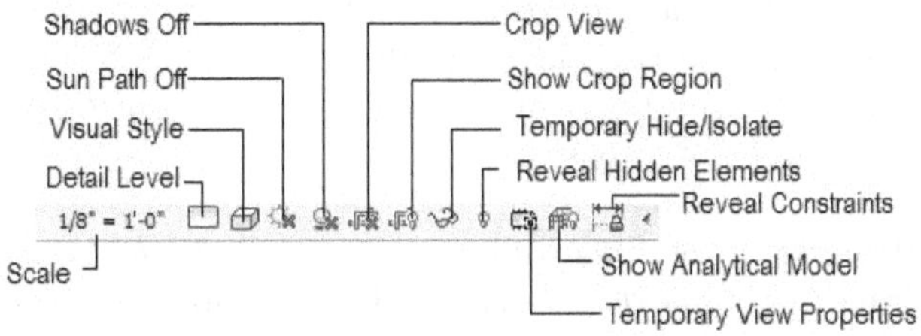

Figure 1-14 The View Control Bar

Options Bar

The **Options Bar** provides information about the common parameters of a component type. It also displays options for creating or editing them. The options displayed in the **Options Bar** depend on the type of component being created and selected for editing. Figure 1-15 displays the options in the **Options Bar** to create a structural column.

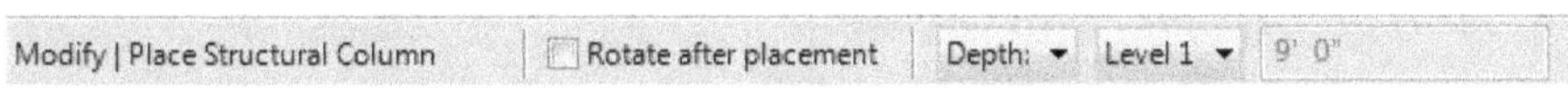

*Figure 1-15 The **Options Bar** with different options to create a structural column*

Type Selector

The **Type Selector** drop-down list is located in the **Properties** palette for the currently invoked tool. On invoking the **Beam** tool, the properties of the beam will be displayed in the **Properties** palette. In this palette, you can use the **Type Selector** drop-down list to select the required type of the beam. The content in the **Type Selector** drop-down list keep changing, depending upon the current function of the tool or the elements selected. The **Type Selector** drop-down list can also be used to specify the type of the element or component while placing an element or a component in a drawing. You can also use this drop-down list to change the type of a selected element.

Drawing Area

The Drawing Area is the actual modeling area where you can create and view the building model. It covers the major portion of the interface screen. You can draw building components in this area. The position of the pointing device is represented by the cursor. The Drawing Area also has the standard Microsoft Windows functions and buttons such as close, minimize, maximize, scroll bar, and so on. These buttons have the same function as that of the other Microsoft Windows-based programs.

PROJECT BROWSER

The **Project Browser** is located below the ribbon. It displays project views, schedules, sheets, families, and groups in a logical tree-like structure, as shown in Figure 1-16 and helps you open and manage them. To open a view, double-click on the name of the view, or drag and drop the view in the Drawing Area. You can close the **Project Browser** or dock it anywhere in the Drawing Area.

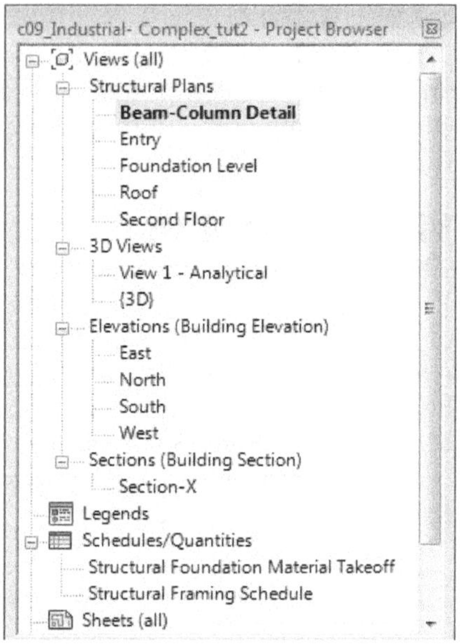

Figure 1-16** The **Project Browser

Note

*If the **Project Browser** is not displayed on the screen,choose the **View** tab from the ribbon and then select the **Project Browser** check box from **View > Windows > User Interface** drop-down.*

The **Project Browser** can be organized to group the views and sheets based on the project requirement. For example, while working on a large project with a number of sheets, you can organize the **Project Browser** to view and access specific sheets.

Note

*In the **Project Browser**, you can expand or contract the view listing by selecting the '+' or '-' sign, respectively. The current view in the drawing window is highlighted in bold letters. The default project file has a set of preloaded views.*

Keyboard Accelerators

In Autodesk Revit, accelerator keys have been assigned to some of the frequently used tools. These keys are shortcuts that you can type through the keyboard to invoke the corresponding tool. Accelerator keys corresponding to a tool appear as a tooltip when you move the cursor over the tool.

Properties Palette

The **Properties** palette, as shown in Figure 1-17, is a modeless interface, which displays the type and element properties of various elements and views in a drawing. The **Properties** palette is dockable and resizable, and it supports multiple monitor configurations. The **Properties** palette is displayed in the Revit interface by default and it shows the instance properties of an active view. When you select an element from a drawing, the **Properties** palette displays its instance properties. You can also access the type properties of the selected element from the **Properties** palette. To do so, choose the **Edit Type** button from the palette; the **Type Properties** dialog box will be displayed. In this dialog box, you can change the type properties of the selected element.

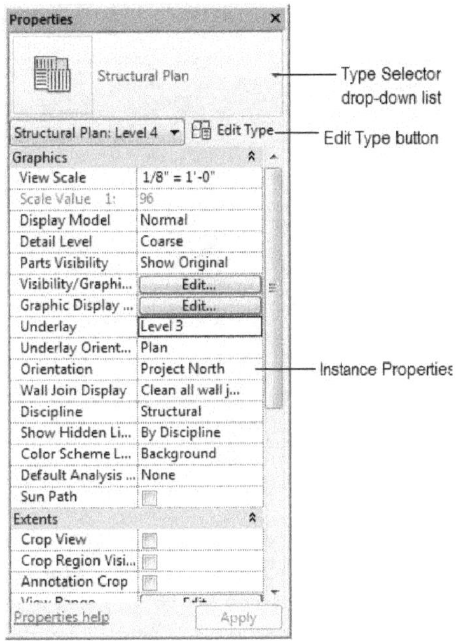

*Figure 1-17 The **Properties** palette*

In the **Properties** palette, you can assign a type to a selected element in a drawing from the **Type Selector** drop-down list. In Revit, you can toggle the display of the **Properties** palette in its interface. Choose the **Properties** button in the **Properties** panel of the **Modify** tab to hide it. Similarly, you can choose the **Properties** button to display the palette if it is not visible in the interface.

Tip

*As you become accustomed to use Autodesk Revit, you will find these **Keyboard Accelerators** quite useful because they save the effort of browsing through the menus.*

DIALOG BOXES

Some Autodesk Revit tools, when invoked, display a dialog box. A dialog box is an interface for accessing, specifying, and modifying the parameters related to that tool. For example, when you choose **Save As > Project** from the **Application Menu**, the **Save As** dialog box is displayed, as shown in Figure 1-18. A dialog box consists of various parts such as dialog label, radio buttons, text or edit boxes, check boxes, slider bars, image box, buttons, and tools, which are similar to other windows-based programs. Some dialog boxes contain the [**...**] button, which displays another related dialog box. There are certain buttons such as **OK**, **Cancel**, and **Help**, which appear at the bottom of most of the dialog boxes. The names of the buttons imply their respective functions. The button with a dark border is the default button.

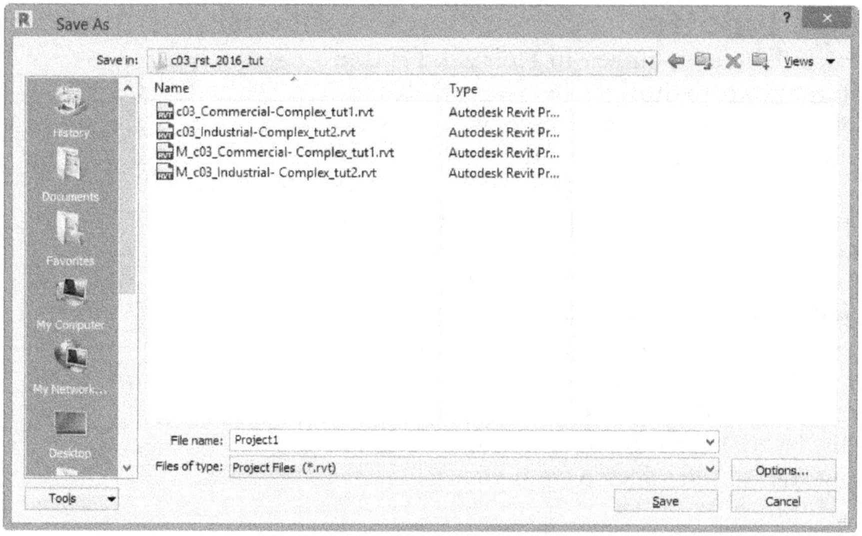

*Figure 1-18 The **Save As** dialog box*

MULTIPLE DOCUMENT ENVIRONMENT

The multiple document environment feature allows you to open more than one project at a time in a single Autodesk Revit session. This is very useful when you want to work on different projects simultaneously and make changes with reference to each other.

Sometimes, you may need to incorporate certain features from one project into the other. With the help of multiple document environment feature, you can open multiple projects and then use the **Cut**, **Copy** and **Paste** tools from the **Clipboard** panel of the **Modify (type of element)** tab to transfer the required components from one project to another. These editing tools can also be invoked by using the CTRL+C and CTRL+V keyboard shortcuts.

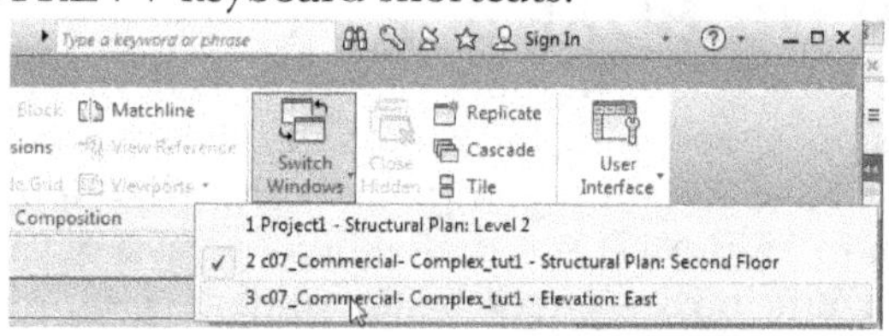

*Figure 1-19 Selecting an option from the **Switch Windows** drop-down list*

To access the opened projects, click on the **Switch Windows** drop-down list in the **Windows** panel of the **View** tab; the options for the names of different opened project files will be displayed, as shown in Figure 1-19. Like other Microsoft Windows-based programs, you can select and view the opened projects using the **Cascade Windows** and **Tile Windows** tools from the **Windows** panel of the **View** tab. The cascaded view of projects is shown in Figure 1-20.

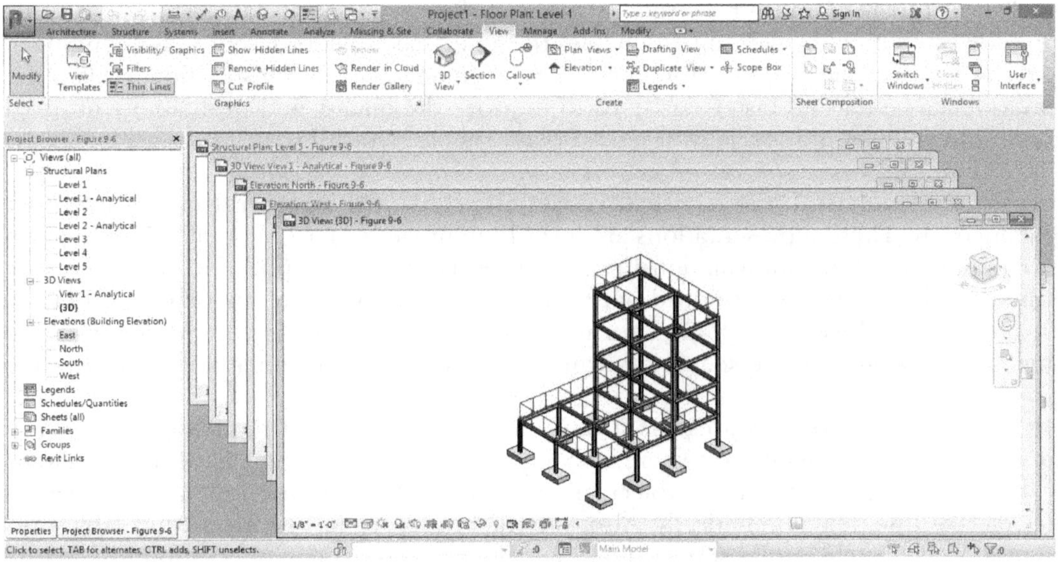

Figure 1-20 The cascaded view of projects

INTEROPERABILITY OF Autodesk Revit

The models or geometries created in Autodesk Revit can be easily exported to AutoCAD and AutoCAD Architecture in the DWG file format. This enables structural engineers to collaborate with Architects.

One of the important aspects of the job of a structural engineer is to collaborate and share information with the rest of the design team including the architect. To facilitate this requirement, Revit follows a wide range of industry standards and supports various CAD file formats such as DWF, DGN, DWG, DGN, IFC, SKP, and SAT. For image files, it supports JPG, TIFF, BMP, PNG, AVI, PAN, IVR, and TGA file formats. Besides these, the formats that are supported by Revit include ODBC, HTML, TXT, XML, XLS, and MDB. Autodesk Revit is compatible with any CAD system that supports the DWG, DXF, or DGN file format. Revit can import the models and geometries as ACIS solids. This enables engineers to import models from AutoCAD Architecture and AutoCAD MEP (Mechanical, Electrical, and Plumbing) software and to link and import 3D information to Revit. This feature makes Autodesk Revit 2017 an efficient, user-friendly, and compatible software.

BUILDING INFORMATION MODELING AND Autodesk Revit 2017

Building Information Modeling (BIM) is defined as a design technology that involves the creation and use of coordinated, internally consistent, and computable information about a building project in design and construction.

Using BIM, you can demonstrate the entire life cycle of a building project starting from the process of construction, facility operation, and information about quantities and shared properties of elements. BIM enables the circulation of virtual information model from the design team to contractors and then to the owner, thereby adding changes and their knowledge to update the model at each stage of transfer. The ability to keep information up-to-date and make it available in an integrated digital environment enables the architects, owners, builders, and engineers to have clear vision of the project before the commencement of actual construction. It also enables them to make better and faster decisions to improve the quality and profitability of projects. Autodesk Revit 2017 is a specially designed platform based on BIM.

In Revit, the analytical and physical representations of a structural model are created simultaneously. These representations are just different views of a computable building model that contains necessary information for a third-party analysis application. Revit provides a common modeling interface for third-party analysis applications. You can use Revit API to move data directly from the Revit building information model to the analysis software. You can further bring back the analysis while keeping the analysis, design, and documentation synchronized.

Revit's parametric model represents a building as an integrated database of coordinated information. In Revit, change anywhere is change everywhere. A change made in your project at any stage is reflected in the entire project, and also, due to the parametric behavior of elements, the project is updated automatically. Also, the integration of Revit with the available in-built commercial tools such as solar studies, material takeoffs, and so on greatly simplifies the project design and reduces the time consumed by these analyses, thereby enabling faster decision making.

WORKSHARING USING REVIT SERVER

Worksharing is a method of distributing work among people involved in a project, and accomplishing it within the stipulated period of time. In worksharing, each person involved in the project is assigned a task that has to be accomplished through proper planning and by coordinating with the other members of the team.

In a large scale building project, worksharing helps in finishing a project in time and meeting the quality requirements that are set during the process. Generally, in a large scale building project, worksharing is based on the specialization of work. The professionals such as structural engineers, architects, interior architects, and MEP engineers are involved in their respective fields to accomplish the project. So, the distribution of work at the primary stage is made on the basis of the area of specialization. Each professional has his own set of work to perform for the accomplishment of the project. Therefore, worksharing is an important process that is required to be implemented efficiently to complete the project in time.

In Autodesk Revit, you can apply server-based worksharing with the help of Revit Server, which is a server application. Revit Server uses a central server and multiple local servers for collaborating across a Wide Area Network (WAN). The central server hosts the central model of a workshared project and remains accessible to all team members over the Wide Area Network. Similarly, the local server is accessible to all team members in a Local Area Network (LAN). The local server hosts a local updated copy of the central model. In the Worksharing environment, the team members are not aware of the local server as it is transparent in their daily operations. Refer to Figure 1-21 for the network model of Revit Server.

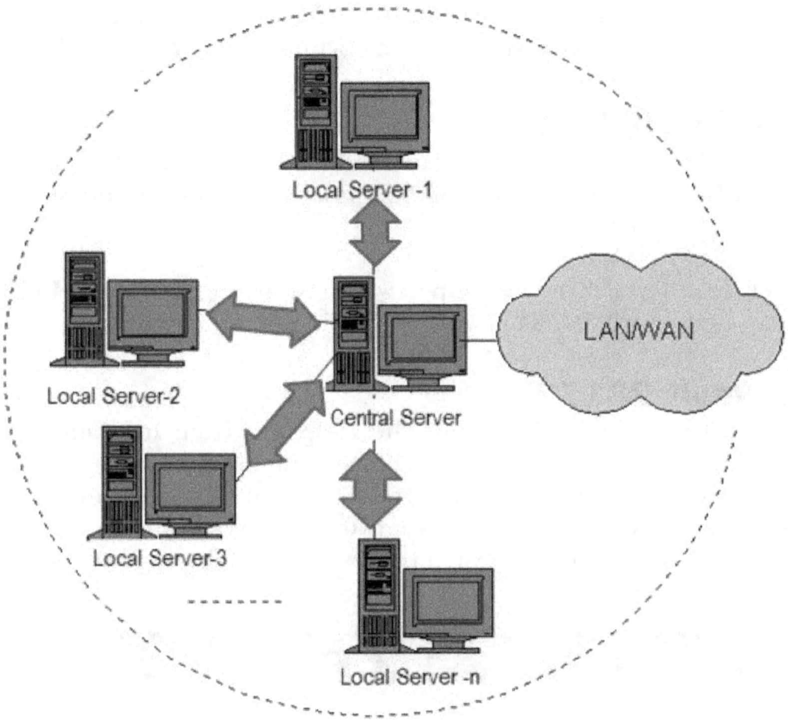

Figure 1-21 *The Network Model of Revit Server*

In Worksharing environment, a team member starts working on the local model of the central model. The local model will be saved in the computer of the team member. As the team member works, the local server requests updated information from the central model on the central server, using available network capacity to transfer the data over the WAN. The updated version of the model is stored on the local server, so the updates are readily available when a team member requests them.

LINKING ANALYTICAL MODEL FOR ANALYSIS

In Autodesk Revit, the physical model represents the physical layout that consists of structural walls, beams, columns, footings, and so on. The analytical model is created along with the physical model. The analytical model is the engineering description of a structural physical model.

The following elements have analytical model associated with them: Structural Columns, Structural Framing elements (such as beams and braces), Structural Floors, Structural Footings, and Structural Walls. The analytical model is a 3D model that is used for structural analysis. In this model, you can add loads and material properties that are required for analysis. To analyze the analytical model, transfer it into a third-party analysis software. You can link the analytical model via application programming interface (API) to several leading industry applications for building analysis, such as ADAPT-Builder, RISA 3D, Fastrack, ROBOT Millennium, Sofistik, and others. After the analysis has been performed, the results of the analysis are returned to the building model (physical and analytical) in Revit. The results that are returned to Revit, dynamically update the building model and the documentation associated with it.

Autodesk Revit HELP

Autodesk Revit 2017 helps you to easily understand the tools and methods used in a project. In Autodesk Revit 2017, you can access online help documentation (Autodesk WikiHelp) as well as local (offline) help documentation.

To access the help feature, click on the **Help** down arrow on the right of the **InfoCenter**; a flyout with various help options will be displayed. The options to access the help are discussed next.

Using the Revit 2017 Help

To access the local Revit 2017 help, choose the **Help** tool from **InfoCenter**; the **Revit Help** page will be displayed, as shown in Figure 1-22. You can also display the **Revit Help** page by pressing the F1 key. You can use this page to get information on various tools and enhancements introduced in Autodesk Revit 2017. There are two tabs provided in this page. The options in these tabs will help you speed up the search process. These tabs are discussed next.

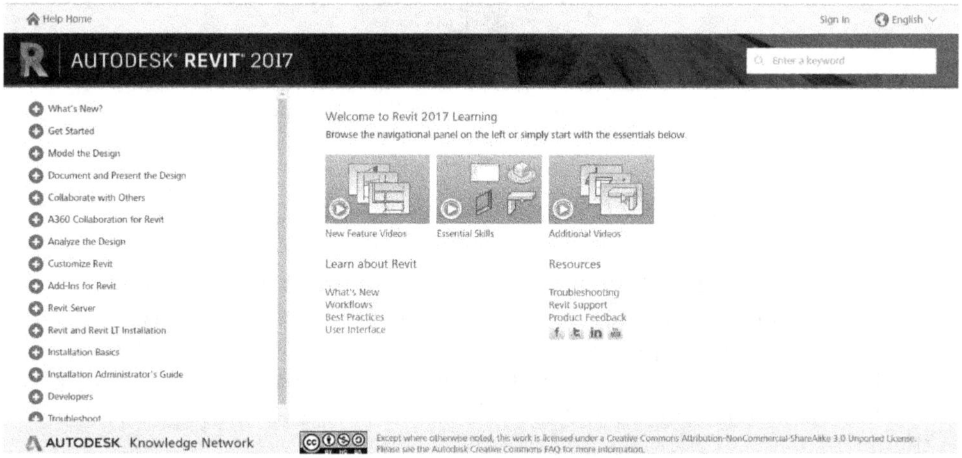

*Figure 1-22 The local **Revit Help** page*

WORKING WITH BIM 360

In Revit 2017, BIM 360 has been introduced as an add-in. BIM 360 is a cloud based BIM management and collaboration solution. It is used for publishing models on the BIM 360 cloud and for sharing latest project information within a team. This add-in allows you to perform several functions such as sharing models with BIM360 Glue, viewing clash results, and adding BIM 360's field equipment properties to your current model.

Sharing Models With BIM 360

In BIM 360, you can use the **Glue** tool to share models with the BIM 360 cloud which can be shared further within a team. To share model with the cloud, first load the desired model in revit and then choose the **Glue** tool from the **BIM 360** panel in the **Add-Ins** tab; the BIM 360 apps page will be displayed. Next, you need to sign in with your Autodesk ID and password, and specify a name for the **BIM 360** page. Follow the instructions mentioned on the apps page and install BIM 360 cloud application.

If you are already signed in, then on choosing the **Glue** tool, the **Select Project** dialog box will be displayed, as shown in Figure 1-23. Select the Sample Project and then choose the **Next** button in the dialog box; the **Select Views** dialog box will be displayed, as shown in Figure 1-24.

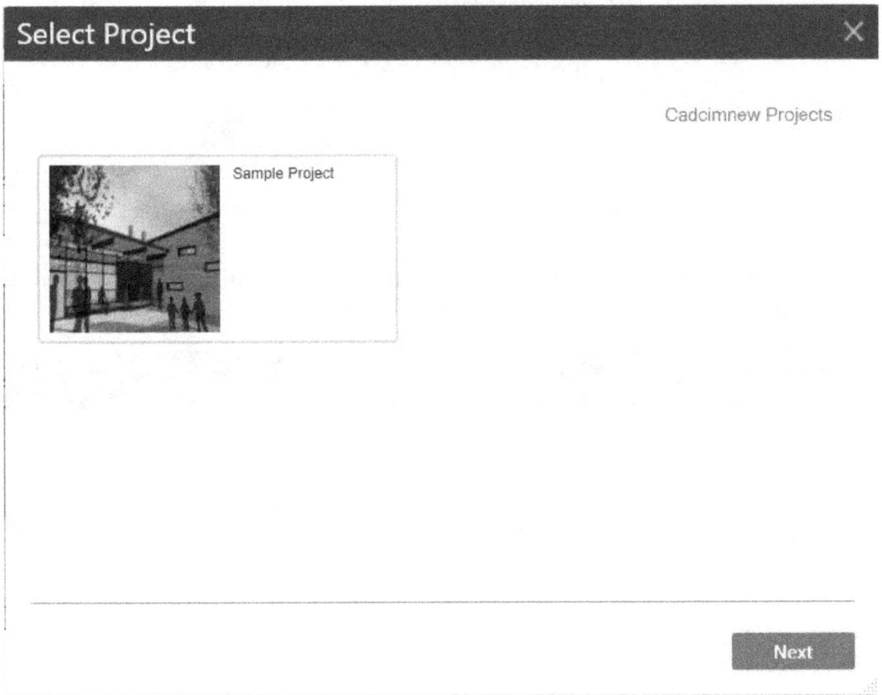

***Figure 1-23** The **Select Project** dialog box*

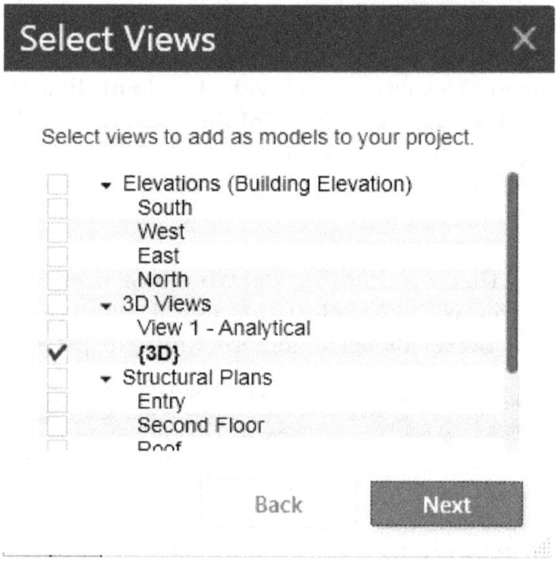

***Figure 1-24** The **Select Views** dialog box*

In this dialog box, select the check box corresponding to the required view and then choose the **Next** button; the **Review and Confirm** dialog box will be displayed, as shown in Figure 1-25.

***Figure 1-25** The **Review and Confirm** dialog box*

In this dialog box, select the required folder from the **Folder** column; the **Add to Folder** dialog box will be displayed. In this dialog box, select the required folder and choose the **Select** button; the **Add to Folder** dialog box will be closed. In the **Review and Confirm** dialog box, choose the **Glue It** button; the **Gluing in Progress** message box will be displayed. Choose the **OK** button to close the message box. After sometime, the **Gluing Complete** message box will be displayed. Choose the **OK** button; the model will be shared with the BIM 360 cloud.

Self-Evaluation Test

Answer the following questions and then compare them to those given at the end of this chapter:

1. You cannot add template file under the **Projects** head. (T/F)

2. You can create analytical model in Revit. (T/F)

3. Worksharing method is used only to hide elements in Revit. (T/F)

4. You can work on four detail levels in Revit. (T/F)

5. Elements of the same family may have different properties. (T/F)

Answers to Self-Evaluation Test

1. T, **2.** T, **3.** F, **4.** F, **5.** T

Chapter 2

Getting Started with a Structural Project

Learning Objectives

After completing this chapter, you will be able to:

- *Start a new structural project*
- *Set the units of various measurement parameters for a project*
- *Understand the concept of snaps, dimensions, and object snaps*
- *Save a project*
- *Close a project and exit Revit 2017*
- *Open an existing project*

INTRODUCTION

In this chapter, you will learn how to create a new structural project. Also, you will learn the methods of creating initial setups such as units and snaps.

STARTING A NEW STRUCTURAL PROJECT

Application Menu: New > Project
Shortcut Key: CTRL+N

In Revit, a structural project is considered as a single database containing information related to a building design. This means that the project file created will contain all the information related to building design, from geometry to construction documentation. Generally, this information include the details of the components used to design building models (like walls, doors, windows, beams, columns, and other relevant information about the building design), different views of the model, working drawings created from model, and documentation related to design of the model.

In Revit, you can create different views such as plan, elevation, and sections from a single model created for a building project. These views are associated with each other. This means, when you change the building design in one view, it is propagated throughout the project.

In Revit, a new project file uses the default settings that are defined in the project template.

To start a new project, choose the **Application** button; the **Application Menu** will be displayed. From the **Application Menu**, choose **New > Project**, as shown in Figure 2-1; the **New Project** dialog box will be displayed, refer to Figure 2-2. In this dialog box, in the **Template File** area, the **Structural Template** is selected by default in the **Template file** drop-down list. As a result, the new project will adopt the settings of the template file. A template file has various project parameters saved in it such as units, views, and so on. When you apply the template file on a new project, it will adopt the same parameters as the template file. The difference between a template file format and a project file format is that the former has a *.rte* extension, whereas the latter has a *.rvt* extension. You can either select any of the template files provided in Revit or create your own template file.

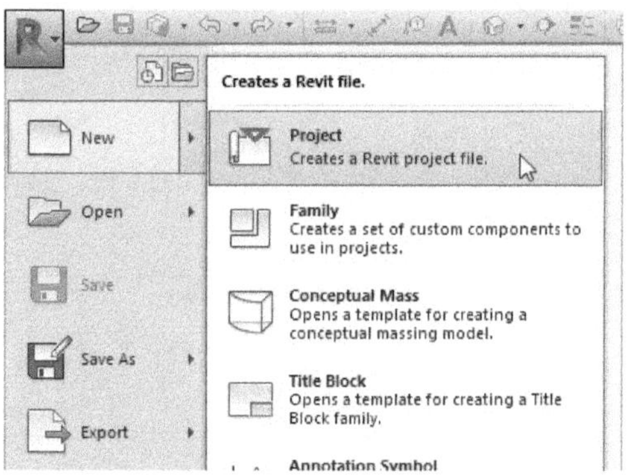

Figure 2-1** Choosing the **Project** option from the **Application Menu

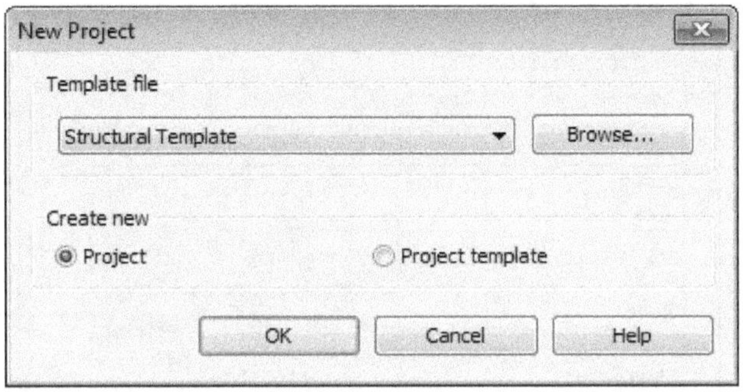

***Figure 2-2** The **New Project** dialog box*

You can also save any project file as a template file. You can select a file as template by first choosing the **Browse** button in the **Template file** area of the **New Project** dialog box and then selecting the required file from appropriate location. To create a new project, select the **Project** radio button and to create a template, select the **Project template** radio button in the **New Project** dialog box.

SETTING PROJECT UNITS

Ribbon: Manage > Settings > Project Units
Shortcut Key: UN

Units are important parameters of a project as they provide a standard for measurement of different entities. While installing Revit, you are prompted to set the Imperial (feet and inches) or Metric (meter) unit as default. Setting a default unit system helps you to start your project with a specific type of unit as per the conventions followed in the industry. To set units, choose the **Project Units** tool from the **Settings** panel of the **Manage** tab; the **Project Units** dialog box will be displayed, as shown in Figure 2-3. Alternatively, type **UN** to display the **Project Units** dialog box. Under the **Units** column in this dialog box, you can specify various units that are relevant to the building project. In the **Project Units** dialog box, units are grouped into six disciplines: **Common**, **Structural**, **HVAC**, **Electrical**, **Piping**, and **Energy**. Each discipline has a set of measurement parameters. You can select any of these disciplines from the **Discipline** drop-down list in the **Project Units** dialog box. The **Format** column in this dialog box displays the current unit format for the corresponding parameter in the **Units** column. You can preview and select the possible digit grouping and decimal separators from the **Decimal symbol/digit grouping** drop-down list, which is at the lower left corner of the dialog box. Different types of units and their parameters are discussed next.

Note

*The values displayed in the **Format** column of the **Project Units** dialog box may differ, depending upon the type of unit system, Imperial or Metric, selected for the project. In this textbook, the Imperial and Metric unit systems have been used in tutorials and illustrations.*

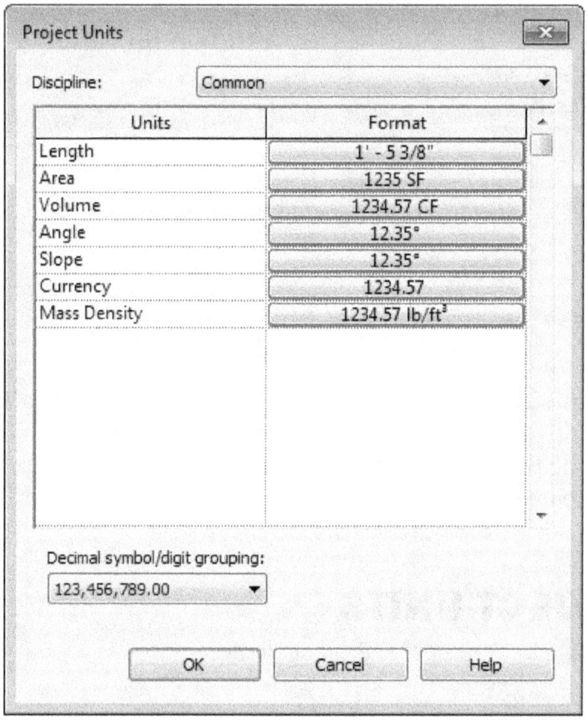

*Figure 2-3 The **Project Units** dialog box*

Common Unit Type

The **Common** unit type used in a structural project includes the parameters such as length, volume, area, angle, slope, mass density, and currency. In the **Project Units** dialog box, the **Common** option is selected by default in the **Discipline** drop-down list, refer to Figure 2-3. The **Common** unit type used in Revit is similar to those used in other Revit platforms. Moreover, the settings of the parameters of common units are similar to those used in other CAD programs. The methods of setting various parameters under the **Common** unit type are discussed next.

Setting Length Units

In a structural project, you can assign a unit for the measurement of length. To do so, click on the field corresponding to the **Length** parameter in the **Format** column of the **Project Units** dialog box; the **Format** dialog box will be displayed. This dialog box displays the units of length and their settings, as shown in Figure 2-4. Select the required unit from the **Units** drop-down list in the dialog box. The options that can be selected from this drop-down list are **Decimal feet**, **Feet and fractional inches**, **Decimal inches**, **Fractional inches**, **Meters**, **Decimeters**, **Centimeters**, **Millimeters**, and **Meters and centimeters**. By default, the **Feet and fractional inches** option is selected in this drop-down list if the **Imperial** is selected as the unit system at the time of installing Revit. But, if you have selected **Metric** as the unit system then the **Millimeters** option is selected in this drop-down list by default. The settings for rounding increment for the length unit can be made from their respective drop-down lists and edit boxes. To do so, select the desired option from the **Rounding** drop-down list. For certain units, you can specify custom rounding value using the **Rounding increment** edit box, which is placed next to the **Rounding** drop-down list. This edit box is inactive by default. To make it active, first select any option from the **Units** drop-down list and then the **Custom** option from the **Rounding** drop-down list. Once the **Rounding increment** edit box is active, you can specify the desired rounding value in it for the selected unit.

Note
*The **Rounding increment** edit box is not available for **Meters and Centimeters**, **Feet and fractional inches**, and **Fractional inches** options.*

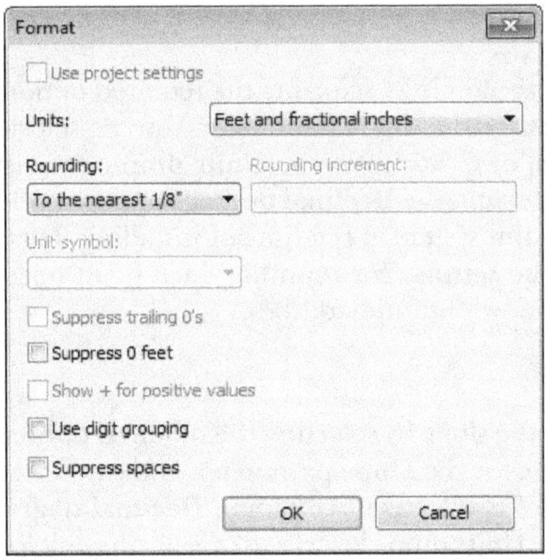

***Figure 2-4** The **Format** dialog box*

The **Unit symbol** drop-down list will be inactive if any of the **Feet and fractional inches**, **Fractional inches**, and **Meters and centimeters** options is selected in the **Units** drop-down list.

Once you make the **Unit symbol** drop-down list active, you can select the desired option from it to specify the measurement symbol to be used along with the unit of length in a project. For example, to use the symbol 'm' after all metric length measurements, you can select **m** option from the **Unit symbol** drop-down list as the measurement symbol. You can select the **Suppress spaces** check box to remove all spaces around the dash from the length strings. For example, you can remove spaces around the dash when a length string is expressed in feet and fractional inches to denote a particular measurement.

Setting the Area and Volume Units

In the **Project Units** dialog box, you can assign a unit for the measurements of areas. To do so, click on the field corresponding to the **Area** parameter in the **Format** column of the **Project Units** dialog box; the **Format** dialog box will be displayed. In this dialog box, you can set the unit for measuring the area by using the options in the **Units** drop-down list. This drop-down list contains various options for the units of area such as **Square feet**, **Square meters**, **Square millimeters**, **Acres**, and so on. By default, the **Square feet** option is selected in this drop-down list, if you had selected the **Imperial** units at the time of installing Revit. But, if you have selected the **Metric** units then the **Square meters** option is selected in this drop-down list by default. The settings for rounding increment and unit symbol for the area units can be made from their respective drop-down lists and edit box. Similar to set the units for the area, you can set the units for volume. To set unit for the volume measurement, click in the field of the **Format** column corresponding to the **Volume** parameter in the **Project Units** dialog box; the **Format** dialog box will be displayed. In this dialog box, click in the **Units** drop-down list and select: **Cubic yards**, **Cubic feet**, **Cubic meters**, **Cubic millimeters**, or **Liters**. After selecting a suitable option from the **Units** drop-down list, choose the **OK** button; the **Format** dialog box will be closed and the selected unit for the volume measurement will be displayed in the field of the **Format** column corresponding to the **Volume** parameter in the **Project Units** dialog box.

Setting Angle Units

You can specify the unit for the angle by selecting the required option from the **Units** drop-down list in the **Format** dialog box for the **Angle** parameter. You can select **Decimal degrees**, **Degrees minutes seconds, Radians**, or **Grads** from the **Units** drop-down list to specify the unit for the angular measurement. By default, the **Decimal degrees** option is selected in this drop-down list, if **Imperial** is selected as a unit system at the time of installing Revit. Similarly, **12.35°** option is selected for Metric unit. The settings for rounding increment for the angle unit can be made from their respective drop-down list and edit box.

Setting Slope Units

You can specify the unit for the slope by selecting the required option from the **Units** drop-down list in the **Format** dialog box for the **Slope** parameter. You can select options such as **Ratio: 12**, **Ratio: 10**, **Rise / 12"**, **Rise / 1'-0"**, **Rise / 1000mm**, **Decimal degrees**, and **Percentage** for the slope measurement from the **Units** drop-down list to specify the unit for the slope measurement. By default, the **Decimal degrees** option is selected in this drop-down list if **Imperial** is selected as the unit system while installing Revit. The settings for rounding increment for the slope unit can be made from their respective drop-down list and edit box.

Setting the Currency Units

In Revit, you can set the unit for the currency as well. To do so, click on the corresponding field for the **Currency** parameter in the **Project Units** dialog box; the **Format** dialog box will be displayed. In this dialog box, select appropriate currency symbols and rounding values from the **Unit symbol** and **Rounding** drop-down lists, respectively.

Setting the Mass Density Units

In Revit, you can set the unit for mass density. To do so, click on the corresponding field for the **Mass Density** parameter in the **Project Units** dialog box; the **Format** dialog box will be displayed. In this dialog box, you can specify the desired unit by selecting it from the **Units** drop-down list. The **Unit** drop-down list contains options such as **Kilogram per cubic meter** and **Pound per cubic foot** for mass density measurement. By default, the **Pound per cubic meter** option is selected in this drop-down list for the Imperial unit system and **Kilogram per cubic foot** option is selected for the Metric unit system. The settings for the parameters of the rounding, rounding increment, and units symbol can be set by selecting the required option from the respective drop-down lists.

Note

You can format only the display of units on the screen or for printout using the ***Project Units*** *dialog box. The actual values for these units in the project may be different. For example, if you set the wall length rounding to the nearest value 1', the wall may show this rounded value, but the actual length of the wall might be in fractional feet. Similarly, for Metric, if you set the wall length rounding to the nearest value 10, the wall may show this rounded value, but the actual length of the wall might be different.*

Structural Unit Type

Structural units include the settings such as Force, Moment, Stress, Unit Weight, and so on. In Revit, you can set the structural units in the **Project Units** dialog box. To do so, invoke the **Project Units** dialog box and then select the **Structural** option from the **Discipline** drop-down list. Figure 2-5 displays the **Project Units** dialog box with the **Structural** option selected in the **Discipline** drop-down list. Some of the structural units are discussed next.

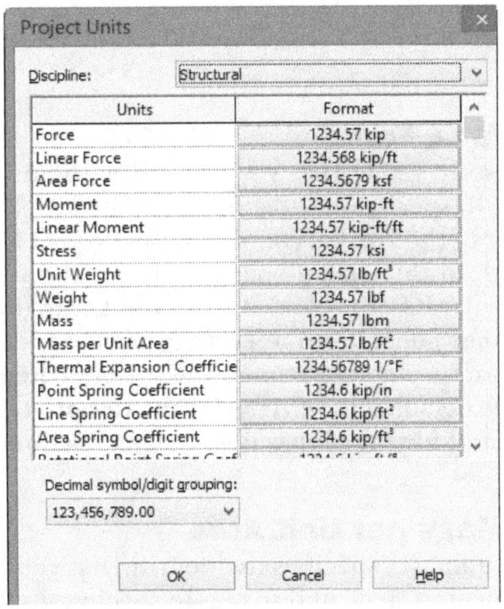

Figure 2-5 The ***Project Units*** *dialog box with the* ***Structural*** *option selected from the* ***Discipline*** *drop-down list*

Setting the Unit for Force

To specify the unit for the concentrated forces that act on the structural model, click on the field corresponding to the **Force** parameter in the **Format** column; the **Format** dialog box will be displayed. In this dialog box, specify a unit by selecting the required option from the **Units** drop-down list. The **Units** drop-down list contains the options such as **Newtons**, **Decanewtons**, **Kilonewtons**, **Meganewtons**, **Kips**, **Kilogram force**, and so on. For the Imperial unit system, **Kips** is the default option selected in the drop-down list. For the Metric unit system, **Kilonewtons** is the default option selected.

Setting the Unit for Linear Force

The linear force implies a uniformly distributed load that you can apply to any structural member. To specify unit for the linear force, click on the field corresponding to the **Linear Force** parameter in the **Format** column of the **Project Units** dialog box; the **Format** dialog box will be displayed. In this dialog box, specify the desired unit for the linear force by selecting an option from the **Units** drop-down list. The **Units** drop-down list contains options such as **Newtons per meter**, **Decanewtons per meter**, **Kilonewtons per meter**, **Meganewtons per meter**, **Kips per foot**, **Kilogram force per meter**, and so on. The default unit selected for the Imperial unit system in is **Kips per foot**. The default unit selected for the Metric unit system in the drop-down list is **Kilonewtons per meter**.

Tip

*While selecting a rounding value from the **Rounding** drop-down list in the **Format** dialog box, you should consider the extent of detailing that may be required for the project, and then select it. For projects that require too much detailing, a lower rounding value may be set. This parameter, however, can be modified at any time during the project development.*

Setting the Unit for Area Force

Area forces denote the pressure forces that you can apply to any structural element (beams, columns, slabs, and so on). To specify unit for the area force, click on the field corresponding to the **Area Force** parameter in the **Format** column of the **Project Units** dialog box; the **Format** dialog box will be displayed. In this dialog box, specify the desired unit for the area force by selecting an option from the **Units** drop-down list. In the **Units** drop-down list, you can select various options such as **Newtons per square meter**, **Decanewtons per square meter**, **Kilonewtons per square meter**, **Meganewtons per square meter**, **Kips per square foot**, **Kilogram force per square meter**, and so on. The default unit selected for the Imperial unit system in the drop-down list is **Kips per square foot** and for the Metric unit system is **Kilonewtons per square meter.**

Setting the Unit for Mass per Unit Area

The **Mass per Unit Area** parameter specifies the density that you can apply on any structural elements such as rebars, structural steel, and so on. To specify a unit for the mass per unit area, click on the field corresponding to the **Mass per Unit Area** parameter in the **Format** column of the **Project Units** dialog box; the **Format** dialog box will be displayed. In this dialog box, specify the desired unit for the mass per unit area by selecting an option from the **Units** drop-down list. In the **Units** drop-down list, you can select various options such as **Kilograms per square meter** and **Pounds per square foot**. The default unit selected in the Imperial unit system for this parameter is **Pound per square foot** and in the Metric unit system is **Kilograms per square meter**.

Similarly, you can set the units for the other options of different discipline types.

STRUCTURAL SETTINGS

Ribbon: Manage > Settings > Structural Settings

In Autodesk Revit, you can specify various structural settings of a project such as settings of the symbolic representation of structural framing components, settings of various load cases to be applied to structural model, settings for the analysis to be performed on model, and settings for providing supports to structural model. You can specify all these settings by using the **Structural Settings** dialog box. The **Structural Settings** dialog box can be invoked by choosing the **Structural Settings** tool from the **Settings** panel of the **Manage** tab, as shown in Figure 2-6. This dialog box contains five tabs, namely **Symbolic Representation Settings**, **Load Cases**, **Load Combinations**, **Analytical Model Settings**, and **Boundary Conditions Settings**. In this dialog box, the **Symbolic Representation Settings** tab is chosen by default, refer to Figure 2-7.

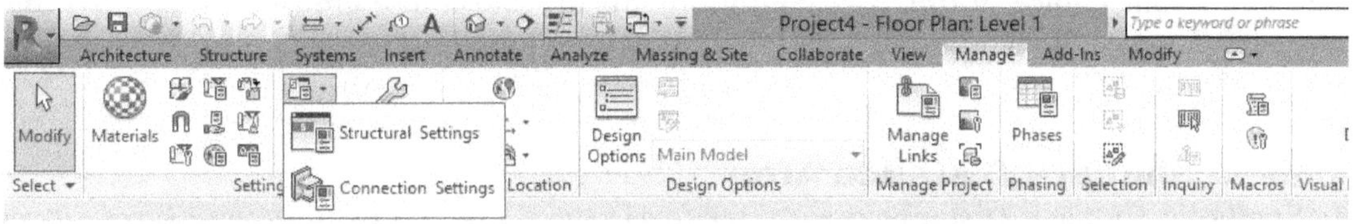

Figure 2-6 *Choosing the* ***Structural Settings*** *tool*

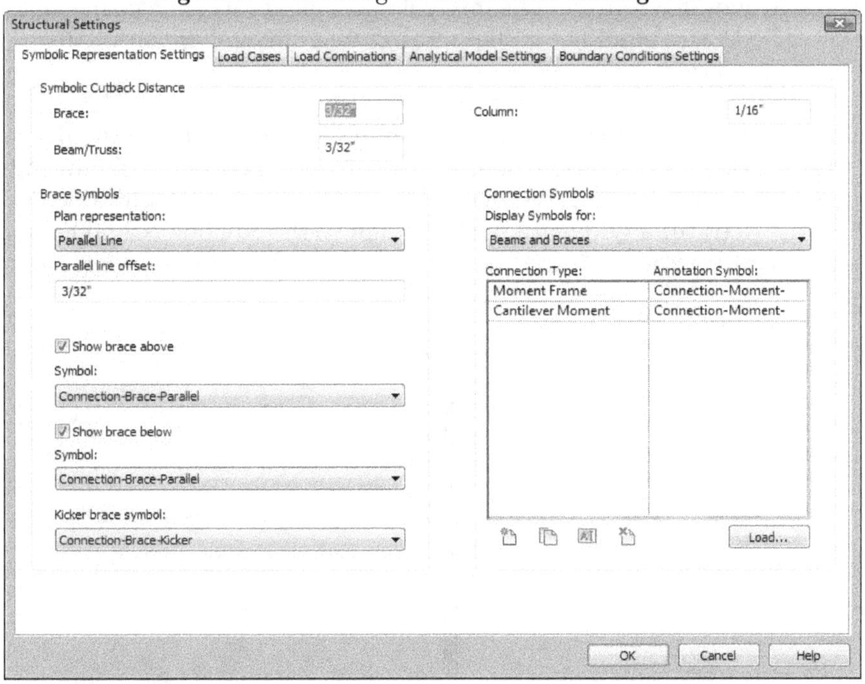

Figure 2-7 *The* ***Symbolic Representation Settings*** *tab chosen in the* ***Structural Settings*** *dialog box*

The settings that you specify in the tabs of the **Structural Settings** dialog box pertain to the appearance of the structural members. In the **Structural Settings** dialog box, you can also specify the load cases, boundary conditions, and the settings for the analytical model. In this chapter, you will learn about various options in the **Symbolic Representation Settings** tab. This will help you adjust the symbolic representation of various structural elements for documentation output.

Symbolic Representation Settings Tab

Whenever you invoke the **Structural Settings** dialog box, the **Symbolic Representation Settings** tab is chosen by default, refer to Figure 2-7. The options in this tab control the display and symbology of different structural conditions in plans, sections, and schedules of steel shapes. These options determine how certain elements of an object will display or allow you to use a symbol of your choice in different conditions. In this tab, there are three areas: **Symbolic Cutback Distance**, **Brace Symbols**, and **Connection Symbols**. The options in these areas are discussed next.

Symbolic Cutback Distance Area

The **Symbolic Cutback Distance** area, refer to Figure 2-7, contains three edit boxes: **Brace**, **Beam/Truss**, and **Column**. The values in these edit boxes control the cutback distances of the structural symbolic lines that represent beams, columns, and braces at a coarse detailed level of a structural project. These values are global settings that affect all beams, trusses, braces, and steel columns in a project.

In the **Symbolic Cutback Distance** area, you can enter a numeric value in the **Brace** edit box to control the symbolic cutback distance between a structural steel beam or column and a structural steel brace at a coarse detail level. In the Imperial unit system, the default value in this edit box is 3/32" whereas in Metric unit system, the default value in the edit box is 2.5000 mm. Figure 2-8 shows the symbolic cutback distance between a brace and a column at a coarse detail level.

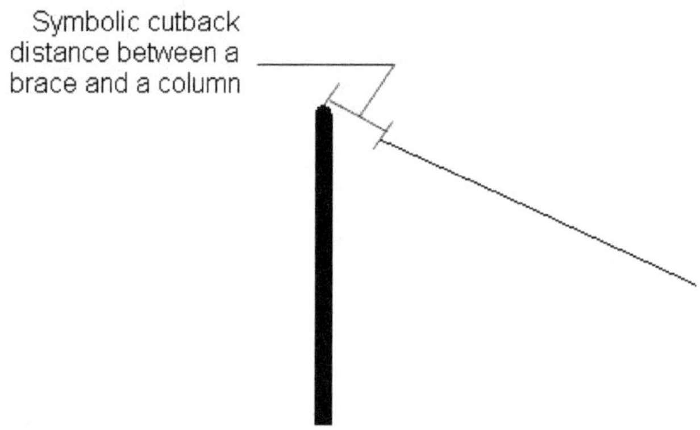

Figure 2-8 *The symbolic cutback distance between a column and brace*

In the **Beam/Truss** edit box of the **Symbolic Cutback Distance** area, you can enter any numeric value to control the symbolic cutback distance between beams and columns, or between beams, trusses, and columns. In the Imperial unit system, the default value in this edit box is 3/32" whereas in Metric unit system, the default value in the edit box is 2.5000mm. Figure 2-9 shows the symbolic cutback distance between two beams and a column. The value in the **Column** edit box is used to increase or decrease the gap between the symbolic representations of joined columns in an elevation view.

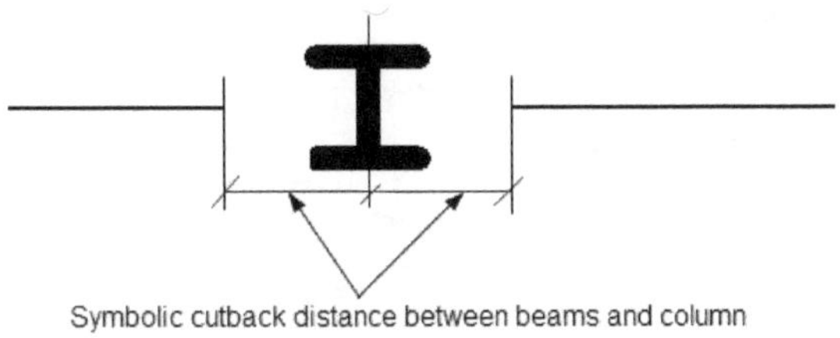

***Figure 2-9** The symbolic cutback distance between beams and column at coarse detail level*

This gap will be visible in a coarse detail elevation view only if the **Top Connection Symbol** instance parameter of the bottom column is set to a value other than **None**. The default value in the **Column** edit box is 1/16" for Imperial unit system and in Metric system, the default value is 1.5000mm. Figure 2-10 shows the symbolic cutback distance between columns in an elevation view.

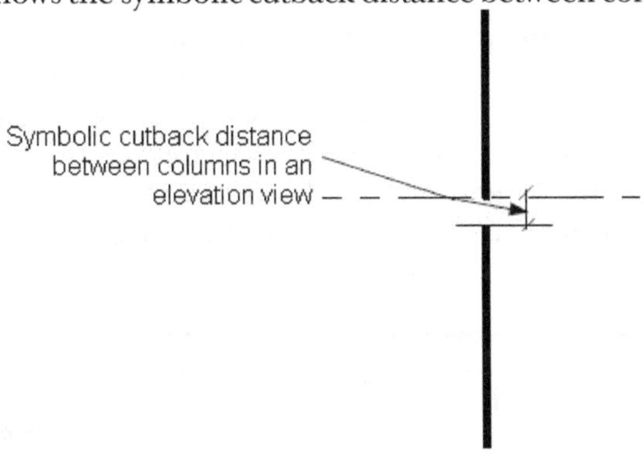

***Figure 2-10** The symbolic cutback distance between columns in an elevation view at a coarse detail level*

The Brace Symbols Area

The settings in the **Brace Symbols** area are used to control the symbolic display of bracing and kicker bracings that exist in a project. Figure 2-11 shows the settings in the **Brace Symbols** area in the **Symbolic Representation Settings** tab of the **Structural Settings** dialog box. The **Plan representation** drop-down list in this area contains two options: **Parallel Line** and **Line with Angle**. The **Parallel Line** option is selected by default. As a result, the symbolic representation of braces in a plan view is displayed with a line parallel to and offset from the actual brace. You can control the offset distance of this parallel line by entering a suitable value in the **Parallel line offset** edit box.

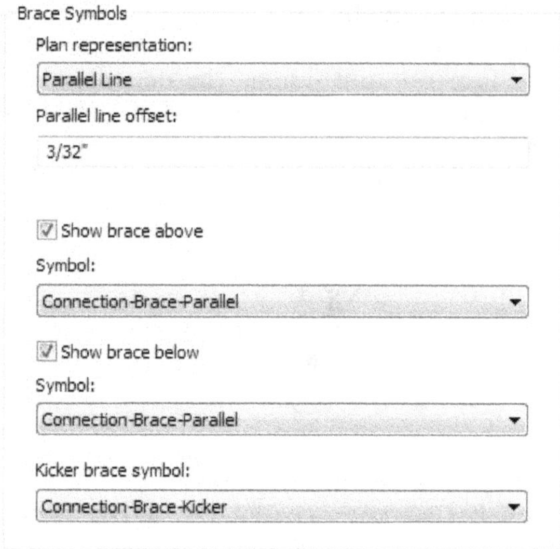

***Figure 2-11** Settings in the **Brace Symbols** area*

In the Imperial unit system, 3/32" is set as the default value in the **Parallel line offset** edit box and in the Metric unit system, default value is 2.5000 mm. Figure 2-12 shows the symbolic representation of a brace in a plan view with default offset distance. On selecting the **Line with Angle** option from the **Plan representation** drop-down list, the symbolic representations of braces in a plan view are displayed with an angled line. The angled line originates from the point of intersection of a brace and level line, if they intersect, or from the closest point to the level line on the brace. The angled line goes upward from the point of origin for the portion of the brace which is above the level or it goes downward for the portion of the brace which is below the level. Figure 2-13 shows a bracing symbol as an angled line.

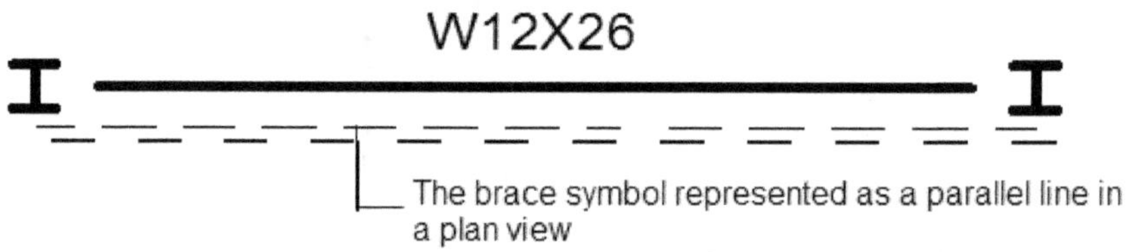

***Figure 2-12** The symbolic representation of a brace in a plan view*

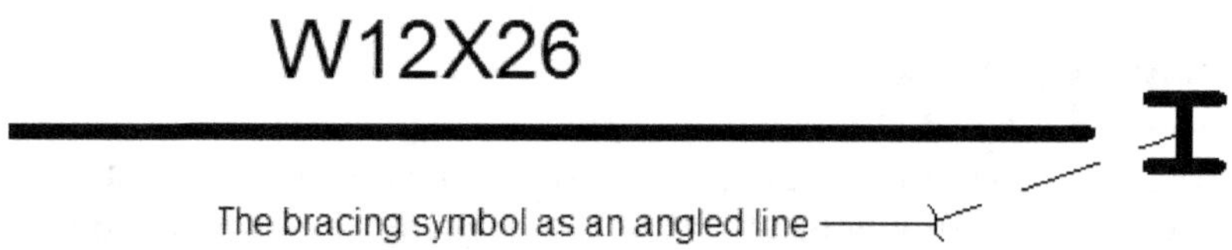

***Figure 2-13** The symbolic representation of a brace as an angled line*

The **Show brace above** check box in the **Brace Symbols** area is selected by default. As a result, the brace symbols representing the braces placed above the current view are displayed. When you select the **Show brace above** check box, the **Symbol** drop-down list below the check box gets activated. From this drop-down list, you can select an option depending upon the option selected from the **Plan representation** drop-down list. You can select the **Connection-Brace -Parallel** option from the **Symbol** drop-down list, if the **Parallel Line** option is selected in the

Plan representation drop-down list. Similarly, if the **Line with Angle** option is selected in the **Plan representation** drop-down list, you can select the **Connection-Brace-Angle** from the **Symbol** drop-down list. On clearing the **Show brace above** check box, the brace option placed above the current view will not be displayed.

Similar to the **Show brace above** check box, the **Show brace below** check box is selected by default. As a result, the brace symbols representing the braces placed below the current view are displayed. The **Symbol** drop-down list below the **Show brace below** check box becomes active when the **Show brace below** check box is selected. The options in the **Symbol** drop-down list located below the **Show brace below** check box are similar to those of the **Symbol** check box located below the **Show brace above** check box.

In the **Kicker brace symbol** drop-down list, the **Connection-Brace-Kicker** option is selected by default. As a result, the symbolic representation of a kicker brace will be a connection-brace in a plan view.

Note

*A normal bracing can be converted into a kicker bracing, if the **Structural Usage** instance parameter in the **Properties** palette of the bracing is set to **Kicker Bracing**, refer to Figure 2-14.*

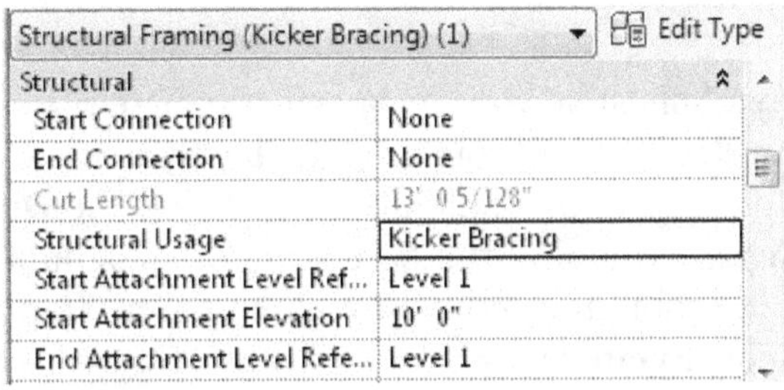

Figure 2-14** The partial view of the **Properties** palette displaying the **Structural Usage** parameter set to **Kicker Bracing

Connection Symbols Area

The options in the **Connection Symbols** area, as shown in Figure 2-15, can be used to control the symbolic display of moment and shear connections for beams, braces, and columns that exist throughout the project. To select the type of framing member for which you want to modify the symbolic display for moment and shear connections, click on the **Display Symbols for** drop-down list and then select any of the three options: **Beams and Braces**, **Column Base**, and **Column Top**. In the **Display Symbols for** drop-down list, the **Beams and Braces** option is selected by default. As a result, the settings for this option will be displayed in a table located below the **Display Symbols for** drop-down list, refer to Figure 2-15. In this table, you will notice two columns: **Connection Type** and **Annotation**

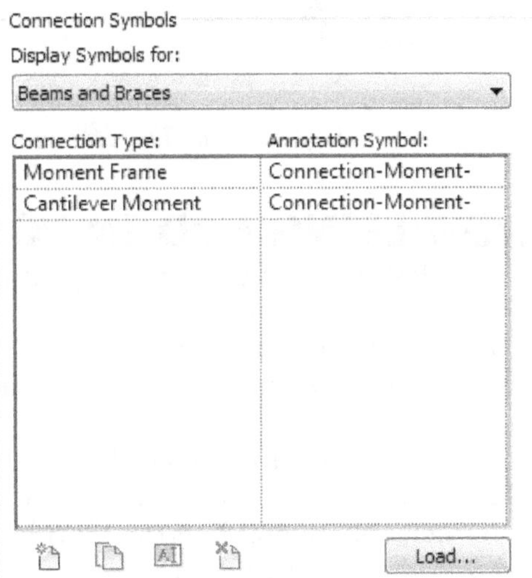

***Figure 2-15** The options in the **Connection Symbols** area*

Symbol. In the **Connection Type** column, two parameters are displayed: **Moment Frame** and **Cantilever Moment**.

The parameters displayed in the **Connection Type** column may vary depending on the option selected from the **Displayed Symbols for** drop-down list. The values for these parameters are displayed in their corresponding fields in the **Annotation Symbol** column. To specify the annotation symbol for a beam or brace of **Moment Frame** parameter, click on the **Annotation Symbol** column corresponding to the **Moment Frame** parameter and select any of the three options: **None**, **Connection-Moment-Filled**, and **Connection-Moment-Not Filled.** Similarly, you can specify an option for **Cantilever Moment** parameter. To add a connection type parameter (for beams and braces) in the table, choose the **Add a connection type** button (first button from the left) displayed below the table, refer to Figure 2-15. On doing so, the **New Connection Type** dialog box will be displayed. In the **Connection Type** edit box of the **New Connection Type** dialog box, enter a name for the new connection type. Next, click in the **Annotation Symbol** drop-down list and select an option to assign an annotation symbol for the connection type. Optionally, in the **New Connection Type** dialog box, you can choose the **Load Symbol** button to load a symbol (with file format *.rfa* or *.adsk)* for the new connection type. Now, choose the **OK** button; the new connection type will be added as parameter in the table. Next, rename, delete, and create a duplicate for the connection type displayed in the table.

In the **Connection Symbols** area, you can specify the base plate symbol for the column. To do so, select the **Column Base** option from the **Display Symbols for** drop-down list and specify a symbol for the base plate from the table displayed below the **Display Symbols for** drop-down list. Similarly, to assign a symbol to the base plate of a column base, you can specify symbols for shear column connection and moment column connection by selecting the **Column Top** option from the **Display Symbols for** drop-down list. Figure 2-16 shows a moment connection and cantilever connection symbols for a steel member.

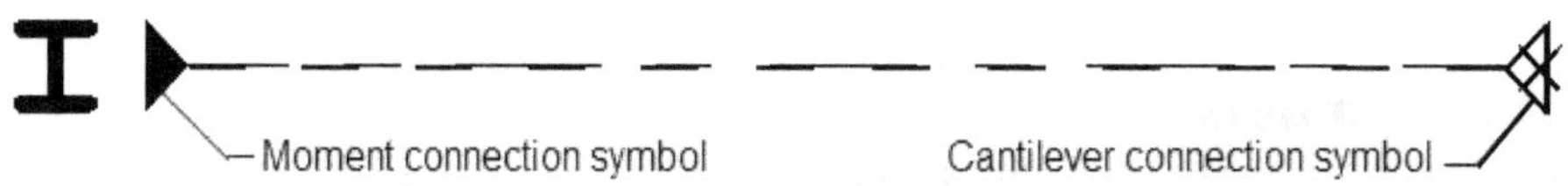

Figure 2-16 *The moment connection and cantilever connection symbols*

INSTALLING ADD-IN FOR STEEL CONNECTION

New

In Autodesk Revit, the Revit API platform can be used to maximum advantage by creating and integrating customized structural connections. Autodesk Revit supports add-ins for user designed parametric structural components. Unless structural connection add-ins are installed, only the symbolic generic connection will be available.

To place the structural connections in Autodesk Revit, you have to first install the **Autodesk Steel Connections** for Revit add-in in your system. The add-in will be available only for the subscribers. The subscribers can download the add-in from their subscription account on *https://manage.autodesk.com/*, refer to Figure 2-17. After the installation is complete, you may need to restart Autodesk Revit to activate the add-in. When the add-in is installed, you have access to use 22 available connections.

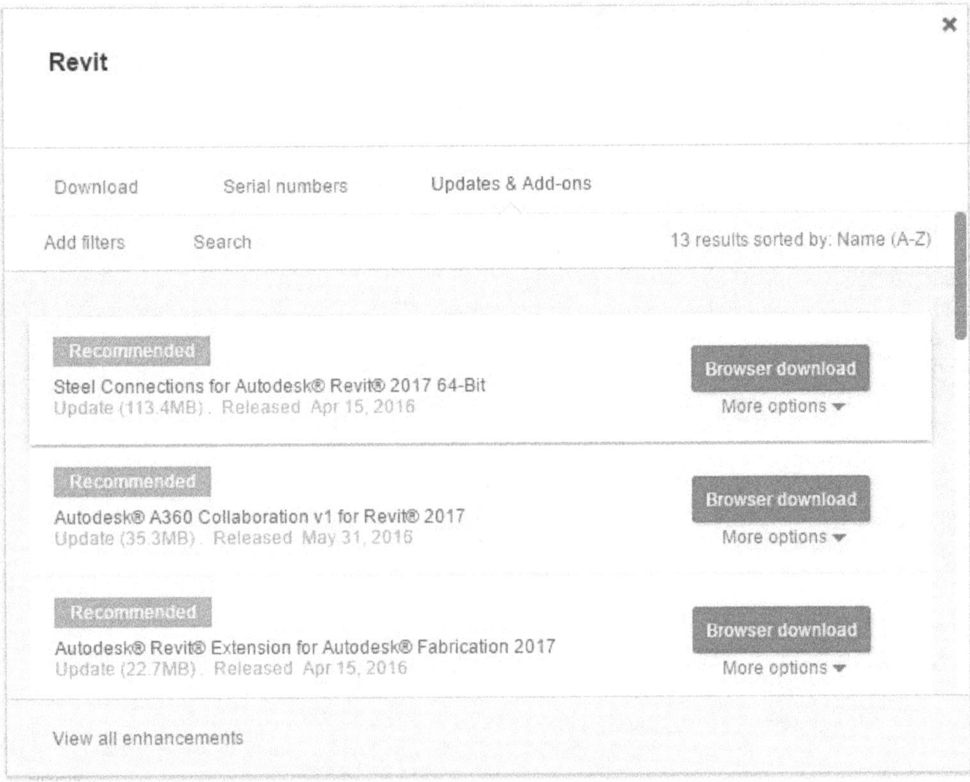

Figure 2-17 The add-in to be downloaded

CONNECTION SETTINGS

New

Ribbon: Manage > Settings > Structural Settings drop-down > Connection Settings

In Autodesk Revit, you can specify various connection settings for structural connections such as the **Beam end to end**, **Column-Beam**, **General bracings**, and so on. You can load different sets of connection in the project by using the **Structural Connection Settings** dialog box. The **Structural Connection Settings** dialog box can be invoked by choosing the **Connection Setting** tool from the **Setting** panel of the **Manage** tab, as shown in Figure 2-18. This dialog box contains two tabs, namely **Connections** and **Parameters**. In this dialog box, the **Connections** tab is chosen by default, as shown in Figure 2-18.

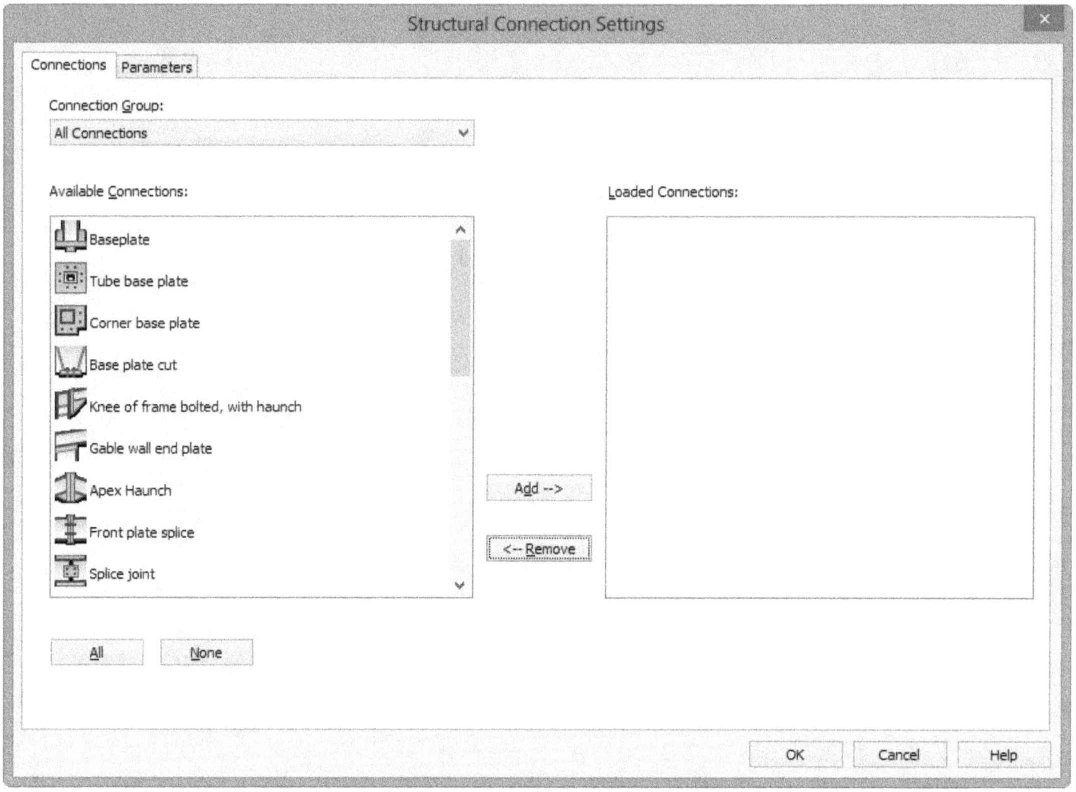

***Figure 2-18** The **Structural Connection Settings** dialog box*

In the **Connections** tab of the **Structural Connection Settings** dialog box, the **All Connections** option is selected in the **Connection Groups** drop-down. In the **Available Connections** area, some of the possible connections between different structural memebers are displayed. To load any of the connection types in the project, select the desired connection type from the **Available Connection** area and choose the **Add** button; the selected type of connection is added in the **Loaded Connections** area. To load more than one connection type in the project from the **Available Connections** area, select the connections and hold the **Ctrl** key to select other connection types and choose the **Add** button.

You can select different types of connection groups from the **Connection Group** drop-down. The **Connection Groups** available in the **Available Connections** area depends on the type of connection selected in the **Connection Group** drop-down. Some of the **Connection Groups** available in this drop-down are **Beam end to end**, **Column-Beam**, **General bracings**, and so on.

Once connections are added in the **Loaded Connections** area, choose the **OK** button, the **Structural Connection Settings** dialog box is closed and the type of connections is loaded in the project.

Now, to set a connection between the structural members, invoke the **Connection** tool from the **Connection** panel of the **Structure** tab. Select the type of connection from the

Type Selector drop-down list in the **Properties** panel and select the structural member between which the connection is to be created. The connection loaded from the **Type Selector** drop-down list is displayed between the member in the 3D view. If the connection type is not displayed in the 3D view, choose the **Visibility/Graphics** option from the **Graphics** panel of the **View** tab; the **Visibility/Graphic Overrides for <View>** will be displayed. Select all the options available in the **Structural Connections** head from the **Visibility** column of the **Model Categories** tab in the displayed dialog box, as shown in Figure 2-19.

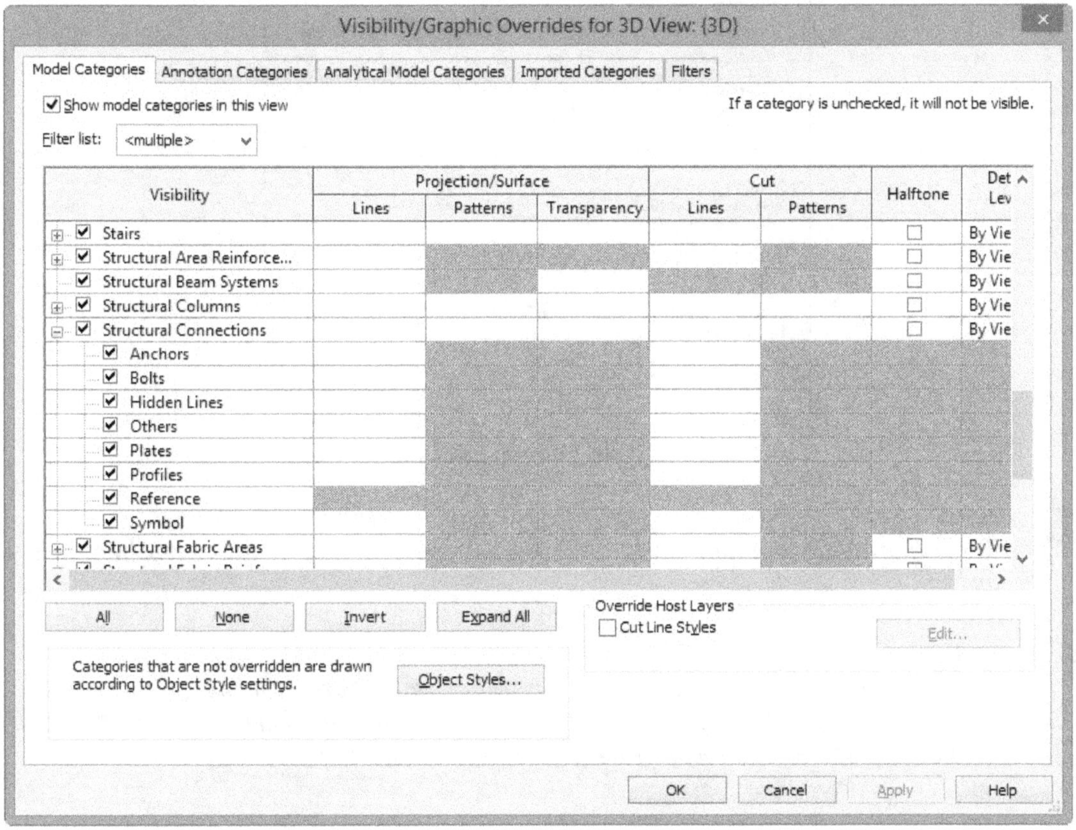

Figure 2-19 The check box selected from the ***Structural Connection*** *head*

SNAPS TOOL

Ribbon: Manage > Settings > Snaps

The **Snaps** tool is one of the important tools used for creating and editing elements in a structural model. This tool is used to make the cursor snap or jump depending on the preset increments or specific object properties of elements such as endpoint, midpoint, and so on. Invoke the **Snaps** tool from the **Settings** panel of the **Manage** tab; the **Snaps** dialog box will be displayed, as shown in Figure 2-20. This dialog box has three areas, **Dimension Snaps**, **Object Snaps**, and **Temporary Overrides**. These areas are discussed next.

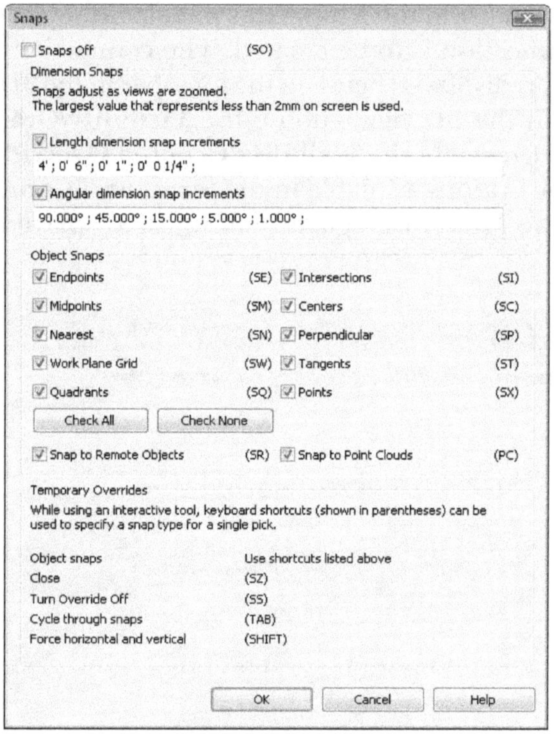

*Figure 2-20 The **Snaps** dialog box*

Note

The settings specified in this dialog box will be applied to all the projects opened in the session and will not be saved.

Tip

*The **Snaps** tool is used frequently not only while creating various building elements but also while editing and placing them. Besides making the modeling simpler, by using this feature efficiently, you can also improve the performance and accuracy of your project.*

Dimension Snaps Area

In the **Dimension Snaps** area, you can set the length and angle dimension snap increments for placing various elements and components in a project. The dimension snap determines the increment by which the cursor will jump/snap along with length or angular dimension while placing elements and components in a project.

The **Length dimension snap increments** check box is selected by default. As a result, you can set the snap increment for the length dimension. On selecting the **Length dimension snap increments** check box, you can enter suitable value(s) in the edit box to set the snap value for length. The default values in this edit box are: 4'; 0'6"; 0'1"; 0'0 1/4", if the **US Imperial** template is selected. While, on selecting the **US Metric** template the default values in this edit box are: 1000; 100; 20; 5. Note that every incremental value is separated by a semicolon (;). You can also set increments by typing the values that are separated by a semicolon. For example, to create an interior layout plan in which the length of the partitions is in 5' (1524mm) modules, counter top width is 2' (610mm), and thickness of partitions is 4" (100mm), you can enter the values for the dimension snaps as **5'; 2';4"** (US Imperial) and **1524; 610; 100** (US Metric). This will enable the cursor to move in these increments and help to create the layout with relative ease. Similarly, you can set the angular dimension snap increments. However, ensure that the **Angular dimension snap increments** check box is selected and then enter suitable value(s) in the edit box below it. This parameter is quite useful for the projects that have radial geometry. To restore the default setting choose the **Restore Defaults** button from the **Snaps** dialog box.

Note

The reason for specifying multiple length and angle snap increment values in the edit boxes is that the priority of increments may change on changing the zoom level of the drawing. Therefore, when you zoom in the drawing, Revit will use smaller increment values, whereas on zooming out, the larger increment values will be used.

Object Snaps Area

Object Snaps refer to the cursor's ability to snap to geometric points on an element such as its endpoint, midpoint, perpendicular, and so on. In the **Object Snaps** area of the **Snaps** dialog box, you can use various options to snap points of elements or objects in a project. The advantage of using these options is that you do not need to specify the exact point in a drawing. When object snaps are enabled, the suitable object snap is displayed as the cursor is moved close to an element. For example, it is virtually impossible to pick the exact endpoint to start a wall from the endpoint of an existing wall. But when you enable the endpoint object snap, the cursor automatically jumps or snaps to the endpoint of this wall. This helps to start the new wall from the endpoint. This, besides making the drawing accurate, later helps add dimensions to the project.

Tip

*The **Dimension Snaps** parameter should be set based on the scale and the amount of detailing required for the project. You may set smaller increments for working on a detailed or small portion of a building.*

Note

The object snaps tool works only with the objects that are visible on the screen. A tooltip, with the same name as the object snap, is also displayed when you bring the cursor close to a snap point.

In a structural model, you can use various **Object Snaps** modes such as **Endpoints**, **Midpoints**, **Nearest**, **Work Plane Grid**, **Quadrants**, **Intersections**, **Centers**, **Perpendicular**, **Tangents**, **Points**, **Snap to Point Clouds**, and **Snap to Remote Objects**.

The name of each object snap option resembles its usage in the project. For example, the **Work Plane Grid** snap option helps you snap the intersection points of grid lines in a work plane grid that is displayed for the current work plane.
Each object snap mode is represented by a geometrical shaped marker to identify it from other object snaps. For example, the endpoint object snap is represented by a square, midpoint by a triangle, nearest by a cross, and so on. To use an object snap mode, move the cursor on the object. As you move it close to the snap point, a marker appears. To select the appropriate snap point, click when the corresponding marker or tooltip is displayed.

In Revit, all enabled object snaps work simultaneously. You can turn off all snap options, including dimension snaps and object snaps by selecting the **Snaps Off** check box located at the top of the **Snaps** dialog box. Alternatively, you can type **SO** on the keyboard to turn them off and on while using a tool. The **Check All** and **Check None** buttons are used to select or clear all check boxes (except the **Snap to Remote Objects**) in the **Object Snaps** area.

Temporary Overrides Area

The options in the **Temporary Overrides** area provide you with an alternative of overriding snaps setting for a single use only. For example, if you have not selected the **Endpoints** check box in the **Snaps** dialog box and you want to use this option while working with a tool, you do not need to open the **Snaps** dialog box to select this option. Instead, you can type the shortcut, **SE** in this case, to temporarily activate the endpoint object snap. Once you have used this object snap option, snapping to the endpoint is automatically turned off.

Using overrides, you can toggle between various object snap options available at the same location. To do so, press the TAB key while snapping the points in the drawing. You can also use other overrides like pressing the SHIFT key to create elements vertically or horizontally. This restricts the movement of the cursor in the orthogonal directions only. Once you release the SHIFT key, the cursor resumes its movement in all directions. You can select the **Snaps Off** check box to disable all types of snapping.

OPTIONS DIALOG BOX

In Autodesk Revit, you can configure global settings by using the **Options** dialog box. This dialog box can be invoked by choosing the **Options** button from the **Application Menu**. The **Options** dialog box, as shown in Figure 2-21, contains nine tabs: **General**, **Graphics**, **File Locations**, **Rendering**, **Check Spelling**, **SteeringWheels**, **ViewCube**, **User Interface**, and **Macros**. These tabs are discussed next.

General Tab

The **General** tab is chosen by default and contains five areas: **Notifications**, **Username**, **Journal File Cleanup**, **Worksharing Update Frequency**, and **View Options**. These areas are discussed next.

Notifications Area

Revit provides an option for setting reminders to save a work at regular intervals. To do so, select the desired option from the **Save reminder interval** drop-down list in the **Notifications** area. By default, 30 minutes is selected in this drop-down list. If you do not want a reminder to save your work in the project, select the **No reminders** option from this drop-down list. Similarly, you can select a value in the **Synchronize with Central reminder interval** drop-down list.

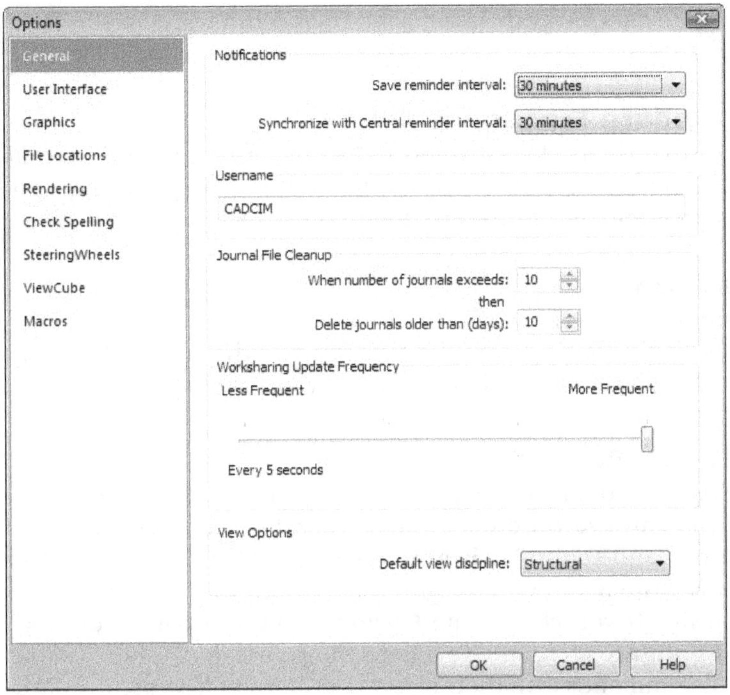

Figure 2-21 *Setting the file options using the* ***Options*** *dialog box*

Username Area

In Autodesk Revit, if you are not signed in as a user of Autodesk account, then the username will not be displayed. To enter a name in the **Username** area, you need to sign in to Autodesk 360. When you sign in, your Autodesk ID will be used as a username.

Note

If you are not logged into an Autodesk account, system username will be taken as the default username.

Journal File Cleanup Area

Journal files are the text files that are used to resolve the technical problems which occur during the Revit session. These files record every step during the session. Whenever you encounter a technical problem with the software, you can run this file to detect the problem or recover the lost files or to know the steps that may have caused the problem. In Autodesk Revit, these files are saved in the following default location: *C:\Users\<Username>\AppData\Local\Autodesk\Revit\Autodesk Revit 2017\Journals* for Windows 7 or Window Vista users and *C:\Documents and Settings\<Username>\Local Settings\Application* for Windows XP users. These files are saved each time you close the Revit session. Therefore, the number of these files keeps on increasing until you remove these files from their location. However to remove these files while retaining some of them, you can use the **Journal File Cleanup** area in the **General** tab of the **Options** dialog box. This area contains two spinners: **When number of journal exceeds** and **Delete journals older than (days)**. You can set the required values in these spinners to retain the recently created files. The default value in both the edit boxes is 10. If you want to delete journal files if their number exceeds 15 and if they were created before 30 days, then in such a situation, set the value in the

When number of journals exceeds spinner to **15** and the value in the **Delete journals older than (days)** spinner to **30**.

Worksharing Update Frequency Area

In this area, you can set the update frequency that indicates the time interval for updating the project in a worksharing environment. To set the limits for worksharing, you can set the slider between the **Less Frequent** and **More Frequent** limits.

View Options Area

In the **View Options** area, you can specify the default view discipline to be used in the Revit project. To do so, select an option from the **Default view discipline** drop-down list in this area. In this drop-down list, the **Structural** option is selected by default.

User Interface Tab

The **User Interface** tab contains two areas: **Configure** and **Tab Switching Behavior**. In the **Configure** area, you can use different options such as **Active theme**, **Keyboard Shortcuts**, **Double-click Options**, and **Tooltip assistance** to change the appearance and operational settings. You can change the display of active theme by selecting appropriate option from the **Active theme** drop-down list. If you choose the **Customize** button corresponding to the **Keyboard Shortcuts** in the **Configure** area, the **Keyboard Shortcuts** dialog box will be displayed. In this dialog box, various commands with their respective shortcut keys are displayed. You can use the **Assign** and **Remove** buttons to add and remove the shortcut keys.

In the **Configure** area, you can select an option from the **Tooltip assistance** drop-down list to set the required level of assistance that a tool tip will display when a cursor is placed over a tool or an element in the project view. From the drop-down list, you can select any of the following options: **Normal**, **None**, **Minimal**, and **High**. By default, the **Normal** option is selected in the **Tooltip assistance** drop-down list.

In the **Tab Switching Behavior** area, you can specify the tab to be displayed once you clear a selection or exit a tool. In this area, the **Project Environment** drop-down list contains two options: **Stay on the Modify tab** and **Return to the previous tab**. Select the **Stay on the Modify tab** option to display the options in the **Modify** tab after exiting a tool or clearing a selection. Alternatively, you can select the **Return to the previous tab** option to display the last used tab after exiting a tool or clearing a selection. Similarly, you can use the settings in the **Family Editor** heading to specify the display behavior of the family editor mode. In the **Tab Switching Behavior** area, the **Display the contextual tab on selection** check box is selected by default. As a result, the contextual tab is displayed once you select a tool from the Revit interface.

Graphics Tab

The options in the **Graphics** tab enable you to configure the display card of your computer to improve the display performance. You can also use this tab to assign colors to selections, highlights and alerts, and enable anti-aliasing for 3D views. In the **Graphics Mode** area of this tab, the **Use Hardware Acceleration** check box is selected by default. As a result, the hardware accelerators are enabled. Hardware accelerators help display the larger models faster on refreshing the views. In addition, the hardware accelerators help you speed up the process of switching between the windows of views. In this area, the **Allow navigation during redraw** check box is selected by default. The selection of check box allows you to navigate the model in a 2D view or a 3D view, and also pan, zoom and orbit while sketching the drawing. On clearing the check box, you will not be able to perform any of the above actions before the completion of drawing. In this area, you can select the **Smooth line with anti-aliasing** check box to improve the quality of lines in 3D views. In the **Graphic** tab area of the **Options** dialog box, you can select the **Use for all views** radio button to apply the settings to all views in Revit. In the **Colors** area of the **Graphics** tab, choose the button corresponding to the **Background** parameter to change the color of the background. The **Selection** parameter refers to a color that an element acquires when it is selected. The default color is **RGB 000-059-189**. To use any other color, click the button on the right of the **Selection** parameter and select the desired color from the **Color** dialog box. The **Pre-selection** parameter is used to specify the color of the highlighted elements. To use any other color, click the button on the right of the **Selection** parameter to display the **Color** dialog box and then select the desired color from this dialog box. Revit uses the **Alert** button to highlight elements when an error occurs. In the **Colors** area, you can select the **Semi-transparent** check box to make the selected elements semi-transparent. Semi-transparency of a selected element allows you to view the elements behind it. In the **Temporary Dimension Text Appearance** area, you can select an option from the **Size** drop-down list to specify the size of the text used in temporary dimensions. In this area, you can set the background of the text in the temporary dimensions. To do so, select the **Opaque** or **Transparent** option from the **Background** drop-down list.

File Locations Tab

By using the **File Locations** tab, you can set the path for various files and directories that are accessed frequently, refer to Figure 2-22. The path for these files is set while installing Revit. However, you can modify the location of a file by choosing the corresponding **Browse** button and specify a new location.You can also change the default template file location, in case, you wish to use a custom made template file for your projects. In Revit, the default path to save or open a project can be specified in the **Default path for user files** edit box. You can also specify the default path for family template files and for the point clouds by using the corresponding **Browse** button.

Rendering Tab

Autodesk Revit uses the mental ray rendering engine for its rendering process. The mental ray has its own library, Render Appearance Library. This library stores information about render appearances for materials, default RPC contents in the software, and other information relevant to the rendering process. The Render Appearance Library is a read-only library and is loaded into the following default location while installing the Revit software: *C:\Program Files (x86)\ Common Files|Autodesk Shared|Materials|2017|assetlibrary_base.fbm.*

In addition to the default Render Appearance Library location, you can also specify paths for the additional image files defining texture, bump map, and custom color for the render appearance that you can use in the project. These image files are not present in the software and therefore, you need to specify their paths to use them. To do so, choose the **Add Value** button in the **Additional Render Appearance Paths** area and specify the required path in the displayed field or choose the **Browse** button; the **Browse for Folder** dialog box will be displayed. In this dialog box, select the desired path and choose the **Open** button to add the path in the field.

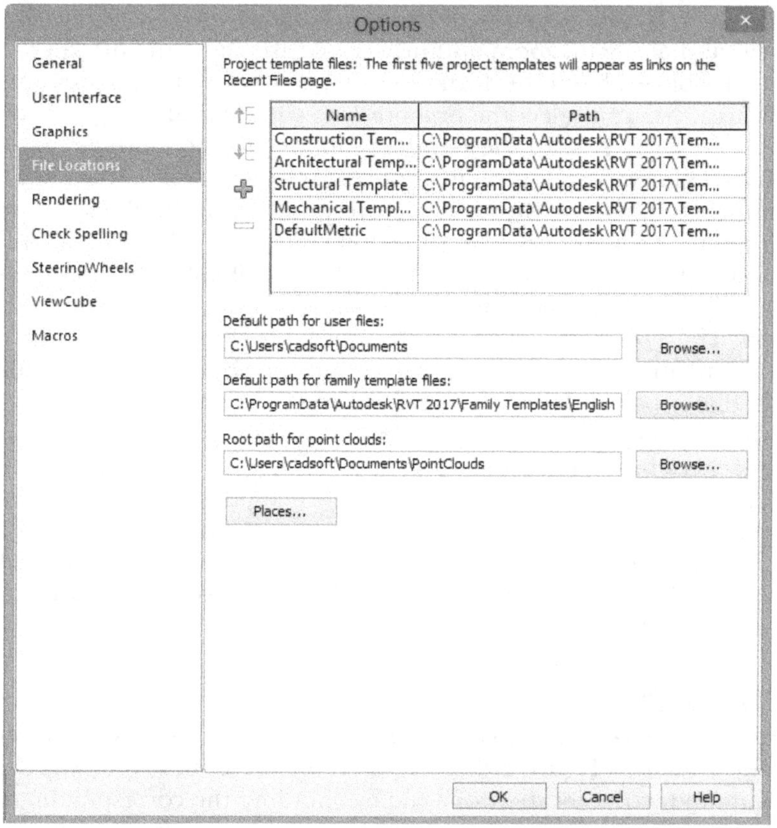

*Figure 2-22 The **File Locations** tab of the **Options** dialog box*

Check Spelling Tab

Revit provides you with the option to run spell check to find out spelling errors in text and then rectify them. You can choose the **Check Spelling** tab from the **Options** dialog box to display its options. Figure 2-23 shows various options in the **Check Spelling** tab. In the **Settings** list box of this tab, you can specify various self-explanatory settings by selecting their respective check boxes. In the **Main Dictionary** area, you can select the type of dictionary to be used as main dictionary for the spell check from the **Autodesk Revit** drop-down list. Apart from the main dictionary, you can also use additional dictionaries available in this dialog box such as the personal and building industry dictionaries. This facilitates the use of various personal and industrial terms in the text. There are many words that are not included in the main dictionary but are frequently used in the building industry. For example, the abbreviation 'conc' for the word concrete is not available in the main dictionary. The additional building industry dictionary has many such words and abbreviations that can be used in the text of the project and therefore, it does not prompt you to check for spelling errors when it finds such terms/words. You can also add or remove words from your personal and building industry dictionary. To do so, choose the **Edit** button next to the required option in the **Additional Dictionaries** area to view the list of words and then enter or remove any word from these lists by using the cursor and keyboard. To run spell check in your drawing, choose the **Check Spelling** tool from the **Text** panel of the **Annotation** tab; the **Spelling** dialog box will be displayed, wherein you can rectify spelling errors in the text by selecting the correct spelling and then choosing the **Change** button in the dialog box. Alternatively, you can press the F7 key to display the **Check Spelling** dialog box.

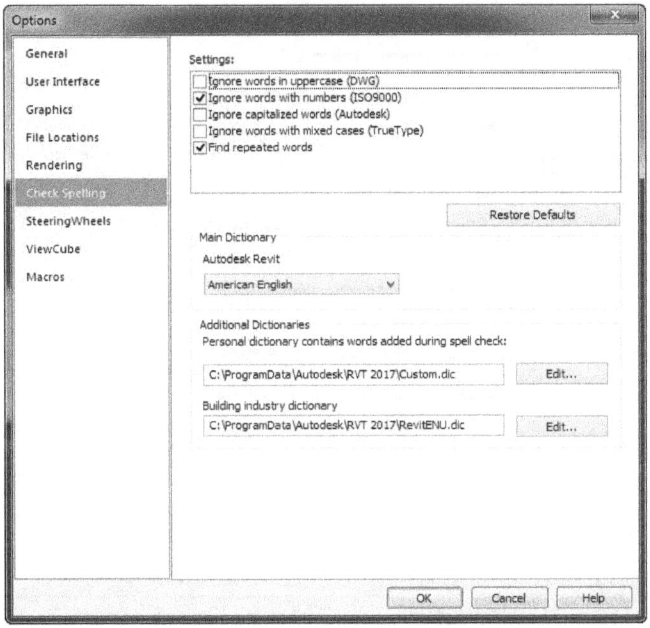

***Figure 2-23** The **Check Spelling** tab of the **Options** dialog box*

SteeringWheels Tab

The **SteeringWheels** tab in the **Options** dialog box has options to control the visibility, appearance, and operational tools of different types of SteeringWheels. This tab has seven different areas, as shown in Figure 2-24. These areas are discussed next.

Text Visibility Area

You can control the display of tool messages, tooltips, and tool cursor of SteeringWheels by using the options in the **Text Visibility** area of the **SteeringWheels** tab. You need to select the **Show tool messages** check box to enable the visibility of tool messages in SteeringWheels. To display tooltips along with SteeringWheels, select the **Show tooltips** check box in this area. Similarly, to control the display of the cursor text when a tool is active, select the **Show tool cursor text** check box.

Big Steering Wheel Appearance and Mini Wheel Appearance Areas

The options in these areas are used to set the size and transparency of the SteeringWheels. To set the size of SteeringWheels, select the required option from the **Size** drop-down list in the corresponding areas and set its size to small, normal, or large. Similarly, you can set the transparency of SteeringWheels by selecting the required option from the **Opacity** drop-down list.

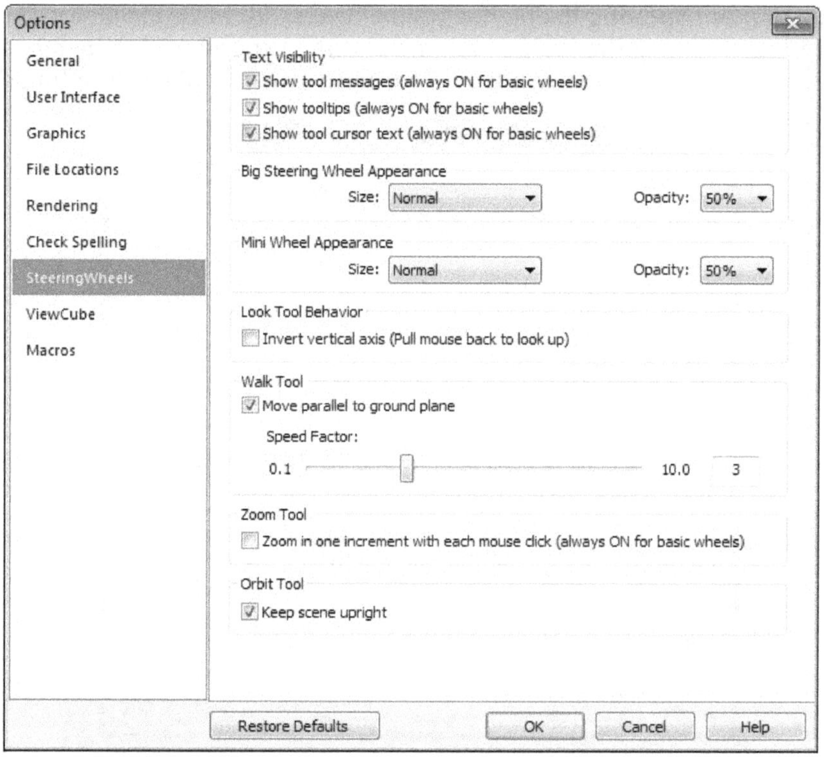

*Figure 2-24 The **SteeringWheels** tab of the **Options** dialog box*

Look Tool Behavior and Walk Tool Areas

In the **Look Tool Behavior** area of the **SteeringWheels** tab, select the **Invert vertical axis** check box to change the vertical axis movement of the view. Selecting this check box enables the view to move in the same direction as the cursor does in the vertical axis.

The **Speed Factor** slider in the **Walk Tool** area is used to change the speed of walk while using the **Walk** tool of the SteeringWheels. You can also select the **Move parallel to ground plane** check box in the **Walk Tool** area to constrain the angular movement of the walk to ground plane.

Zoom Tool and Orbit Tool Areas

Select the **Zoom in one increment with each mouse click** check box in the **Zoom Tool** area to enable the zooming operation with a single click.

In the **Orbit Tool** area, select the **Keep scene upright** check box to maintain perpendicularity between the sides of the model and the ground plane while using the **Orbit** tool.

ViewCube Tab

The **ViewCube** tab in the **Options** dialog box is used to edit various settings of the ViewCube. It has four different areas: **ViewCube Appearance**, **When Dragging the ViewCube**, **When Clicking on the ViewCube**, and **Compass**, as shown in Figure 2-25. These areas are discussed next.

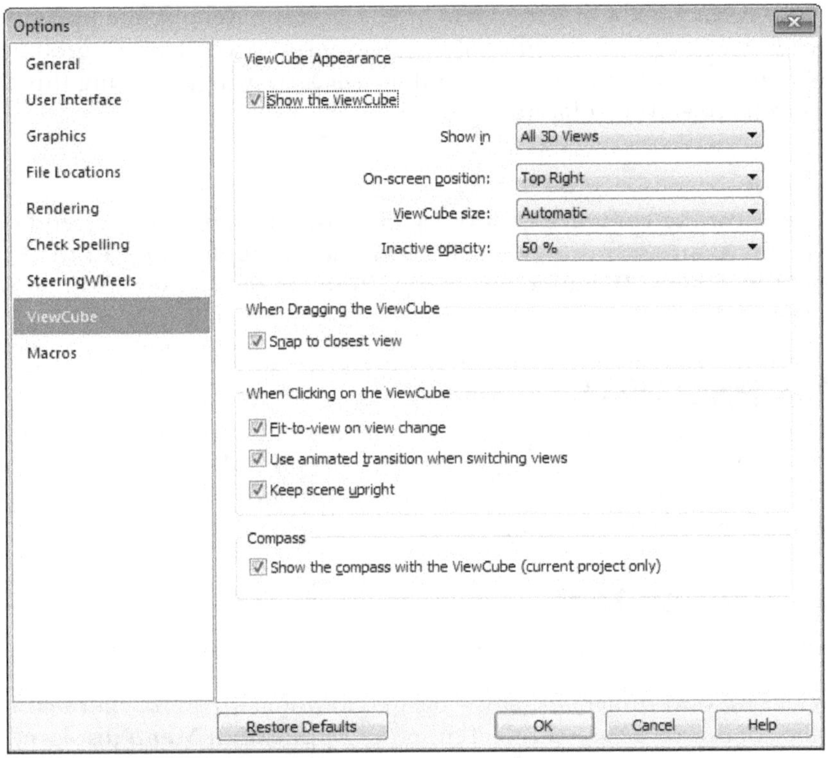

*Figure 2-25 The **ViewCube** tab of the **Options** dialog box*

ViewCube Appearance Area

This area is used to control the appearance and display of the **ViewCube**. In this area, the **Show the ViewCube** check box is selected by default. As a result, the ViewCube will be visible. If you clear this check box, the ViewCube will disappear and all options in the **ViewCube** tab will be deactivated. In the **ViewCube Appearance** area, you can use various drop-down lists to align, resize, and change the transparency of the ViewCube.

Select the options from the **On-screen position** drop-down list to align the ViewCube on the screen. Similarly, if you want to resize the ViewCube, select the required option from the **ViewCube size** drop-down list. You can also set the opacity of the inactive ViewCube by selecting an option from the **Inactive opacity** drop-down list.

When Dragging the ViewCube Area

Select the **Snap to closest view** check box in this area to enable the snap to select the closest view in the ViewCube.

When Clicking on the ViewCube Area

Select the **Fit-to-view on view change** check box in the **When Clicking on the ViewCube** area to fit the view on the screen while changing the viewing direction. In this area, the **Use animated transition when switching views** check box is selected by default. As a result, the animated transition occurs while switching the views. If you clear this check box, the animated transition will not occur while using the ViewCube. Select the **Keep scene upright** check box to keep the sides of ViewCube and the sides of the view perpendicular to the ground plane. Clear the check box to turn around the model in full 360-degree swing. Clearing this check box can be useful when you are editing a family.

Compass Area

In this area, the **Show the compass with the ViewCube** check box is selected by default. As a result, the compass along with the ViewCube is visible in the drawing. In the **ViewCube** tab, you can choose the **Restore Defaults** button to restore the default settings that were changed in its different areas.

SAVING A PROJECT

Before you close or exit a Revit session, it is recommended to save the project file. You can save a project file in a permanent storage device, such as a hard disk or a removable storage device like CD or USB. Also, you must save your work at regular intervals to avoid data loss due to any error in the computer's hardware or software.

Using the Save As Tool

Application Menu: Save As > Project

In Revit, you can save your project file at the desired location. To do so, choose the **Application** button and then choose **Save As > Project** from the **Application Menu** displayed; the **Save As** dialog box will be displayed, refer to Figure 2-26. Alternatively, you can save the project file by choosing the **Save** button from the **Quick Access Toolbar**.

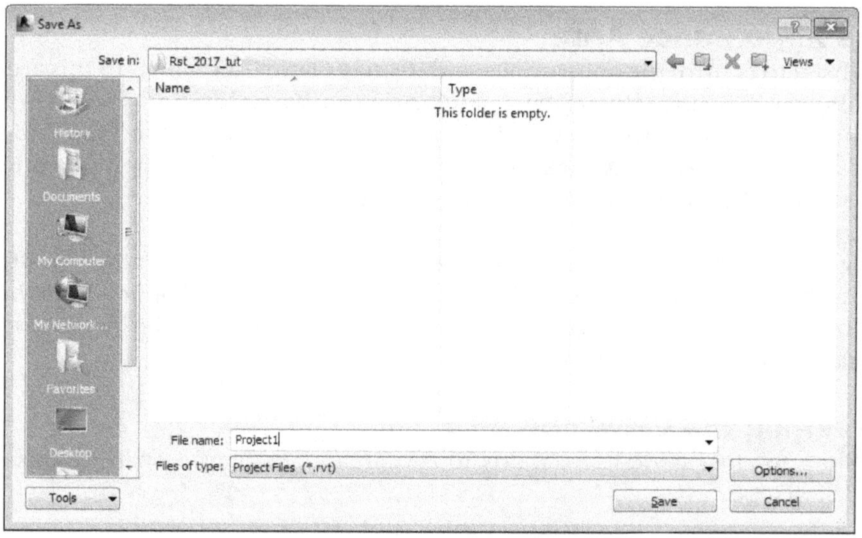

*Figure 2-26 The **Save As** dialog box*

In the **Save As** dialog box, the **Save in** drop-down list displays the current drive and path in which the project file will be saved. The list box below the **Save in** drop-down list shows all folders available in the current directory. The **File name** edit box is used to specify the name of the file to be assigned to the project. The **Places List** area on the left of the **Save As** dialog box contains shortcuts for the folders that are frequently used.

You can use different file saving features by choosing the **Options** button from the **Save As** dialog box. On choosing this button, the **File Save Options** dialog box will be displayed, as shown in Figure 2-27. Using the **Maximum backups** edit box from this dialog box, you can specify the maximum number of backup files that you need to store for a project. In Autodesk Revit, the non-workshared projects have three backup files and the workshared projects have twenty backup files by default. The options in the **Worksharing** area are inactive for non-workshared projects. You can use the options in this area to make the current workshared file central compact and select the default workset. The options in the **Thumbnail Preview** area enable you to specify the image to be used as the preview of the project file. This image is used at the time of opening a project file. You can specify a view of the model as a preview image by selecting the corresponding option from the **Source** drop-down list in the **Thumbnail Preview** area of the **File Save Options** dialog box. By default, the **Active view/sheet** option is selected in the **Source** drop-down list of the **Thumbnail Preview** area for the preview of a project file. For example, to make **Structural Plan: Level 1** as the preview image, select it from the drop-down list. As a result, when you invoke the **Open** dialog box and select a file to open, the view displayed in the **Preview** area will correspond to the selection that is made from the **Source** drop-down list in the **File Save Options** dialog box.

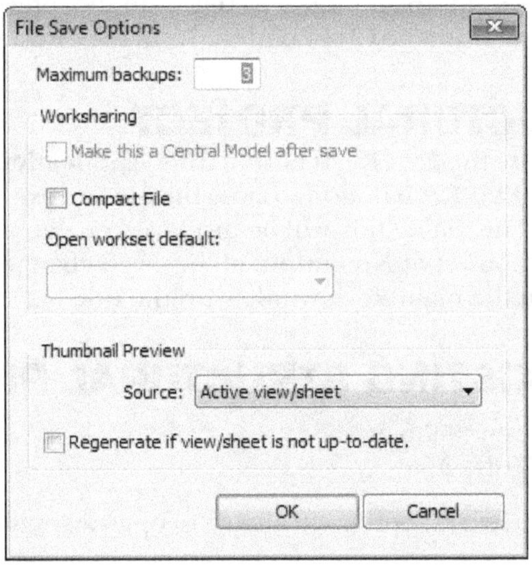

Figure 2-27 *The* ***File Save Options*** *dialog box*

In the **File Save Options** dialog box, you can select the **Regenerate if view/sheet is not up-to-date** check box to see the preview with the latest modifications. On selecting this check box, the preview image will be updated when you close the project file.

Note

Revit updates the preview image continuously. Therefore, selecting the ***Regenerate if view/sheet is not up-to-date*** *check box can consume considerable resources.*

Using the Save Tool

Once a project has been saved using the **Save As** tool, you do not need to re-enter file parameters to save it again. To save a project to the hard disk, choose the **Application** button and then choose the **Save** option from **Application Menu**. If you are saving the project for the first time, the **Save As** dialog box will be displayed, even if you invoke the **Save** tool.

> **Tip**
> *By default, the preview of a file is the last active view or sheet at the time it was last saved. You can set the preview to a particular view using the **Options** button in the **Save As** dialog box.*

Alternatively, you can save your project by choosing the **Save** button from the **Quick Access Toolbar**. As you save your project file, Revit updates it automatically without prompting you to re-enter the file name and path.

CLOSING A STRUCTURAL PROJECT

To close a project, choose the **Application** button and then select the **Close** option from **Application Menu** displayed. If you have already saved the latest changes, the project file will be closed. Otherwise, Revit will prompt you to save the changes through the **Save File** dialog box. You can save the changes by choosing the **Yes** button or discard them by choosing the **No** button. You can also choose the **Cancel** button to return to the interface and continue working on the project file. You can also use the **Close** button (X) in the drawing area to close the project.

EXITING A STRUCTURAL PROJECT

To exit a Revit session, choose the **Exit Revit** button from **Application Menu**. Even if the project is open, you can choose the **Exit Revit** button to close the file and exit Revit. If the project has not been saved once, the **Save File** dialog box will be displayed on choosing the **Exit Revit** button. In this dialog box, if you choose the **No** button, all unsaved changes will be lost. You can also use the **Close** button (**X**) in the main Revit window (in the title bar) to end the Revit session.

OPENING AN EXISTING STRUCTURAL PROJECT

Application Menu:	Open > Project
Shortcut Keys:	CTRL+O

In Autodesk Revit, there are several options available to open an existing project. These options are discussed next.

Opening an Existing Project Using the Open Tool

To open an existing project file, choose **Open > Project** from **Application Menu**, as shown in Figure 2-28. Alternatively, you can open the project file by choosing the **Open** button from the **Quick Access Toolbar** or by pressing the CTRL+O keys. On invoking the **Open** tool, the **Open** dialog box will be displayed, as shown in Figure 2-29. Using the **Look in** drop-down list in this dialog box, you can access the desired folder and open the desired file.

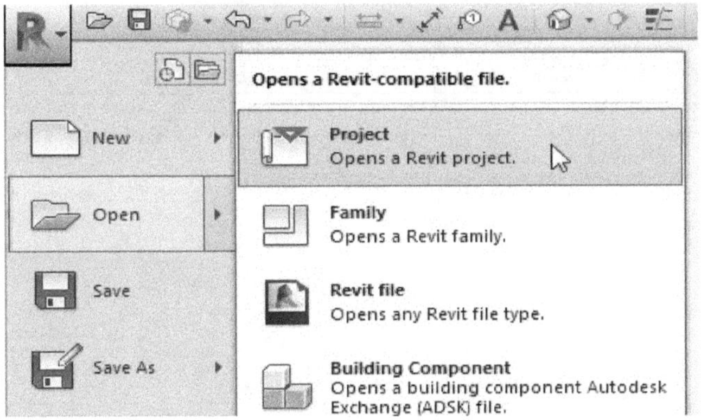

*Figure 2-28 Choosing the **Project** option from **Application Menu***

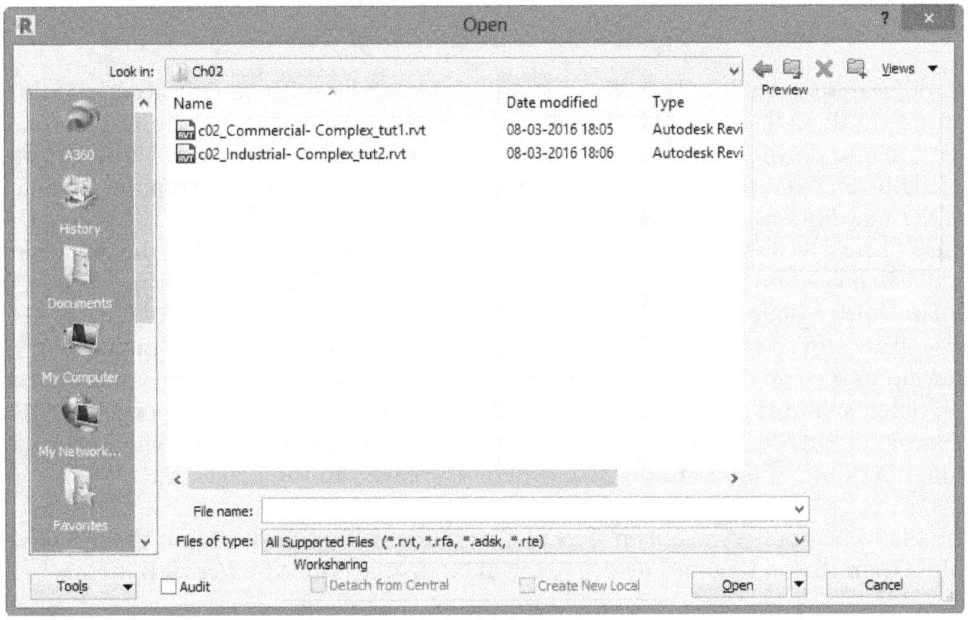

*Figure 2-29 The **Open** dialog box*

The **Preview** area of the **Open** dialog box shows the preview of the selected project file. It helps you select a particular file by viewing its contents, even if you are not sure about the name of the file. The window icons such as the **Views** menu, placed on the right of the **Look in** drop-down list, helps you select a project file based on its size, type, or date when it was last saved. On choosing the **Thumbnails** option from the **Views** menu, you can preview the contents of the project files inside the selected folder in the file list area. In the **Open** dialog box, you can browse to important locations from the **Places** list. The **Places** list is located on the left side of the **Open** dialog box. In this list, you can add or remove folders as per your requirement. To do so, choose the **Options** button from **Application Menu**; the **Options** dialog box will be displayed. In this dialog box, choose the **File Locations** tab and then the **Places** button from it; the **Places** dialog box will be displayed, as shown in Figure 2-30. The **Places** dialog box contains two columns: **Library Name** and **Library Path**.

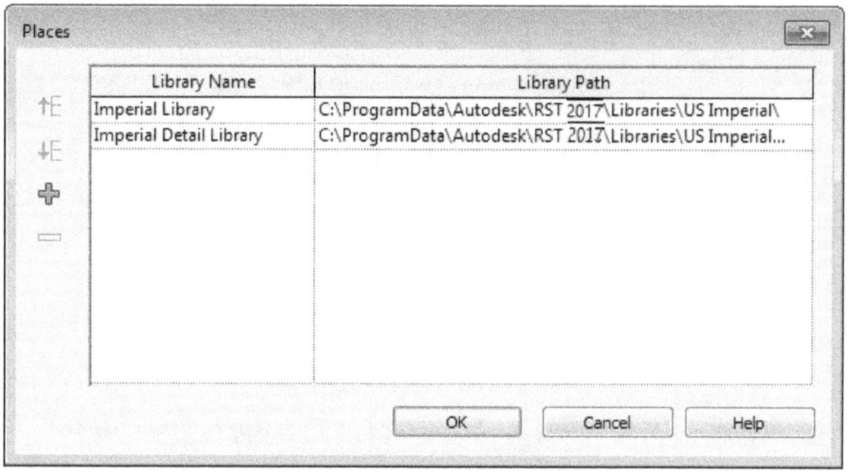

***Figure 2-30** The **Places** dialog box*

You can add or remove folders in the libraries list to create a list of frequently accessed folders. The four buttons on the left side of the **Places** dialog box can be used to create or delete a library, or move it up and down in the list. To create a new library, choose the **Add Value** button, which is the third button from the top; a new library will be added to the defined path. By default, the name of the new library in the **Library Name** section will be **New Library1**. Change the name of the new library and then click in the **Library Path** column to display the **Browse** button. Choose the **Browse** button and select the folder to be added in the libraries list by using the **Browse For Folder** dialog box. Next, choose the **Open** button; the new folder will get added to the list. If required, choose the upward arrow button from the **Places** dialog box to move the folder up to the top of the list. Similarly, you can choose the down arrow button to move it down. To delete a library, select it and choose the **Remove Value** button. Choose the **OK** button in the **Places** dialog box to exit, and then close the **Options** dialog box. When you invoke the **Open** tool next time, the new folder icon will be displayed in the places list.

Once the file to be opened has been selected, its name will be displayed in the **File Name** edit box of the **Open** dialog box and its preview will be displayed in the **Preview** area.

Note

*If you try opening an already opened file that has been modified in the Revit session, a message box appears, prompting you to close the file first and then reopen it. In case, you open a file that has been created using an older version of Revit, the **Program Upgrade** message box is displayed. This message box informs that the file is being upgraded to the latest file format and that this is a onetime process. Once the file is opened, choose the **Save** button from **Application** menu to save it to upgraded Autodesk Revit version.*

Using the Windows Explorer to Open an Existing Project

Apart from using the **Open** tool from the Revit interface to open a file, you can also open files directly from the **Windows Explorer** by using the methods discussed next.

A file can be opened by double-clicking on its icon in the **Windows Explorer**. It opens the project file in the latest Revit session. If Revit is not running, double-click on the file icon to start Revit and then open the file.

Another method of opening a project file is by dragging the project file icon from the **Windows Explorer** and dropping it in the drawing window of the Revit interface. You can also select, drag, and drop more than one file in the drawing window. In this case, Revit prompts you to open the files in separate windows. Choose the **OK** button to open all files in the same Revit session.

TUTORIALS

In this tutorials, you will create two structural projects as tutorials: a commercial project of a concrete structure and an industrial project of steel and concrete structure. These tutorials will be created in a sequence according to the topics discussed in the chapter.

Tutorial 1 Commercial Complex

In this tutorial, you will create a revit file for the *Commercial Complex* project with the following parameters and project specifications: **(Expected time: 15 min)**

1. Template file:
   For Imperial **US Imperial > Structural Analysis-Default.rte**
   For Metric **US Metric > Structural Analysis-DefaultMetric.rte**

2. Project Units:
   For Imperial **Feet and fractional inches**, Rounding- **To the nearest 1/2"**
   For Metric **Millimeters**, Rounding- **To the nearest 10**
3. Length dimension snap increment:
   For Imperial **5' ; 2' 6" ; 0' 3" ; 0' 01/2"**
   For Metric **1500 ; 700 ; 70 ; 10**

4. File name to be assigned:
   For Imperial *c02_Commercial-Complex_tut1.rvt*
   For Metric *M_ c02_Commercial-Complex_tut1.rvt*

The following steps are required to complete this tutorial:

a. Start a Revit session.

b. Use the template file for the project,
   For Imperial **Structural Analysis-Default.rte**.
   For Metric **Structural Analysis-DefaultMetric.rte**

c. Set **Endpoint, Midpoint**, **Nearest**, **Perpendicular**, **Work Plane Grid**, and **Intersection** as the object snaps in the **Snaps** dialog box for both the systems (Imperial & Metric).

d. Save the project by using the **Save** tool.
   For Imperial *c02_Commercial-Complex_tut1.rvt*
   For Metric *M_c02_Commercial-Complex_tut1.rvt*

e. Close the project by using the **Close** tool.

Starting Autodesk Revit 2017

1. Start Autodesk Revit by choosing **Start > All Programs > Autodesk > Revit 2017**. On doing so, the Revit interface window is displayed.

Opening a New Project

1. Choose the **Application** button; the **Application Menu** is displayed. Choose **New > Project** from this menu; the **New Project** dialog box is displayed.

Selecting the Template File

To select the template file for the project, you need to access the appropriate folder and select the required template file from it.

1. In the **New Project** dialog box, choose the **Browse** button from the **Template file** area; the **Choose Template** dialog box is displayed. This dialog box shows a list of template files available in the **US Imperial** folder, if **Imperial** unit system is selected at the time of Revit installation. (On selecting **Metric** unit system, list of metric template is displayed).

2. In the **Choose Template** dialog box, the template file **Structural Analysis-Default.rte**, as shown in Figure 2-31, is selected as the default template file in the **US Imperial** folder. In **US Metric** folder, default template file is **Structural Analysis-DefaultMetric.rte**. Choose the **Open** button to assign the selected template file to the **New Project** dialog box.

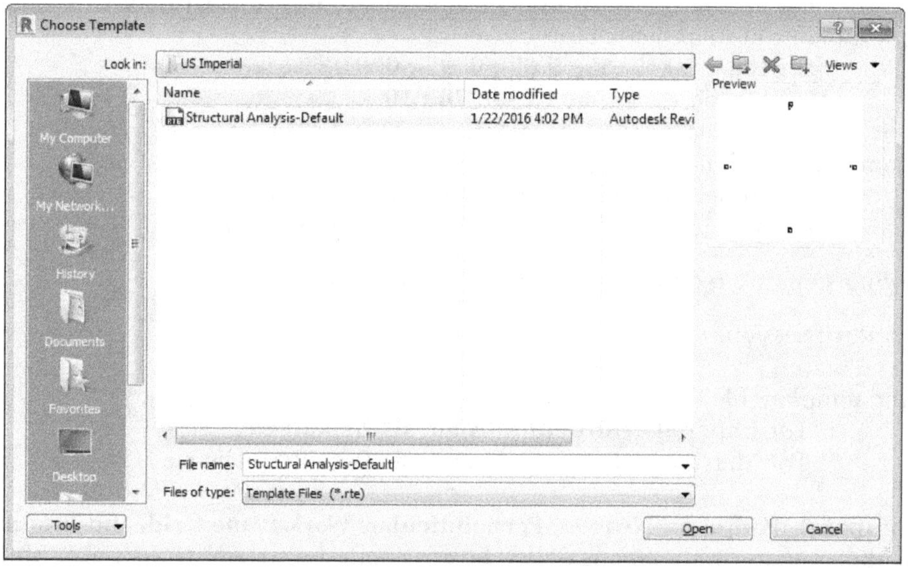

***Figure 2-31** The **Choose Template** dialog box*

3. Select the **Project** radio button (default) to create a project file and then choose the **OK** button; the selected template file of selected unit system is loaded. Note that the **Project Browser** now shows different levels and views that have already been created in the selected template file.

Setting Project Units

1. To set units for the project, choose the **Project Units** tool from the **Settings** panel of the **Manage** tab; the **Project Units** dialog box is displayed.

2. Make sure that the **Common** option is selected in the **Discipline** drop-down list of the **Project Units** dialog box. Now, click in the value field corresponding to the **Length** parameter; the **Format** dialog box is displayed.

3. In this dialog box, select the option by clicking on the **Units** drop-down list.
   For Imperial **Feet and fractional inches**
   For Metric **Millimeters**

4. Select the option from the **Rounding** drop-down list.
   For Imperial **To the nearest 1/2"**
   For Metric **To the nearest 10**

5. Choose the **OK** button in the **Format** dialog box and then in the **Project Units** dialog box to close them.

Setting Dimension and Object Snaps

You need to set dimension snaps and object snaps by using the **Snaps** tool. These settings are based on the type of the project and the amount of detailing required.

1. Choose the **Snaps** tool from the **Settings** panel of the **Manage** tab; the **Snaps** dialog box is displayed.

2. In the **Dimension Snaps** area of this dialog box, ensure that the **Length dimension snap increments** check box is selected.

3. Now, click in the edit box below the **Length dimension snap increments** check box and delete the existing values and then enter the new values.
   For Imperial **5' ; 2' 6" ; 0' 3" ; 0' 0 1/2"**
   For Metric **1500 ; 700 ; 70 ; 10**

   Figure 2-29 shows the value entered for length snaps increment.

4. In the **Object Snaps** area, clear the **Quadrants**, **Centers**, **Tangents**, and **Points** check boxes to disable them. Leave the other options such as **Endpoints**, **Midpoints**, **Nearest**, **Perpendicular**, **Work Plane Grid**, **Snap to Remote Objects**, **Snap to Point Clouds**, and **Intersections** enabled, refer to Figure 2-32. Choose the **OK** button to apply settings and to close the **Snaps** dialog box.

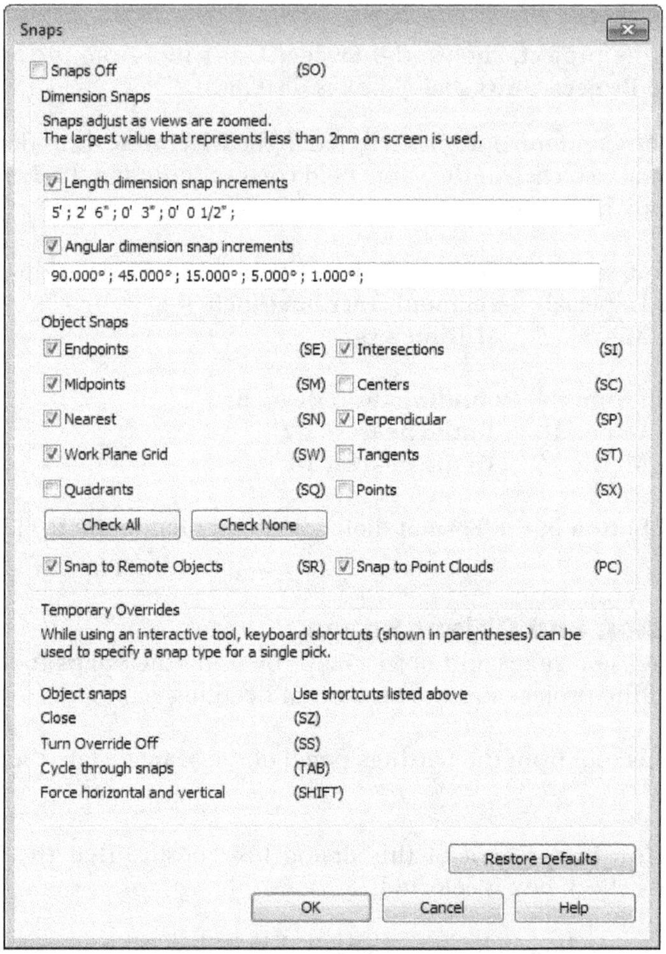

*Figure 2-32 The **Snaps** dialog box*

Saving the Project

The project parameters set in the previous steps are the global settings of the project file. To save this project file, use the **Save As** tool.

1. To save the project, choose the **Save** option from **Application Menu**. As you are saving the project for the first time, the **Save As** dialog box is displayed.

2. In this dialog box, browse to the *C* drive and then create a folder with the name **rst_2017**.

3. Open the *rst_2017* folder and then create a sub-folder with the name *c02_rst_2017_tut*. Next, open the created folder and save the file with the name *c02_Commercial-Complex_tut1.rvt* (For Imperial), refer to Figure 2-33. For Metric, save the file with the name *M_c02_Commercial-Complex_tut1.rvt*.

4. Choose the **Save** button; the **Save As** dialog box closes and the project file is saved.

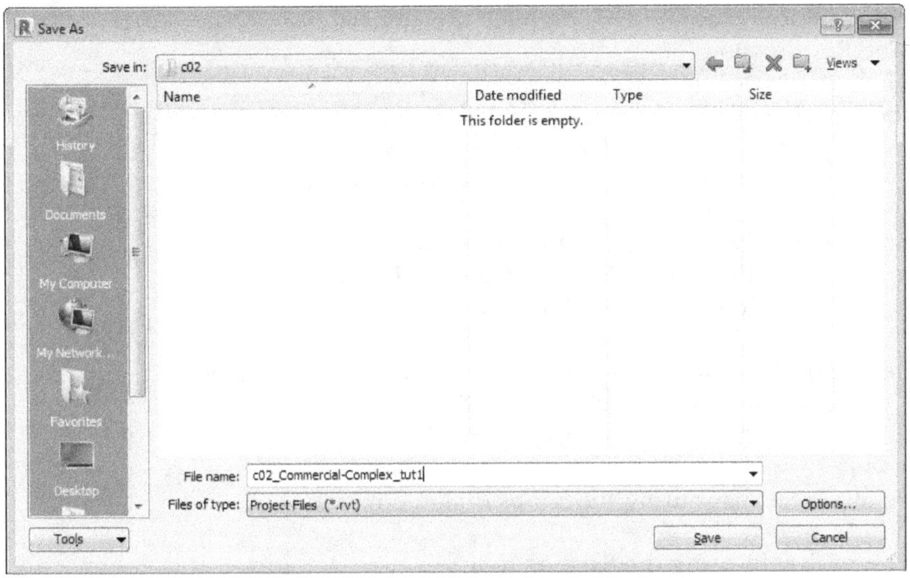

Figure 2-33 The ***Save As*** *dialog box*

Closing the Project

1. To close the project, choose the **Close** option from **Application Menu** in the Ribbon.

Tutorial 2 — Industrial Complex

In this tutorial, you will create a new project file for the *Industrial Complex* project by using the following parameters and project specifications. **(Expected time: 15 min)**

1. Template file:
   For Imperial **US Imperial > Structural Analysis-Default.rte**
   For Metric **US Metric > Structural Analysis-DefaultMetric.rte**
2. Project Units:
   For Imperial **Feet and fractional inches**
   For Metric **Millimeters**

3. Length dimension snap increment:
   For Imperial **10';2'6";1'; 0'3"**
   For Metric **3000;760;300; 70**

4. Object snaps to be set- all available object snaps.

5. File name to be assigned:
   For Imperial *c02_Industrial-Complex_tut2.rvt*
   For Metric *M_c02_Industrial-Complex_tut2.rvt*

The following steps are required to complete this tutorial:

a. Start the Revit session.

b. Use the template file.
For Imperial *Structural Analysis-Default.rte*
For Metric *Structural Analysis-DefaultMetric.rte*

c. Set the project units by using the **Project Units** dialog box.
For Imperial **Feet and fractional inches**
For Metric **Millimeters**

d. Set the length dimension snap increment in the **Snaps** dialog box.
For Imperial **10';2'6";1'; 0'3"**
For Metric **3000;760;300; 70**

e. Enable all object snaps using the **Snaps** dialog box.

f. Save the project by using the **Save As** tool.
For Imperial *c02_Industrial-Complex_tut2.rvt*
For Metric *M_c02_Industrial-Complex_tut2.rvt*

g. Close the project by using the **Close** tool.

Starting Revit and Opening a New Project

1. Start a new Revit session by double-clicking on the Autodesk Revit shortcut icon on the desktop. The program is loaded and the user interface screen is displayed. If the Revit session is already running, you can ignore the step and open the project file directly.

2. Choose **New > Project** from the **Application Menu**; the **New Project** dialog box is displayed.

Using the Template File

As given in the project parameters, you need to use the template file for this project. You can select this template file from the **New Project** dialog box.

1. In the **New Project** dialog box, choose the **Browse** button in the **Template file** area; the **Choose Template** dialog box is displayed. Select the **Structural Analysis-Default.rte** template file from the **US Imperial** folder or the **Structural Analysis-DefaultMetric.rte** template file from the **US Metric** folder and then choose the **Open** button; the **Choose Template** dialog box is closed. Next, choose **OK** to exit the **New Project** dialog box; the template file is loaded for this project.

Setting Project Units

You can set units for various measurement parameters by using the **Project Units** dialog box.

1. Choose the **Project Units** tool from the **Settings** panel of the **Manage** tab; the **Project Units** dialog box is displayed.

2. Click in the **Format** column corresponding to the **Length** unit; the **Format** dialog box is displayed. In this dialog box, select the option from the **Units** drop-down list.

   For Imperial **Feet and fractional inches**
   For Metric **Millimeters**

3. Click on the **Rounding** drop-down list and select the desired option, as shown in Figure 2-34.

   For Imperial **To the nearest 1/4"**
   For Metric **To the nearest 10**

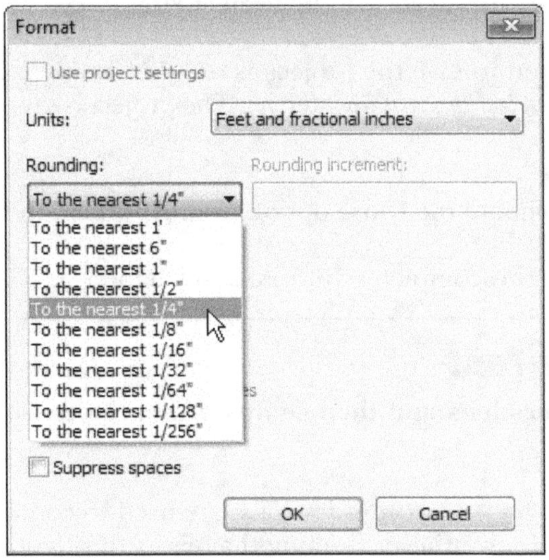

***Figure 2-34** The **Format** dialog box*

4. Choose **OK** to return to the **Project Units** dialog box. Next, choose the **OK** button; the dialog box is closed and the user interface screen is displayed.

Setting Dimensions and Object Snaps

Next, you need to access and modify settings in the **Snaps** dialog box. You need to specify the dimension snap increment and enable all object snap options.

1. Choose the **Snaps** tool from the **Settings** panel of the **Manage** tab; the **Snaps** dialog box is displayed.

2. In this dialog box, enter values in the edit box below the **Length dimension snap increments** check box in the **Dimension Snaps** area.

   For Imperial **10';2'6";1'; 0'3"**
   For Metric **3000;760;300; 70**

3. In the **Object Snaps** area, ensure that all the object snaps options are selected and then choose the **OK** button to apply settings and exit the **Snaps** dialog box.

Saving the Project

In this section, you need to save the project and the settings using the **Save As** tool.

1. To save the project with the settings, choose **Save As > Project** from the **Application Menu**. As you are saving the project for the first time, the **Save As** dialog box is displayed.

2. Browse to the *C:\rst_2017\c02_rst_2017_tut* and for imperial enter **c02_Industrial-Complex_tut2** in the **File name** edit box (for Metric enter **M_c02_Industrial-Complex_tut2**). Notice that in the **File of type** drop-down list, the **Project Files (*.rvt)** option is selected by default.

3. Choose the **Save** button to save the project as *c02_Industrial-Complex_tut2.rvt* for Imperial *(M_c02_Industrial-Complex_tut2.rvt* for Metric*)*. The project is saved at the specified location.

Closing the Project

1. To close the project, choose the **Close** option from **Application Menu**.

   The file is closed and this completes Tutorial 2 of Chapter 2.

Self-Evaluation Test

Answer the following questions and then compare them to those given at the end of this chapter:

1. The options in the __________ drop-down list are used to control the symbolic display of moment connections for cantilever framing that exists throughout a project.

2. The __________ drop-down list in the **Structural Settings** dialog box contains two options: **Parallel Line** and **Line with Angle**.

3. In Autodesk Revit, you can set the symbolic representation of structural framing components by using the __________ dialog box.

4. To specify the location of the default template file, you can use the _________ tab in the **Options** dialog box.

5. The __________ button in the **Save As** dialog box can be used to specify the maximum numbers of backup(s) for a project file.

6. The __________ option of Revit enables you to override snap settings for a single pick only.

7. You can add folders to the __________ in the **Save As** dialog box to access the frequently used folders directly.

8. You can open only one Revit project at a time. (T/F)

9. In Revit, all enabled object snaps work together. (T/F)

10. A project file can be opened by double-clicking on its file name in the Windows Explorer. (T/F).

Review Questions

Answer the following questions:

1. Which of the following options is not an object snap option?

   a) **Endpoints** b) **Work Plane Grid**
   c) **Dimension** d) **Centers**

2. Which of the following keys is used to toggle between the object snap options available at the same point?

   a) TAB b) CTRL
   c) ALT d) F3

3. You can specify the settings for **ViewCube** and **SteeringWheels** in the ______________ dialog box.

4. You can select any of the three options, namely **Beams and Braces**, **Column Base**, and **Column Top** from the___________drop-down list.

5. You can modify the project unit settings anytime during a project. (T/F)

6. The name and path of a project file should be specified each time you save the project. (T/F)

7. You cannot control the visibility of the tooltip assistant. (T/F)

8. The **Save reminder interval** drop-down list available in the **General** tab of the **Options** dialog box is used to specify the time interval between reminder prompts to save a project file. (T/F)

9. If you choose the **Close** button without saving the changes made in a project file, Revit will prompt you to save the changes before closing it. (T/F).

10. You can add words to additional dictionaries for checking spelling errors. (T/F)

EXERCISES

Exercise 1 Academic Institution

In this exercise, you will create a new project file for the *Academic Institution* project with the following parameters. **(Expected time: 15 min)**

1. Template file-

   For Imperial **US Imperial > Structural-Analysis Default.rte**
   For Metric **US Metric > Structural-Analysis DefaultMetric.rte**

2. Project Units-

   For Imperial **Feet and fractional inches**
   For Metric **Millimeters**

3. File name to be assigned:

   For Imperial *c02_Academic-Institution_exer1.rvt*
   For Metric *M_c02_Academic-Institution_exer1.rvt*

Exercise 2 Factory Shed

In this exercise, you will create a new project file for the *Factory Shed* project by using the following parameters. **(Expected time: 15 min)**

1. Template file:

   For Imperial **US Imperial > Structural-Analysis Default.rte**
   For Metric **US Metric > Structural-Analysis DefaultMetric.rte**

2. Project Units:

   For Imperial **Feet and fractional inches**
   For Metric **Millimeters**

3. File name to be assigned:

   For Imperial *c02_Factory-Shed_exer2.rvt*
   For Metric *M_c02_Factory-Shed_exer2.rvt*

Answers to Self-Evaluation Test

1. **Cantilever connection**, **2.** **Plan representation**, **3.** **Structural Settings**, **4.** **File Locations**, **5.** **Options**, **6.** **Temporary Overrides**, **7.** **Places list**, **8.** F, **9.** T, **10.** T

Chapter 3

Setting up a Structural Project

Learning Objectives

After completing this chapter, you will be able to:

- *Create project templates*
- *Understand and create grids*
- *Understand and create levels*
- *Understand and create work planes*
- *Sketch reference planes*
- *Create scope box*

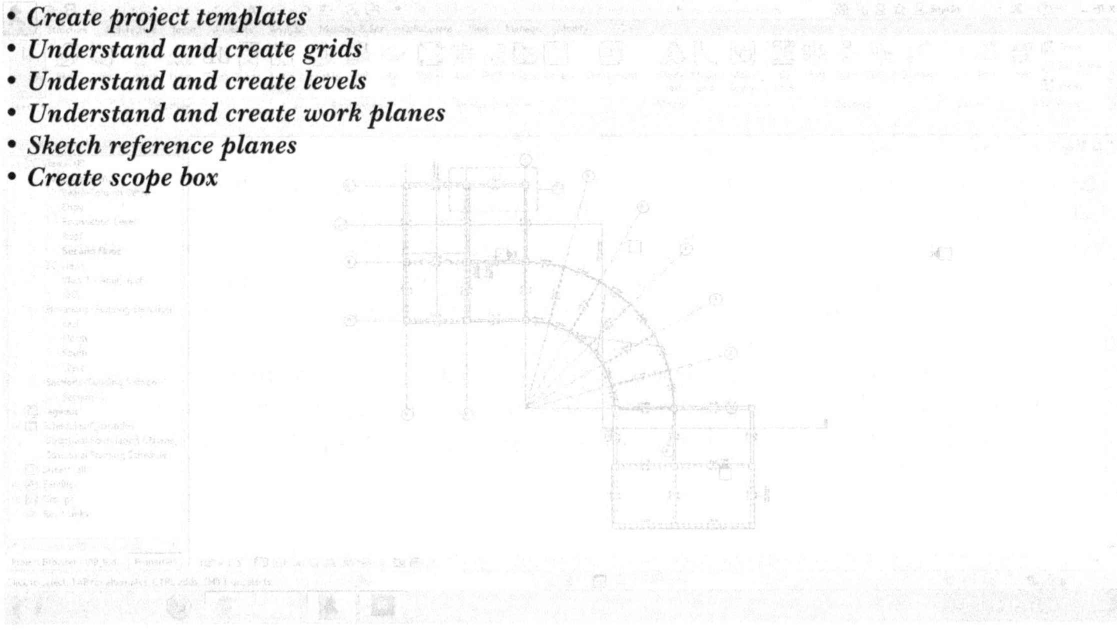

In the previous chapter, you learned to open a new file and create initial setups. Now, after opening a new file and creating initial setups such as units and snaps, you need to learn about creating advanced setups such as creating the project templates, levels, and grids for your project. These advance setups help you to start your project more efficiently and accurately.

In this chapter, you will learn to create a project template, save the project template and then create structural grids and levels for a structural project. Further, you will learn the use of Work Planes and Reference Planes in a project.

PROJECT TEMPLATE

Project templates are files that can be opened while starting a new project. These files, commonly known as template files, contain predefined settings for projects, display of annotations, display of graphics, and so on. Similarly, for the display of graphics, the template files contain predefined settings for materials, line styles, line weights, line patterns, and structural symbols.

Project templates provide initial conditions for a project. In Autodesk Revit, you will find in-built templates that are saved with *.rte* file extension when you install the software. You can also create your own template based on the project requirement. In Revit, any new template-based project inherits all families, settings (such as units, fill patterns, line styles, line weights, and view scales), and geometry from the template.

Template files carry project-specific information. You may have different project template files for different types of projects. For example, for a concrete structure, you may not use steel members for your design purpose. For this kind of project, you may require a template that carries families with more concrete sections and less steel members. A template file can also carry predefined levels and grids and thus saves a lot of time while starting up a new project. Schedules for footings, columns, piers, and other structural elements can also be a part of a project template.

In the next sections, you will learn how to create custom templates and use them in a structural project.

Creating a Custom Project Template

In Autodesk Revit, there are various methods to create a custom project template. The common method is to open an existing template file and modify its settings based on the project requirement and then save it as a different template file. You can also create a custom project template by starting a blank project file, defining all settings such as naming the viewports, creating levels, adding grids, and others, and then saving it as a template (.rte) file. For certain projects, you can create a template file which includes geometry that can be used repeatedly as a base for the new projects. For example, if you have defined geometry for a shopping complex and want to include this geometry whenever you start a new project, you can save the file that includes this geometry as a template. Each time you open a project with this template, the geometry is included.

Creating a New Template from a Blank Project File

Application Menu: New > Project

In Revit, you can create a new template file from a blank project file or use the default template file (*Structural Analysis-Default.rte* for Imperial and *Structural Analysis-DefaultMetric.rte* for Metric). To create a template file from a blank project file, choose **New > Project** from the **Application Menu**; the **New Project** dialog box will be displayed. In the **Template file** area of this dialog box, select the **<None>** option from the drop-down list. Next, select the **Project template** radio button in the **Create new** area and choose the **OK** button; the **Undefined System of Measurement** dialog box will be displayed, as shown in Figure 3-1. This dialog box prompts you to select a system of measurement for your project.

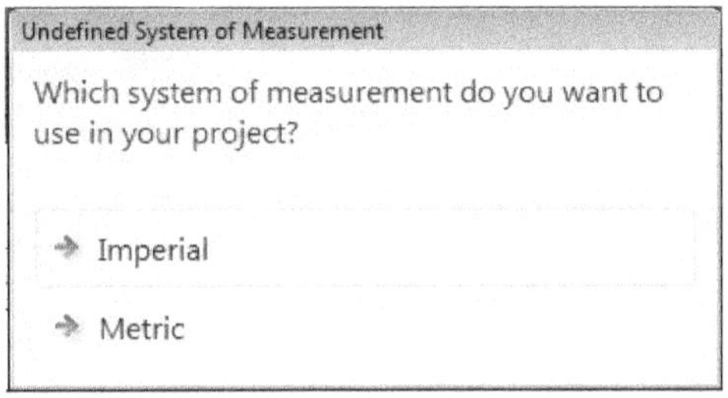

***Figure 3-1** The **Undefined System of Measurement** dialog box*

You can choose the **Imperial** or **Metric** option from this dialog box. If you choose the **Imperial** option, a template file containing all default unit settings in Imperial unit system will open. Similarly, if you choose the **Metric** option, the template file containing default units in the Metric unit system will open. After opening the template file, you can modify its existing settings based on your project environment and then save the modified file as a template file. To do so, choose **Save As > Template** from the **Application Menu**; the **Save As** dialog box will be displayed. In this dialog box, select a folder from the **Save in** drop-down list and enter a name for the template file in the **File name** edit box. Note that the **Template Files (*.rte)** option is selected from the **Files of type** drop-down list. After entering the file name, choose the **Save** button; the template file will be saved with the settings defined in the blank project file.

Note

*You can also select the template file, Structural Analysis-Default.rte for Imperial and Structural Analysis-DefaultMetric.rte for Metric, to create a new template file from a blank project. To do so, select the **Structural Template** option from the drop-down list in the **Template file** area of the **New Project** dialog box, and then choose the **OK** button.*

Tip

*You can change the default template file displayed in the **New Project** dialog box. To do so, invoke the **Options** dialog box by choosing the **Options** button from the **Application Menu**. Next, in the **Project template file** area of the **File Locations** tab, click in the **Path** column corresponding to the **Structural Template** name; a browse button will be displayed. Choose the browse button; the **Browse For Template File** dialog box will be invoked. In this dialog box, select the file that is to be made default template file and then choose the **Open** button. Next click **OK** button to save changes.*

Creating a New Project Template from an Existing Project Template

To speed up your project, you may be required to use predefined template files. These template files contain predefined information or settings pertaining to the project you need to start. To use these template files, choose **New>Project** from the **Application Menu**; the **New Project** dialog box will be displayed. In this dialog box, ensure that the **Project template** radio button in the **Create new** area is selected. Now, to select the desired template file for your project, choose the **Browse** button; the **Choose Template** dialog box will be displayed, as shown in Figure 3-2. In this dialog box, browse to the desired folder to locate the template file. Next, select the template file from the folder and choose the **Open** button; the **Choose Template** dialog box will close and the **New Project** dialog box will be displayed again. In this dialog box, keep the default setting and choose the **OK** button; a new project file will open, which inherits all project settings from the selected template file.

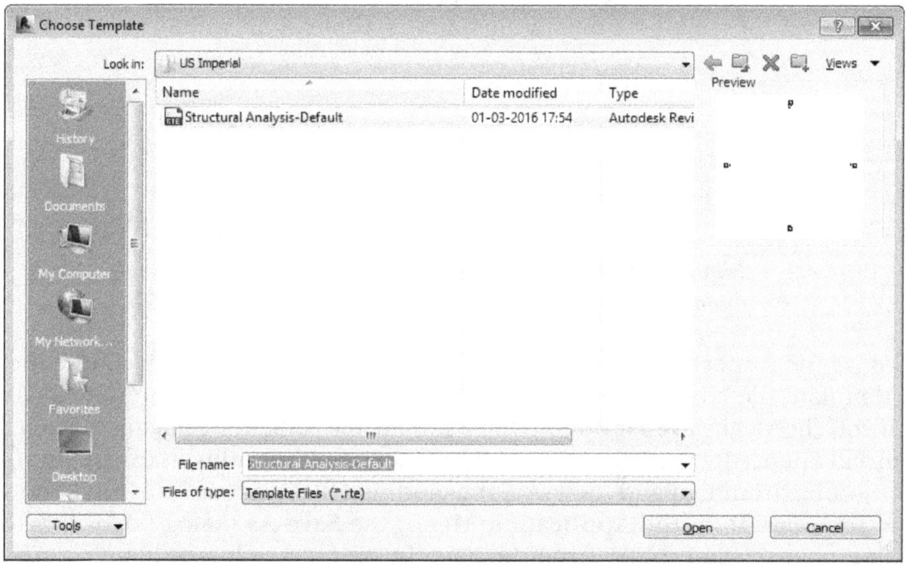

Figure 3-2 The ***Choose Template*** *dialog box*

Settings for the Project Template

While creating a project template, you can predefine certain settings based on your project requirement. To start a new project template, you need to fill in the information specific to the project. The information includes the name of the project, project number, client's name, and so on. These information are useful while publishing or plotting the drawing. Next, you need to enter the project settings. These settings include units, snaps, the line styles for components and lines, fill patterns for materials, and more. Some of the settings such as units and snaps have already been discussed in Chapter 2.

After modifying the project settings, you can create settings for families. The families in a project template can be system families and loaded families. While defining the settings for the project template, you can modify or duplicate system families (for example, walls) as required for the project. You can also load the commonly used families, user-defined families, and title blocks.

After setting families in the project template, you can modify or create settings for project views. These settings include plan views (Structural Plans), levels, schedules, sheets, and so on. There are other settings that can be made for the project template other than those discussed. These are visibility/graphics settings, and the plot (Print) settings. The settings for the project information are discussed in the next section.

Setting the Project Information

When you create a project template, you can also set the project information. To do so, choose the **Project Information** tool from the **Settings** panel of the **Manage** tab; the **Project Information** dialog box will be displayed, refer to Figure 3-3. In this dialog box, you can specify various settings related to the project information displayed in the **Instance Parameters** area. To enter information regarding the **Organization Name**, **Organization Description**, **Building Name**, and **Author** of the project, click on the **Value** fields of their corresponding parameters and enter appropriate values in them. Similarly, to edit the energy settings of the project, choose the **Edit** button in the **Value** field of the **Energy Settings** parameter; the **Energy Settings** dialog box will be displayed. In this dialog box, you can specify various settings related to the type of the building, site location of the building, and the level that will represent the ground plane of the site. To do so, click on the **Value** fields corresponding to the **Building Type**, **Location**, and **Ground Plane** parameters and specify the desired values in them. You can also choose the **Edit** button corresponding to the **Other Options** parameter from the **Advanced** area of the **Project Information** dialog box. On doing so, **Advanced Energy Settings** dialog box will be displayed. You can specify various advance settings like HVAC system, building type, building operating schedule and so on from the **Advanced Energy Settings** dialog box. Choose the **OK** button twice; the settings will be applied and the **Energy Settings** dialog box will be closed and the **Project Information** dialog box will be displayed.

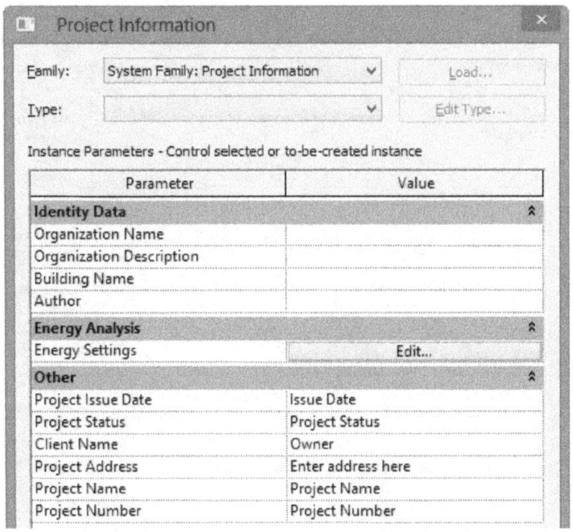

Figure 3-3 *Partial view of the* ***Project Information*** *dialog box used for setting the project information*

In the **Project Information** dialog box, you can enter the information regarding the start or issue date of the project. To do so, click on the **Value** field of the **Project Issue Date** parameter and enter a valid date in the **Project Information** dialog box. Similarly, to specify the status of the project, click on the **Value** field of the **Project Status** parameter and specify a valid status of the project.

To enter the address of location of the project, choose the browse button in the **Value** field of the **Project Address** parameter; the **Edit Text** dialog box will be displayed. In the text area of this dialog box, enter a suitable address and choose the **OK** button; the address of the project will be updated.

Next, specify the name and number of the project in the **Value** fields of the **Project Name** and **Project Number** parameters, respectively. After entering the appropriate information in the **Project Information** dialog box, choose the **OK** button; the **Project Information** dialog box will be closed and the project information entered will be updated.

In addition to the parameters discussed in the **Project Information** dialog box, you can add more parameters for entering project information. To do so, choose the **Project Parameters** tool from the **Settings** panel of the **Manage** tab; the **Project Parameters** dialog box will be displayed. Choose the **Add** button from this dialog box; the **Parameter Properties** dialog box will be displayed, as shown Figure 3-4. Ensure that the **Project parameter** radio button is selected in the **Parameter Type** area of this dialog box. Next, select the **Project Information** check box from the **Categories** area. In the **Parameter Data** area, enter the name of the parameter data in the **Name** edit box. For example, you can specify **Engineer's Name** as the name of the parameter. Next, you need to assign a discipline for the new parameter. To do so, select an appropriate option from the **Discipline** drop-down list. For example, to assign a discipline for the **Engineer's Name** parameter, ensure that the **Common** option is selected in the **Discipline** drop-down list.

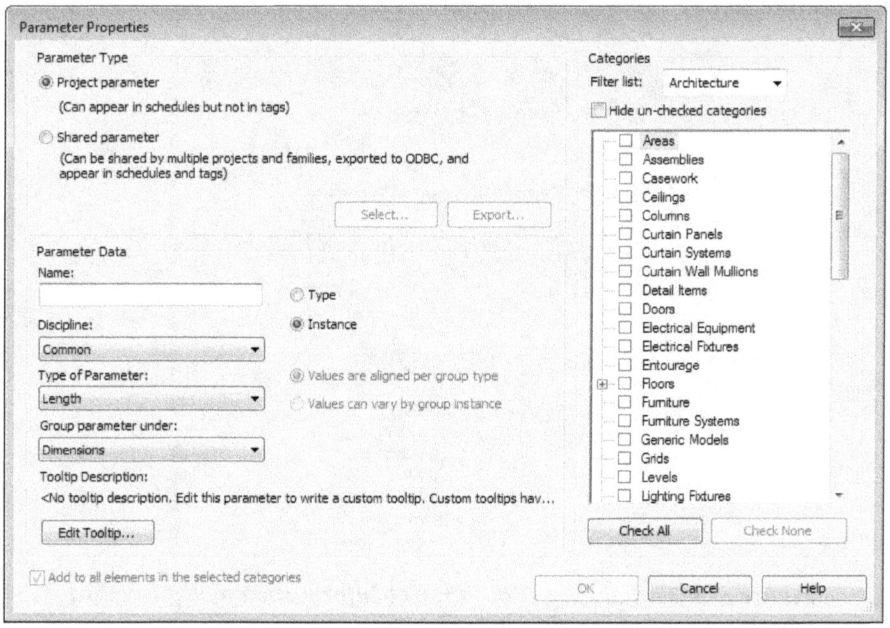

*Figure 3-4 The **Parameter Properties** dialog box*

To assign type and group for the **Engineer's Name** parameter, select the **Text** options from both the **Type of Parameter** and **Group parameter under** drop-down lists. Next, choose the **OK** button; the **Project Parameters** dialog box will be displayed with the added parameter displayed in the list box. Choose the **OK** button from this dialog box; the newly created project parameter will be added in the **Project Information** dialog box. Figure 3-5 shows the partial view of the **Project Information** dialog box for the project information with **Engineer's Name** as the added parameter.

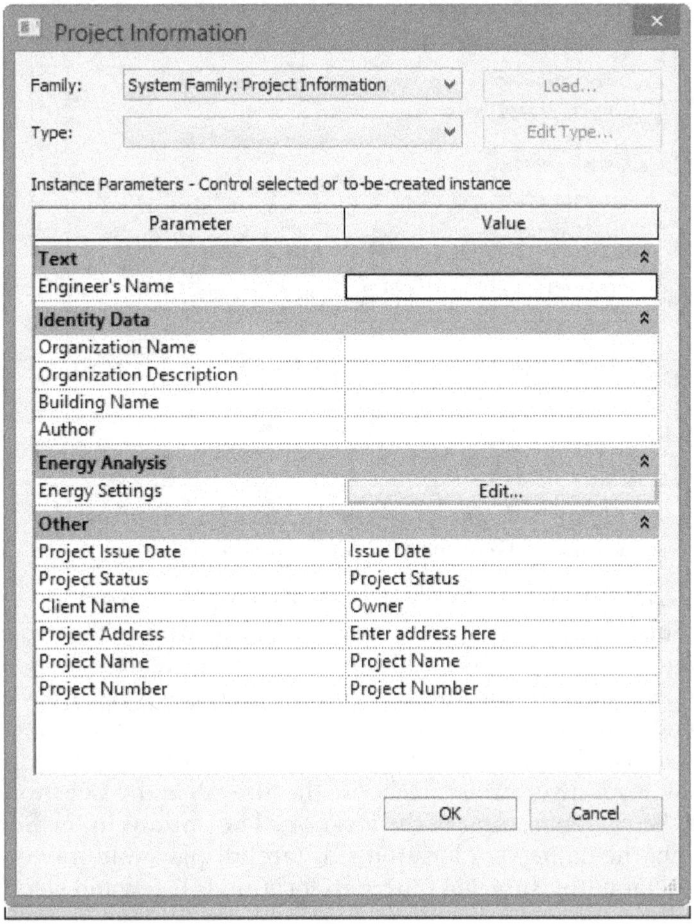

***Figure 3-5** The **Project Information** dialog box displaying the newly added parameter*

Setting the Project Location

For each project in Autodesk Revit, you need to define a location with respect to a coordinate system. This definition includes the geographical location (azimuth) of a project, which needs the latitude and longitude of the nearest city and the coordinate system of the host model and the linked model (if present).

To define the geographical location of a project, choose the **Location** tool from the **Project Location** panel of the **Manage** tab; the **Location Weather and Site** dialog box will be displayed, as shown in Figure 3-6. This dialog box contains three tabs: **Location, Weather** and **Site**. The **Location** tab is chosen by default. In the **Define Location by** drop-down list of this tab, the **Internet Mapping Service** option is selected by default. As a result, you can use the bing map service to find out the geographical location of a desired place. To find out the geographical location, type it in the **Project Address** edit box and then choose the **Search** button; the location searched will be displayed in a map.

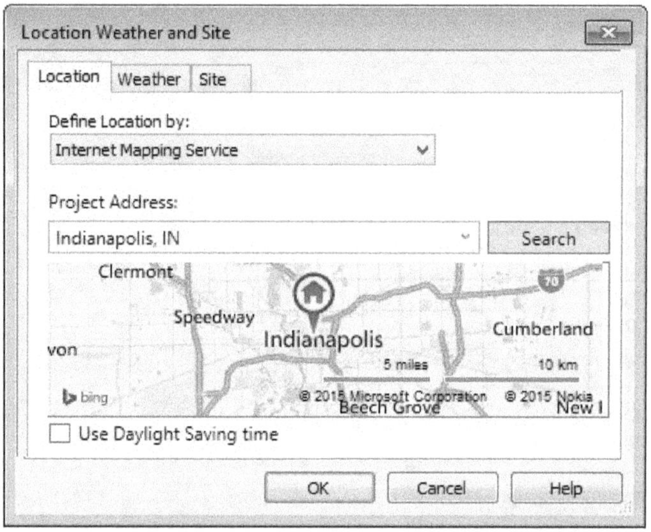

Figure 3-6 *The options in the* ***Location*** *tab of the* ***Location Weather and Site*** *dialog box*

Alternatively, you can select the **Default City List** option from the **Define Location by** drop-down list. This option is selected to specify a city nearest to the project location and to define the latitude and longitude of the location. To specify the nearest city, select an option from the **City** drop-down list; the **Latitude** and **Longitude** edit boxes will display the corresponding values for the latitude and longitude of the selected city. If the name of a city is not available in the **City** drop-down list, then enter the values of latitude and longitude in the **Latitude** and **Longitude** edit boxes, respectively. You can choose the **Site** tab in the **Location Weather and Site** dialog box to name the current setting of the location. The options in the **Site** tab are shown in Figure 3-7. In this tab, the name(s) of location(s) is (are) displayed in the **Sites defined in this project** list box. By default, the **Internal (current)** location is listed and selected in the list box.

To define a different location for the project, choose the **Duplicate** button from the **Location Weather and Site** dialog box; the **Name** dialog box will be displayed. Enter a name in the **Name** edit box and choose the **OK** button; the **Name** dialog box will be closed and the name of the new location will be listed in the **Sites defined in this project** list box. After you have added a location, you will notice that the **Delete** and **Make Current** buttons are active. You can choose the **Delete** button to delete location(s) apart from the current location. A current location is a location being currently used. To make a location current, select the location from the **Sites defined in this project** list box and choose the **Make Current** button. Note that after you have made a location current, the name of the location is suffixed with the word **(current)**.

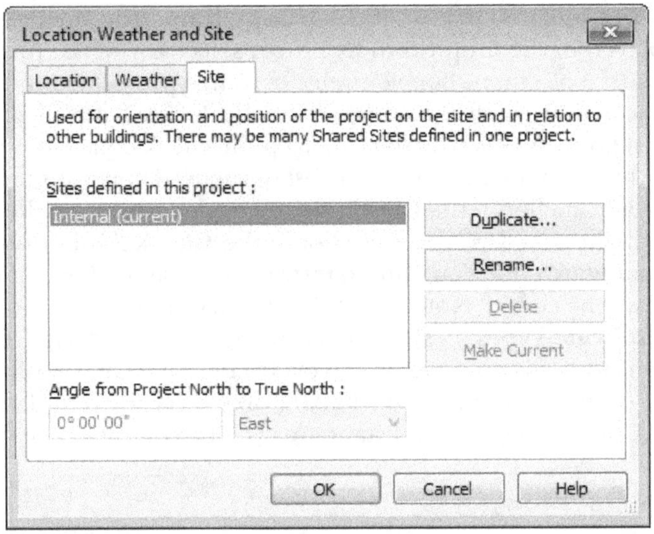

*Figure 3-7 The options in the **Site** tab of the **Location Weather and Site** dialog box*

Transferring Project Standards

Ribbon: Manage > Settings > Transfer Project Standards

When you create a template file, you can copy project standards from another project to the current file. These standards include various project settings such as Family Types (only system families, not loaded families), Line Weight, Line Styles, Line Patterns, Materials, and View Templates. To transfer project standards to the template file (target file), open the source file from which standards are to be copied. Next, choose the **Transfer Project Standards** tool from the **Settings** panel of the **Manage** tab; the **Select Items To Copy** dialog box will be displayed, as shown in Figure 3-8.

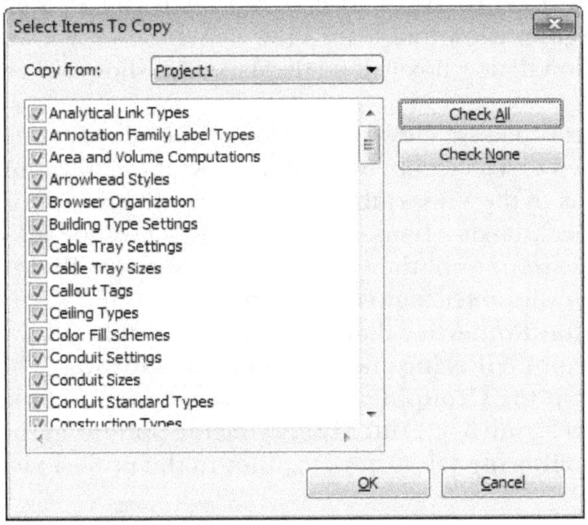

*Figure 3-8 The **Select Items To Copy** dialog box*

In the **Select Items To Copy** dialog box, click on **Copy from** drop-down list; the list of opened projects will be displayed in the drop-down list. Next, select any of the projects from which you want to copy standards. Note that when you select a source project from this list, the standards included in the project along with their respective check boxes are displayed. By default, all check boxes in the list box are selected. You can keep the check boxes selected for the standards you want to copy to the template file and clear rest of them. If you want to clear all check boxes for fresh selection, you can choose the **Check None** button located on the right of the list box. Similarly, to select all check boxes, you can choose the **Check All** button. After selecting the check boxes for the standards that you want to transfer, choose the **OK** button; the **Select Items To Copy** dialog box will be closed. Note that if the selected standards are already present in the template file, the **Duplicate Types** message box will appear, refer to Figure 3-9. You can choose the **Overwrite** button from this message box to overwrite the common standards or choose the **New Only** button to transfer the standards that are present in the template file. After you have chosen the required options, the selected standards will be copied to the destination file.

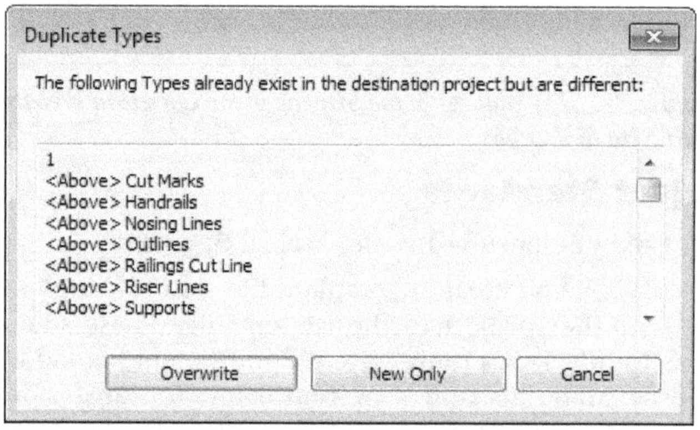

Figure 3-9 *The* ***Duplicate Types*** *message box*

Setting the Browser Organization

While creating a project template, you can organize the **Project Browser**. To do so, choose the **Browser Organization** tool from **View > Windows > User Interface** drop-down; the **Browser Organization** dialog box will be displayed, as shown in Figure 3-10.

This dialog box contains two tabs: **Views** and **Sheets**. The **Views** tab is chosen by default. The options in this tab are used to select, edit, or create a browser organization for the views present in the project. In the list box of the **Views** tab, the list of default browser organizations is displayed with their respective check boxes. By default, the check box for **all** browser organization is selected. You can edit the settings of the check boxes displayed. To do so, click on the name of the browser organization whose settings you want to change, and then choose the **Edit** button; the **Browser Organization Properties** dialog box will be displayed, refer to Figure 3-11. This dialog box contains two tabs: **Filtering** and **Grouping and Sorting**. The **Filtering** tab is chosen by default. The options in the **Grouping and Sorting** tab are used to create group by sorting the project views, refer to Figure 3-11. After specifying the options in the **Grouping and Sorting** tab, you can choose the **Filtering** tab to apply a filter to the project views.

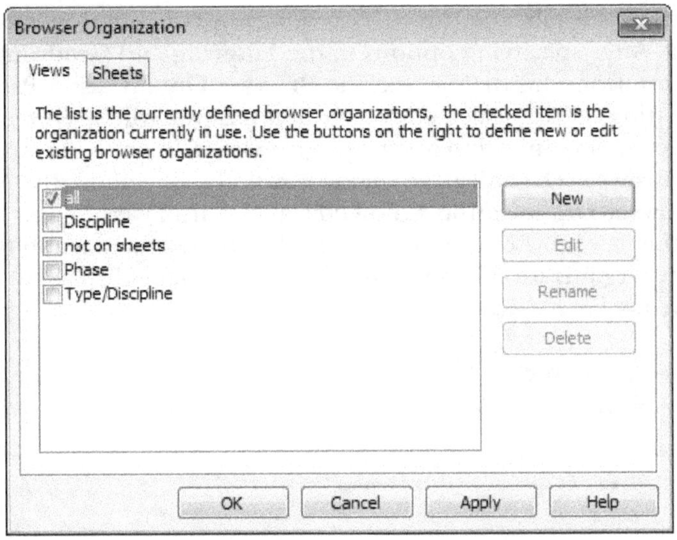

***Figure 3-10** The **Browser Organization** dialog box*

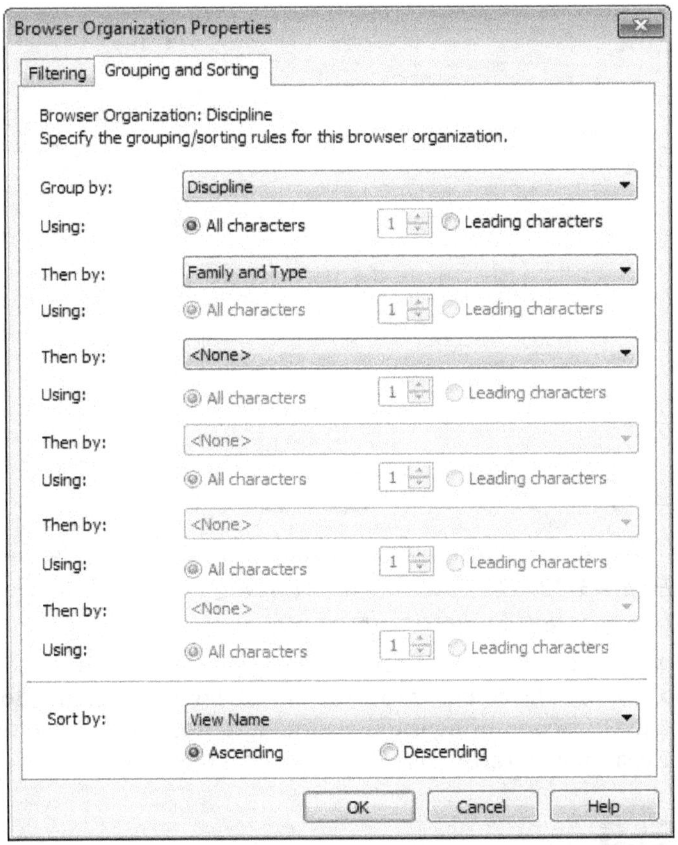

***Figure 3-11** The options in the **Grouping and Sorting** tab of the **Browser Organization Properties** dialog box*

In the **Filter by** area of this tab, you can specify view property, filter operator, and filter value, refer to Figure 3-12. After specifying options in the **Filtering** and **Grouping and Sorting** tabs, choose the **OK** button from this dialog box; the **Browser Organization Properties** dialog box will close and the settings of the selected browser organization will be edited. Similarly, you can also create a new browser organization for the project views. To do so, ensure that the **Views** tab is chosen from the **Browser Organization** dialog box and then choose the **New** button from it; the **Create New Browser Organization** dialog box will be displayed. In this dialog box, you can enter the name of the new browser organization in the **Name** edit box and then choose the **OK** button; the **Browser Organization Properties** dialog box will be displayed. The options in this dialog box have already been discussed.

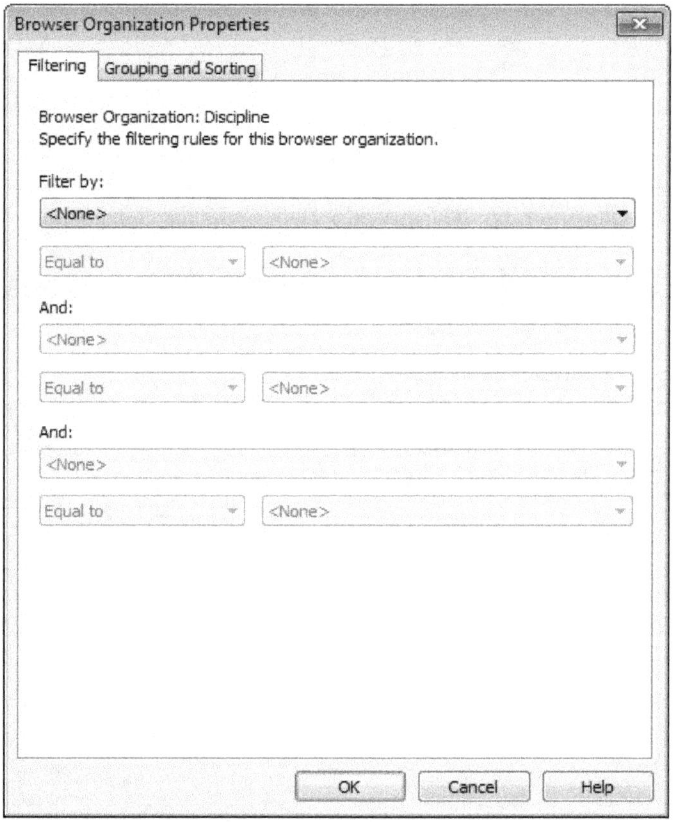

Figure 3-12 *The options in the **Filtering** tab of the **Browser Organization Properties** dialog box*

As you created browser organization for a project view, you can also create browser organization for sheets. To do so, choose the **Sheets** tab from the **Browser Organization** dialog box. The **Sheets** tab displays a list box containing the list of default browser organizations for sheets. The options in this tab are similar to those discussed for the **Views** tab.

USING LEVELS

Creating levels is one of the important tasks in a BIM project. In any structural model of a building project, levels define horizontal planes for placing the level-hosted elements, such as walls, roofs, foundations, floors, slabs, beams, columns, and ceilings.

In Revit, you can use the **Level** tool to add new levels by sketching the required level lines in the elevation or section views. By default, all level lines have associated labels that display the name and elevation of the level. There are two types of level that you can create: story and nonstory. The procedure to create these levels is discussed later in this chapter. A story level defines a floor, roof, or ceiling of a building. You can create a level for each known story in a structural model. The main characteristic of story levels is that they have corresponding plan views. In a project view, the level bubble for a story level will be displayed in blue.

On the other hand, a nonstory level does not hold any plan view but can act as host for placing objects and information. The examples of a nonstory level are top of wall, top of foundation, and top of slab. The level bubble for a nonstory level is displayed in black. Figure 3-13 shows both types of levels in the **East** elevation view of a structural model. In this figure, the **T.O. Columns** and **Column Splice** levels are nonstory levels and the rest of the levels are story levels.

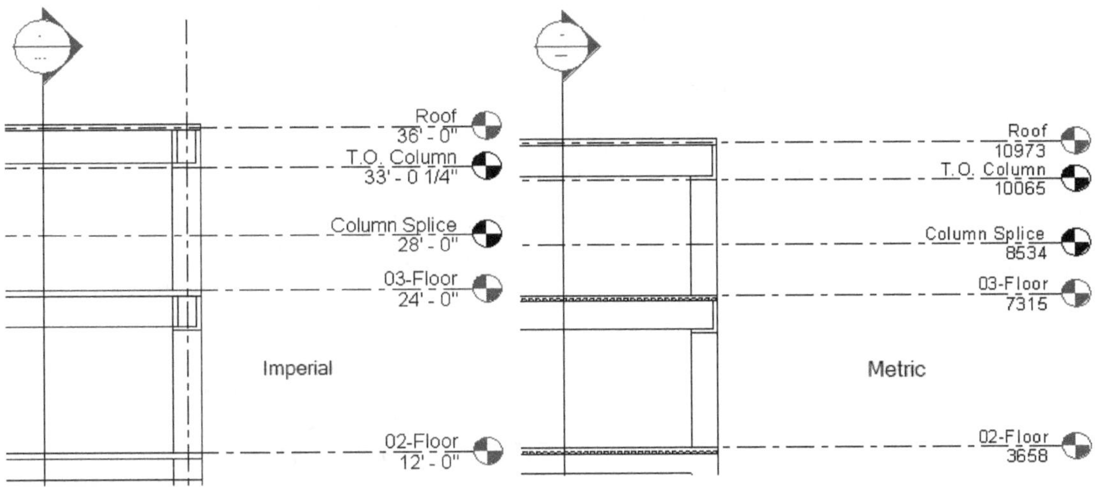

***Figure 3-13** Different types of level in the **East** elevation view*

In Autodesk Revit, when you use the default template file for creating a new project file, it loads two predefined levels: **Level 1** and **Level 2**. These levels are displayed in the elevation or section view. You can view any of the elevations or section views using the **Project Browser**. Levels can be added, renamed, and modified at any time during the project development. In the following sections, you will learn more about properties of levels, creating levels, modifying levels, and working with levels.

Understanding Controls in a Level

A typical level is represented by a level line, level bubble, level name, level elevation, and so on, as shown in Figure 3-14. You can modify the appearance of a level by using these controls and parameters. The level name is an editable parameter which refers to different levels.

The level elevation is the distance of the level from the base level. The visibility of level bubble on either sides of the level line can be controlled by using the bubble display control. The length alignment control can be used to align level lines. Autodesk Revit provides the 2D or 3D extent control for datums when you select them. As a result, you can change their extents in one or multiple views in which they are visible. When a datum is in the 3D mode, any modification made in the 3D view is propagated in all views of the structural model. On the other hand, the 2D mode can be used to modify the datum in a specific view, thereby making it view-specific. In the next section, you will learn about adding levels in your project.

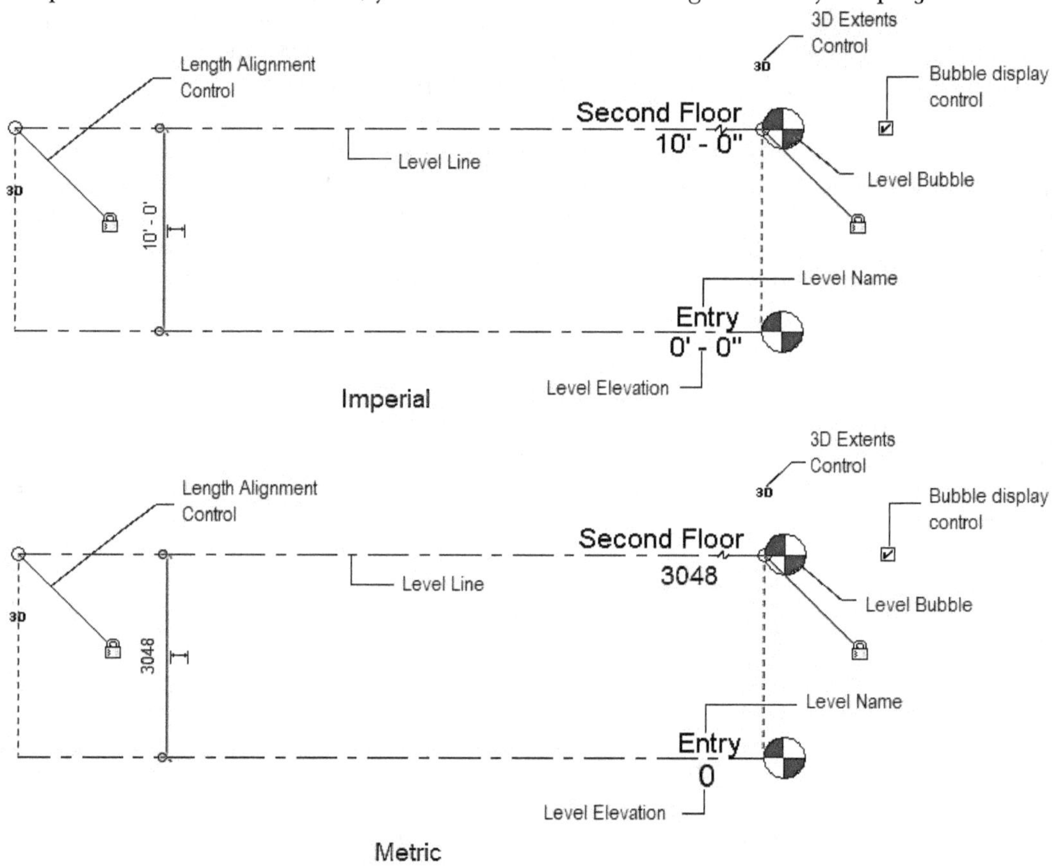

Figure 3-14 *Various controls of a typical level*

Adding Levels

Ribbon: Structure > Datum > Level
Shortcut Keys: LL

In Autodesk Revit, you can create multiple levels in an elevation or a section view. To create a level, invoke the **Level** tool from the **Datum** panel of the **Structure** tab, as shown in Figure 3-15; the **Modify | Place Level** tab will be displayed. In this tab, choose any of the sketching options displayed in the **Draw** panel to create levels in your project. Alternatively, you can invoke the **Level** tool by pressing **LL**. After invoking the **Level** tool, you can select a level type from the **Type Selector** drop-down list in the **Properties** palette. There are three options in this drop-down list: **Level: 1/4" Head**, **Level: Story Level**, and **Level: Story Level-no head** in the **Imperial** unit system and **Level: 8mm Head** option for the **Metric** unit system. If you select the **Level: 1/4" Head** option of **Imperial** or **Level: 8mm Head** of **Metric** from the **Type Selector** drop-down list, level lines will be added as dashed lines along with the level bubbles. If you select the **Level: Story Level** option from the drop-down list, level lines will be added with continuous line along with the level bubbles. Similarly, by selecting the **Level: Story Level-no head** option, you can add level lines as continuous lines without level bubbles.

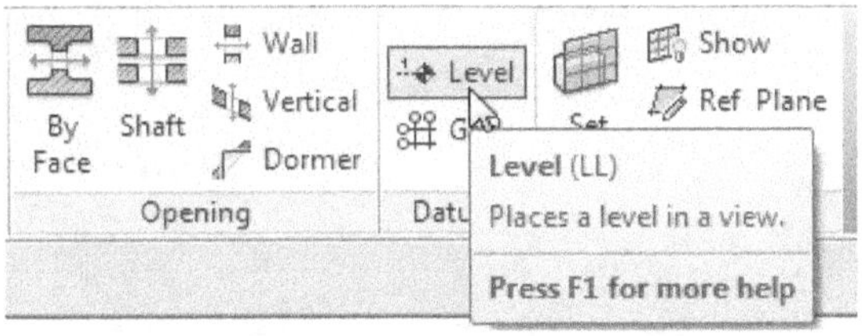

Figure 3-15 *Invoking the **Level** tool*

You can choose the **Line** (default) or **Pick Lines** tool from the **Draw** panel of the **Modify | Place Level** tab to sketch the level line. Notice that the **Make Plan View** check box is selected in the **Options Bar**. This implies that when a level is added, Autodesk Revit treats it as a story level under the **Structural Plans** head in **Project Browser**. As a result, the level created will be displayed under the **Structural Plans** head in **Project Browser**. While working with the **Level** tool, if you clear the **Make Plan View** check box from **Options Bar**, a nonstory level will be created. As a result, the associated views will not be created. The **Offset** edit box in the **Options Bar** can be used to add a level at a specified distance from the selected point or element.

To add a level, invoke the **Level** tool and move the cursor near the existing level line. The temporary dimension displayed indicates the perpendicular distance between the nearest level and the cursor. To add a level at the specified distance from the existing level line, enter the perpendicular distance value in the **Imperial** and **Metric** template, as shown in Figure 3-16. Next, click to specify the first point and move the cursor to left or right. You will notice that the level line, level name, and elevation appear and move with the cursor. On moving the cursor above the endpoint of the existing level line, a dashed alignment line appears, indicating its alignment with the existing level, as shown in Figure 3-17. When the alignment line appears, click to specify the endpoint of the level line.

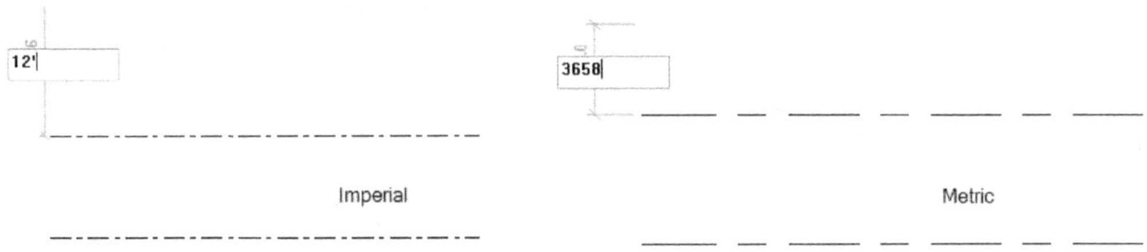

Figure 3-16 *Entering the perpendicular distance value*

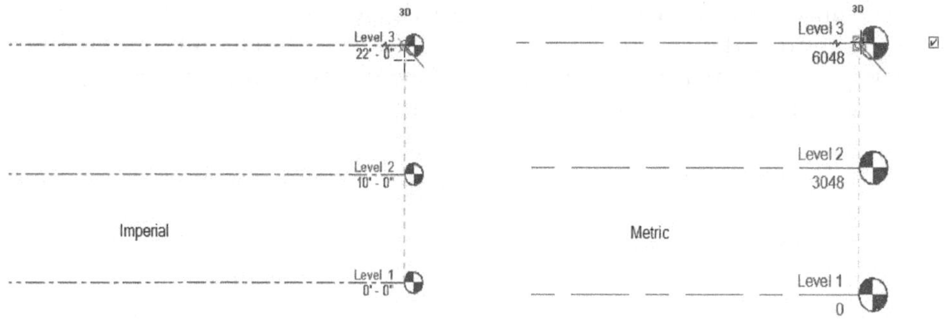

Figure 3-17 The appearance of a dashed alignment line along with the cursor

Note

Although levels are infinite horizontal planes, you are required to specify the start point and endpoint of level line. However, the placement of a level line can be useful in elevation and section views.

The recently added level when selected will be highlighted and will display two square boxes, one on either side of the level, as shown in Figure 3-18 in both **Imperial** and **Metric** unit system. These boxes are used to control the visibility of the level bubble. They can be checked or cleared to make the bubble visible or invisible, respectively, at the desired side(s). The two small circles representing the drag controls for the level line can be used to increase or decrease its length. The padlocks act as the length alignment control for the alignment of all level lines. When the padlock is locked, if you change the length of a level line, all level lines aligned to it will increase or decrease simultaneously. To modify the length of a single level line, unlock the control and then modify its length.

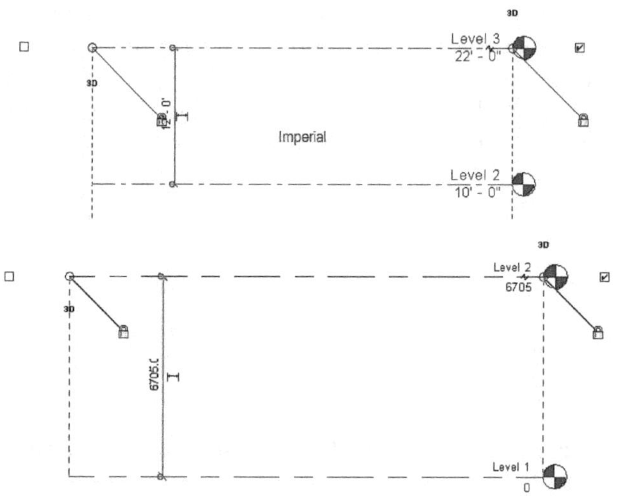

*Figure 3-18 The selected level displaying one check box on its right and left in both the **Imperial** and **Metric** template files*

Instance and Type Properties of a Level

Like other structural and building elements used in Autodesk Revit project, levels also have associated types and instance properties. To view and modify the instance properties of a level, select the level from the drawing; the instance properties of the selected level will be displayed in the **Properties** palette, as shown in Figure 3-19. You can use this palette to view and modify the type properties of the selected level. To do so, choose the **Edit Type** button; the **Type Properties** dialog box will be displayed. Alternatively, you can display this dialog box by choosing the **Type Properties** tool from the **Properties** panel of the **Modify | Levels** contextual tab. You can use the **Type Properties** dialog box to modify and view the type properties of the level selected from the drawing. The parameters in the **Properties** palette and the **Type Properties** dialog box are discussed next.

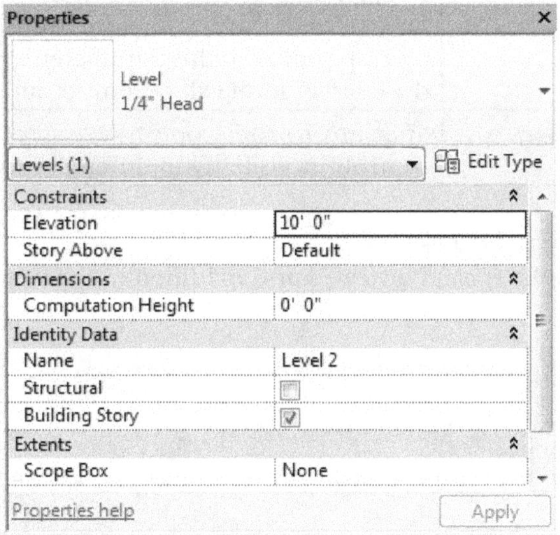

Figure 3-19 *The instance properties of a level displayed in the* ***Properties*** *palette*

Instance Properties of a Level

When you change the instance property of a selected level, the properties of the selected instance are changed. Different instance properties of a level are described in the following table:

Parameter Name	**Description**
Elevation	Specifies to the vertical height of a level from the elevation base.
Story Above	Specifies the building story above the current level. By default, the next highest level will be specified for this parameter. To change the value of this parameter, click on the value field corresponding to this parameter; a drop-down list will be displayed. The drop-down list will display all the levels above the current level. You can select any of these levels to specify it as the story above the current level.
Computation Height	Specifies the computation height for a level. This value is entered to calculate the area, perimeter, and volume of a room.

Name	Specifies the name assigned to the selected level in the model.
Structural	Specifies that the current level is a structural level. To specify this parameter, select the check box displayed in the value field corresponding to this parameter.
Building Story	Specifies whether the current level will be a building story or not. A building story is a floor level such as first, second, third, and more. By default, the check box corresponding to this parameter is selected; As a result, the current level will become a building story.
Scope Box	Specifies to the scope box assigned to the level that controls its visibility in different views.

Type Properties of a Level

The type properties of a level can be viewed and modified in the **Type Properties** dialog box, refer to Figure 3-20.

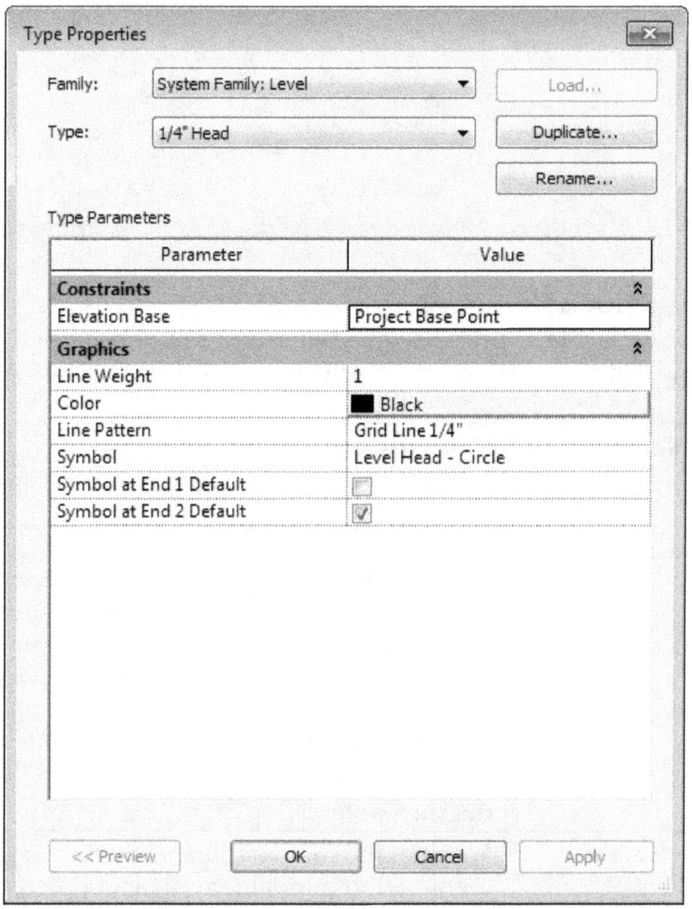

***Figure 3-20** The **Type Properties** dialog box of the selected level*

When you change the type properties of a level, all instances of the level are modified. In the **Type Properties** dialog box, you can modify the value of a parameter by clicking on its **Value** field and selecting a new value from the drop-down list or entering a new value in the field. Different level type properties are described in the table given next.

Parameter Name	Value and Description
Elevation Base	Describes the elevation base value with respect to project or the shared origin.
Line Weight	Refers to the weight of a line defining the level. It can be selected from the drop-down list.
Color	Refers to the color of a level line and can be selected from the available colors. The default color is black.
Line Pattern	Used to set the linetype of a level line.
Symbol	Refers to the symbol indicating level and can be selected from the drop-down list. The **None** option can be selected if the level head is not required.
Symbol at End 1 Default	Check box is used if a bubble is required at the left end of a level line.
Symbol at End 2 Default	Check box is used if a bubble is required at the right end of a level line.

Tip
It is recommended that you name the building story level according to the floor such as first floor, second floor, or roof. This helps in referring to their corresponding plan views. This will also help in editing the level-hosted elements. The names assigned to the levels are also reflected in the schedules created for the project.

Changing the Level Parameters

While working in a structural project, you may need to change various parameters of a level based on your project requirement. The type and instance parameters can be changed by using the **Properties** palette and the **Type Properties** dialog box. The options in this dialog box have already been discussed in the previous sections. Further, you can change the level type by selecting a level and then selecting a different level type from the **Type Selector** drop-down list. In Revit, some of the parameters can also be modified by clicking on the level element in the drawing window and entering a new value. For example, after selecting a level, you can click on its name. On doing so, an edit box will appear with the present name highlighted. Remove the present name, type a new name, and then click again or press ENTER. Note that if you are renaming the level for the first time, Autodesk Revit will prompt you to specify whether you want to rename all the corresponding views. If you choose to rename the views, the name of the associated views will also be changed in the **Project Browser**. Similarly, you can also modify the elevation of a level by selecting the elevation value below the elevation name and entering a new elevation value in the edit box displayed on clicking it, as shown in Figure 3-21. When you enter a new value, the level automatically moves to the specified elevation.

Alternatively, you can move levels by simply dragging them to the desired location. To do so, select a level line and drag it to the desired location. As you drag the level line, the elevation level changes dynamically with respect to the cursor location.

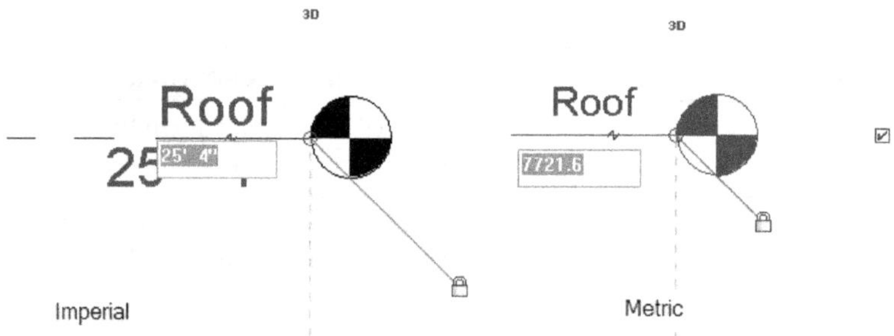

***Figure 3-21** Entering a value for the new elevation in the edit box*

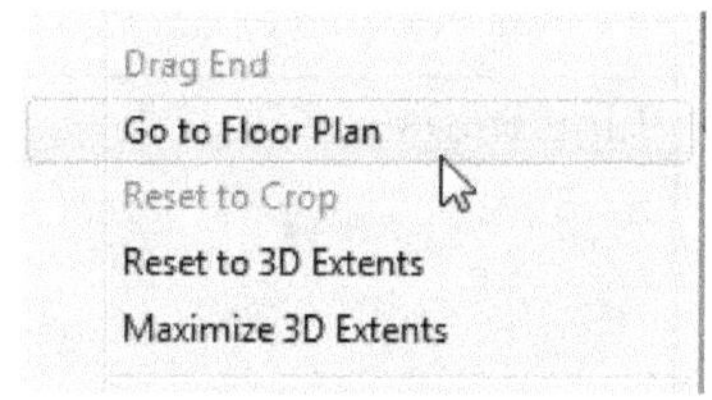

***Figure 3-22** Choosing the **Go to Floor Plan** option from the shortcut menu*

It is recommended to hold the SHIFT key to constrain the cursor to move vertically. After dragging the level line to the desired location, release the mouse button; the level will be relocated to a new position. While working on a project, you may need to quickly view the structural plan of a level that is displayed in the elevation. To do so, open an elevation view and move the cursor near a level line until it gets highlighted. Next, right-click on the highlighted level line; a shortcut menu will be displayed. From the shortcut menu, choose the **Go to Floor Plan** option to open the corresponding floor plan for the level, as shown in Figure 3-22.

Tip

When you select a level, its distance from the adjacent levels is displayed. Click on the corresponding temporary dimension and enter a new value to move the level to a new location.

For certain levels, you may want to move the level bubble to a different location. On selecting a level, you will notice that a break control appears below its name, next to the level bubble. The break control can be used to break the level line and move the level bubble away from it. On clicking this control, you will notice that the level name and the level bubble are also moved to the new location and an extension line is created. Figure 3-23 illustrates the appearance of the bubble dots after they are moved. Notice that as the bubble is moved, it appears with two blue dots. You can use the blue dots to place the level bubble at the appropriate location.

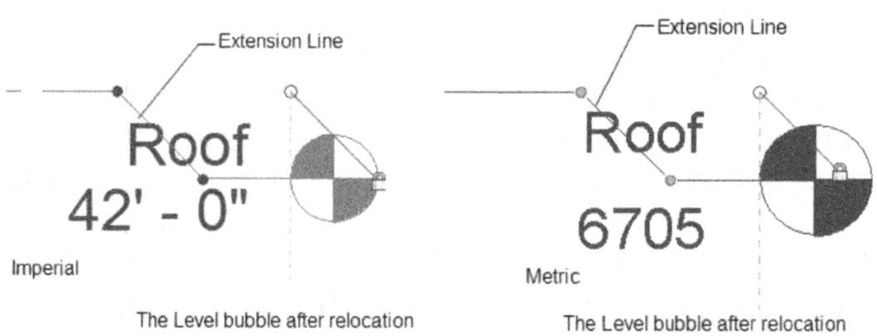

***Figure 3-23** The level bubble after its relocation*

Controlling the Visibility of Levels

While working in an elevation view or a section view, sometimes you may need to hide some or all level lines to avoid congestion of construction lines. This helps in viewing the detail lines for sections and elevations properly. To do so, in the elevation view or section view, select the level line that you need to hide, and then right-click; a shortcut menu will be displayed. From this shortcut menu, choose **Hide in View > Elements**, as shown in Figure 3-24; the selected level line will be hidden.

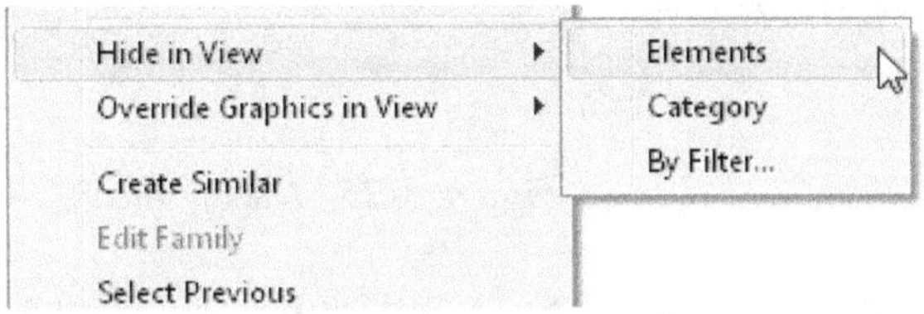

***Figure 3-24** Choosing the **Elements** option from the shortcut menu*

If you need the hidden level line to be visible, choose the **Reveal Hidden Elements** button from the **View Control Bar**; the hidden level line will appear red in the drawing area. Next, right-click on the level appearing in red color in the drawing area; a shortcut menu will be displayed. Choose **Unhide in View > Element** from the shortcut menu. Next, choose the **Close Reveal Hidden Elements** button from **View Control Bar**; the hidden level will appear.

Similar to hiding a single level line, you can also hide all levels that are displayed in an elevation or a section view. To do so, select any level from the view and right-click; a shortcut menu will be displayed. Choose **Hide in View > Category** from the shortcut menu; all levels in the view will hide.

GRIDS

Grids are the essential features that will help you to align and place structural elements such as columns, beams, and walls in a structural plan. With grids, you can easily trace the exact location of a desired column or any other structural elements. Grids are definite vertical planes represented as lineworks in the plan, elevation, and section views.

You can add grids as lines or arcs. You can also create them from existing lines, or link/import them from other software such as AutoCAD Architecture, Revit Architecture, and AutoCAD and then convert them to revit grids. In the following sections, you will learn how to create, modify, and understand the grids.

Creating Grids

Ribbon: Structure > Datum > Grid
Shortcut Key: GR

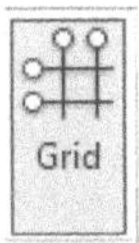

To create a grid, invoke the **Grid** tool from the **Datum** panel of the **Structure** tab; the **Modify | Place Grid** tab will be displayed, as shown in Figure 3-25. The options in the **Modify | Place Grid** tab can be used to assign and change the element properties (instance and type) of the grid, to select a type for grid, and to draw grid in the desired plan or elevation.

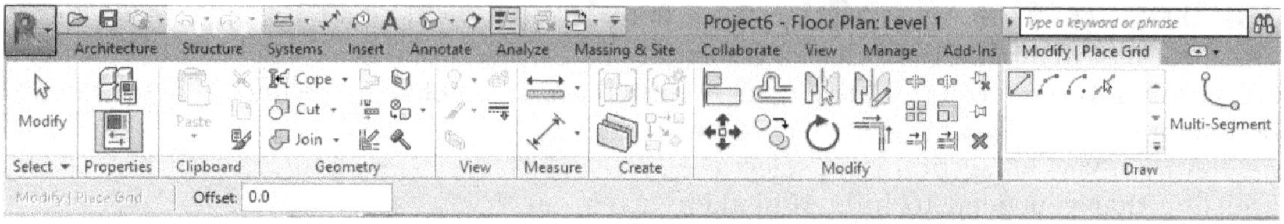

Figure 3-25 The options in the ***Modify | Place Grid*** *tab*

Before you start sketching grids, you need to assign their element properties (instance or type) using the **Properties** palette. To do so, invoke the **Grid** tool; the **Properties** palette will display the parameters that can be added to the grid. In the palette, you will notice that the value assigned to the **Scope Box** parameter is **None**, which implies that no scope box will be assigned to the grid to be added. Next, in the palette, choose the **Edit Type** button; the **Type Properties** dialog box will be invoked. In this dialog box, you can change the type parameters of grids, as discussed next.

Type Parameters of a Grid

Before sketching a grid, you can assign its type parameters. To do so, choose the **Type Properties** tool from the **Properties** panel in the **Modify | Place Grid** tab; the **Type Properties** dialog box will be displayed. Figure 3-26 shows the partial view of the dialog box. In this dialog box, you can change values for various type parameters, which are given in the following table:

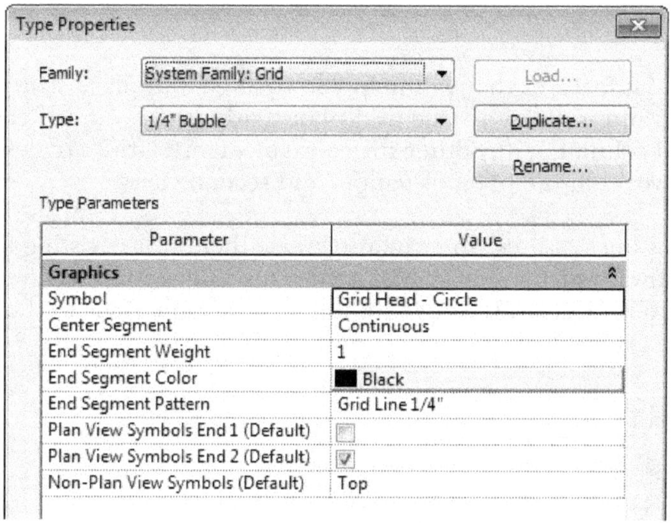

Figure 3-26 Partial view of the ***Type Properties*** *dialog box*

Parameter Name	**Value and Description**
Symbol	Refers to the display of symbol at the end grid.
Center Segment	Refers to the display type of the center segment of a grid line. You can select **Continuous**, **None**, or **Custom** from the drop-down list.
Center Segment Weight	Refers to the line weight of the center segment, if the **Center Segment** parameter is set to **Custom**.
Center Segment Color	You can assign a color to the center segment, if the **Center Segment** parameter is set to **Custom**.
Center Segment Pattern	Refers to the line type of the center segment if the **Center Segment** parameter is set to **Custom**. You can select various line types from the drop-down list.
End Segment Weight	Refers to the line weight of a grid line, if the **Center Segment** parameter is set to **Continuous, None** or **Custom**.
End Segment Color	Refers to the color assigned to a grid line, if the **Center Segment** parameter is set to **Continuous, None** or **Custom**.
End Segment Pattern	Refers to the line type of the grid line if the **Center Segment** parameter is set to **Continuous, None** or **Custom**.
End Segments Length	Refers to the length of each end segment as measured in the sheet, if only the **Center Segment** parameter is set to **None** or **Custom**.
Plan View Symbols End 1 (Default)	Refers to the default visibility status of the symbol at end 1 of the grid line in the plan view. By default, the check box is cleared. If you select the check box, the visibility of the symbol at end 1 in the plan view will be turned on.
Plan View Symbols End 2 (Default)	Refers to the default visibility status of the symbol at end 2 of grid line in the plan views. By default, the check box is selected. If you clear the check box, the visibility of symbol at end 2 in the plan view will be turned off.
Non-Plan View Symbols (Default)	Refers to the default visibility status of the grid line in sections and elevations other than in the plan views. You can control the visibility of the symbol at the top and bottom of grid line by selecting various options from the drop-down list.

Sketching the Grids

After assigning the type parameters in the **Type Properties** dialog box, you can now start sketching the grid line. To sketch the grid line, invoke any of the sketching tools from the **Draw** panel of the **Modify | Place Grid** tab. The **Draw** panel of the **Modify | Place Grid** tab contains four tools: **Line**, **Start-End-Radius Arc**, **Center-ends Arc**, and **Pick Lines**, as shown in Figure 3-27. The use of these sketching tools is discussed next.

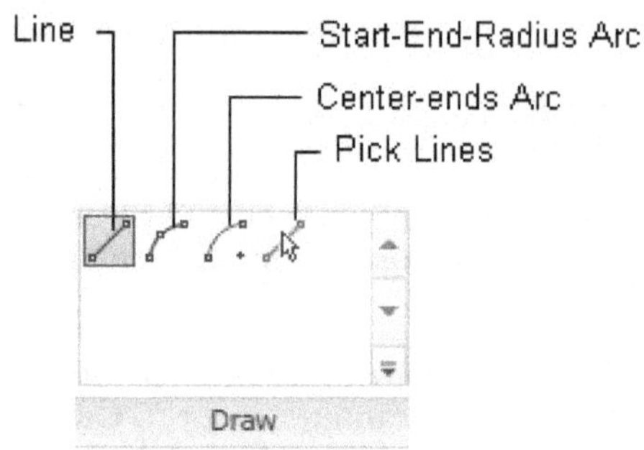

***Figure 3-27** The sketching tools in the **Draw** panel*

Sketching a Grid Using the Line Tool

The **Line** tool in the **Draw** panel of the **Modify | Place Grid** tab is invoked by default. To start sketching the grid line, click at the desired location in the drawing area to specify its start point. Notice that as you move the cursor, a grid line with one end fixed at the specified point and the other end attached to the cursor is created. A temporary angular dimension, indicating the angle of the grid line with the horizontal axis is also displayed, as shown in Figure 3-28. Click to specify the endpoint of the grid line when the appropriate angular dimension is displayed. You can also sketch an arbitrary inclined grid line and then click on the angular dimension to enter a new value of the angle. To draw orthogonal grids, hold the SHIFT key and restrain the movement of the cursor to the horizontal and vertical axes.

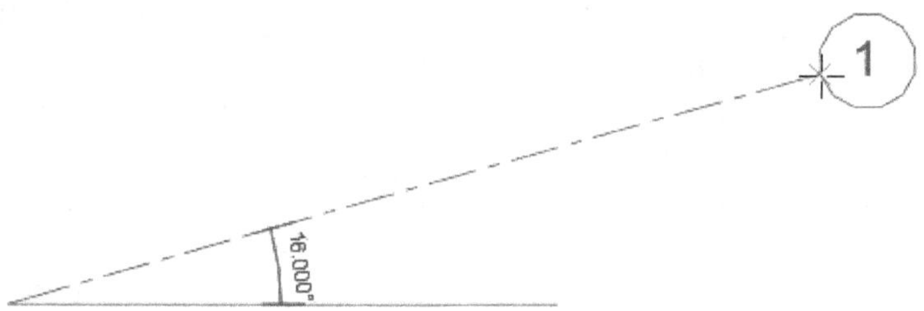

***Figure 3-28** Temporary angular dimension displayed while sketching the grid*

When you click to specify the endpoint, a grid is created and its controls are highlighted, as shown in Figure 3-29. Also, the recently added grid is highlighted and displays one square box on either side. These boxes are used to control the visibility of grid bubble. They can be selected or cleared to make the grid bubble visible or invisible on the desired side(s), respectively. The two circles on the start point and the endpoint can be dragged to extend or reduce the extent of the grid line.

***Figure 3-29** The highlighted controls of the grid created*

Similarly, when you sketch a new grid line near an existing one, a temporary dimension indicating the distance between them is displayed. You can enter a value in the edit box to specify the distance, as shown in Figure 3-30. Alternatively, you can move the cursor to the desired location using the temporary dimensions and click to specify the start point of the second grid line. Then, move the cursor horizontally to the right, and when the alignment line appears, click to specify the endpoint of the grid line.

Tip

It is recommended that you draw grid lines with the endpoints aligned to each other. You can use the alignment line that is displayed when the cursor is moved to its proximity. The ***Lock*** *constraints can be used to drag the extents of the single or multiple grid line.*

On doing so, the second grid line will be created. Notice that the name of this grid is 2. Autodesk Revit automatically numbers the grid lines as they are created. Similarly, you can draw more parallel grid lines. The **Offset** edit box in the **Options Bar** is used to specify the offset distance of a grid line from the desired point. To create vertical grid lines, specify the start point below the first grid line. Next, move the cursor vertically upward and click below the last horizontal grid line to specify the endpoint. You can create multiple vertical grid lines by using the method used for creating multiple horizontal lines.

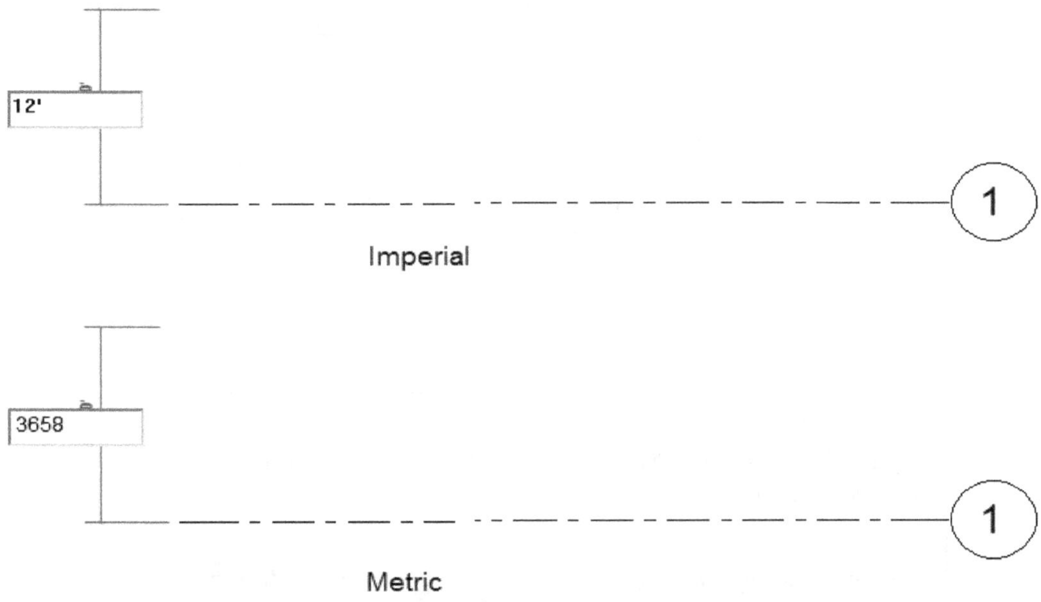

Figure 3-30 *Distance value entered in the edit box for another grid*

You can also create a rectangular grid pattern that is aligned at a given angle to the horizontal axis. You can also specify different angles for grid lines and create grid patterns based on your project requirement.

Sketching a Curve Grid

You can sketch curved or radial grid patterns based on your project requirement. To do so, invoke the **Start-End-Radius Arc** or **Center-ends Arc** tools from the **Draw** panel of the **Modify | Place Grid** tab.

After you have invoked the **Start-End-Radius Arc** tool, click at a point in the drawing area to specify the start point of the curved grid. Next, click at an appropriate location to locate the endpoint of the grid. On doing so, the curved grid will emerge with the temporary dimension displayed for the radius. Enter a suitable value for the radius or click in the drawing area; the curved grid will be created. Alternatively, after invoking the **Start-End-Radius Arc** tool, you can create multiple curved grids by selecting the **Radius** check box from the **Options Bar**. To do so, enter a suitable value in the edit box next to the **Radius** check box and then pick the start point and endpoint in the drawing to create an arc. You can also use the **Center-ends Arc** tool to create arcs by defining its center point, start point, and endpoint.

While sketching a structural grid, generally a combination of radial and linear grid patterns is used. Figure 3-31 shows a grid pattern created by using both the line and arc tools.

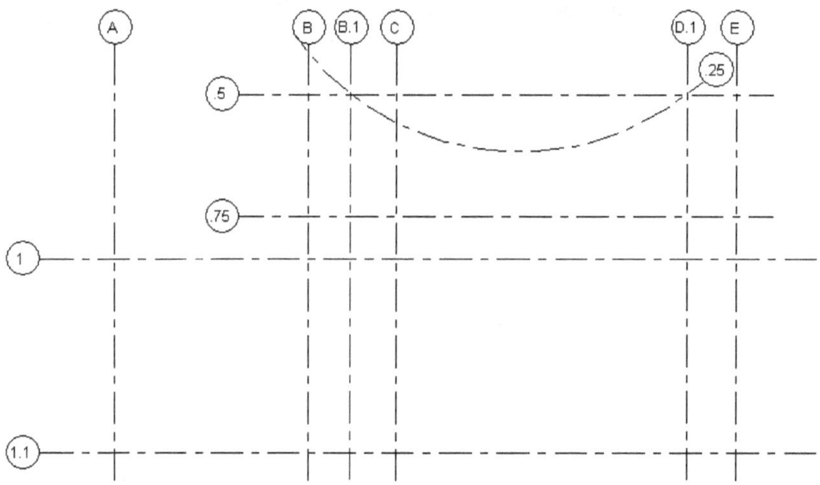

Figure 3-31 *An example of a grid pattern created using an arc and straight grid lines*

> **Tip**
> *You can also use different editing tools such as* ***Copy, Array, Mirror, Move,*** *and so on to create grid patterns. Autodesk Revit numbers grids intuitively. For example, if you create an array of a grid named A, the new grids will sequentially be named as B, C, D, E, and so on.*

Creating Grids Using the Pick Lines Tool

The **Pick Lines** tool is useful for creating grid lines that are aligned to existing elements. When you choose this tool from the **Draw** panel of the **Modify | Place Grid** tab and move the cursor near an existing element, the cursor snaps to certain object properties such as the interior face of walls, centerlines of columns, and so on. Click when the desired property of the element is highlighted; Autodesk Revit automatically creates a grid line along the specified alignment of element.

For example, to create grid lines for the structural wall profile, invoke the **Pick Lines** tool from the **Draw** panel and move the cursor near the curved exterior wall. As you move the cursor over the wall, it will snap to the interior and exterior faces. Click when the interior face is highlighted, as shown in Figure 3-32; a grid line aligned with the interior face of the wall will be displayed, as shown in Figure 3-33. Similarly, you can click on the other wall segments to create a grid pattern required for the building.

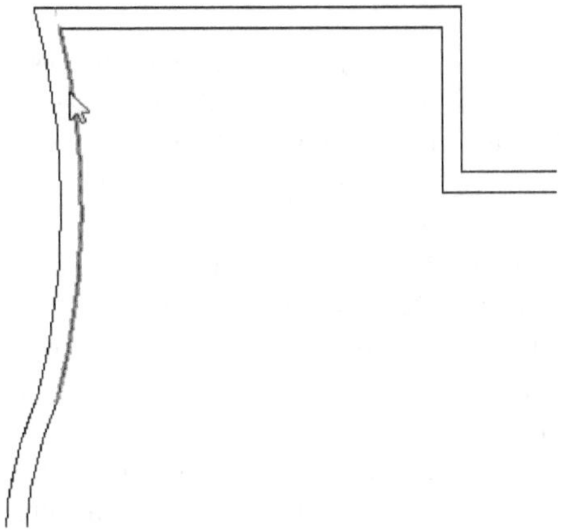

Figure 3-32 *The interior face of the wall highlighted*

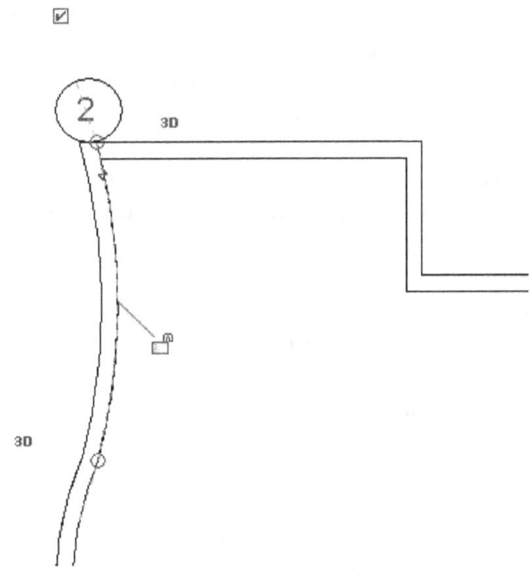

Figure 3-33 *Grid line created from the edge of the interior face of the wall*

If the cursor does not snap to the centerline of the wall, you can align the grid lines along the center of the wall by specifying the distance value of the centerline from the interior face in the **Offset** edit box and then use the **Pick Lines** tool to select the interior face.

Creating Multi-Segmented Grids

In Autodesk Revit, you can create grids having more than one segment. To create a multi-segmented grid, choose the **Grid** tool from the **Datum** panel of the **Structure** tab; the **Modify | Place Grid** tab will be displayed. In the **Draw** panel of this tab, choose the **Multi-Segment** tool; the **Modify | Edit Sketch** tab will be displayed. In the **Draw** panel of the tab, the **Line** tool is chosen by default. As a result, the grid segment that you will draw will be a straight line. In the **Draw** panel, you can choose other sketching tools such as **Start-End-Radius Arc**, **Center-ends Arc**, **Tangent-end Arc**, and others depending on the requirement of the sketch of the grid lines in the project. In the **Options Bar** the **Chain** check box is selected by default. As a result, the grid will be created from end-to-end in a chain. If you clear this check box you need to specify start and end points for each grid segment that will be created. In the **Options Bar**, you can enter a suitable value in the **Offset** edit box to specify an offset distance by which the grid segment will be offset from the point you specify in the drawing area.

In the **Options Bar**, the **Radius** check box is cleared by default. As a result, no fillet or curve will be created at the join of the grid segment. On selecting the **Radius** check box, the edit box next to it will be enabled. As a result, you can specify a suitable value in it to set the fillet radius of the curve that will be created at the join of two grid segments. Next, after specifying the desired options in the **Options Bar**, click in drawing area to start the multi-segment grid line. Now, click in the drawing area to specify the endpoint of the grid line.

Modifying Grids

Grids can be modified before or after they are created. To modify a grid, you need to select the grid and modify its properties using the **Properties** palette. Some of the parameters can also be modified by selecting the grid in the drawing window and entering the new values. For example, after selecting the grid, you can click on its name and assign it a new name. Similarly, you can modify the distance between the grids by selecting the temporary dimension and entering a new value. When you enter a new value, the grid automatically moves to the specified distance.

You can also move multiple grids by simply dragging them to the desired locations. To do so, click and select a single grid. Then, hold the CTRL key and click to select multiple grids. Next, you can drag the grid(s) to the desired location. Hold the SHIFT key to restrain the movement of the cursor in the orthogonal direction. When you move the cursor near a grid, it gets highlighted. At this stage, right-click and then choose the **Properties** option from the shortcut menu to display the **Properties** palette, if it is not displayed in the interface of Revit. You can also choose other options from the shortcut menu such as **Select Previous**, **Select All Instances**, and so on to select other grids. For certain grids, you may need to move or offset grid bubble to a different location. To do so, select a grid; a blue circle will appear on each of its endpoints. Also, the grid line break control will appear near the grid bubble. This control is used to create a grid bubble offset. You can click on this control and use the drag controls displayed to move the grid bubble to the desired location. The grid name also moves to the new location. Note that an extension line is also created along it.

Customizing the Grid Display

In Revit, you can customize the grid display. You can change the color, line weight, and line type of the entire grid line or part of the grid line such as center segment and end segment. You can also change the display of symbols at the end of the grid line.

Changing the Continuous Grid Line

A grid line is said to be continuous when the **Center Segment** parameter in the **Type Properties** dialog box is set to **Continuous**. You can change the line type, line weight, and color of the end segments using various options available in the **Type Properties** dialog box. Similarly, you can change the display of a symbol at the ends of a grid by using the other options available in the **Type Properties** dialog box.

Creating a Grid Line with Central Gap

You can create a grid line with central gap between its two end segments. To do so, select the **1/4" Bubble Custom Gap** or **1/4" Bubble Gap** option from the **Type Selector** drop-down list for Imperial unit system in the **Properties** palette. In Metric, you can select the option **6.5mm Bubble Custom Gap** or **6.5mm Bubble Gap** option from the **Type Selector** drop-down list.

Then, set the **Center Segment** parameter in the **Type Properties** dialog box to **None**. Figure 3-34 shows the display of grid line with gap between the two segments. You can also change the display properties of the end segments, as discussed above.

Figure 3-34 *The display of a grid line with a central gap*

Creating a Grid Line with Center Segment

You can create a grid line with the center and end segments containing different display properties such as line color, line weight, and line pattern. To do so, select the **1/4" Bubble Custom Gap** or **1/4" Bubble Gap** option from the **Type Selector** drop-down list in the **Properties** palette for the Imperial unit system. In Metric, you can select the option **6.5mm Bubble Custom Gap** or **6.5mm Bubble Gap** option from the **Type Selector** drop-down list. Then, set the **Center Segment** parameter in the **Type Properties** dialog box to **Custom**. You can change the display properties of the center segment and end segments using various options available in the **Type Properties** dialog box.

Controlling the Visibility of Grids

You can control the visibility of each grid or all grids in any of the project views. To do so, select a grid and right-click; a shortcut menu will be displayed. Choose **Hide in View > Category** from the shortcut menu; the grid category will be hidden in the current view. Similarly, you can also hide one particular grid by choosing **Hide in View > Elements** from the shortcut menu. On doing so, the selected grid will be hidden in the current view. However, the grid will be displayed in all the other views.

Alternatively, you can control the visibility of the grids by using the visibility/graphics feature of Autodesk Revit. To do so, select a grid from the drawing and choose the **Override by Category** tool from **Modify | Grids > View > Override Graphics in View** drop-down; the **View-Specific Category Graphics** dialog box will be displayed. Choose the **Open the Visibility Graphics dialog** button; the **Visibility/Graphic Overrides for Structural Plan: <view name>** dialog box for the current project view will be displayed. In the **Annotation Categories** tab of this dialog box, clear the check box for the visibility of grids; the visibility of all grids will turn off in the current view. You can also control the visibility of grids using the **Scope Box** feature.

WORKING WITH REFERENCE PLANES

Ribbon: Structure > Work Plane > Ref Plane
Shortcut Keys: RP

Reference planes help sketching and adding structural elements to a design. They can be used as datum planes to act as guidelines to create elements. They can also be used for creating new family elements. To create a reference plane, invoke the **Ref Plane** tool from the **Work Plane** panel of the **Structure** tab. Alternatively, enter **RP**; the **Modify | Place Reference Plane** tab will be displayed. In the **Draw** panel of this tab, the **Line** and **Pick**

Line tools are present, the **Line** tool is invoked by default. You can use this tool to draw a straight line or a chain of connected line segments to represent a reference plane. Before drawing a reference plane using the **Line** tool, you can use the **Offset** edit box in the **Options Bar** to specify a suitable offset distance for it. To draw a line for the reference plane, click at the desired location in the drawing window. Next, drag the cursor to a new location and release the left button to specify the endpoint of the reference line; the reference plane will be created. Select the reference plane and assign a name to it in the corresponding field of the **Name** parameter in the **Properties** palette.

Alternatively, you can draw a reference plane in your project view by selecting the edge of a wall, beam, column, or an existing model line in the drawing. To do so, invoke the **Ref Plane** tool as discussed earlier and then choose the **Pick Lines** option from the **Draw** panel of the **Modify | Place Reference Plane** tab. On doing so, you will be prompted to select an edge or a line from the drawing. Move the cursor toward the desired edge of an element; the edge of the element will be highlighted. Click on the highlighted edge to add a reference plane aligned to it.

WORKING WITH WORK PLANES

A work plane is a plane which is used to sketch elements. In Autodesk Revit, you can create and edit only those elements that are in the current work plane. A work plane can be horizontal, vertical, or inclined at any specified angle. Each generated view has an associated work plane. This plane is automatically defined for some standard views such as floor plans. For others, such as sections, elevations, and 3D views, you can set the work plane based on the location of the elements to be created or edited. The concept of work planes is especially useful for creating elements in elevations, sections, or inclined planes. You will learn more about the work planes in the following sections.

Setting a Work Plane

Ribbon:	Structure > Work Plane > Set

You can set a work plane based on your project requirement. To set a work plane, invoke the **Set** tool from the **Work Plane** panel of the **Structure** tab; the **Work Plane** dialog box will be displayed, as shown in Figure 3-35. This dialog box shows the current work plane. You can also specify parameters for setting a new work plane in this dialog box. In the **Specify a new Work Plane** area, you will notice drop-down list is available on the right of the **Name** radio button. This drop-down list contains the options that can be levels, grids, and named reference planes. Select an option from the drop-down list to set the current work plane.

The **Pick a plane** radio button can be used to set a work plane along an existing plane. To do so, select this radio button and then an existing plane in the drawing. You can select the **Pick a line and use the work plane it was sketched in** radio button to create a work plane that is coplanar with the plane on which the selected line was created.

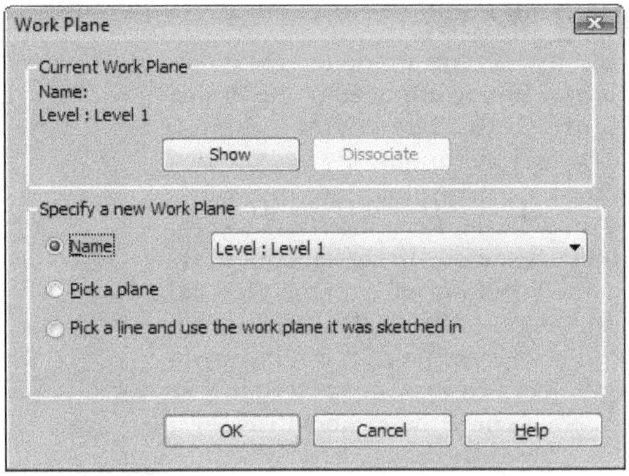

*Figure 3-35 The **Work Plane** dialog box*

Controlling the Visibility of Work Planes

Ribbon:	Structure > Work Plane > Show

You can control the visibility of the current work plane by invoking the **Show** tool from the **Work Plane** panel of the **Structure** tab. The work plane appears as a grid in the current view. To hide it, choose the **Show** tool again.

You can also set the grid spacing for a work plane. To do so, select a work plane from the drawing; the work plane will be highlighted. Specify the spacing by entering a new value in the **Spacing** edit box in the **Options Bar**. You can also enter the spacing in the corresponding value field of the **Work Plane Grid Spacing** parameter in the **Properties** palette. You can snap to the work plane grid using the object snap tools. These tools can be used to create basic locations for columns and footings in a project view.

Using the Workplane Viewer Window

While working on a project you may require to modify elements at a specified workplane. To do so, you can use the **Workplane Viewer** window. This window displays temporary views of models present in the Project Environment and helps in editing the profile in forms, sweeps, blends, and swept blends. You can invoke the **Workplane Viewer** window by choosing the **Viewer** tool from the **Workplane** panel of the **Structure** tab.

WORKING WITH PROJECT VIEWS

While working on a structural model, you may need to view its different exterior and interior portions in order to add or edit elements in the design. Autodesk Revit provides various features and techniques that can be used to view the structural model. In this section, you will learn about the tools that help in working with views.

Viewing a Building Model

The default template file has certain predefined standard project views displayed in the **Views (all)** head of the **Project Browser**, refer to Figure 3-36. These views include structural plans, 3d views, and elevations. To open a view, double-click on its name; the corresponding view will be displayed in the viewing area. In the Revit interface, you can hide or unhide the **Project Browser**. To do so, choose the **User Interface** tool from the **Windows** panel of the **View** tab; a cascading menu will be displayed. By default, the **Project Browser** check box is selected in this cascading menu, which indicates that the **Project Browser** is displayed in your drawing. Clear this check box to hide the **Project Browser** in your drawing.

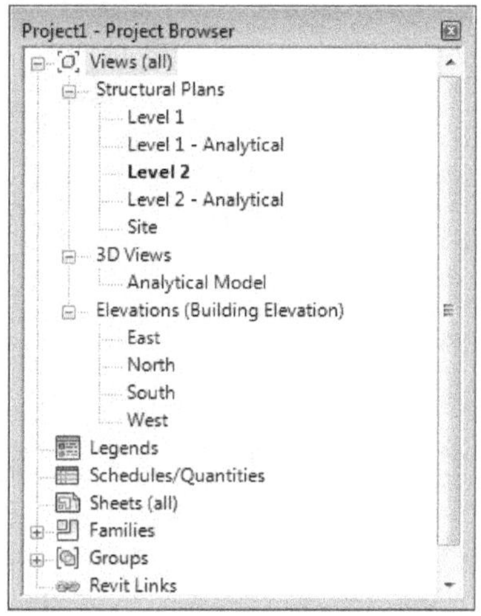

Figure 3-36** The options displayed in the **Project Browser

Autodesk Revit also enables you to add elements to a building model by simply dragging and dropping them directly from the **Project Browser** instead of using the **Type Selector** drop-down list.

When you open a new project, the viewing area displays four inward arrow symbols in the floor plan view, which indicates the four-side elevations: North, East, South, and West. You can use these symbols to view appropriate building elevation by double-clicking on them.

Controlling the Visibility of Elements in Views

To control the visibility of certain categories of elements, select the element from the drawing; the contextual tab related to it will be displayed in the ribbon. From the **View** panel of the displayed contextual tab, choose the **Override Graphics in View** drop-down menu; a list of available override tool will be displayed. Next, select the **Override by Category** option from the list; the **View-Specific Category Graphics** dialog box will be displayed. Choose the **Open the Visibility Graphics dialog** button; the **Visibility/Graphic Overrides for Structural Plan: <Level name>** dialog box for the specific view will be displayed. You can use the various tabs in this dialog box to hide different categories of elements.

SCOPE BOX

Ribbon: View > Create > Scope Box

The **Scope Box** tool helps you to control the visibility of the datum elements (grids, reference lines, and levels) in the project views. As described earlier in this chapter, these datum elements have infinite scope and they extend throughout the project. Using the **Scope Box** tool, you can create boundary limit for the visibility of these datum elements. You can also specify the views in which these datum elements become visible.

Creating a Scope Box

A scope box can be created in the plan view by invoking the **Scope Box** tool. You can invoke this tool from the **Create** panel of the **View** tab. Once this tool has been invoked, the **Options**

Bar displays the **Name** and **Height** edit boxes. You can enter the name and height of the scope box in their respective edit boxes. To create a scope box, move the cursor in the viewing area; the cursor will change into a cross symbol and will prompt you to draw scope box in the plan view. To define the scope box, click on its upper left corner, move the cursor to the lower right corner, and then click to specify the diagonally opposite ends. Note that the rectangle should be drawn in such a way that the elements that need to be visible are enclosed in it. The scope box with the assigned name will be created. When you select the scope box, drag controls are visible on it, as shown in Figure 3-37. These drag controls can be used to resize the scope box.

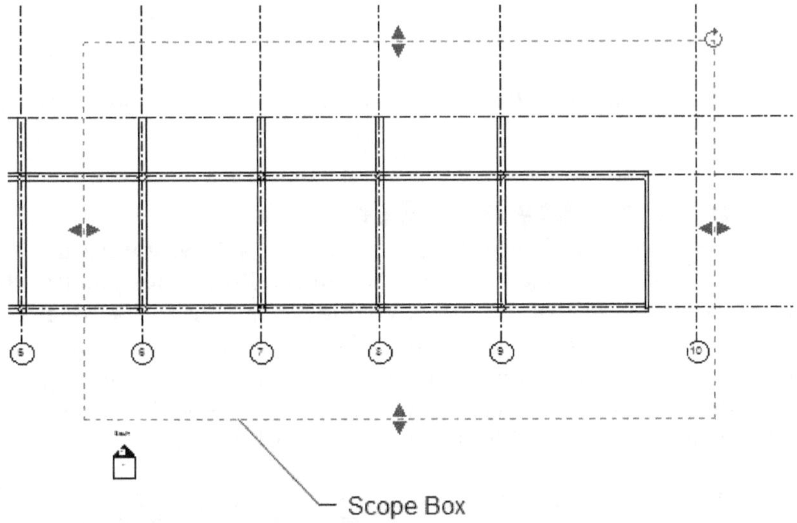

Figure 3-37 The scope box displayed in the drawing

Applying a Scope Box to Datum Elements

The visibility of datum elements can be controlled by associating them with a scope box. To associate the scope box created to the desired datum elements, select the scope box from the drawing; the properties of the selected scope box will be displayed in the **Properties** palette. Click on the value field corresponding to the **Scope Box** parameter and select the name of the scope box from the drop-down list. Next, choose the **Apply** button to apply property to the selected datum element. On doing so, you will notice that the datum elements that intersect the boundary of the assigned scope box will restrict its extent throughout the boundary of the scope box. Figures 3-38 and 3-39 illustrate the extent of the grid elements before and after applying the scope box to them.

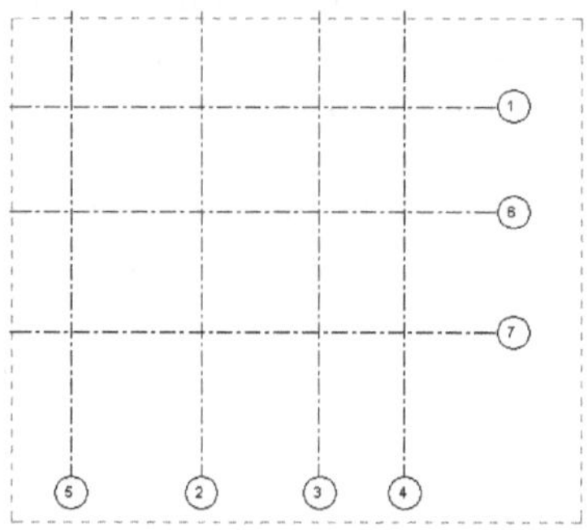

Figure 3-38 *Grids before applying the scope box*

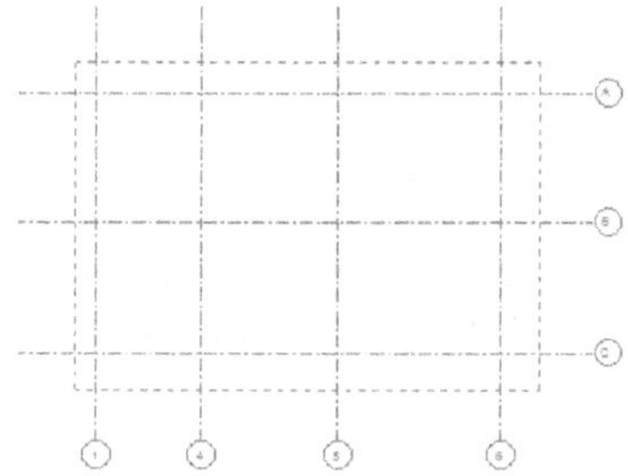

Figure 3-39 *Grids after applying the scope box*

Controlling the Visibility of a Scope Box

A scope box can be resized to limit its visibility for certain views. Its visibility can also be controlled for each view. To do so, select a scope box, and then in the **Properties** palette, choose the **Edit** button displayed in the value field for the **Views Visible** parameter; the **Scope Box Views Visible** dialog box will be displayed. This dialog box lists all views types and view names available in the project. The **Automatic visibility** column shows the current visibility of scope boxes. You can click on the **Override** column for a specific view and select the **Visible** or **Invisible** option from the drop-down list displayed to override the automatic visibility setting, refer to Figure 3-40.

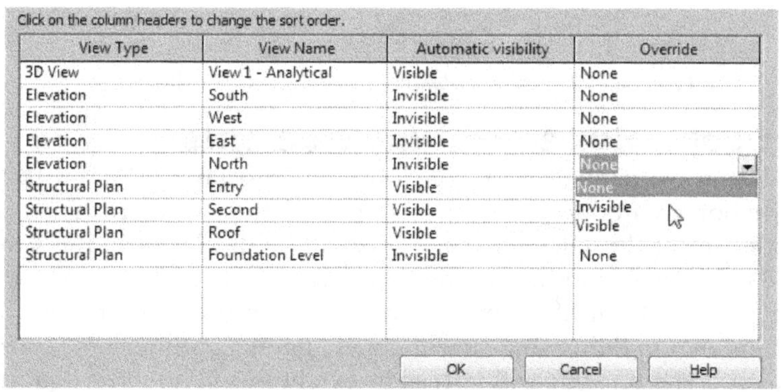

Figure 3-40 *Selecting an option from the drop-down list in the* ***Override*** *column*

TUTORIALS

Tutorial 1 Commercial Complex

In this tutorial, you will create various project settings such as project information, project location, and browser organization for the project created in Tutorial 1 of Chapter 2. Further, you will add levels and grids to the project file. Use the following project parameters and specifications:

(Expected time: 1 hr 15 min)

1. Rename Level 1 and Level 2 as **Below Grade** and **Entry**, respectively.

2. Specify the elevation of **Below Grade** and **Entry** levels as follows.
   For Imperial **-4' 0"** and **0' 0"**
   For Metric **-1219** mm and **0** mm

3. Levels to be added:
   For Imperial **Second Floor - Elevation 12' 6"**
   Roof - Elevation 25' 0"
   For Metric **Second Floor - Elevation 3810 mm**
   Roof - Elevation 7620 mm

4. File name to be assigned:
   For Imperial *c03_Commercial-Complex_tut1.rvt*
   For Metric *M_c03_Commercial-Complex_tut1.rvt*

The following steps are required to complete this tutorial:

a. Open Tutorial 1 of Chapter 2.
b. Set the project information, project location, and browser organization.
c. Modify existing levels and add new levels.
d. Add grid lines.
e. Save the project using the **Save As** tool.
f. Close the project using the **Close** tool.

Opening an Existing Project

In this section, you will open the project file created in Tutorial 1 of Chapter 2.

1. To open the project file, choose **Open > Project** from the **Application Menu**; the **Open** dialog box is displayed.

2. In this dialog box, browse to *C:\rst_2017\c02_rst_2017_tut* folder and then choose the *c02_Commercial-Complex_tut1.rvt* (for Imperial) or *M_c02_Commercial-Complex_tut1.rvt* (for Metric) project file. You can also download this file from *http://www.cadcim.com*. The path of the file is as follows: *Textbooks > Civil/GIS > Revit Structure > Exploring Autodesk Revit 2017 for Structure*

3. Choose the **Open** button from the **Open** dialog box to open the selected project file.

Setting the Project Information

In this section, you will add project information to the opened file. Further, you will define a parameter for the project information and add it to the project information of the project file.

1. To set the project information for the project, choose the **Project Information** tool from the **Settings** panel of the **Manage** tab; the **Project Information** dialog box is displayed.

2. In this dialog box, specify the project parameters as follows:

Parameter	**Value**
Project Issue Date	**01/01/2016**
Project Status	**Started**
Client Name	**Tickoo Institute of Emerging Technologies**
Project Address	**525, Andrews Drive Schererville**
Project Name	**Commercial Complex**
Project Number	**TIET/02/16**

3. In the **Project Information** dialog box, choose the **OK** button; the dialog box closes and the project parameters are saved in the project file.

4. Next, you need to create an additional instance parameter for the project information. To do so, choose the **Project Parameters** tool from the **Settings** panel of the **Manage** tab; the **Project Parameters** dialog box is displayed.

5. Choose the **Add** button from the **Project Parameters** dialog box; the **Parameter Properties** dialog box is displayed.

6. In the **Parameter Properties** dialog box, scroll the list in the **Categories** area and then select the **Project Information** check box.

7. In the **Parameter Data** area, ensure that the **Instance** radio button is selected. Now, enter the name **Project Engineer** in the **Name** edit box in the **Parameter Data** area.

8. Next, select the **Common** option from the **Discipline** drop-down list, if it is not selected by default.

9. Select the **Text** and the **Other** options from the **Type of Parameter** and **Group parameter under** drop-down lists, respectively.

10. Choose the **OK** button in the **Parameter Properties** dialog box; the **Project Parameters** dialog box is displayed, as shown in Figure 3-41.

Figure 3-41 *The **Project Parameters** dialog box*

11. Choose the **OK** button; the **Project Parameters** dialog box is closed.

12. To enter the value for the new parameter, choose the **Project Information** tool from the **Settings** panel of the **Manage** tab; the **Project Information** dialog box is displayed.

13. In this dialog box, click in the **Value** field corresponding to the **Project Engineer** parameter and enter **Sham Tickoo**.

14. Next, choose the **OK** button; the **Project Information** dialog box closes and the value for the newly added parameter is saved in the project file.

Setting the Project Location

In this section, you will specify the geographical location of the project.

1. To specify the geographical location of the project, invoke the **Location** tool from the **Project Location** panel of the **Manage** tab; the **Location Weather and Site** dialog box is displayed.

2. In the **Location Weather and Site** dialog box, choose the **Site** tab and then the **Duplicate** button; the **Name** dialog box is displayed.

3. Enter **TIET-IN** in the **Name** edit box and then choose the **OK** button; the **Name** dialog box closes and the name entered is displayed and highlighted in the **Sites defined in this project** list box of the **Location Weather and Site** dialog box, refer to Figure 3-42.

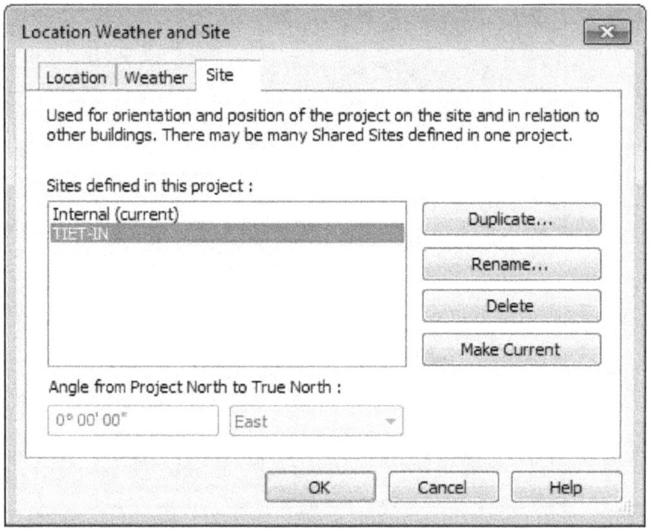

***Figure 3-42* *TIET-IN* option selected in the *Sites defined in this project* list box of the *Site* tab**

4. Next, choose the **Location** tab in the **Location Weather and Site** dialog box; the options in this tab are displayed.

5. Select the **Default City List** option from the **Define Location by** drop-down list.

6. Next, select the **Indianapolis, IN** option from the **City** drop-down list.

7. Next, select the **Use Daylight Saving time** check box and then choose the **OK** button; the dialog box closes and the settings for the project location are saved in the project file.

Setting the Browser Organization

In this section, you will specify the settings for customizing the **Project Browser** displayed in the drawing area of the project.

1. To customize the **Project Browser**, choose the **Browser Organization** tool from **View > Windows > User Interface** drop-down; the **Browser Organization** dialog box is displayed.

2. Ensure that the **Views** tab is chosen, and then select the **Discipline** check box in this tab.

3. Next, choose the **Sheets** tab in the **Browser Organization** dialog box; the options in this tab are displayed.

4. In this dialog box, select the **Issue Date** check box.

5. Now to apply the modified settings, choose **Apply** and then **OK** from the **Browser Organization** dialog box; the modified settings are applied to the **Project Browser** and the **Browser Organization** dialog box is closed.

Modifying the Existing Levels

In this section, you will modify the existing levels in the project file. To do so, you need to select an elevation view from the **Project Browser** and then use the **Level** tool.

1. Before modifying the levels, move the cursor to the **Project Browser** and expand the Structural head and then the **Elevations (Building Elevation)** node. Now, double-click on the **North** node; the current view is changed to the North elevation view.

> **Note**
> *In the north elevation view, make sure that the **Hidden Line** option is chosen for the **Visual Style** in the **View Control Bar**.*

2. Next, zoom in the drawing area to make the text on the level lines visible. Click on the **Level 1** level line and move the cursor over the text **Level 1**; the text is highlighted.

3. Click on the text; an edit box is displayed.

4. Enter **Below Grade** in the edit box and press ENTER; the **Revit** message box is displayed. This message box prompts you to confirm whether or not to rename the corresponding view of the selected level.

5. Choose the **Yes** button; the name of the level changes in the **Project Browser**. Also, notice that the name of the view corresponding to the changed level has been changed under the **Structural Plans** head in the **Project Browser**.

6. Next, move the cursor over the elevation value of the **Below Grade** level; the elevation value is highlighted.

7. Click on the highlighted elevation value; an edit box is displayed.

8. Enter value in the edit box as given below:
   For Imperial **-4' 0"**
   For Metric **-1219 mm**

   Press ENTER and then press ESC to exit from the modification controls.

9. Repeat the procedure followed in steps 2 through 7 to rename the **Level 2** level to **Entry** and modify its elevation as given below:
   For Imperial **0'0"**.
   For Metric **0** mm.

10. Next, right-click in the drawing area; a shortcut menu is displayed.

11. Choose the **Zoom To Fit** option from the shortcut menu; the extents of the level lines are displayed in the drawing area.

12. Next, select the **Below Grade** level line and then place the cursor over the grid length control displayed on the left.

13. Press the left mouse button and drag the mouse to the right; the level lines move along the cursor.

14. Click in the drawing area when the length of the level line reduces to half of their existing length.

15. Press ESC to exit from the edit controls.

16. Next, right-click in the drawing area and then choose **Zoom To Fit** from the shortcut menu; the current view fits the extent of the length of level lines.

Adding New Levels

In this section, you will add new levels to the project file using the **Level** tool.

1. To add a new level, choose the **Level** tool from the **Datum** panel of the **Structure** tab; the **Modify | Place Level** tab is displayed. Alternatively, you can press LL to activate the **Level** tool.

2. Move the cursor near the left endpoint of the **Entry** level line, and when the alignment line appears, type value in the edit box displayed as given below:

   For Imperial **12'6"**, refer to Figure 3-43.
   For Metric **3810 mm**

   Press ENTER.

3. Next, move the cursor toward the right side until the alignment line appears above the **Entry** level bubble and then click to complete the level line, as shown in Figure 3-44. A new level line is created, displaying the level name as **Level 3** and the elevation as **12'6"** for imperial and for metric the elevation is **3810** .

Figure 3-43 *Specifying the distance of the new level line*

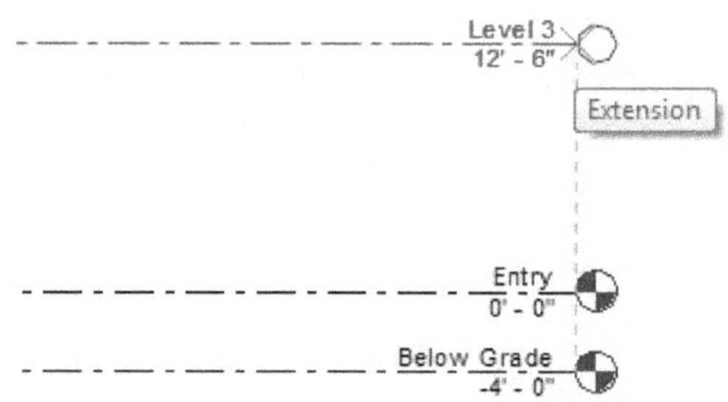

Figure 3-44 *Creating the new level line using the alignment line*

4. Without exiting the **Level** tool, repeat steps 2 and 3 to create a level line with the name **Level 4** at a distance from the **Level 3** level line as given below:

   For Imperial **12'6"**
   For Metric **3810**

   Note that the elevation of the **Level 4** level line is displayed as shown below:

   For Imperial **25' 0"**
   For Metric **7620**

5. Press ESC twice to exit from the **Level** tool.

6. Select the **Level 3** level line; the properties of the selected level line are displayed in the **Properties** palette.

7. In the **Properties** palette, click in the value field corresponding to the **Name** parameter and replace the existing name **Level 3** with **Second Floor**.

8. Choose the **Apply** button in this palette; the **Revit** message box is displayed.

9. In the **Revit** message box, choose the **Yes** button; the name of the level line **Level 3** changes to **Second Floor**. Also, notice that the name of the corresponding view in the **Project Browser** has changed to the new name.

10. Next, to rename the **Level 4** level view, select it from the drawing area; the properties of the selected level are displayed in the **Properties** palette.

11. Repeat steps 7 to 9 and rename **Level 4** to **Roof**.

12. Press ESC to exit from the tool.

13. Choose the **Level 1 - Analytical** option from the **Structural Plans** subhead in the Project Browser.

Note
*The **Structural Plans** head is a subhead in the **Structural** head of the **Views (Discipline)** hierarchy in the **Project Browser**.*

14. Right-click and then choose **Delete** from the displayed shortcut menu; the selected view name is deleted from the **Project Browser**.

15. Repeat steps 13 and 14 to delete the **Level 2- Analytical** option from the **Structural Plans** head.

Adding Grid Lines

In this section, you will add grids to the structural project. To add grids, you need to change the current view to the **Entry** level structural plan and then use the **Grid** tool.

1. Double-click on **Entry** from the **Structural Plans** head in the **Project Browser**; the **Entry** structural plan view will be displayed in the drawing area.

2. Invoke the **Grid** tool from the **Datum** panel of the **Structure** tab; the **Modify | Place Grid** tab is displayed.

3. In this tab, ensure that the **Line** tool is invoked from the **Draw** panel. Next, move the cursor toward the lower left corner of the drawing area and then click to specify the start point of the first grid line; a grid line with temporary dimensions for angle emerges, refer to Figure 3-45 for location.

4. Move the cursor vertically up (at an angle of 90° from the horizontal plane) and click when it reaches near the **North** elevation arrow, refer to Figure 3-46 for location.

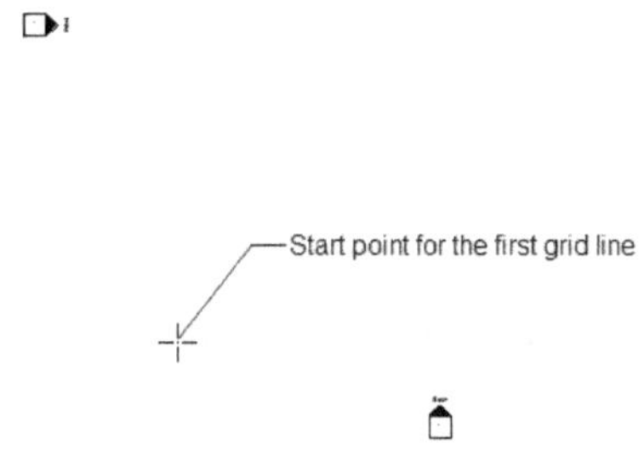

Figure 3-45 *Specifying the start point of the first grid line*

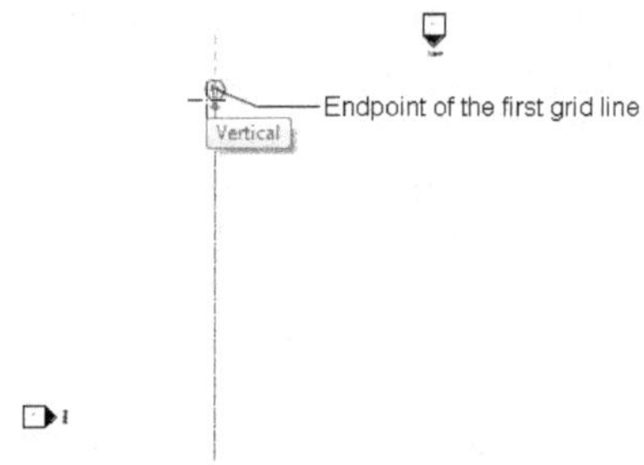

Figure 3-46 *Specifying the endpoint of the first grid line*

5. Now, move the cursor downward and position it near the right side of the endpoint of the grid line created previously, as shown in Figure 3-47.

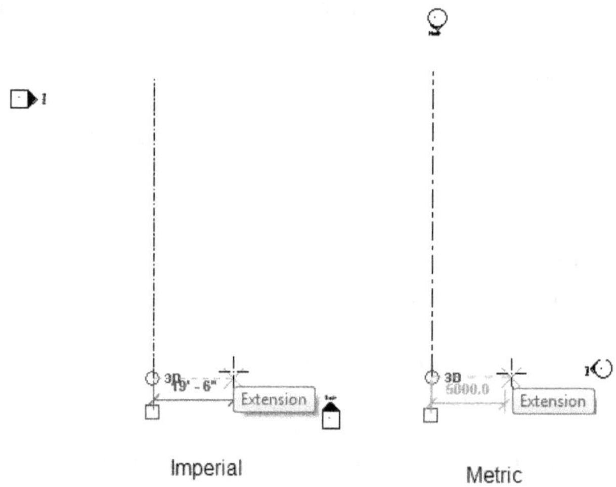

Figure 3-47 *Specifying the start point of the second vertical grid line*

6. As you place the temporary dimension near the endpoint, a temporary dimension appears with an extension line and the **Extension** snap. Enter value as given below:
   For Imperial **10'**
   For Metric **3048**

   Press ENTER; the second grid line starts at a distance of entered value from the start point of the first grid line.

7. Move the cursor vertically up and place it near the bubble of the first grid line.

8. Click when the **Vertical and Extension** snap is displayed; the second grid line is created.

9. Repeat steps 5 to 8 to create five more vertical grids with the distance specified in Figure 3-48.

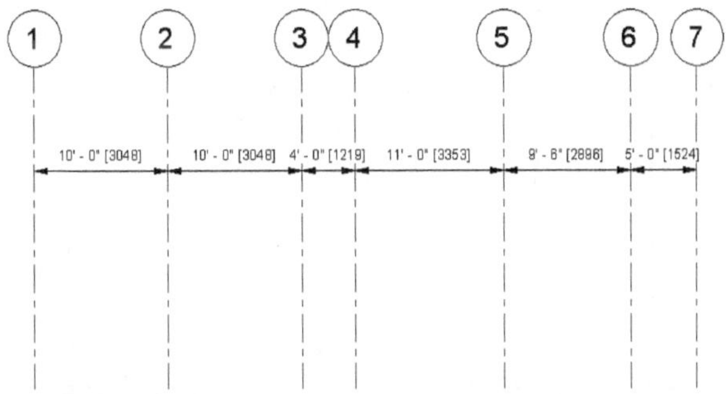

Figure 3-48 *Distance between the vertical grid lines*

10. After creating all the vertical grids, press ESC twice to exit the **Grid** tool.

11. To rename grids, select the vertical grid line marked **4**. Now, place the cursor over the text in the grid bubble of the selected grid line and click; an edit box is displayed, as shown in Figure 3-49.

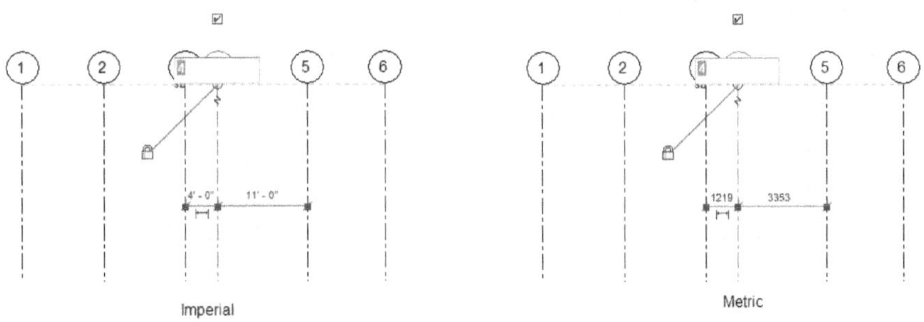

Figure 3-49 *Renaming the grid line using the edit box*

12. Enter **3.1** in the edit box and press ENTER; the selected grid line is renamed.

Note

While renumbering the grid line, you can use the ***Zoom in Region*** *and* ***Zoom Out(2x)*** *tools to view the grid numbers easily.*

13. Without exiting the editing mode, select the next grid marked **5** and then move the cursor over its grid bubble and click to display the edit box.

14. Enter **4** in the edit box and press ENTER to rename the selected line.

15. Repeat steps 11 and 12 to rename the grid lines marked **6** and **7** to **5** and **5.1**, respectively, as shown in Figure 3-50.

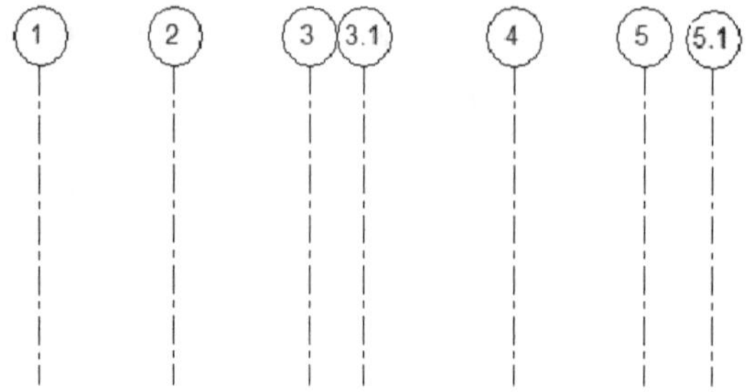

Figure 3-50 The display of vertical grids after they are renumbered

16. After renumbering all required vertical grids displayed in the drawing, press ESC to exit the editing mode.

17. To add horizontal grids, invoke the **Grid** tool from the **Datum** panel of the **Structure** tab.

18. Next, move the cursor toward the top right of the vertical grid line marked **5.1** and click at the location shown in Figure 3-51.

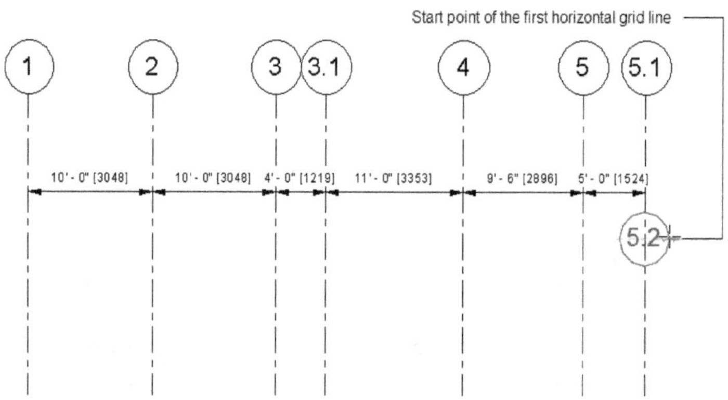

Figure 3-51 The location of the start point of the first horizontal grid line

19. Move the cursor horizontally toward the left until it has crossed the vertical grid line marked **1**. Click in the drawing area to create the first horizontal grid line, as shown in Figure 3-52.

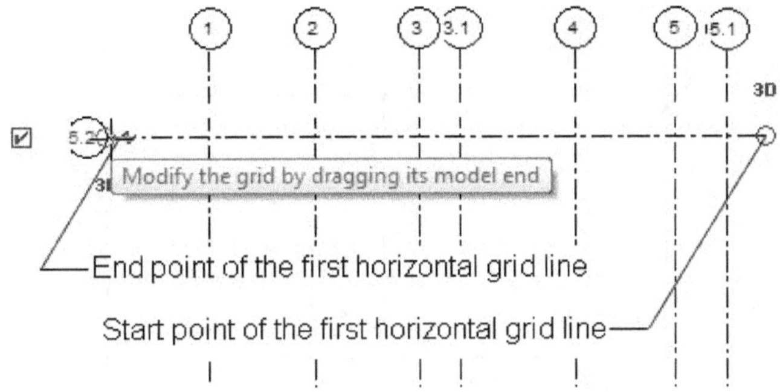

Figure 3-52 *The location of the start point and endpoint of the first horizontal grid line*

20. Move the cursor below the grid line alignment control of the start point of the first horizontal grid line (grid line created previously); an alignment line with a temporary dimension is displayed.

21. Type **8'** (or **2438 mm** for Metric); an edit box is displayed with the entered value. Press ENTER; the second horizontal grid line will start below the start point of the first horizontal grid line at a distance of **8'** (or **2438** mm in Metric) from it.

22. Move the cursor horizontally toward the left and place it below the grid bubble (marked **5.2**) of the first horizontal grid line when the alignment line appears.

23. Click when the **Horizontal and Extension** snap appears; the second horizontal grid line is created.

24. Repeat steps 20 to 23 to create the rest of horizontal grid lines and to rename them, refer to Figures 3-53 and 3-54.

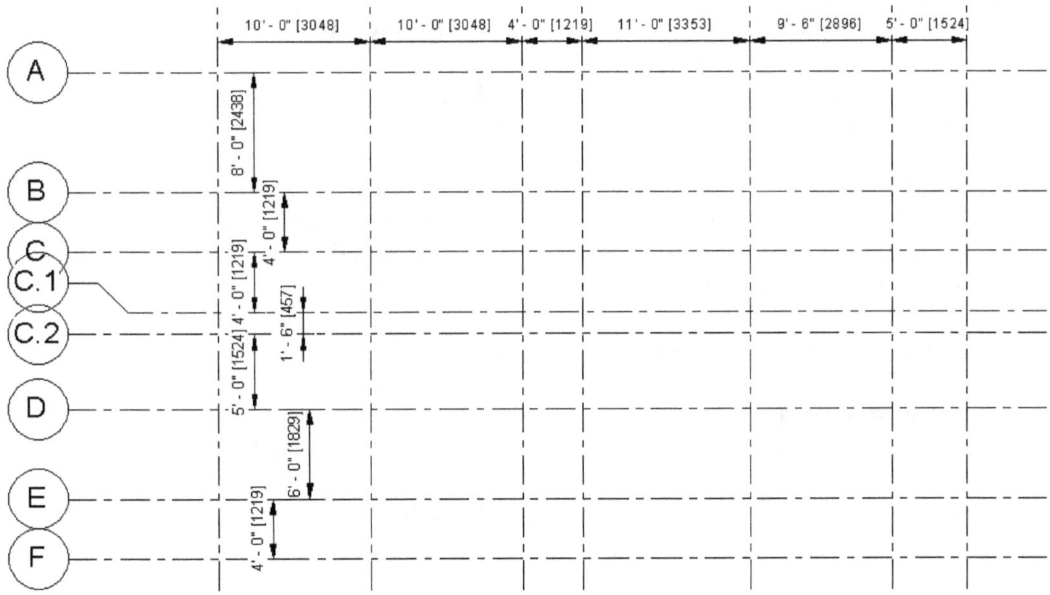

Figure 3-53 *The horizontal grid lines to be created with the specified names and distances*

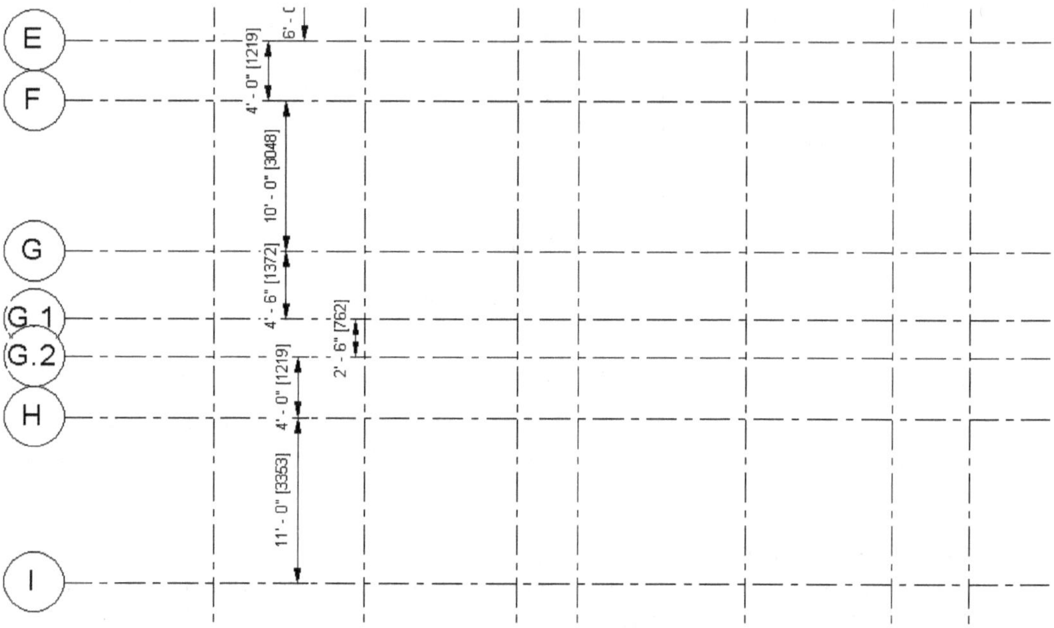

Figure 3-54 *The remaining horizontal grid lines in continuation of Figure 3-53*

Note

*To create the remaining horizontal grids from grid **B** to grid **I** refer to Figures 3-53 and 3-54. Note that both these figures are part of a single grid plan view. Figure 3-54 is in continuation with Figure 3-53 and is displayed separately to show the grid dimensions clearly. Further, the horizontal grids will be renamed using the same method used for vertical grids.*

Saving the Project

In this section, you will save the project file by using the **Save As** tool.

1. To save the project, choose the **Save As > Project** option from the **Application Menu**. On doing so, the **Save As** dialog box is displayed.

2. In this dialog box, browse to the C:*rst_2017* location and create a folder with name *c03_rst_2017_tut*. Now, open the *c03_rst_2017_tut* folder and enter **c03_Commercial-Complex_tut1** (for Imperial) **or M_c03_Commercial-Complex_tut1** (for Metric) in the **File name** edit box.

3. Choose the **Save** button; the **Save As** dialog box closes and the project file is saved.

Closing the Project

1. To close the project, choose the **Close** option from the **Application Menu**.

   The file is closed and this completes Tutorial 1 of Chapter 3.

Tutorial 2 Industrial Complex

In this tutorial, you will create various project settings such as project information, project location, and browser organization for the project created in the Tutorial 2 of Chapter 2. Further, you will add levels and grids to the project file. Use the following project specifications:

(Expected time: 1 hr 15 min)

1. Rename Level 1 and Level 2 as **Foundation Level** and **Entry**, respectively.

2. Specify the elevation of **Foundation Level** and **Entry** levels.
   For Imperial **-6' 0"** and **0' 0"**
   For Metric **-1829 mm** and **0 mm**

3. Levels to be added:
   For Imperial **Second Floor - Elevation 10' 0"**
   Roof - Elevation 24' 0"
   For Metric **Second Floor - Elevation 10' 0"**
   Roof - Elevation 7315 mm

4. File name to be assigned:
   For Imperial *c03_Industrial-Complex_tut2.rvt.*
   For Metric *M_c03_Industrial-Complex_tut2.rvt.*

   The following steps are required to complete this tutorial:

a. Open the Tutorial 2 of Chapter 2.
b. Set project information and project location.
c. Modify existing levels and add new levels.
d. Add grid lines.
e. Resizing and renaming the grid lines.
f. Save the project using the **Save As** tool.
g. Close the project by using the **Close** tool.

Opening the Existing Project

In this section, you will open the project file created in the Tutorial 1 of Chapter 2.

1. To open the project file, choose **Open > Project** from the **Application Menu**; the **Open** dialog box is displayed.

2. In this dialog box, browse to *C:\rst_2017\c02_rst_2017_tut* folder and then choose the *c02_Industrial-Complex_tut2.rvt* (for Imperial) *or M_c02_Industrial-Complex_tut2.rvt* (for Metric) project file. You can also download this file from *http://www.cadcim.com*. The path to download this file is as follows: *Textbooks > Civil/GIS > Revit Structure > Exploring Autodesk Revit 2017 for Structure.*

3. Choose the **Open** button from the **Open** dialog box; the selected project file opens in the drawing window.

Setting the Project Information

In this section, you will add the project information to the industrial complex project file.

1. To set the project information, choose the **Project Information** tool from the **Settings** panel of the **Manage** tab; the **Project Information** dialog box is displayed.

2. In the **Project Information** dialog box, specify the project parameters as follows:

Parameter	**Value**
Project Issue Date	**01/01/2016**
Project Status	**Started**
Client Name	**CADCIM Technologies**
Project Address	**525, Andrews Drive Schereville**
Project Name	**Industrial Complex**
Project Number	**CIT/02/16**

3. Choose the **OK** button; the **Project Properties** dialog box closes and the project parameters are saved in the project file.

4. Next, create an additional instance parameter for the project information. To do so, choose the **Project Parameters** tool from the **Settings** panel in the **Manage** tab; the **Project Parameters** dialog box is displayed.

5. Choose the **Add** button from the dialog box; the **Parameter Properties** dialog box is displayed.

6. In this dialog box, scroll the list in the **Categories** area and then select the **Project Information** check box.

7. In the **Parameter Data** area of the **Parameter Properties** dialog box, enter **Project Manager** in the **Name** edit box.

8. Next, in the same area, ensure that the **Instance** radio button is selected and then ensure that the **Common** option is selected in the **Discipline** drop-down list.

9. Select the **Text** and **Other** options from the **Type of Parameter** and **Group parameter under** drop-down lists, respectively.

10. Next, choose the **OK** button from the **Parameter Properties** dialog box; the **Project Parameters** dialog box is displayed.

11. Choose the **OK** button from the **Project Parameter** dialog box; the dialog box closes.

12. To enter the value for the new parameter, choose the **Project Information** tool from the **Settings** panel of the **Manage** tab; the **Project Information** dialog box is displayed.

13. In this dialog box, click in the **Value** field corresponding to the **Project Manager** parameter and enter **Sham Tickoo**.

14. Now, choose the **OK** button; the **Project Information** dialog box closes and the value for the newly added parameter is saved in the project file.

Setting the Project Location

In this section, you will specify the geographical location of the project.

1. To specify the geographical location of the project, choose the **Location** tool from the **Project Location** panel of the **Manage** tab; the **Location Weather and Site** dialog box is displayed.

2. Ensure that the **Location** tab is chosen in this dialog box. Then, select the **Internet Mapping Service** option from the **Define Location by** drop-down list.

Note

*While selecting the **Internet Mapping Service** option from the **Define Location by** drop-down list, ensure that the internet is connected to your system. If the internet is not connected, a message box will be displayed below the **Project Address** edit box, informing that the connection could not be established to the Internet Mapping Service.*

3. In the **Project Address** edit box, delete the default text, and enter **Indianapolis IN**, and then choose the **Search** button; the desired location is displayed on the map in the area below the edit box. Note that the search uses the bing map service to display the location, refer to Figure 3-55.

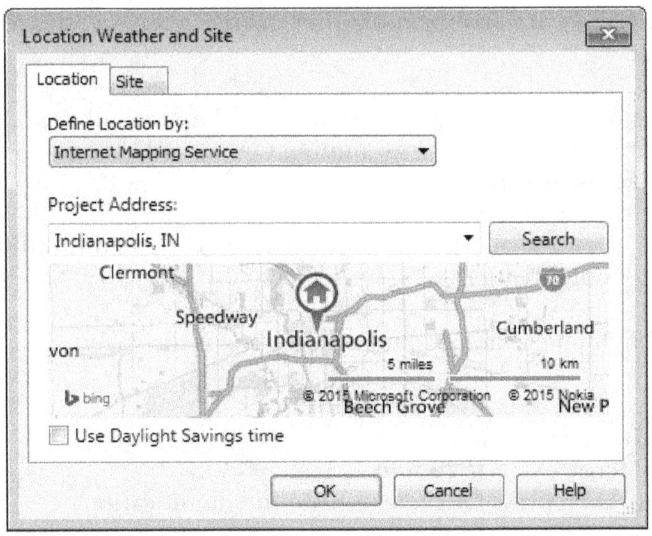

***Figure 3-55** The desired location displayed in the **Location Weather and Site** dialog box*

4. Next, choose the **Site** tab from the **Location Weather and Site** dialog box and then choose the **Duplicate** button; the **Name** dialog box is displayed.

5. Enter **CADCIM-IN** in the **Name** edit box and then choose the **OK** button; the **Name** dialog box closes and the name entered is displayed and highlighted in the **Sites defined in this project** list box.

6. Now, choose the **Make Current** button and then choose the **OK** button; the **Location Weather and Site** dialog box closes and the settings for the project location are saved in the project file.

Modifying the Existing Levels

In this section, you will modify the existing levels in the project file. To do so, you need to select an elevation view from the **Project Browser** and then use the **Level** tool.

1. Before modifying levels, move the cursor to the **Project Browser** and then double-click on the **South** option displayed under the **Elevation (Building Elevation)** head; the current view is changed to the **South** elevation view.

Note

*In the **South** elevation view, make sure that the **Hidden Line** option is chosen for the visual style in the **View Control Bar**.*

2. Next, zoom in the drawing area to make the text on the level lines visible. Click on the **Level 1** level line and move the cursor over the text **Level 1**; the text is highlighted.

3. Again, click on the text; an edit box is displayed.

4. Enter **Foundation Level** in the edit box and press ENTER; the **Revit** message box is displayed, prompting you to confirm whether or not to rename the corresponding view of the selected level.

5. In the **Revit** message box, choose the **Yes** button; the name of the level is changed. Also, notice that the name of the view corresponding to the level changes in the **Structural Plans** head of the **Project Browser**.

6. Next, move the cursor over the elevation value of the **Foundation Level** level; the elevation value gets highlighted.

7. Click on the highlighted elevation value; an edit box is displayed.

8. Enter value in the edit box as given below:

   For Imperial **-6' 0"**
   For Metric **-1829 mm**

   Press ENTER and then press ESC to exit from the modification.

9. Repeat steps 2 to 8 to rename **Level 2** level to **Entry** and modify its elevation as given below:

   For Imperial **0' 0"**
   For Metric **0 mm**

10. Next, right-click in the drawing area; a shortcut menu is displayed.

11. Choose the **Zoom to Fit** option from the shortcut menu; the extent of the level lines are displayed in the drawing area.

12. Next, select the **Foundation Level** level line and place the cursor over the grid length control displayed on the left.

13. Press and hold the left mouse button and drag the mouse to the right; the level lines move along the cursor.

14. Click in the drawing area when the length of the level line reduces to almost half of its existing length.

15. Next, press ESC to exit from the edit controls.

16. Right-click in the drawing area and then choose **Zoom to Fit** from the shortcut menu displayed; the current view fits the extent of the length of the level lines.

Adding New Levels

In this section, you will add two levels to the project file using the **Level** tool.

1. To add a new level, invoke the **Level** tool from the **Datum** panel of the **Structure** tab.

2. To create a nonstory level, clear the **Make Plan View** check box in the **Options Bar**.

3. Now, move the cursor near the left endpoint of the **Foundation Level** level line, and when the alignment line appears, enter the following values:

   For Imperial **6"**
   For Metric **152 mm**

   Press ENTER.

4. Move the cursor toward the right until the alignment line appears above and across the **Foundation Level** level bubble. Next, click to complete the level line. The new level line is created. It displays the level name as **Level 3** and the elevation as given below:

   For Imperial **-5' 6"**
   For Metric **-1676 mm**

5. Without exiting the **Level** tool, select the **Make Plan View** check box in the **Options Bar**.

6. Move the cursor above the left endpoint of the **Entry** level line and when the alignment line appears, enter value as given below:
   For Imperial **10'**
   For Metric **3048 mm**
   Next, press ENTER.

7. Move the cursor toward the right till the alignment line appears above the **Entry** level bubble, and click to complete the level line. The new level line is created. It displays the level name as **Level 4** and the elevation.
   For Imperial **10' - 0"**
   For Metric **3048 mm**

8. Repeat steps 6 and 7 to create a level line with the default name **Level 5** at a distance of given distance from **Level 4**.
   For Imperial **14'**
   For Metric **4267 mm**
   Note that the elevation of the **Level 5** level line is displayed as:
   For Imperial **24' 0"**
   For Metric **7315 mm**

9. Next, press ESC twice to exit the **Level** tool.

10. Select the **Level 3** level line; the properties of the selected level line are displayed in the **Properties** palette.

11. In the **Properties** palette, click in the value field corresponding to the **Name** parameter and replace the existing name **Level 3** with **T.O.F**.

12. Choose the **Apply** button in the palette; the **Revit** message box is displayed

13. Choose the **Yes** button from the message box; the **Revit** message box closes and the name of the level line changes in the project view.

14. To rename the **Level 4** level, select it from the drawing area; the properties of the selected level are displayed in the **Properties** palette.

15. In the **Properties** palette, click in the value field corresponding to the **Name** parameter and replace the existing name **Level 4** with **Second Floor**.

16. Choose the **Apply** button in the palette; the **Revit** message box is displayed.

17. Choose the **Yes** button from the message box; the **Revit** message box closes. Notice that the name of the selected level line as well as its corresponding view name have been changed.

18. Repeat steps 13 to 16 and rename the **Level 5** to **Roof**.

19. Press ESC to exit the **Level** tool.

20. Now, move the cursor to the **Project Browser** and choose the **Level 1 - Analytical** option from the **Structural Plans** head.

21. Right-click and then choose **Delete** from the shortcut menu; the name of the selected view is deleted from the **Project Browser**.

22. Repeat steps 19 and 20 to delete the **Level 2 - Analytical** option from the **Structural Plans** head.

Adding Grid Lines

In this section, you will add grids to the project using the **Grid** tool.

1. To add grids, change the current view to the **Entry** structural plan view. To do so, double-click on **Entry** from the **Structural Plans** head in the **Project Browser**; the **Entry** structural plan view is displayed in the drawing area.

2. Invoke the **Grid** tool from the **Datum** panel of the **Structure** tab; the **Modify | Place Grid** tab is displayed.

3. Ensure that the **Line** tool is invoked from the list box in the **Draw** panel of this tab. Then, move the cursor toward the lower left corner of the drawing area (near the **South** elevation arrow) and click; a grid line emerges, as shown in Figure 3-56.

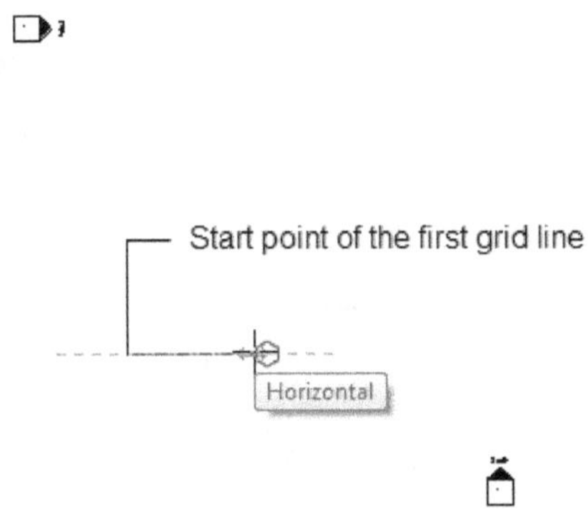

Figure 3-56 *The location for the start point of the first horizontal grid line*

4. Move the cursor horizontally to the right and click when it reaches a point below the **East** elevation arrow; refer to Figure 3-57 for location.

5. Next, right-click in the drawing area; a shortcut menu is displayed. Choose **Snap Overrides > Endpoints** from the shortcut menu.

6. Move the cursor near the left endpoint of the first grid (grid created earlier) and place the cursor on it until it (the first grid line) is highlighted and the **Endpoint** snap appears.

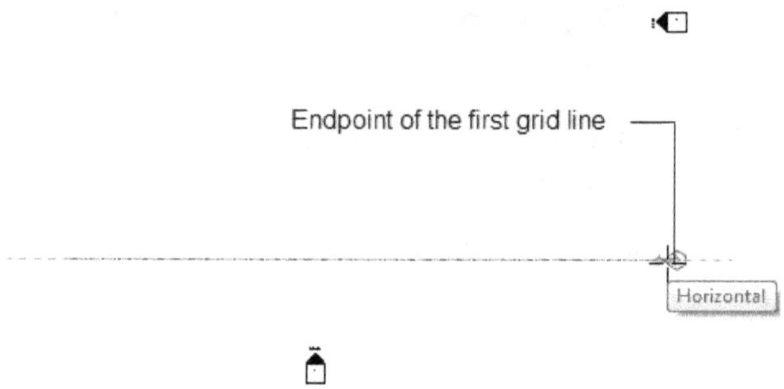

Figure 3-57 *The location for the endpoint of the first horizontal grid line*

7. Next, click as the **Endpoint** snap appears and the first grid line is highlighted; a grid line appears that displays the temporary dimension of its angle measured from the horizontal axis.

8. Next, move the cursor toward the right and above the grid bubble of the grid line created earlier.

9. Click when the temporary angular dimension displays **15.00°** and an alignment line appears from the endpoint of the previous grid line, refer to Figure 3-58.

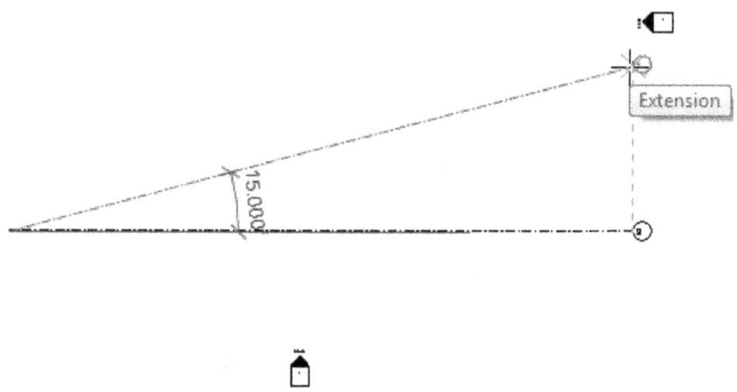

Figure 3-58 *The grid line displaying the angular dimension and an alignment line*

10. Repeat steps 5 to 9 to create the remaining five grid lines that emerge from the start point of the first grid line. The temporary dimensions that you will consider to create the five radial grids are : **30°**, **45°**, **60°**, **75°**, and **90°**, refer to Figure 3-59.

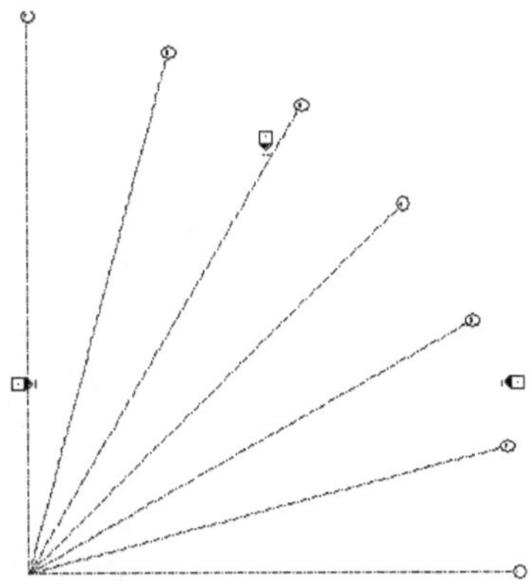

***Figure 3-59** The radial grid lines with varying angles*

Note
*Whenever a grid is created, you need to use the **Endpoints** snap override option to align the endpoint of each grid with the endpoint of the grid created earlier.*

11. Press ESC twice to exit the **Grid** tool.

12. Right-click in the drawing area and then choose **Zoom to Fit** from the shortcut menu.

13. Now, invoke the **Grid** tool from the **Datum** panel of the **Structure** tab.

14. Choose the **Center-ends Arc** tool from the list box in the **Draw** panel of the **Modify |Place Grid** tab.

15. Now move the cursor toward the start point of the first grid line and click when the **Endpoint** snap appears; a circle appears with temporary dimensions denoting its radius and the start angle of the arc with the first grid line.

16. Move the cursor vertically upward along the last grid line created (the vertical grid line).

17. When the linear temporary dimension displays **36'** (or 10973 mm in Metric) and the angular temporary dimension displays **90°**, enter **30'** (or 9144 mm in Metric) and press ENTER; a curved arc emerges and starts from a distance of **30'** (or 9144 mm in Metric) from the start point of the first grid line.

18. Move the cursor clockwise until it is placed below the first grid line and click when the first alignment line with the **Extension** snap emerges from the start point of the first grid line.

19. Now, repeat steps 15 to 18 to create a curved grid line with a radial distance of **50'** (or **15240 mm** in Metric) from the start point of the first grid, refer to Figure 3-60.

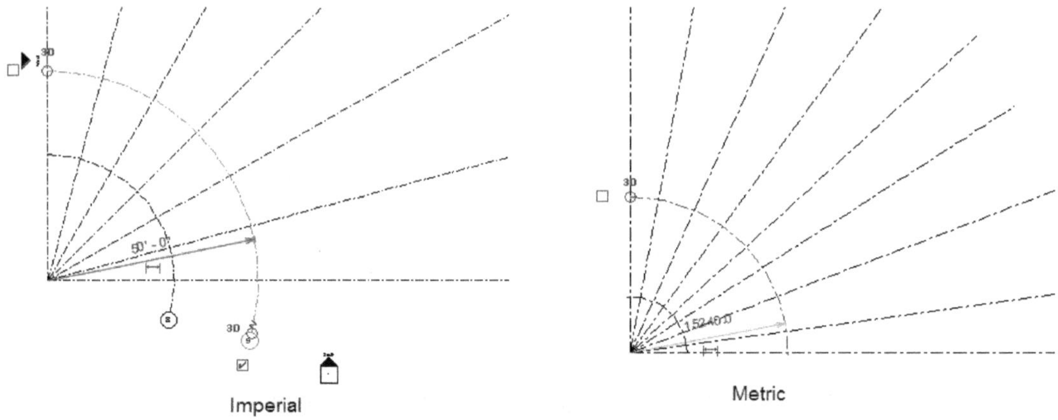

Figure 3-60 The curved grid lines created

20. Next, press ESC to exit the **Grid** tool.

21. To complete the grid pattern for the project, invoke the **Grid** tool and use the **Line** tool from the list box in the **Draw** panel of the **Modify | Place Grid** tab, to create the remaining grids with the specified distance, as shown in Figure 3-61.

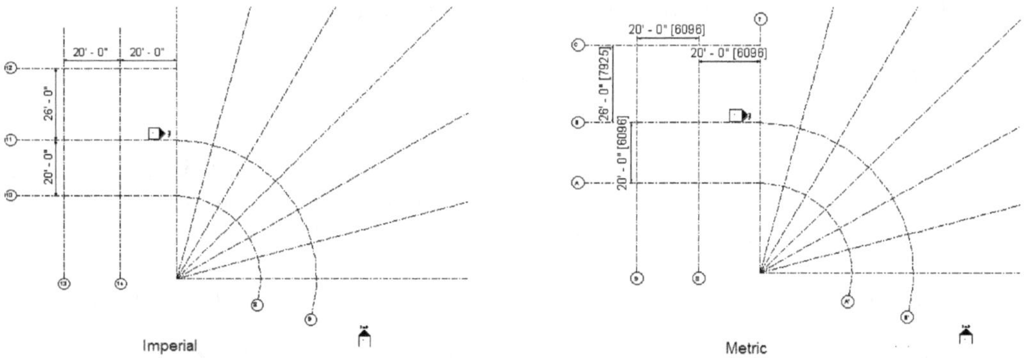

Figure 3-61 The dimensions of the remaining grid lines

Resizing the Grid Lines

In this section, you will resize the grids using the drag controls.

1. To start resizing, select the first grid line of the project and then move the cursor toward its endpoint near the grid bubble, as shown in Figure 3-62.

2. Now, press and drag the cursor toward the left and release the mouse when the cursor along with the grid endpoint reaches to the location shown in Figure 3-63.

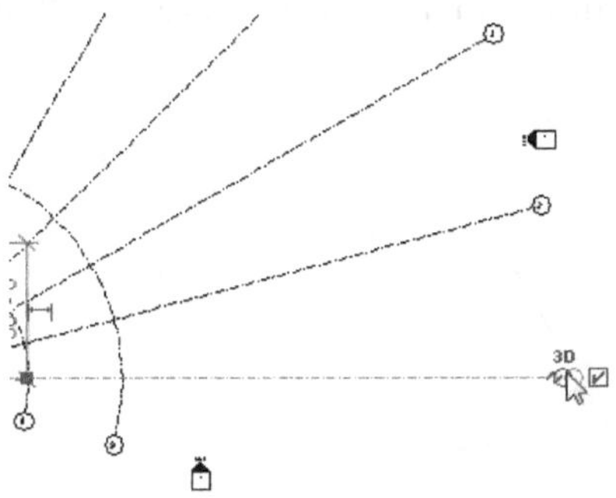

Figure 3-62 Moving the cursor toward the endpoint

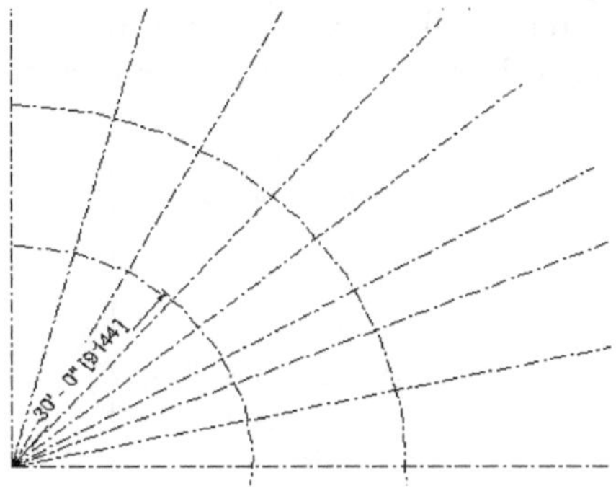

Figure 3-63 Dragging the grid endpoint to resize

3. Press ESC to exit from the editing mode.

4. Select the other grids individually and repeat steps 1 to 4 to resize all the grids in the project view.

Note

While resizing the grids, you can align them in reference to their previous grids with the help of the alignment control.

Renaming Grids

In this section, you will rename all grids created in the previous section.

1. To rename grids, select the curved grid, as shown in Figure 3-64.

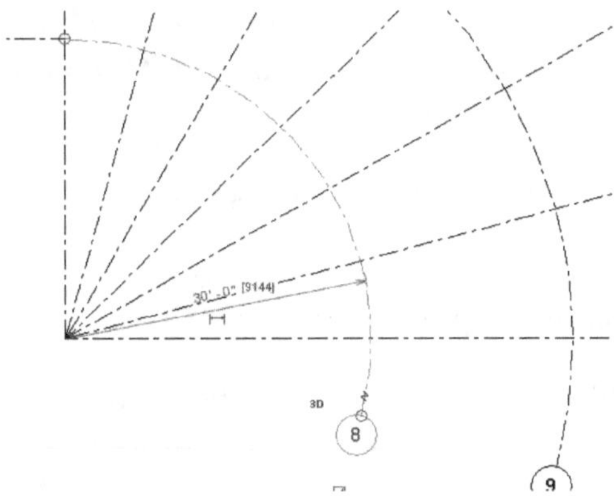

Figure 3-64 Selecting the curved line to rename

2. Move the cursor over the grid bubble of the selected grid and click; an edit box is displayed. Enter **A'** in the displayed edit box and press ENTER. The selected curved grid is renamed.

3. Press ESC to exit the editing mode.

4. To rename all the grids in the project, select them individually and repeat steps 2 and 3. Refer to Figure 3-65 for new names of the grids.

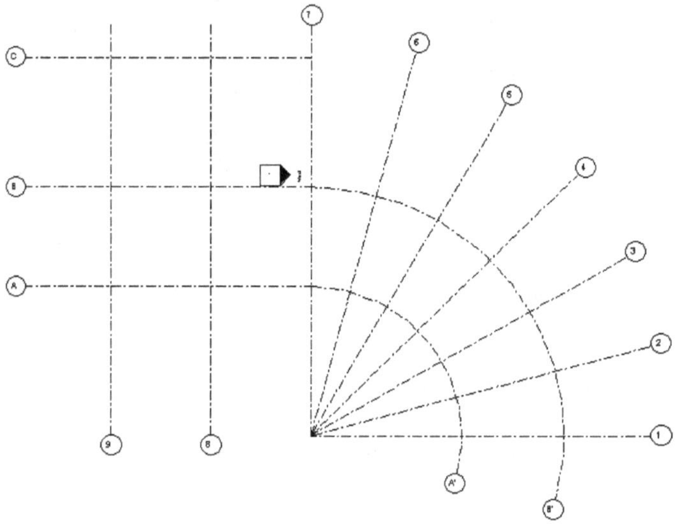

Figure 3-65 The grid lines with new names

Saving the Project

In this section, you will save the project file using the **Save As** tool.

1. Choose **Save As > Project** option from the **Application Menu**; the **Save As** dialog box is displayed.

2. In this dialog box, browse to the *C:\rst_2017\c03_rst_2017_tut* location and enter **c03_Industrial-Complex_tut2** (for Imperial) or **M_c03_Industrial-Complex_tut2** (for Metric) in the **File name** edit box.

3. Now, choose the **Save** button; the **Save As** dialog box closes and the project file is saved.

Closing the Project

1. Choose the **Close** option from **Application Menu**.

   The file is closed and this completes Tutorial 2 of Chapter 3.

Self-Evaluation Test

Answer the following questions and then compare them to those given at the end of this chapter:

1. ________________ is the default template file in Autodesk Revit 2017.

2. To define the geographical location of a project, you need to invoke the ________________ tool from the **Manage** tab.

3. To transfer project standards from the source file to the target file, you need to invoke the ________________ tool from the **Manage** tab.

4. You can use the ________________ tool to manage the **Project Browser**.

5. In a grid line, you can create a gap between two end segments by setting the________________ parameter in the **Type Properties** dialog box to **None**.

6. You can use the________________option in the **Options Bar** to create a grid line that starts at a specified offset distance from a point defined in the drawing area.

7. ________________can be used as datum planes to act as guideline for creating elements.

8. The visibility of a work plane grid can be controlled by using the ________________tool from the **Work Plane** panel of the **Structure** tab.

9. In Revit, you can include project information to the schedules of linked models. (T/F)

10. The visibility of level bubble on either sides of a level line cannot be controlled. (T/F)

11. To add the name of the author in a project, you need to invoke the ________________ tool from the **Manage** tab.

Review Questions

Answer the following questions:

1. The visibility of datum elements can be controlled by using the__________ tool.

2. The **Project Browser** displays the __________ folder, which contains the element families loaded into the project file.

3. To create browser organization for sheets, choose the __________ tab from the **Browser Organization** dialog box.

4. To add levels to a project view, you need to invoke the **Level** tool from the __________ panel.

5. You can use the__________tool to create curved grids.

6. To make a hidden level line visible, choose __________from the **View Control Bar**.

7. A __________level does not hold any plan view but acts as a host for placing objects and information.

8. You can clear the __________check box in the **Options Bar** of the **Modify | Place Level** tab to create a nonstory level.

9. You cannot copy project standards from a project to the current project. (T/F)

10. A work plane can be horizontal, vertical, or inclined at any specified angle. (T/F)

EXCERCISES

Exercise 1 Academic Institution

In this exercise, you will create project information for the project created in the Exercise 1 of Chapter 2. Further, you will add levels and grids to the project file. Refer to Figure 3-66 for grid dimensions and names. Use the following project specifications:

(Expected time: 1 hr 15 min)

1. Project file to be used:
   For Imperial *c02_Academic-Institution_exer1.rvt.*
   For Metric *M_ c02_Academic-Institution_exer1.rvt.*

2. Rename Level 1 and Level 2 as **Foundation Level** and **Entry**, respectively.

3. Specify the elevation of **Foundation Level** and **Entry** levels as:
   For Imperial **-5'6" 0"** and **0' 0"**.
   For Metric **-1676 mm** and **0 mm**, respectively.

4. Story levels to be added:
   For Imperial **Second Floor - Elevation 12' 6"**
   Roof - Elevation 25' 0"
   For Metric **Second Floor - Elevation 3810 mm**
   Roof - Elevation 7620 mm

5. Nonstory level to be added:
   For Imperial **T.O.F- Elevation -5'**
   For Metric **T.O.F- Elevation -1524 mm**

6. Project information to be added:

Project Issue Date	**01/01/2017**
Project Status	**Started**
Client Name	**CADCIM Technologies**
Project Address	**525, Andrews Drive Schereville**
Project Name	**Academic Institution**
Project Number	**AI/03/13**

7. Grids to be added to the **Entry** project view using the dimensions and names as shown in Figure 3-66.
8. File name to be assigned:
   For Imperial *c03_Academic-Institution_exer1.rvt.*
   For Metric *M_c03_Academic-Institution_exer1.rvt.*

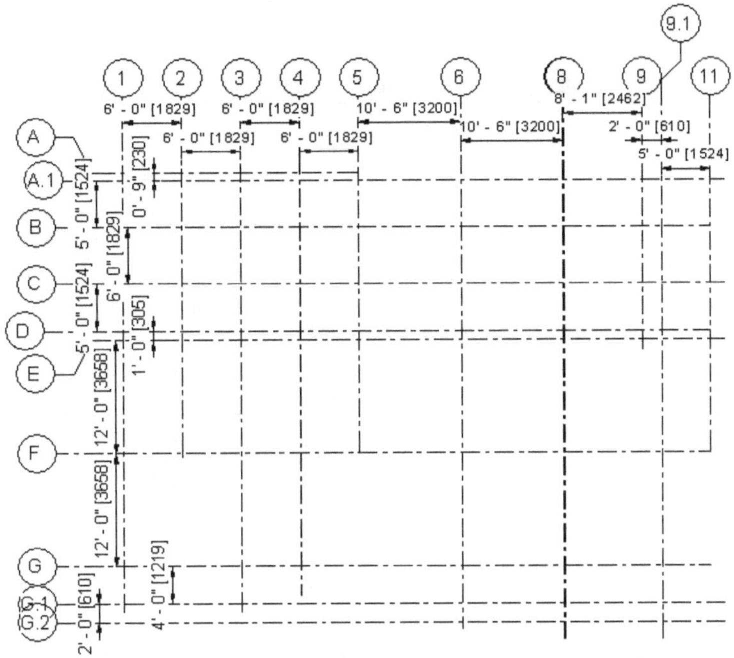

***Figure 3-66** The grid lines with new names for the Academic Institution structure*

Exercise 2 Factory Shed

In this exercise, you will add levels and grids to the project file created in Exercise 2 of Chapter 2. Refer to Figure 3-67 for grid dimensions and names. Use the following project specifications:

(Expected time: 1 hr 15 min)

1. Project file to be used:

   For Imperial *c02_Factory-Shed_exer2.rvt.*
   For Metric *M_c02_Factory-Shed_exer2.rvt.*

2. Rename Level 1 and Level 2 as **Below Ground** and **Plinth**, respectively.

3. Specify the elevation of **Below Ground** and **Plinth** levels.

   For Imperial **-2'0"** and **1'0"**, respectively
   For Metric **-610 mm** and **305 mm**

4. Story levels to be added:

   For Imperial **Second Floor - Elevation 14'**
   Roof - Elevation 24' 0"
   For Metric **Second Floor - Elevation 4200 mm**
   Roof - Elevation 7315 mm

5. Grids to be added in the **Plinth** project view using dimensions and names, as shown in Figure 3-67.

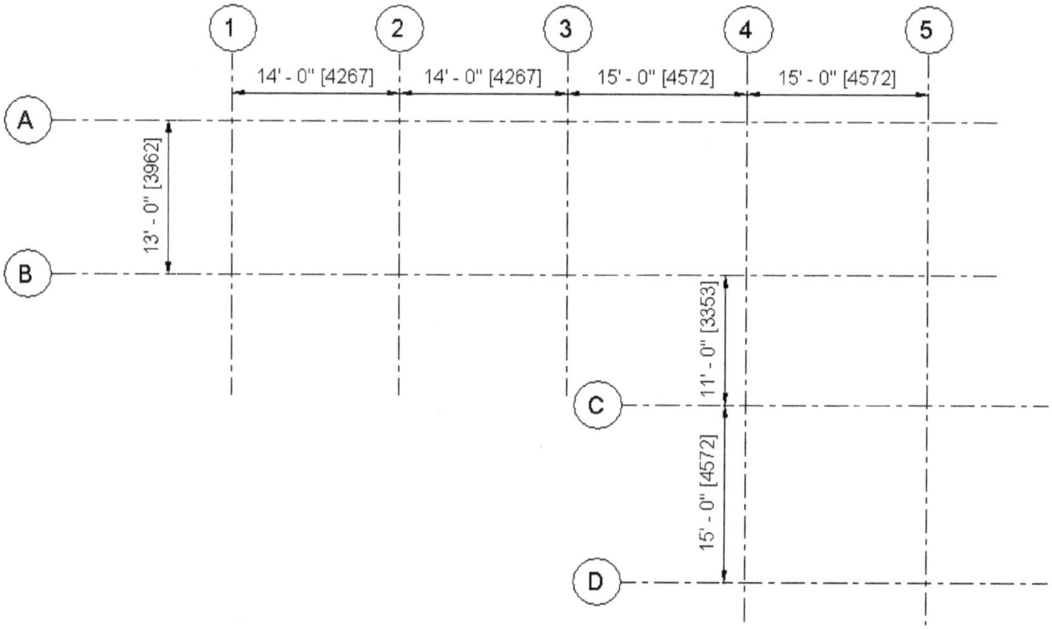

Figure 3-67 *The grid lines with new names for the Factory Shed structure*

6. File name to be assigned: *c03_Factory-Shed_exer2.rvt*.

Exercise 3 Residential Building

Download the *c03_Residential_Build_exer3* file from *http://www.cadcim.com*. The path of the file is as follows: *Textbooks > Civil/GIS > Revit Structure > Exploring Autodesk Revit 2017 for Structure*. In this exercise open the *c03_Residential_Build_exer3* file (for Imperial) and *M_c03_Residential_Build_exer3* file (for Metric) and add levels and grids to it. Refer to Figures 3-68, 3-69, and 3-70. Further, you will add project information to the project. Use the following specifications to complete the exercise: **(Expected time: 45 min)**

1. Rename Level 1 and Level 2 as **Subgrade** and **GL**, respectively.

2. Specify the elevation of **Subgrade** and **GL** levels.
   For Imperial **-4'6" 0"** and **0' 0"**.
   For Metric **-1372 mm** and **0 mm**

3. Story levels to be added: For Imperial **FL- Elevation 11'0"**
   TL- Elevation 19'-6"
   For Metric **FL- Elevation 3353 mm**
   TL- Elevation 5944 mm

4. Nonstory level to be added:
   For Imperial **T.O.F- Elevation -4'**
   For Metric **T.O.F- Elevation -1219 mm**

5. Project information to be added:

Project Issue Date		**01/01/2015**
Project Status		**Started**
Client Name		**CADCIM Technologies**
Structural Consultant		**CADCIM-Structural Engineers**
Mechanical Consultant		**Sham Tickoo**
Quantity Surveyors		**CADCIM-Quantity Surveyors**
Project Name		**Residential Building**
Documented Contract Commencement Date		**20/09/2015**
Documented Contract Completion Date		**31/07/2017**
Provisional Period Allowed in Contract		**30 Days**
Gross Area of the Project	For Imperial	**1200 sq. ft.**
	For Metric	**112 sq. meter**

6. File name to be assigned:
   For Imperial *c03_Residential_Build_exer3_final.*
   For Metric *M_ c03_Residential_Build_exer3_final.*

Figure 3-68 *Elevation of the project displaying the name of the levels and their respective elevations*

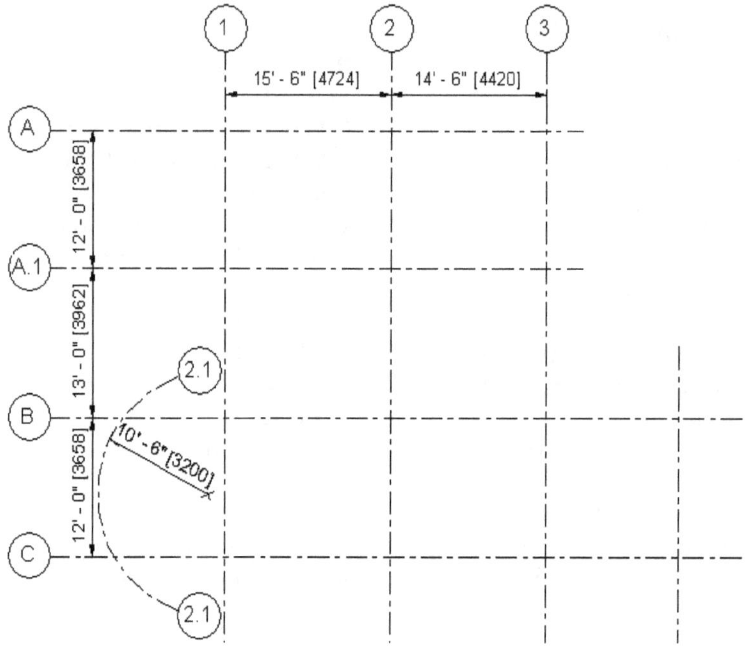

Figure 3-69 The grid lines with the specified names and distances

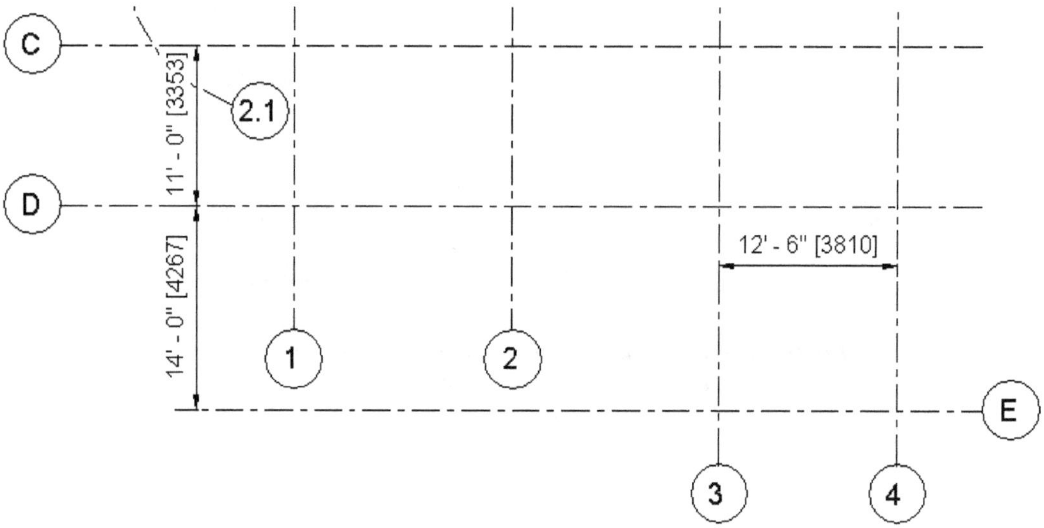

Figure 3-70 The remaining grid lines in continuation of Figure 3-69

Answers to Self-Evaluation Test

1. **Structural Analysis-Default, 2. Location, 3. Transfer Project Standards, 4. Browser Organization, 5. Center Segment, 6. Offset, 7. Reference Planes, 8. Show, 9.** T, **10.** F, **11. Project Parameters**

Chapter 4

Structural Columns and Walls

Learning Objectives

After completing this chapter, you will be able to:

- *Create Structural Columns*
- *Modify Structural Columns*
- *Create Structural Walls*
- *Modify Structural Walls*

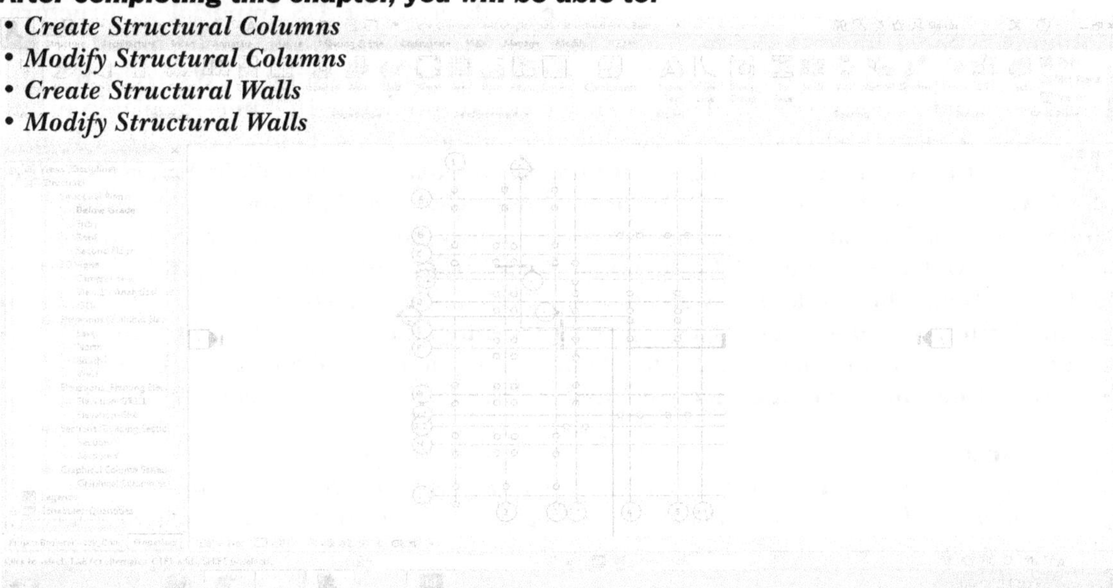

In the previous chapter, you learned to add datums such as structural levels and grids to a project and also learned to add project information to a project. In this chapter, you will learn to add structural elements such as columns, walls, beams, and foundations. Also, you will learn to work with structural columns and walls in a project in detail.

STRUCTURAL COLUMNS

Structural column is an integral part of the primary structural framing of a building system. The basic function of structural columns is to carry vertical load from the roof system, floor slab, and beam to the foundation. Functionally, columns are subjected to axial loads, which form the basis of its design.

In Revit, you need to add columns in a plan view. But before adding columns, it is recommended that you create a grid system. The grid lines and grid intersections of the grid system help you place columns at appropriate locations. When you add columns at grid intersections, they are anchored to it. As a result, whenever there is a change in the grid spacing, the columns move with grid intersections. Before you add structural columns to a project, you need to learn about different types of structural columns. In the following sections, you will learn about different types of structural columns, loading structural columns, adding structural columns and much more.

Types of Structural Column

In Revit, there are three types of structural columns that can be added to a building. These columns are made of three types of material: steel, concrete, and wood. Out of these, steel and concrete columns are commonly used in the construction of a building. While working in Revit, you will get to know that some of the families of these column types are pre-loaded in the project template. You can load other families from the Imperial or Metric structural column libraries that become available on installing Autodesk Revit .

In a project, you can add two types of predefined steel columns depending on your requirement. These columns are rolled steel columns and light gauge steel columns. For rolled steel columns, there are eleven predefined families which are available in the **US Imperial > Structural Columns > Steel** folder location. These families are **Double C-Channel Column**, **HSS-Hollow Structural Section-Column**, **L-Angle-Column**, **Pipe-Column**, **WWF-Welded Wide Flange-Column**, and more. Similarly, there are other predefined families available in the **US Metric > Structural Columns > Steel** folder location. These families are **CHS-Circular Hollow Section-Column**, **M_L-Angle-Column**, **M_M Miscellaneous Wide Flange-Column**, **M_W-Wide Flange Column**, **PFC-Parallel Flange Channel-Column**, and **SHS- Square Hollow Section-Column**. For light gauge steel columns, there are seven predefined families which are available at the **US Imperial > Structural Columns > Light Gauge Steel** folder location. The predefined light gauge steel families for columns are **Light Gauge-Channel Studs-Column**, **Light Gauge-Furring Channels-Column**, and **Light Gauge-Studs-Column**. The predefined light gauge steel families for columns available at the **US Metric > Structural Columns > Light Gauge Steel** folder location are **M_Light Gauge-Zeds-Column**.

Note

Apart from using the predefined sections available in Autodesk Revit, you can also customize a section with different physical properties for your project. To use customized sections, first you need to create families and then load them into your project.

There are two types of concrete columns in both US Imperial and US Metric that can be added to a project: casted (in-situ) concrete columns and pre-cast concrete columns. For casted concrete columns in US Imperial, there are several pre-defined families that can be used such as **Concrete-Rectangular-Column with Drop Caps**, **Concrete-Round-Column**, and more. For casted concrete columns in **US Metric**, several pre-defined families available are **M_Concrete-Rectangular-Column with Drop Caps**, **M_Concrete-Rectangular-Column**, and more. For pre-cast concrete columns in **US Imperial**, you can use any of the given predefined families such as **Precast-Rectangular Column with Bearing**, **Precast-Rectangular Column with Multi Layer Corbels**, **Precast-Rectangular Column with Rectangular Fork bearing**, and **Precast-Square Column**. For pre-cast concrete columns in **US Metric**, you can use any of the given predefined families such as **M_Precast-Rectangular Column with Bearing**, **M_Precast-Rectangular Column Cone-Shaped Fork bearing**, and **M_Precast-Square Column**.

To add wood (timber) columns for Imperial project, you can use any of the following families such as **Glulam-Southern Pine-Column**, **Timber-Column**, and more. To add wood (timber) columns in a Metric project, you can use any of the following families: **M_Dimension Lumber-Column**, **M_PSL-Parallel Strand Lumber-Column**, **M_Timber-Column**, and more.

Loading Structural Column Families

As discussed in the previous section, Autodesk Revit comes with some predefined families. When you start a new project with the default template, you can access the **W Shapes-Column** family. Note that before adding structural columns to a project, you need to access more families that are present in the software. To do so, choose the **Load Family** tool from the **Load from Library** panel of the **Insert** tab; the **Load Family** dialog box will be displayed, as shown in Figure 4-1.

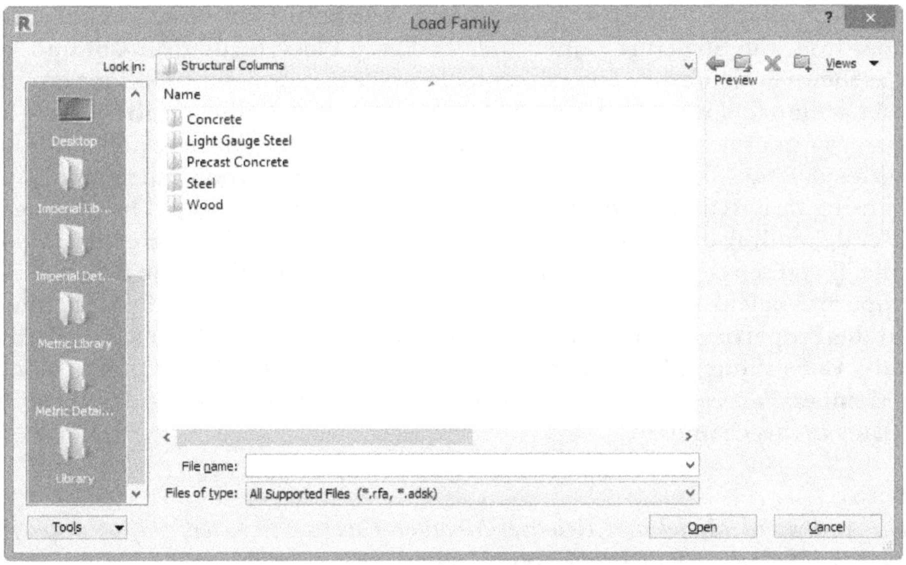

***Figure 4-1** The **Load Family** dialog box*

In this dialog box, for Imperial unit system browse to **US Imperial > Structural Columns** folder location and for Metric unit system browse to **US Metric > Structural Columns** folder location and then select any of the desired sub-folders such as **Concrete**, **Light Guage steel, Precast Concrete**, **Steel**, and **Wood** from the list. On choosing a sub-folder, the list of available family files (*.rfa*) will be displayed. Click on the desired file or press SHIFT and then click on multiple files in the folder for multiple selection. Then, choose the **Open** button; the selected family (ies) will be loaded in the project.

Note

In case the files that you are trying to load using the ***Load Family*** *dialog box are already present in the project file, the* ***Family Already Exists*** *window will be displayed. You can choose* ***Overwrite the existing version*** *or* ***Overwrite the existing version and its parameter values*** *from the window based on your project requirement.*

Adding Structural Columns to a Project

After loading families for columns, you can add them to a project based on the design requirement. Before adding columns, the grid patterns and levels should be created. This will help you in placing the column on grids or grid intersections. For adding columns, it is recommended that they should be modeled and placed on your project as they are constructed at site. The advantage of following this practice is that the column placed this way will give you a more accurate BIM model and you will spend less time in creating sections, schedules, quantity take-offs, and estimation of materials. You can place structural columns in your project, both vertically and inclined. You can add them individually (without grids) on grids, or on existing architectural columns. The methods of placing vertical and slanted structural columns are discussed next.

Placing Vertical Columns

Ribbon: Structure > Column

To place vertical columns, select an appropriate structural plan view and then choose the **Column** tool from the **Structure** panel; the **Modify | Place Structural Column** tab will be displayed, as shown in Figure 4-2. This tab contains twelve panels and the **Options Bar**. By default, the **Vertical Column** tool is invoked from the **Structural** panel. In the **Modify | Place Structural Column** tab, you can select the desired column type from the **Type Selector** drop-down list in the **Properties** palette. Also, you can change the instance and type properties of the selected element by using various parameters displayed in the **Properties** palette. The parameters in this palette differ depending on the type of column selected from the **Type Selector** drop-down list. For example, if you select **Concrete-Rectangular-Column** from the **Type Selector** drop-down list, the **Properties** palette will display those parameters that are required for designing concrete column. In the **Properties** palette, the properties of the concrete columns are classified under seven heads: **Constraints, Materials and Finishes, Structural, Dimensions, Identity Data, Releases/Member Forces**, and **Analytical Model**. The parameters under these heads are discussed later in this chapter.

Note

The ***Analytical Model*** *and* ***Releases/Member Forces*** *parameters will be displayed only for Analytical Columns in the* ***Properties*** *palette.*

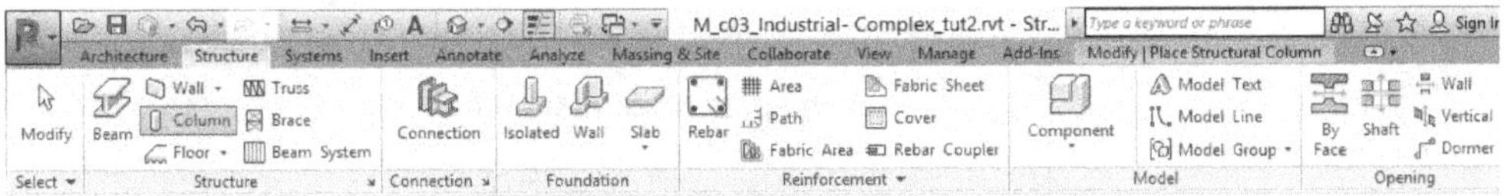

*Figure 4-2 The **Modify / Place Structural Column** tab*

To change the type properties of a selected column type, choose the **Edit Type** button from the **Properties** palette; the **Type Properties** dialog box will be displayed. This dialog box displays different parameters for different column types. You can select column types from the **Type Selector** drop-down list. This dialog box displays parameters under three heads: **Structural**, **Dimensions**, and **Identity Data**. Various parameters under these heads are discussed later in this chapter.

Note

*The **Structural** head and its parameters will not be displayed in the **Type Properties** dialog box for concrete columns.*

After setting the instance and type properties for a selected column type, you need to specify whether to place columns at the desired location without grids, at grids, or at existing architectural columns. To place structural columns anywhere in the drawing area (without grids), click the left mouse button in the drawing area; a column of the selected type will appear at the location where you have clicked.

If you want to place a column at the intersection of grids, then choose the **At Grids** tool from the **Multiple** panel in the **Modify | Place Structural Column** tab; the **Modify | Place Structural Column > At Grid Intersection** contextual tab will be displayed, and in the **Status Bar**, you will be prompted to select grid lines to place the structural column at the intersection of grids. Select the first grid line and then the other grid line that intersects the first grid line; a column will be inserted at the intersection of the two selected grid lines. You can add more columns in a drawing. To do so, press the CTRL key; a "+" symbol will appear with the cursor. Select the other grid lines that intersect with grid line (highlighted) selected last; multiple columns will be inserted at grid intersections. Next, choose the **Finish** button from the **Multiple** panel in the **Modify | Place Structural Column > At Grid Intersection** tab; the columns will be inserted at the specified grid intersections. Figures 4-3 and 4-4 show the examples of **Concrete-Rectangular-Column** column type placed at the intersection of two grid lines and at multiple grid line intersections.

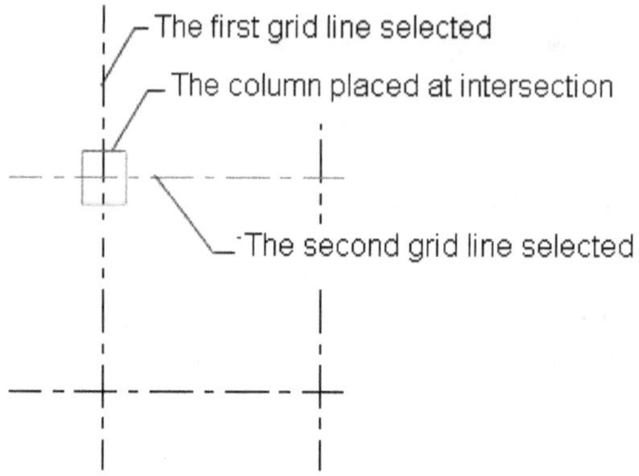

Figure 4-3 *Concrete column placed at grid intersection*

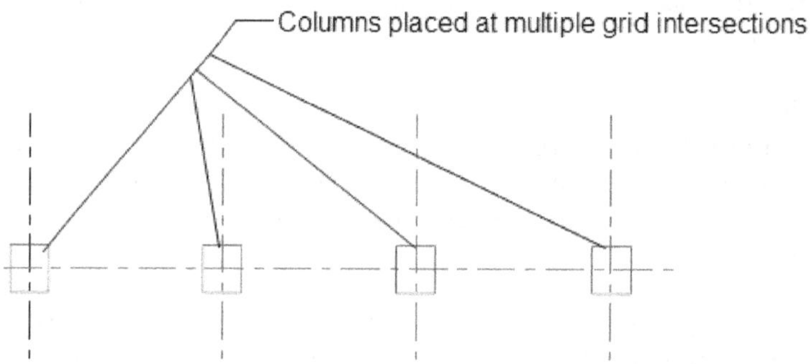

Figure 4-4 *Concrete columns placed at multiple grid intersections*

Note

*You can tag a structural column while placing it in structural model. To do so, choose the **Tag on Placement** tool from the **Tag** panel of the **Modify | Place Structural Column** tab; a tag will appear with the column while adding to the drawing. Refer to Figure 4-5 for an example of a tag added to a concrete column.*

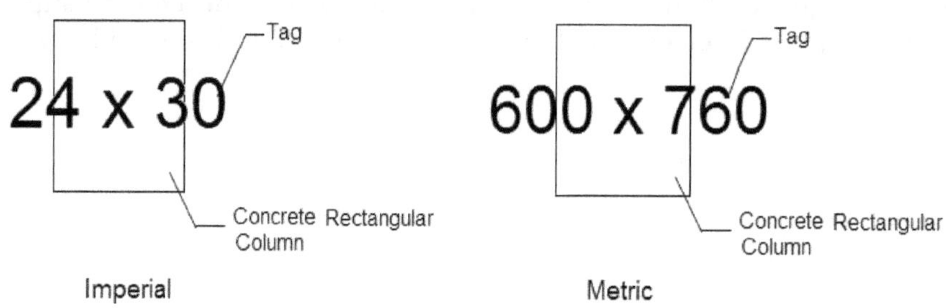

Figure 4-5 *Concrete rectangular column with a tag*

Using the Options in Options Bar

When you invoke the **Column** tool from the **Structure** panel of the **Structure** tab, the **Modify | Place Structural Column** tab is displayed. At the same time, various options related to the alignment and annotation of column are also displayed in the **Options Bar** below it, as shown in Figure 4-6.

Figure 4-6 *The* ***Options Bar*** *displaying various options for columns*

In the **Options Bar**, the **Rotate after placement** check box is clear by default. As a result, the column will be placed in the model in the same alignment as it was created in the family. To rotate the column before it has been inserted in the model, select the **Rotate after placement** check box and click on the screen; the column will be inserted and displayed in blue color along with the rotation line. Next, move the cursor; you will notice that a temporary angular dimension is displayed, as shown in Figure 4-7. Click on the screen to achieve the desired alignment or enter the required angle. On doing so, the column will be placed at the desired alignment. Note that you can also enter the value for the rotation angle in the **Angle** edit box that will be displayed in the **Options Bar** while defining the rotation angle.

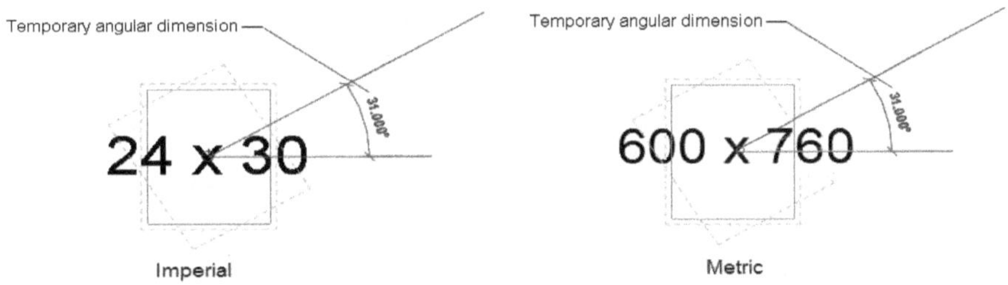

Figure 4-7 *Temporary angular dimension displayed*

In the **Options Bar**, there are two drop-down lists and one edit box displayed next to the **Rotate after placement** check box, refer to Figure 4-8. The options in the two drop-down lists are used to align the column vertically along levels or at a particular height. For example, if your current view is the structural plan for level 1 and you need to place a column whose base is at level 1 and its top is constrained to another level, say level 2, which is at a distance of 12' in Imperial system and in Metric system level is at a distance of 3048 mm above level 1, then select the **Height** and **Level 2** options from drop-down lists marked as **1** and **2** respectively in the **Options Bar**, refer to Figure 4-8.

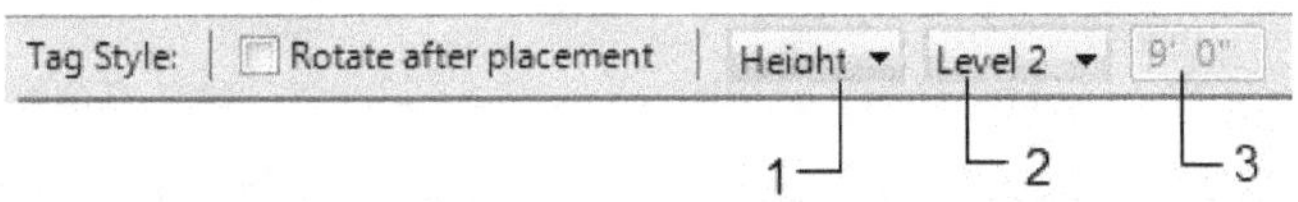

Figure 4-8 *The* ***Height*** *and* ***Level 2*** *options displayed in the drop-down lists marked 1 and 2*

Note

*If you select the **Unconnected** option for height from the drop-down list marked as **2**, the edit box marked as **3** will be activated. In the edit box, you can enter value to assign the unconnected height of the column from the current level.*

Similarly, if you need to place another column with its base below the current level at a certain depth, select the **Depth** option from the drop-down list marked as **1**, refer to Figure 4-8. Next, select an option to assign a level below the current level from the drop-down list marked as **2**. Alternatively, you can select the **Unconnected** option from the drop-down list marked **2** and enter a value in the edit box located next to it to assign an unconnected depth to column.

Figure 4-9 displays two columns in the North elevation view. The columns are placed in the plan view with different level constraints for their top and base. The column marked as **1** has been placed in the structural plan view of level 1 with the **Height** option assigned to **Level 2** from the **Options Bar**. The column marked **2** has also been placed in level 1 but with the **Depth** option assigned to an unconnected depth of **4'6"**(Imperial) and **1372 mm** (Metric) from the **Options Bar**.

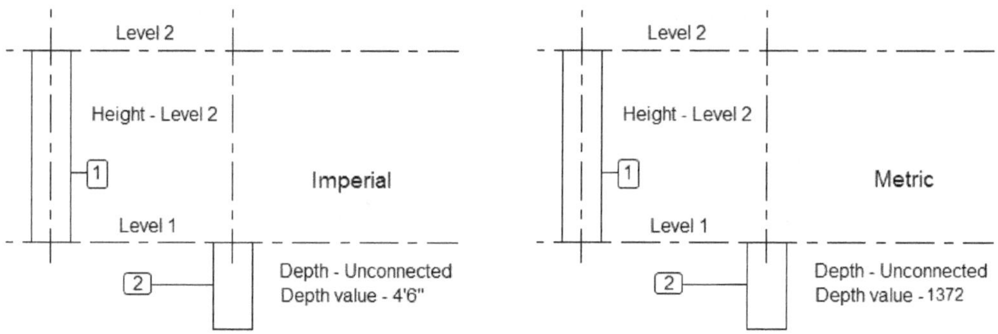

Figure 4-9 *Two columns with different level constraints displayed in the North Elevation view*

Placing Slanted Structural Columns

Slanted columns are commonly used in a structural design. In Revit, you can add these columns to a structural model. Unlike vertical columns, you can add these columns in a 3D view, an elevation view, or in section view. To add a slanted column, you first need to change the view of the structural model to an elevation, section, or 3D view. To place a slanted column in a 3D view, invoke the **Column** tool and then select the desired column type from the **Type Selector** drop-down list in the **Properties** palette. Next, choose the **Slanted Column** tool from the **Placement** panel in the **Modify | Place Structural Column** tab; the options to place the slanted column will be displayed in the **Options Bar**, refer to Figure 4-10.

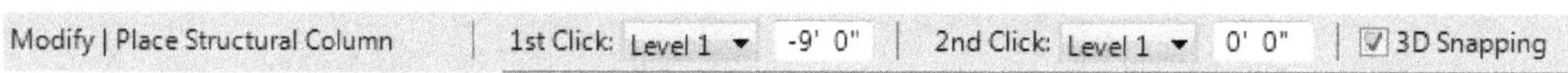

Figure 4-10 *The **Options Bar** displaying options for placing slanted structural column*

Tip

*After placing the column, you can further rotate it. To do so, choose the **Rotate** tool from the **Modify / Place Structural Column** tab and rotate the column at the desired angle.*

In the **Options Bar**, select the level options from the **1st Click** and **2nd Click** drop-down lists to define the associated levels and offset the endpoints of a slanted column. In the **1st Click** drop-down list, you can select a level for the endpoint of the column that you will define first. Similarly, you can select a level for the other end point of the column by selecting a level option from the **2nd Click** drop-down list. You can also use the edit boxes next to the **1st Click** and **2nd Click** drop-down lists to define offset values, if required, for the levels that you will define for the elevations of the column ends. After selecting the levels for the column end elevations, you can select the **3D Snapping** check box in the **Options Bar** to snap the points of the existing structural elements.

After setting the options in the **Options Bar**, move the cursor to a location in the structural model to define the first endpoint of the slanted column. Click on the desired location in the structural model in reference to the level option selected from the **1st Click** drop-down list in the **Options Bar**. Next, move the cursor to another location and click to specify the other endpoint of the slanted column based on the selection made in the **2nd Click** drop-down list in **Options Bar**. Figures 4-11 to 4-13 illustrate the insertion of a slanted column in a structural model.

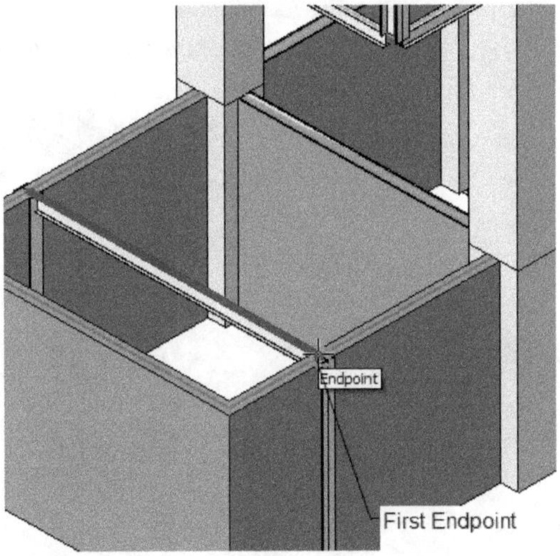

Figure 4-11 *Specifying the first endpoint of the slanted structural column*

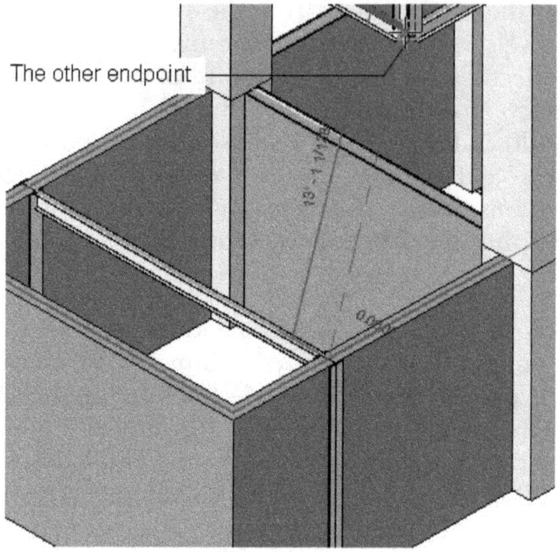

Figure 4-12 *Specifying another endpoint of the slanted structural column*

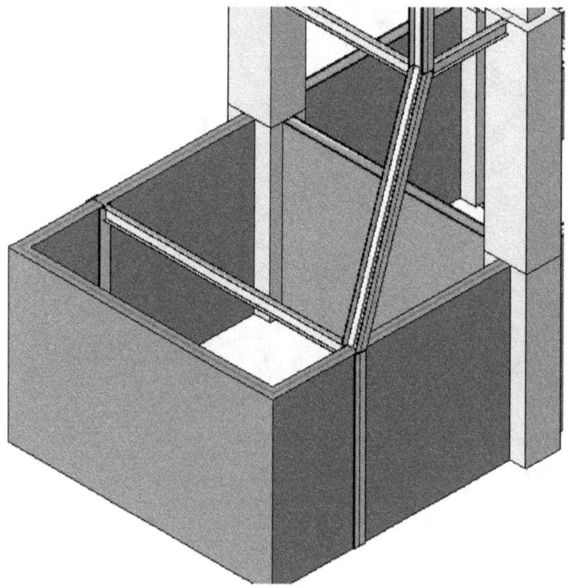

Figure 4-13 *The slanted column after insertion*

Creating Openings in Structural Columns

Ribbon: Structure > Opening > By Face

While working with columns in a structural project, you may need to cut openings through it to allow other structural elements such as beams and girders to pass through. To create an opening in a column, first align the view preferably to 3D view and zoom in the column in which you want to create the opening. Next, choose the **By Face** tool from the **Opening** panel of **Structure** tab, refer to Figure 4-14; the cursor will change into a selection cursor. Move the selection cursor toward the face of the column where you want to cut the opening; a plane will be displayed on that face, as shown in Figure 4-15. Click on the plane; the **Modify | Create Opening Boundary** tab will be displayed.

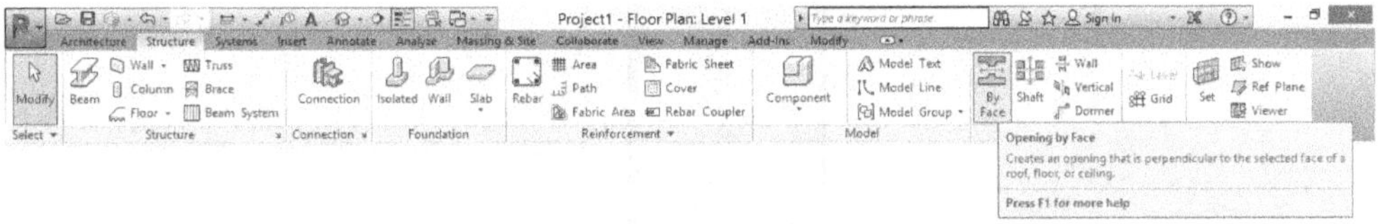

Figure 4-14 *Choosing the **By Face** tool to cut an opening in the column*

In the **Draw** panel of this tab, you will notice that the sketching tools to sketch the opening profile are displayed in a list box. You can choose any of these tools to create the opening in columns. After choosing a tool, align the view using the **ViewCube**. Next, sketch the desired profile, using the tool invoked from the **Draw** panel. After sketching the desired profile, choose the **Finish Edit Mode** button from the **Mode** panel; an opening of the desired profile will be created on the selected face. Figure 4-16 shows a column with the sketch of the opening profile and Figure 4-17 shows an opening cut through the face of the column.

Note

While selecting the face for sketching the opening profile, Revit allows you to select any of the perpendicular planes that align with the major or minor axis of the column section.

You can modify the opening on the face of the column by using the options in the **Modify | Structural opening cut** tab. Note that the **Modify|Structural opening cut** tab is displayed on creating the opening in the column. In the **Modify| Structural opening cut** tab, you can use various tools to edit the sketch. In this tab, you can use various modification tools such as **Move**, **Rotate**, **Array**, and **Mirror** to modify the profile. In this tab, you can also use other tools related to clipboard operations and visibility graphics. To edit the sketched profile of the opening, choose the **Edit Sketch** tool from the **Mode** panel; the **Modify | Structural opening cut > Edit Boundary** tab will be displayed and the profile will be highlighted in the sketch mode. You can modify the profile by using different tools from various panels in this tab. After editing the sketch of the opening profile, choose the **Finish Edit Mode** button from the **Mode** panel, the opening cut will be modified as per the revised sketch. After exiting the **Modify | Structural opening cut** tab, if you want to display it again, select the created opening.

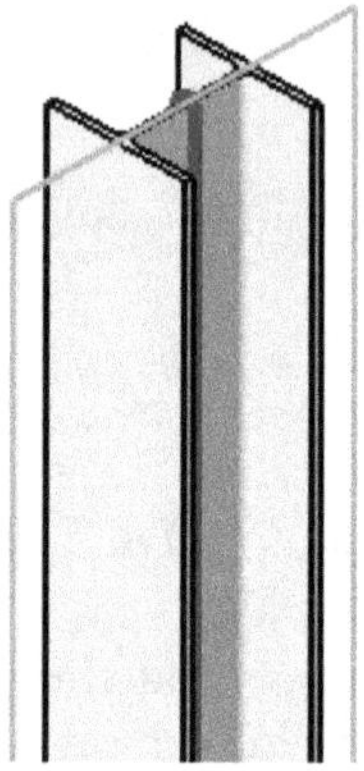

Figure 4-15 A plane displayed on the face of the column

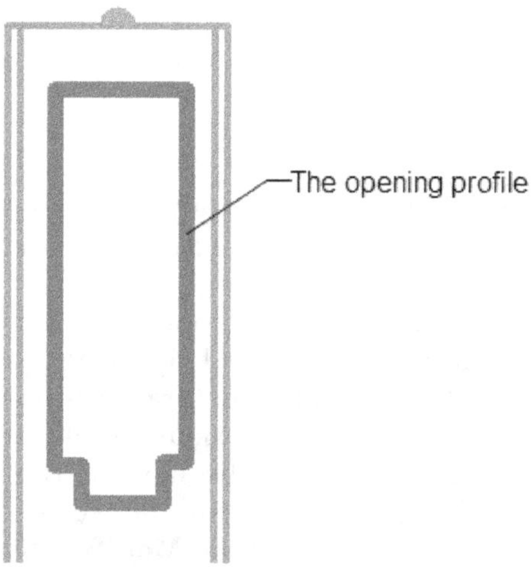

Figure 4-16 The opening profile created on the face of the column

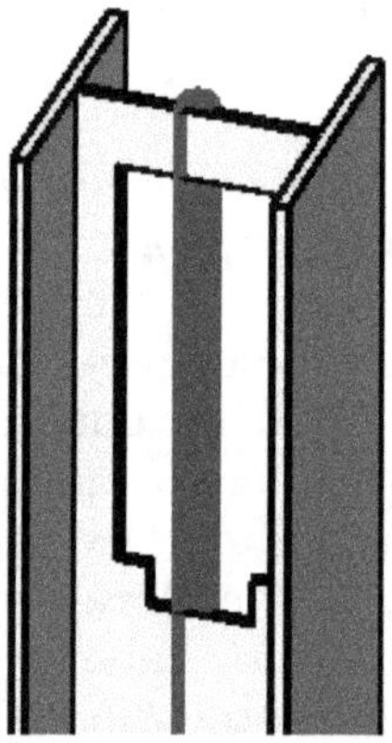

Figure 4-17 An opening cut through the face of the column

Modifying the Element Properties of Structural Columns

Element properties for a structural column include its instance and type properties. The instance properties of a column refer to the individual instances of the column placed in a project. For example, if you select a column and modify its instance properties, the change in the properties will be reflected only on the selected column.

The type properties of a column may vary from one column to another. If there is a change in the type property of a column, it may affect other columns of the same type. The type properties for steel columns are different from the concrete columns. The type properties of these columns will be discussed individually in detail in the later sections of this chapter.

To modify the instance properties of a column, select it from the project. For example, you can select a concrete column type from the project. On doing so, the **Properties** palette will display the instance properties of the selected column, as shown in Figure 4-18.

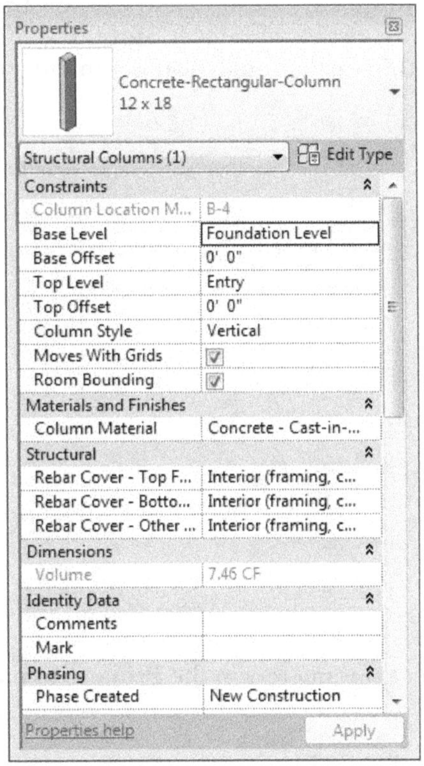

Figure 4-18 *The* ***Properties*** *palette displaying the instance properties in a column*

You can modify various properties of a column of a particular instance, such as its location in the elevation, its height or height constraint, material assigned, room bounding properties, and so on. You can also modify the structural properties such as cover distances for rebars at the top, bottom, and other faces. To modify the type properties of a selected column, select the column from the project and then choose **Edit Type** from the **Properties** palette; the **Type Properties** dialog box will be displayed. In this dialog box, you can modify the type properties of a column such as its cross-sectional dimensions and schedule data. Note that the parameters in the **Type Properties** dialog box for a column differ depending upon the type of column selected from the project. For instance, the parameters in the **Type Properties** dialog box for a steel column may differ from that of a concrete column. Some of the instance and type properties of different columns are discussed in the next sections.

Note

In Family template, the steel section shape parameter is added to ***Steel*** *columns in the* ***Properties*** *palette.*

The Instance Properties of the Steel and Concrete Columns

The instance properties of a column may vary depending upon the column families specified during the installation of Revit software. The instance properties can be specified from the properties displayed in the **Properties** palette, when you select the column. For example, in the **Properties** palette, you can specify the top and base constraints of a column by specifying a level. Figure 4-19 shows some of the references of the instance properties of a column in a project view. You can also modify various analytical parameters of the column.

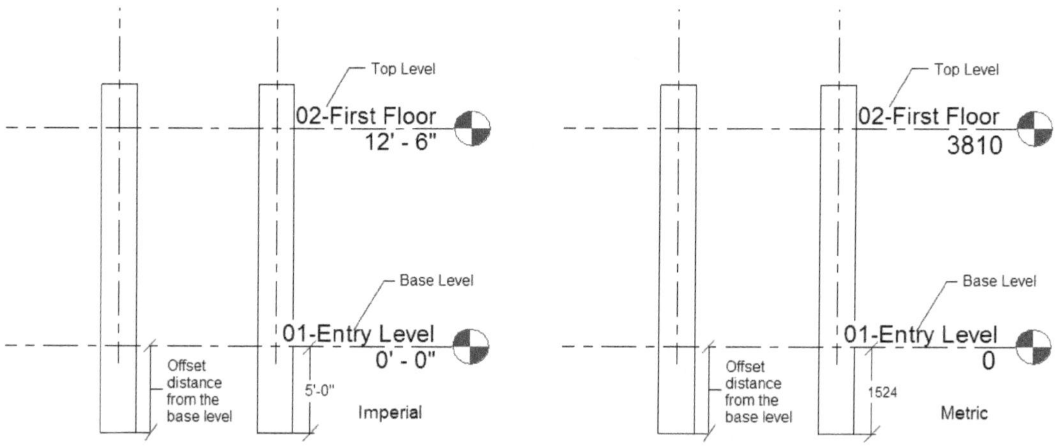

Figure 4-19 *References of the instance properties of a column in a project view*

Brief description of the instance parameters in the **Properties** palette is given in the table next:

Parameter	**Description**
Column Location Mark	Specifies the coordinate location of a vertical column on the project grid.
Base Level	Sets a level to constrain the column base. For example, refer to Figure 4-19, **01-Entry Level** is the base level for a particular column.
Base Offset	Specifies the offset distance of the base of the column from the base level assigned. The value specified for this parameter is negative. For example, in Imperial unit system the base of the column is 5' and in Metric unit system, the base of the column is 1500mm down from the **01-Entry Level** level (base level). Refer to Figure 4-19.
Top Level	Sets a level to constrain the top of a column. For example, in Figure 4-19, **02-First Floor** is the top level for a particular column.

Top Offset	Specifies the offset distance of the top of a column from the top level assigned to it. The value specified for this parameter is positive. For example, in Figure 4-19, in Imperial system, the top of the column is 2'-6" and in Metric system, the top of the column is 760mm up from the **02-First Floor** level (top level).
Moves With Grids	Specifies whether a column associated with a particular grid intersection will move when the grid intersection moves. If the check box corresponding to this parameter is selected, the column will move when the associated grid intersection moves. This parameter will be available only for vertical columns.
Move Top With Grids	Specifies a constraint at the top endpoint of the slanted column to the grid. Select the check box corresponding to this parameter to assign constraint so that the top endpoint moves when the corresponding grid intersection will move.
Move Base With Grids	Specifies a constraint at the base endpoint of the slanted column to the grid. Select the check box corresponding to this parameter to assign the constraint so that the base endpoint moves when the corresponding grid intersection will move.
Column Style	Specifies the style of a column. Select any of the options such as **Vertical**, **Slanted - End Point Driven**, or **Slanted - Angle** from the drop-down list corresponding to this parameter.
Cross-Section Rotation	Specifies the clockwise or counter-clockwise rotation angle for a slanted column at its longitudinal axis. This parameter can have both positive and negative values. A positive value indicates a clockwise rotation angle and a negative value indicates a counter-clockwise rotation angle.
Top Connection	Turns on the visibility of a moment or shear connection symbol of a steel column. This parameter is applicable only for steel columns. Select any of the options from the drop-down list corresponding to this parameter. The options available in the drop-down list are **None**, **Shear Connection**, and **Moment Connection**. Note that the top connection symbol is visible only in the elevation views and sectional views that are cut parallel to the main axis of the column.

Base Connection	Turns on the visibility of base plate symbols. Select the check box corresponding to this parameter to display the base plate symbols. The symbols are visible only in the elevations that are cut parallel to the main axis of the column.
Structural Material	Specifies the material to be assigned for a column. To specify a material for a column, invoke the **Material Browser** dialog box by choosing the **Browse** button displayed in the value field of this parameter.
Rebar Cover - Top Face	Sets the rebar cover distance of a concrete column from the top face of its section. This parameter is applicable for concrete columns only.
Rebar Cover - Bottom Face	Sets the rebar cover distance from the bottom face of the column. This parameter is applicable to the concrete columns only.
Rebar Cover - Other Faces	Sets the cover distance of rebars in the column to the faces of other elements.
Volume	Specifies the volume of a selected column. The value displayed for this parameter is non-editable or read-only.
Image	Specifies the image of the column. Select the Browse button to add the image of column.
Comments	Specifies a comment that a user can assign to a column.
Top Release	Specifies the top release condition for a column. Select any of the options such as **Fixed**, **Pinned**, **Bending Moment**, or **User-Defined** from the drop-down list corresponding to this parameter.
Top Fx	Specifies the translational release at the top end of a column along the X-axis. The option to set this parameter will be available only if the **Top Release** parameter is set to **User-Defined**.
Top Fy	Specifies the translational release at the top end of the column along the Y-axis. The option to set this parameter will be available only if the **Top Release** parameter is set to **User-Defined**.
Top Fz	Specifies the translational release at the top end of a column along the Z-axis. The option to set this parameter will be available only if the **Top Release** parameter is set to **User-Defined**.

Top Mx	Specifies the translational release at the top end of a column along the Z-axis. The option to set this parameter will be available only if the **Top Release** parameter is set to **User-Defined**.
Top My	Specifies the rotational release at the top end of the column along the Y-axis. The option to set this parameter will be available only if the **Top Release** parameter is set to **User-Defined.**
Top Mz	Specifies the rotational release at the top end of a column along the Z-axis. The option to set this parameter will be available only if the **Top Release** parameter is set to **User-Defined**.
Base Release	Specifies the bottom release condition for a column. Select any of the options such as **Fixed**, **Pinned**, **Bending Moment**, and **User-Defined** from the drop-down list corresponding to this parameter.
Base Fx	Specifies the translational release at the bottom end of a column along the X-axis. Select the check box corresponding to this parameter to enable this parameter. Note that this parameter will be available only when the **Bottom Release** parameter is set to **User-Defined**.
Base Fy	Specifies the translational release at the bottom end of a column along the Y-axis.
Base Fz	Specifies the translational release at the bottom end of a column along the Z-axis.
Base Mx	Specifies the rotational release at the bottom end of a column along the X-axis.
Base My	Specifies the rotational release at the bottom end of a column along the Y-axis.
Base Mz	Specifies the rotational release at the bottom end of a column along the Z-axis.
Analyze As	Specifies whether a column will be analyzed as Hanger, Gravity, Lateral, or Not for analysis. This parameter is used by external analysis applications to determine whether the column contributes to lateral analysis.
Analytical Links	Specifies whether to keep a analytical link between the analytical model of a beam and the analytical model of a column. Select the check box corresponding to this parameter to enable the rigidity in the linkage between a beam and a column. When the analytical link is enabled, an analytical segment is added in the model between the ends of the analytical models of a beam and a column.

Base Alignment Method	Specifies a method of justifying the horizontal and vertical reference of a column base. Click on the value field corresponding to this parameter and select the **Auto-detect** or **Projection** option from the drop-down list displayed.
Base x-Projection	It is enabled on assigning the **Projection** option to the **Base Extension Method** parameter. The **Base x-Direction** specifies the horizontal limits of the analytical model at the base of the column.
Base y-Projection	It is enabled on assigning the **Projection** option to the **Base Alignment Method** parameter. The **Base y-Direction** specifies the vertical limits of the analytical model at the base of the column.
Base z-Projection	It is enabled on assigning the **Projection** option to the **Base Alignment Method** parameter. The **Base z-Direction** specifies the vertical extension location of the analytical model at the base of the column.
Base Extension Method	Specifies the method for justification of the vertical extension of the column base.
Top Alignment Method	Specifies a method of justifying the horizontal and vertical reference of a column top. Click on the value field corresponding to this parameter and select the **Auto-detect** or **Projection** option from the drop-down list displayed.
Top x-Projection	It is enabled on assigning the **Projection** option to the **Top Extension Method** parameter. The **Top x-Direction** specifies the horizontal limits of the analytical model at the top of the column.
Top y-Projection	It is enabled on assigning the **Projection** option to the **Top Alignment Method** parameter. The **Top y-Direction** specifies the vertical limits of the analytical model at the top of the column.
Top z-Projection	It is enabled on assigning the **Projection** option to the **Top Alignment Method** parameter. The **Top z-Direction** specifies the vertical extension location of the analytical model at the top of the column.
Top Extension Method	Specifies the method for justification of the vertical extension of the column base.

The Type Properties of Steel Columns

The type properties of steel columns include its area, weight flange width, actual depth of section, name of the manufacturer, and so on. Some of the relevant type properties, which can be assigned to a steel column, in the **Type Properties** dialog box are described in the table given next.

Parameter	Description
A	Specifies the section area of the steel section.
W	Specifies the nominal weight of the steel column.
bf	Specifies the flange width of a section.
d	Specifies the actual depth of a section.
tf	Specifies the thickness of a flange.
tw	Specifies the web thickness of the steel column.

Note

*After assigning type parameters in the **Type Properties** dialog box, you can preview the changes made to a model. To do so, choose the **Preview** button in the **Type Properties** dialog box; the preview of the selected type will be described in the expanded area of this dialog box.*

The Type Properties of Concrete Columns

The type properties of a concrete column include its depth, width, and so on. Some of the type properties of a concrete column are described in the following table:

Parameter	Description
b	Specifies the width of a column.
h	Specifies the depth of a column.
Keynote	Specifies the keynote associated to a column. To assign a keynote, click in the **Value** field of this parameter and then choose the Browse button; the **Keynotes** dialog box will be displayed. In this dialog box, you can assign a keynote to the desired column.
Cost	Specifies the cost of a column.
Description	Specifies the description of a column.

ARCHITECTURAL COLUMNS

Ribbon: Architecture > Build > Column drop-down > Column: Architectural

Architectural columns are non load bearing members that are placed in a structural project to add ambience to the structure. Generally, very few architectural columns are placed in a structural model. The placing of architectural columns in a model depends upon the design requirement, both architectural and structural.

You can add architectural columns only in plan view. To add an architectural column, activate the desired plan view and then invoke the **Column: Architectural** tool from the **Build** panel; the **Modify | Place Column** tab will be displayed, as shown in Figure 4-20.

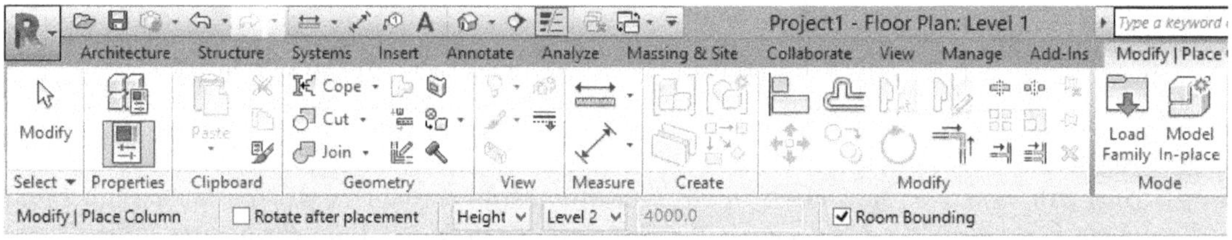

*Figure 4-20 The **Modify / Place Column** tab*

Note

*On invoking the **Column: Architectural** tool, the **Revit** message box may appear informing that no column family is loaded in the project. The message box will prompt you to load a column family in the project. To load a column family, choose the **Yes** button; the **Open** dialog box will be displayed. Browse to the **Column** sub-folder in the **US Imperial** folder for Imperial (for Metric browse to **Column** sub-folder in the **US Metric**) in this dialog box to load the desired column family in your project.*

The tools in the **Modify | Place Column** tab are the same as those discussed in the **Modify | Place Structural Column** tab. After using the tools from the **Modify| Place Column** tab, you can use options in the **Options Bar** to set the alignment and location of the column that you need to place. After setting the options in the **Options Bar**, click on the desired location in the structural plan view; the architectural column will be placed at the point you click. The placed column can now be used for aesthetic purpose as well as to add structural column to it. Figure 4-21 shows a three-dimensional view of an architectural column placed in a structural model. The method of adding structural column to a architectural column is discussed next.

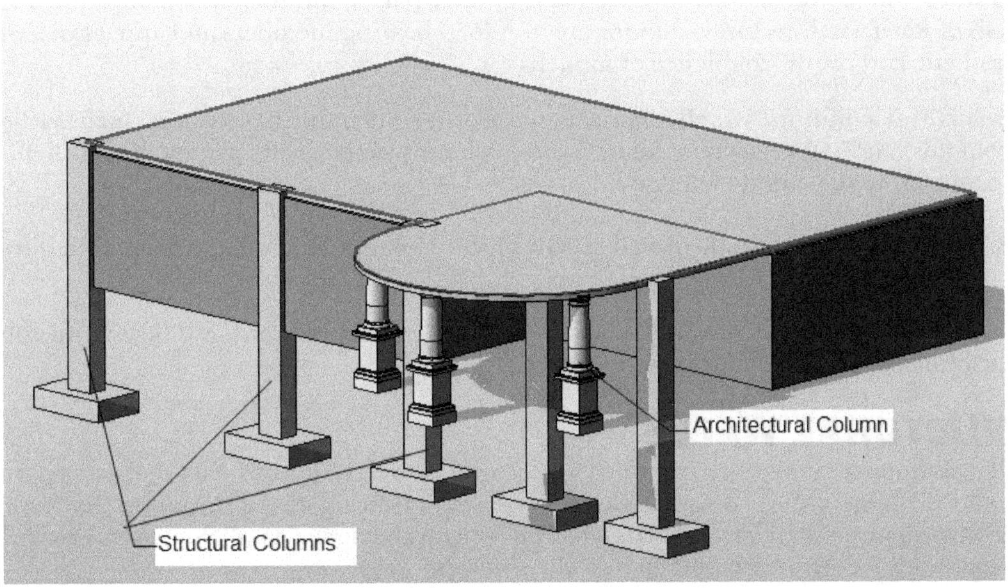

Figure 4-21 *Three-dimensional view of a structural model displaying architectural columns*

Adding Structural Column to an Architectural Column

After adding an architectural column to a project, you can add a structural column to it so that it can become a part of the analytical model. To do so, choose the **Column** tool from **Architecture > Build > Structural Column** drop-down; the **Modify |Place Structural Column** tab will be displayed. In this tab, choose the **At Columns** tool from the **Multiple** panel; the **Modify | Place Structural Column > At Architectural Columns** tab will be displayed. In the **Type Selector** drop-down list in the **Properties** palette, you can select the desired type of column and then click on any existing architectural column from the structural model in the drawing area; the selected structural column type will be placed at the center of the selected architectural column. You can add more structural columns to other architectural columns that are present in the project. To do so, select the architectural columns, press and hold the CTRL key and then click on the architectural columns one by one; the structural columns will be placed at the center of each selected architectural columns. While adding architectural columns to the selection, you can also remove the elements from the selection. To do so, press and hold the SHIFT key and click on the selected column that you need to remove; the selected structural column will be removed from the selected columns.

After placing the structural column(s) in the architectural column(s), choose the **Finish** button from the **Multiple** panel; the structural column(s) will be placed at the desired architectural columns.

DIFFERENCE BETWEEN A STRUCTURAL COLUMN AND AN ARCHITECTURAL COLUMN

In Autodesk Revit, you can add two types of columns that are classified based on their functionality. These are: Architectural and Structural Columns. The major differences between Structural columns and Architectural columns are given next.

Structural columns are load-bearing members and are part of the analytical model for structural analysis of Revit. Architectural columns are non-load bearing members and appear in a project for aesthetic and overall ambience of model.

- Structural columns attach themselves to other structural members such as beams, columns, and slabs, whereas Architectural columns functionally cannot attach themselves to any of the structural elements.

- Structural columns are displayed in Graphical Column Schedule, whereas Architectural columns are not.

Despite differences in the functional behavior, Architectural columns and Structural columns may appear same in a structural model.

STRUCTURAL WALLS

A wall constitutes an important feature in a structural model. Walls can be modeled as a partition, boundary element, a shear resistant element, or a load bearing element. Walls differ from each other on the basis of their structural usage. The structural usage of a wall defines its load-bearing functions used for structural and architectural design.

A building project contains different types of walls depending on their structural usage. These walls can be exterior walls, interior walls, basement walls, and partition wall studs. In Revit, walls are defined as system families and therefore, they can be created and modified within the current structural model. The walls are implicit features in a structural model. This means that they cannot be added in the model as separate entities and also they change with the change in the structural model.

In Revit, you can add structural walls that are load-bearing in function. This implies that they play an active role in transferring the load from one structural element to the other. They can bear both vertical forces and shear forces depending upon the function they are supposed to perform. The function of a structural wall is determined by its structural usage that can be defined in its instance property.

Adding Structural Walls

You can add a structural wall either in a structural plan view or in a 3D view, based on your project requirement. To add a structural wall, choose the **Wall: Structural** tool from **Structure > Structure > Wall** drop-down; the **Modify | Place Structural Wall** tab along with other options in the **Options Bar** will be displayed. In this tab, you can use various options to sketch the wall and assign different instance and type properties to the structural wall, as shown in Figure 4-22. On invoking the **Wall: Structural** tool, the **Properties** palette displays various instance properties and wall types. To select a type of structural wall, select a wall type from the **Type Selector** drop-down list. The wall types that you can select from the drop-down list are **Basic Wall**, **Curtain Wall**, and **Stacked Wall**. These wall types have various sub-types. In Imperial system, by default the **Generic-8"** wall sub-type is selected and in Metric system, by default the **Generic-200mm** is selected in this drop-down list.

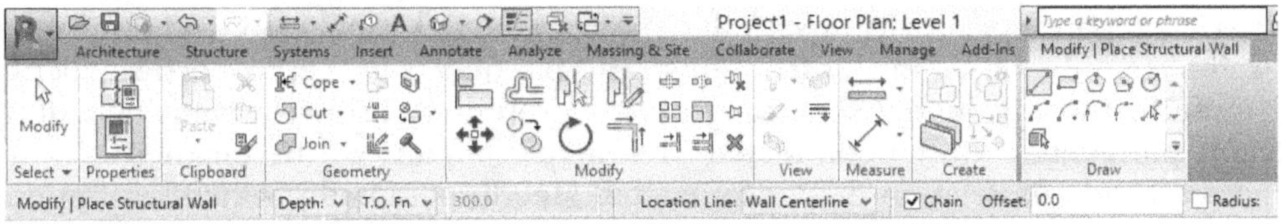

Figure 4-22 Various options and tools in the ***Modify / Place Structural Wall*** *tab*

> **Note**
>
> *At the initial phase of a structural design, you can add generic structural walls to the structural model. These walls act as placeholders for the height, width, and wall type of actual walls.*

After selecting a type from the **Type Selector** drop-down list, you can assign instance and type properties to the selected type. You can also change the instance properties of the selected wall type such as its top and base constraints, and rebar cover by using the parameters displayed in the **Propertie**s palette, as shown in Figure 4-23. Different instance parameters for a structural wall in the **Properties** palette are discussed in the next section.

Instance Parameters of Structural Wall

After invoking the **Wall: Structural** tool, you can set the instance parameters of a structural wall in the **Properties** palette. The **Location Line** parameter in the palette is used to specify the location line of the structural wall to be placed in a model. The location line represents the vertical plane in the structural wall that does not change on changing the wall type. You can select any of the options such as **Wall Centerline**, **Core Centerline**, **Finish Face: Exterior**, **Finish Face: Interior**, **Core Face: Exterior**, and **Core Face: Interior** from the drop-down list in the value field corresponding to the **Location Line** parameter. The **Base Constraint** parameter is used to specify the level to constrain the base of the wall. From the drop-down list in the value field of this parameter, you can select a level to constraint the base of the structural wall. After setting the **Base Constraint** parameter of the structural wall, you can set the offset distance that can be used to extend or offset the base of the wall from the base level. To do so, click in the value field of the **Base Offset** parameter and specify a numeric (positive or negative) value for the offset distance of the base of the wall. Similar to constraining the base of the wall, you can also constrain the top of the wall. To do so, you can use the **Top Constraint** and the **Top Offset** parameters from the **Properties** palette. To set the value for the **Top Constraint** parameter, click on its value field and select an option from the drop-down list displayed. The option thus selected will represent a level in the structural model. After setting the top constraint for the wall, click on the value field corresponding to the **Top**

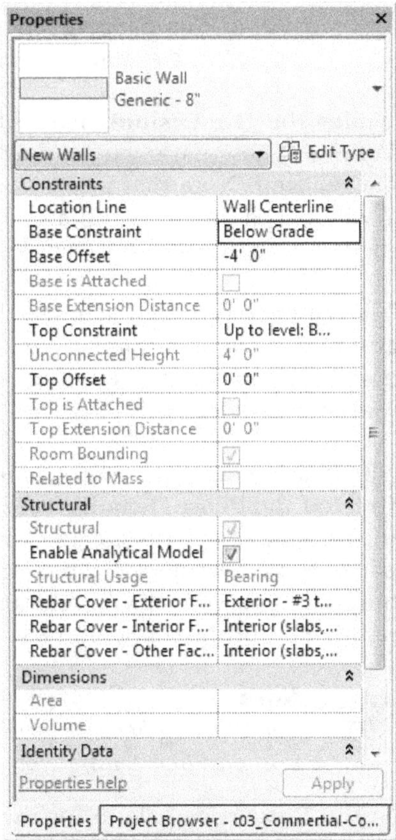

Figure 4-23 The instance properties of a structural wall displayed in the ***Properties*** *palette*

Offset parameter and specify the offset value. You can use this value to extend or offset the top of the wall from the level set in the **Top Constraint** parameter in the **Properties** palette. Note that if you select the **Unconnected** option from the drop-down list in the value field of the **Top Constraint** parameter, the **Unconnected Height** parameter will be enabled. The default unconnected height in Revit in the **Unconnected Height** parameter is **20'0"** in Imperial and **4000** mm in Metric. In the value field of this parameter, you can enter a numeric value to specify the unconnected height of the wall measured from its base.

In the **Properties** palette of the structural wall, the structural and dimensional parameters are specified respectively under the **Structural** and **Dimensions** heads, refer to Figure 4-23. Under the **Structural** head, the following structural parameters of the wall are available: **Rebar Cover - Exterior Face**, **Rebar Cover - Interior Face**, **Rebar Cover - Other Faces**, and **Structural Usage**. Note that the value corresponding to the **Structural Usage** parameter is set to **Bearing** and is read-only. The other parameters are used to specify the rebar cover distance from the exterior or interior face of the wall, or from the face of the adjacent element.

Note

*The value of the **Structural Usage** parameter can be changed from **Bearing** to other values such as **Non-Bearing**, **Shear**, and **Structural combined** only when you modify the instance property of the wall.*

Under the **Dimensions** head, the following instance parameters are available: **Area** and **Volume**. These parameters are read-only and display the volume and area of walls when they are selected. Note that when you select a structural wall and use the **Properties** palette, an additional parameter, **Length**, is displayed under the **Dimensions** head. This parameter displays the length of the wall selected.

The parameters displayed under the **Identity Data** head in the **Properties** palette are **Comments** and **Mark**. These parameters are used to specify comments and marks that can be assigned to a wall. The value assigned in the value field of the **Mark** parameter should be unique for each wall in the project as it serves as a label for the wall.

The **Phasing** head in the **Properties** palette of the structural wall displays two parameters: **Phase Created** and **Phase Demolished**. In these parameters, you can specify the information related to the phasing of the wall created and the wall demolished. This means that in the **Phase Created** parameter, you can specify the information regarding the phase in which the wall was created. Similarly, you will specify relevant information in the value field of the **Phase Demolished** parameter for the phase of the wall in which it was demolished.

Note

*The **Phasing head** parameter is displayed in the **Properties** palette only when the created wall is selected from the drawing.*

In the **Properties** palette for the structural wall, select the check box corresponding to the **Structural** parameter to enable the visibility of the analytical model of a wall. In the project view, you can view and edit the instance properties of the analytical model of the wall. Note that the instance properties of the analytical model of a wall are different from the instance properties of the physical model of the wall. To enable the display of the analytical model of the wall, choose the **Show Analytical Model** button from the **View Control Bar.** Using this button, you can toggle between the two options: **Show Analytical Model** and **Hide Analytical Model**. By default, the **Show Analytical Model** is selected in the options. As a result, the analytical model will be displayed along with the physical model. To highlight the analytical model in the project view, select the structural wall from the project view and then choose the **Highlight Analytical** option from the **Analytical** panel in the **Modify | Walls** tab.

Alternatively, you can choose the **Hide Analytical Model** option from the **View Control Bar**. As a result, only the analytical model will be displayed in the project view. You can select the analytical model of the structural wall to modify its instance properties. The instance properties for the analytical model of a structural wall are different from its associated physical model. As you select the analytical model of the structural wall, its instance properties are displayed in the **Properties** palette. The instance properties for the analytical model are: **Alignment Method**, **Top Extension Method**, **Base Extension Method** and **Analyze As**. After setting various instance parameters for the structural wall in the **Properties** palette, you can set the type parameters. To do so, choose the **Edit Type** button from the palette; the **Type Properties** dialog box will be displayed, as shown in Figure 4-24. In this dialog box, you can set various type parameters related to the structural wall. The type parameters for structural walls are discussed next.

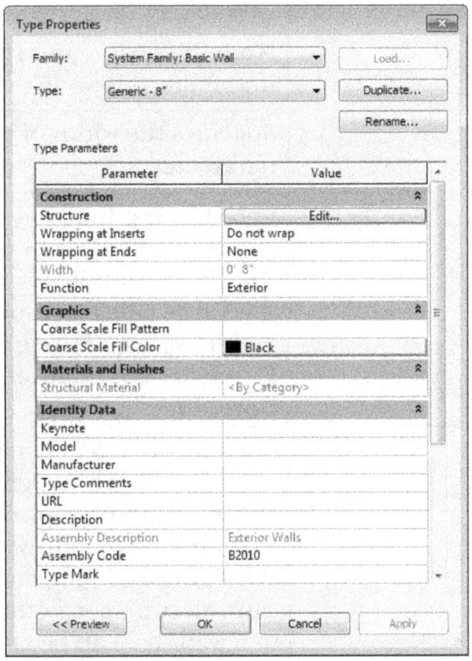

***Figure 4-24** The **Type Properties** dialog box for structural walls*

Type Properties of Structural Walls

The type properties of a structural wall refer to the properties assigned to a particular type of wall. To define the type properties of a wall, invoke the **Type Properties** dialog box and specify the parameters in their specific fields. Some of the type properties that can be specified in the **Type Properties** dialog box are described in the table given next.

Parameter	Description
Structure	Specifies the layers that exist in a structural wall. To edit wall layers, choose the **Edit** button from the **Value** field of the **Structures** parameter; the **Edit Assembly** dialog box will be displayed. You can use this dialog box to specify different layers for a structural wall.
Wrapping at Inserts	Specifies the layer wrapping of walls at the insertion of doors, windows, and openings. To specify an option for this parameter, click on its **Value** field and select any of these options, namely **Do not Wrap**, **Exterior**, **Interior**, and **Both** displayed in the drop-down down list.
Wrapping at Ends	Specifies layer wrapping of wall end caps. To specify an option for this parameter, click on its **Value** field and select any of the options from **None**, **Exterior**, and **Interior** options displayed in the drop-down list.
Width	Specifies the width of a wall. This is a read only parameter.
Function	Specifies the function of a wall that provides a way to schedule walls in a structural model. To specify an option for this parameter, click on the **Value** field corresponding to this parameter and select any of these options, namely **Interior**, **Exterior**, **Foundation**, **Retaining**, **Soffit**, and **Core-Shaft**.
Coarse Scale Fill Pattern	Specifies the hatch pattern for the wall in a coarse view. To assign a pattern, click in the **Value** field corresponding to this parameter and choose the Browse button to invoke the **Fill Patterns** dialog box. In this dialog box, you can select a specific hatch pattern to assign it to a wall that can be viewed in the coarse view.
Coarse Scale Fill Color	Specifies the color assigned to the hatch pattern applied to a wall. To assign a different color, choose the button displayed in the **Value** field corresponding to this parameter; the **Color** dialog box will be invoked. In this dialog box, you can select a color and assign it to the hatch fill pattern applied to the structural wall.

Structural Material	Specifies the material composition of wall.
Fire Rating	Specifies the fire rating of a wall in compliance with the building codes. The fire rating determines the duration which a fire resistant wall can withstand in case of fire.
Cost	Specifies the cost of a wall.

After you have specified the instance and type properties of a structural wall, you can select a sketching tool from the list box in the **Draw** panel to sketch the wall. The tools available in the list box to sketch the wall are discussed in the next section.

Sketching Structural Walls

You can sketch a structural wall by using the tools available in the list box in the **Draw** panel of the **Modify | Place Structural Wall** tab, refer to Figure 4-25. To sketch the wall, you can also use the options available in the **Options Bar** along with the tools from the **Draw** panel. The sketching tools available in the **Draw** panel of the **Modify | Place Structural Wall** tab and their applications are discussed next.

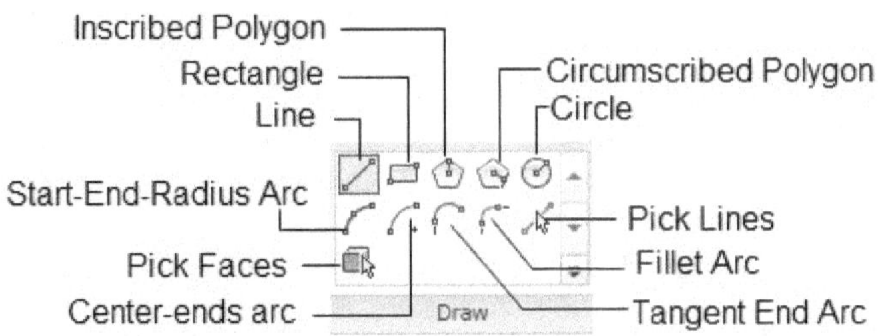

Figure 4-25 *Various sketching tools in the* ***Draw*** *panel*

Sketching Structural Walls Using the Line Tool

You can use the **Line** tool to sketch straight walls in a drawing. You can invoke this tool from the **Draw** panel of the **Modify | Place Structural Wall** tab. After invoking this sketching tool, specify the location of the start point and endpoint of the wall segment. To specify the location of the start point, click in the drawing area. Next, move the cursor in the drawing area. On doing so, you will notice that a wall segment has started from the specified point and displays a dimension on it. This dimension is called a temporary dimension or listening dimension, and it shows the length and angle of the wall segment in Imperial unit system at the cursor location, as shown in Figure 4-26. Also, you will notice that the length of the wall segment is displayed in Metric system, refer to Figure 4-26. Note that the cursor is moving in fixed increments. This increment is determined by the value set in the **Dimension Snaps** option of the **Snaps** dialog box. The options in this dialog box have already been discussed in Chapter 2.

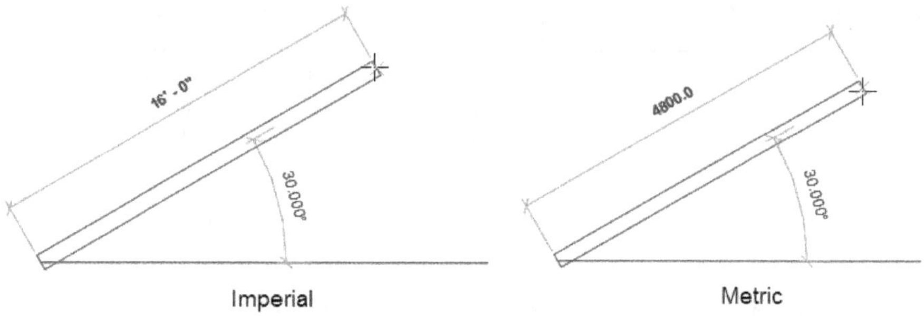

Figure 4-26 *Structural wall with its temporary dimensions*

The angle subtended by the wall on the horizontal axis is also displayed and it keeps on changing dynamically as you move the cursor to modify the inclination of the wall. On moving the cursor near the horizontal or the vertical axis, a dashed line will appear on the wall segment. This line is called alignment line and helps you sketch components with respect to the existing components. You will also notice that a tooltip is displayed at the endpoint of the wall, indicating that the wall segment being sketched is horizontal, as shown in Figure 4-27.

Figure 4-27 *Sketching a horizontal wall*

Autodesk Revit provides you with the option to specify the length of walls in different ways. The first option is to specify the start point of the wall and then moving the cursor in the desired direction and then click when the angle and the temporary dimension attain the required values. The second option is to sketch the wall and then modify its length and angle to the exact value. For example, to sketch a 18'0" (or 5400 mm) long horizontal wall after specifying the starting point, you need to move the cursor to the right until you see a dashed horizontal line parallel to the sketched wall. Click when the temporary dimension shows approximately 20'0" (or 6000 mm). Note that the length of the wall may not be exactly 18'0"(or 5400 mm). You can now use the wall controls to modify the dimensions of the wall.

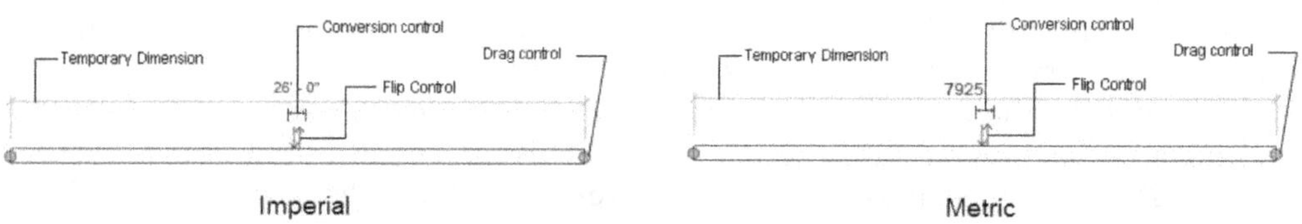

Figure 4-28 *Wall controls displayed on wall segment*

To modify a wall, select the wall segment and view its control and properties. When you select the wall segment, it gets highlighted in blue and three symbols appear above the wall segment in blue. These highlighted symbols are called wall controls, as shown in Figure 4-28. The exact dimension of the sketched wall is visible in the dimension text of the temporary dimension. The conversion control symbol, which appears below the dimension value, is used to convert a temporary dimension into a permanent dimension. The two blue arrows, which also appear on the upper face of the wall, indicate the flip control symbols for the sketched walls. They appear on the side interpreted as the exterior face of the wall. By default, the walls drawn from the left to right have the external face on the upper side and the walls drawn from the top to bottom have external face on the right side. You can flip the orientation of the wall by clicking on the arrow symbol. Alternatively, you can place the cursor on the flip control symbol and notice the change in its color. After the color of the flip control is changed, press SPACEBAR to flip the wall. The two blue dots that appear at the two ends of the wall segments are the drag control symbols. You can use them to stretch and resize the walls. To set the wall to the exact length, click on the temporary dimension; an edit box will be displayed, showing the current dimension of the wall segment. You can replace this value by specifying the exact length, for example **14' 8"** (or 4470), in the edit box, as shown in Figure 4-29. The length of the wall will be modified to **14' 8"** (or 4470). Alternatively, you can create a straight wall by typing the dimension of the length before choosing its endpoint. As soon as you enter the value of the length, an edit box appears above the dimension line. Enter the value of the length and press ENTER to create a wall segment of the specified length. To sketch a wall at a given angle, sketch it at any angle and then click on the angular dimension symbol; an edit box will be displayed. In the edit box, you can enter the exact angular dimension.

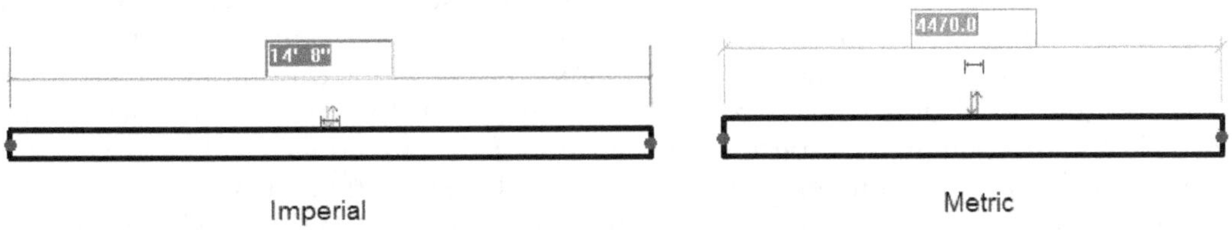

Figure 4-29 *Specifying the exact length of the wall*

Note

*The **Project Browser** shows **Level 1** in bold letters. This indicates that the wall has been sketched in that level.*

Sketching a Structural Wall Using the Circle Tool

The **Circle** tool is used to sketch a circular wall profile. To sketch one such profile, invoke this tool from the **Draw** panel of the **Modify | Place Structural Wall** tab. Next, click in the drawing area to specify the center point of the circular wall. As you move the cursor away from this point, you will notice that a circular wall profile has been created. This circle extends dynamically, taking the specified points as the center and the cursor location as the other end. The temporary radial dimension will also be displayed, as shown in Figure 4-30. Click when the desired value for the radius is displayed. Alternatively, before clicking on the second point, type the value for the radius of the circular profile. On doing so, an edit box with the type value will be displayed as the temporary dimension. Press ENTER to complete the profile. Note that the dimension that you entered is the distance of the center point to the location line of the profile.

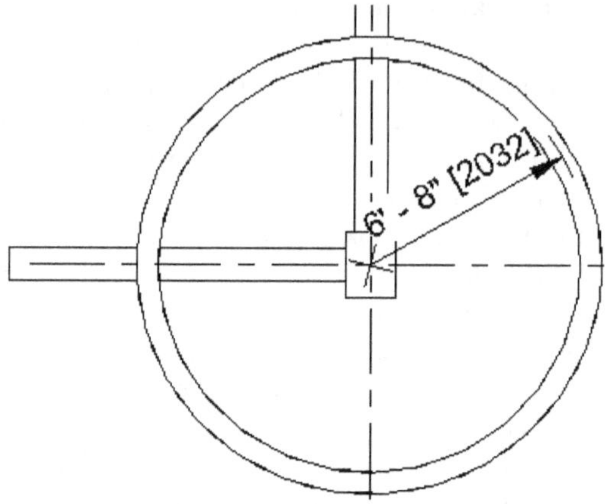

Figure 4-30 Temporary dimensions displayed on the circular wall

Sketching a Wall Using the Start-End-Radius Arc Tool

The **Start-End-Radius Arc** sketching tool enables you to sketch an arc wall by specifying the start point, the endpoint, and the intermediate point that determines the radius of the arc. To create an arc wall, invoke the **Start-End-Radius Arc** tool from the **Draw** panel of the **Modify | Place Structural Wall** tab; you will be prompted to specify the start point of the proposed arc wall. Specify the start point by clicking in the drawing area; you will be prompted to specify the endpoint of the curved wall. Specify the endpoint of the curved wall; a curved wall with a variable radius will stretch dynamically between the two specified points. Specify the location of the third point between the two specified points to specify the subtended angle or the radius of the arc. As you specify the third point, the temporary angular and radial dimensions are displayed along with the curve on the screen. Next, click in the drawing area to specify the third point; the curved wall will be sketched. You can also modify the parameters of the sketched walls such as their curvature, radius, angle subtended, orientation, and so on. To modify the curvature, select the sketched wall, click on the parameter such as radial dimension, and enter the new value in the edit box, as shown in Figure 4-31. You can also use the drag controls to increase or decrease the extents of the wall. The central blue dot can be used to stretch the wall, keeping the subtended angle constant.

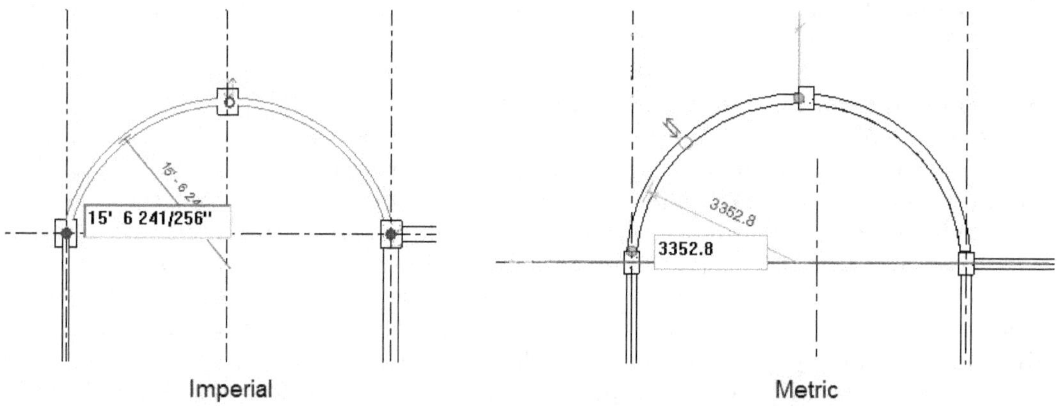

Figure 4-31 Editing the dimension of the arc wall

Sketching a Wall Using the Tangent End Arc Tool

To sketch a curved wall profile that starts tangentially from an existing wall, invoke the **Tangent End Arc** tool from the **Draw** panel in the **Modify | Place Structural Wall** tab; the start point of the wall will be specified as the endpoint of the existing wall. After specifying the start point, move the cursor to the desired distance and click to define the curved wall profile.

Sketching a Wall Using the Fillet Arc Tool

To create a curved fillet wall between two existing walls, choose the **Fillet Arc** tool from the **Modify | Place Structural Wall** tab. Next, click on the walls one by one to create a fillet close to the desired fillet end. On doing so, a fillet wall will appear, showing its possible locations, as shown in Figure 4-32. Click to specify the location of the fillet. Once the fillet wall has been sketched, you can modify its radius by clicking on it and typing its value. Notice that the walls are automatically trimmed after placing the fillet arc.

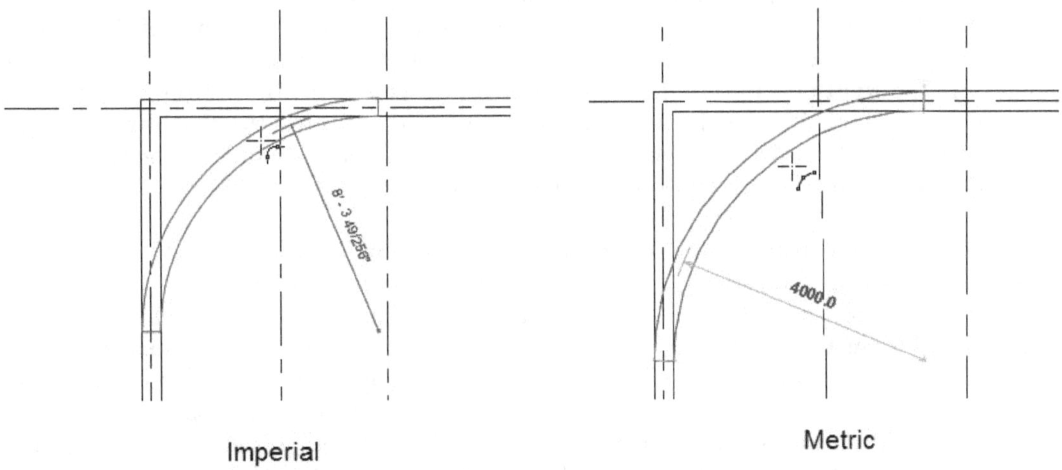

Figure 4-32 *Fillet wall created between two structural walls*

Creating a Wall Using the Pick Lines Tool

The **Pick Lines** tool can be used to create a wall from the existing model lines, edge, or grid lines. Invoke this tool from the **Draw** panel of the **Modify| Place Structural Wall** tab; you will be prompted to select an edge or a line. Move the cursor over the desired line or edge; the desired edge or line will be highlighted. Next, click on the highlighted edge or line, a wall will be created. You can keep on selecting multiple lines and edges to add more walls.

Creating a Wall Using the Pick Faces Tool

The **Pick Faces** tool can be used to create a wall from an existing face of a massing object. Invoke this tool from the **Draw** panel in the **Modify | Place Structural Wall** tab; you will be prompted to select a face from a massing object. Move the cursor over the desired face of the massing object and click over it, a wall will be created on the desired face.

Using the Options from the Options Bar

You can also use the options from the **Options Bar** to sketch a wall. The options in the **Options Bar**, as shown in Figure 4-33, can be used to set the height of a wall, alignment of the location line, offset value of the location line of the wall, and the radius of the circular or arc wall, if required.

Figure 4-33 *Various options in the* ***Options Bar***

The first drop-down list in the **Options Bar** is used to set the reference to which the height of the wall is to be measured. The drop-down list contains two options: **Height** and **Depth**. On selecting the **Height** option from the drop-down list, you can assign a constraint for the top of the wall from the drop-down list available next to the first drop-down list. In this drop-down list, you can select a level that is at a higher elevation from the current level or select the **Unconnected** option to assign an unconnected height which is not constrained to any level. On selecting the **Unconnected** option from the drop-down list, an edit box next to it will be activated. In this edit box, you can enter the unconnected height of the wall. Similarly, while adding a wall, if it is required to constrain the base of the structural wall, then select the **Depth** option from the first drop-down list of the **Options Bar**. After selecting the **Depth** option from the drop-down list, select any option from the drop-down list next to it to specify a level to constrain the base. You can select the **Unconnected** option from this drop-down list to specify the unconnected depth. To specify the unconnected depth, enter a value in the edit box displayed next to the drop-down list.

After setting the constraint for the top or base of the wall, you can select an option from the **Location Line** drop-down list to set the justification of the wall. In this drop-down list, you can select any of these options: **Wall Centerline**, **Core Centerline**, **Finish Face: Exterior**, **Finish Face: Interior**, **Core Face: Exterior**, and **Core Face: Interior**. Note that when you select an option from the **Location Line** drop-down list, the spacing between other structural element and the structural wall is also affected. So, you need to carefully select an option from the **Location Line** drop-down list, based on the project requirement. If you select the **Wall Centerline** option from the drop-down list, the structural wall will be justified to the center of the wall section that is calculated on the basis of its two outermost faces. You should select this option while adding a structural wall for single element wall type such as concrete foundation wall or concrete masonry unit. Figure 4-34 shows the example of a structural wall where the **Wall Centerline** option is selected from the **Location Line** drop-down list while adding a **Foundation - 12" Concrete** (for Imperial) and **Foundation - 300mm Concrete** (for Metric) type of wall in a project. Similarly, you can select the **Finish Face: Exterior** option or **Finish Face: Interior** option from the **Location Line** drop-down list to justify the location line of the structural wall either to the exterior finished face or to the interior finished face of the wall.

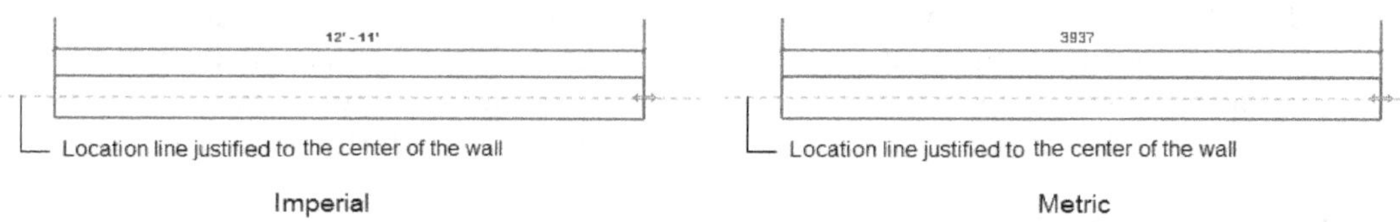

Figure 4-34 *Structural wall justified to the center of the wall*

Refer to the example in Figure 4-35 which shows the location of the location line for justification to the exterior and interior finished faces of the **Brick on CMU** type of wall. It is recommended to select the **Core Face: Exterior** or **Core Face: Interior** option from the **Location Line** drop-down list whenever a compound structural wall type is to be placed in a project. Refer to the example in Figure 4-36 which shows the placement of the location line at the exterior and interior faces of the core layers of the walls.

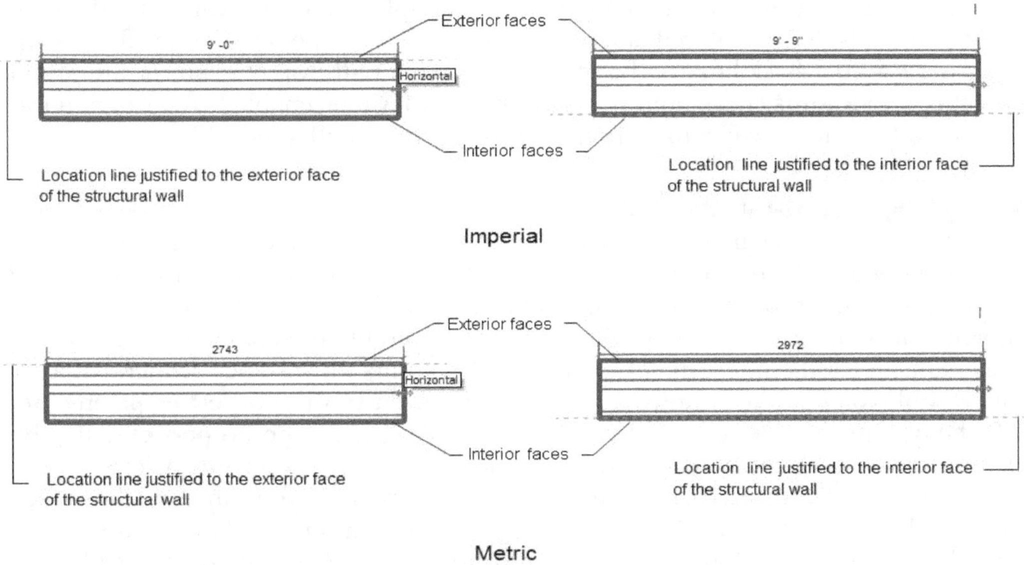

Figure 4-35 *Structural walls justified to the exterior and interior finished faces*

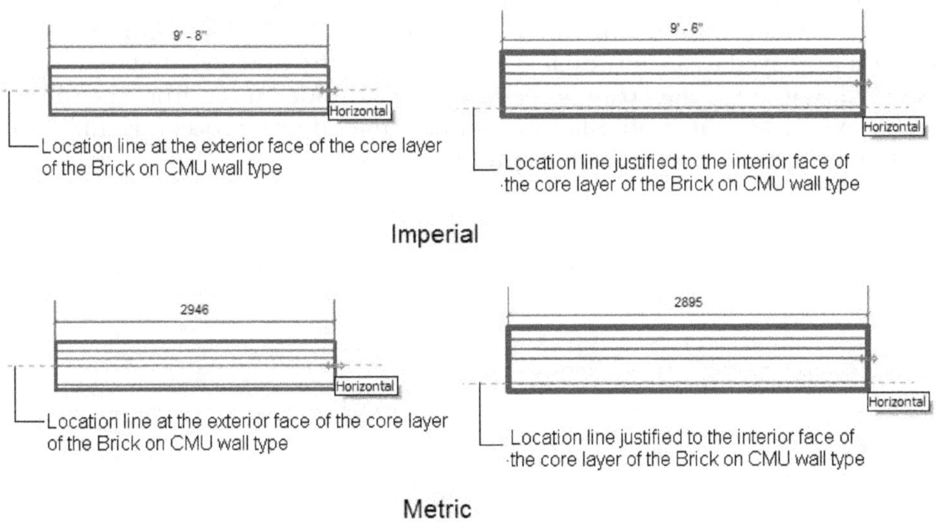

Figure 4-36 *Structural walls justified to the exterior and interior finished faces of the core layers*

In the **Options Bar**, you can select the **Chain** check box to create a continuous wall with wall segments connected end-to-end. The endpoint of the previous wall becomes the start point of the next wall. In the **Options Bar**, you can also enter a value in the **Offset** edit box to create a structural wall that starts at a specified offset distance from a point defined in an existing element. You can enter the offset distance value in the edit box provided in the **Options Bar**. After entering the offset value in the **Offset** edit box, click near the element to define the offset distance. When you move the cursor, the wall will start at the specified distance from the selected point. For example, this option can be used for creating boundary walls that are placed at a specific distance from the building profile. In **Options Bar**, the **Radius** check box can be selected to sketch walls with circular or arc profile. On selecting the **Radius** check box, an edit box next to will be enabled. You can enter a value in this edit box to specify a fixed radius for the circular wall to be added.

Modifying Structural Walls

You can modify profile, element properties (type and instance), and attachments at the top or base of a structural wall. To do so, select the wall that you want to modify; the **Modify | Place Wall** tab will be displayed. In this tab, you can use the options in the **Mode**, **Modify Wall**, and **Reinforcement** panels to edit the profile of the wall, modify the attachments of the top and base of a wall, and add reinforcement to the wall. To modify the instance property of a selected structural wall, you can use various parameters displayed in the **Properties** palette. You can modify the type properties of a structural wall by invoking the **Type Properties** dialog box. To invoke this dialog box, choose the **Edit Type** button from the **Properties** palette. Apart from modifying the element properties of the selected structural wall, you can also edit its elevation profile. To edit the elevation profile, select a wall in a structural floor plan view and choose the **Edit Profile** tool from the **Mode** panel in the **Modify | Place Walls** tab; the **Go To View** dialog box will be displayed. Select an option from this dialog box to specify a suitable view. Now, choose the **Open View** button from the **Go To View** dialog box; the specified view with the wall profile of the selected wall will be displayed in an editable mode, as shown in Figure 4-37. Next, you can use various sketching and editing tools from the **Modify| Walls > Edit Profile** tab. Note that this tab is displayed as a contextual tab on selecting the **Edit Profile** tool from the **Mode** panel. After editing the profile, choose the **Finish Edit Mode** button from the **Mode** panel in the **Modify | Walls > Edit Profile** tab to complete the editing process. Figure 4-38 shows a structural wall before and after editing its profile.

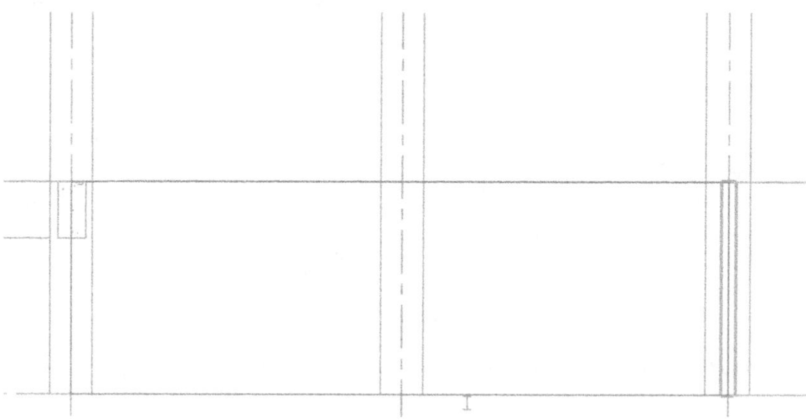

Figure 4-37 *Structural wall in editing mode as a wall profile*

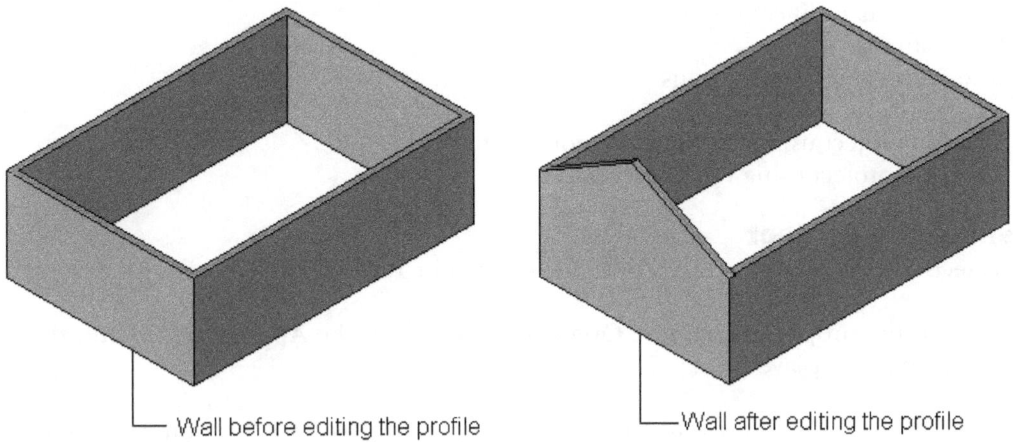

Figure 4-38 *Structural wall before and after editing its profile*

TUTORIALS

Tutorial 1 Commercial Complex

In this tutorial, you will add structural columns and structural walls to the *Commercial Complex* project created in Tutorial 1 of Chapter 3. Use the following project parameters and specifications for completing the project. **(Expected time: 1 hr)**

1. Structural column:
   For Imperial Type: **Concrete-Rectangular-Column**
   Dimension: 12" x 24"
   For Metric Type: **M_Concrete-Rectangular-Column**
   Dimension: 300 x 600mm

2. Structural concrete walls:
   For Imperial Width - **12"**
   Function - **Core-shaft**
   For Metric Width - **300mm**
   Function - **Core-shaft**

3. File name to be assigned:
   For Imperial *c04_Commercial-Complex_tut1.rvt.*
   For Metric *M_c04_Commercial-Complex_tut1.rvt.*

The following steps are required to complete this tutorial:

a. Start a Revit session and open the project file.
   For Imperial *c03_Commercial-Complex_tut1.rvt*
   For Metric *M_c03_Commercial-Complex_tut1.rvt*

b. Create a structural column type.
c. Add columns.
d. Modify the height of columns.
e. Add structural walls.
f. Save the project using the **Save As** tool.
g. Close the project using the **Close** tool.

Opening the Project

In this section, you will open the project file created in Tutorial 1 of Chapter 3.

1. To open the project file, choose **Open > Project** from the **Application Menu**; the **Open** dialog box is displayed.

2. In this dialog box, browse to the *C:\rst_2017\c03_rst_2017_tut* location and then choose the *c03_Commercial-Complex_tut1.rvt* (for Imperial) *or M_c03_Commercial-Complex_tut1.rvt* (for Metric) project file. You can also download this file from *http://www.cadcim.com*. The path of the file is as follows: *Textbooks > Civil/GIS > Revit Structure > Exploring Autodesk Revit 2017 for Structure*.

3. Choose the **Open** button from the **Open** dialog box; the selected project file opens in the drawing window in the **Entry** structural plan view.

Creating a Structural Column Type

In this section, you will create a concrete rectangular column type to add to the project view.

1. To create a rectangular column type, choose the **Column** tool from the **Structure** panel of the **Structure** tab; the **Modify | Place Structural Column** tab is displayed.

2. In the **Properties** palette, select the specified option from the **Type Selector** drop-down list.

   For Imperial **Concrete-Rectangular-Column: 12 x 18**
   For Metric **M_Concrete-Rectangular-Column: 300 x 450mm**

3. Choose the **Edit Type** button; the **Type Properties** dialog box is displayed.

4. In this dialog box, choose the **Duplicate** button; the **Name** dialog box is displayed.

5. In the **Name** dialog box, enter **12 x 24** (in Imperial unit system) and **300 x 600mm** (in Metric unit system) in the **Name** edit box and choose **OK**; the **Name** dialog box is closed.

6. In the **Type Properties** dialog box, click in the **Value** field corresponding to the **h** parameter; the existing value, **1' 6"** (for Imperial) and **450.0** (for Metric), is highlighted.

7. Replace the existing value with **2' 0"** (for Imperial) and **600.0** (for Metric), and then choose the **OK** button; the **Type Properties** dialog box closes and preview of the new column type attached with the cursor is displayed in the drawing area.

Adding Structural Columns

In this section, you will add structural columns to the model.

1. In the drawing area, right-click and then choose the **Zoom In Region** option from the shortcut menu; the cursor changes to a zoom icon.

2. Move the cursor and place it between the grid lines marked **1** and **A**. Click at an appropriate location, as shown in Figure 4-39; a rectangular selection window appears.

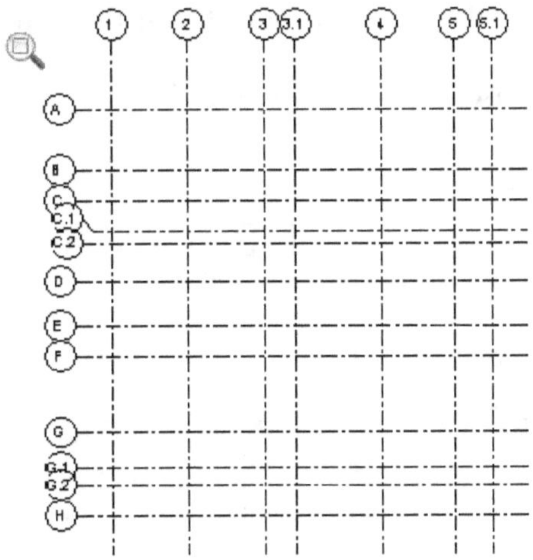

Figure 4-39 *Placing the cursor to mark the first corner of the zoom window*

3. Move the cursor downward toward right. Click between the grid lines marked 5.1 and C.2, refer to Figure 4-40. On doing so, the project view zooms in so that the columns can easily be added at the desired location.

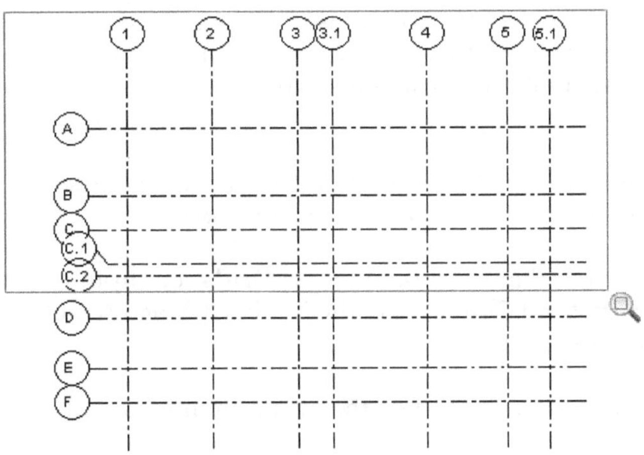

Figure 4-40 *Placing the cursor to mark the other corner of the zoom window*

4. In the **Options Bar**, select the **Rotate after placement** check box and then select the **Height** option from the first drop-down list next to it.

5. Ensure that the **Room Bounding** check box is selected in the **Options Bar**. Place the cursor at the intersection of grid 1 and grid A; the grid **1** and grid **A** are highlighted.

6. Click when both grids are still highlighted; the preview of the column with its center aligned to the intersection of grid 1 and grid A is displayed.

7. Move the cursor up toward the right; a rotation alignment line appears with temporary angular dimension.

8. Type **90**; an edit box with typed value is displayed near the temporary dimension.

9. Press ENTER; a column is inserted at the intersection of grid 1 and grid A, as shown in Figure 4-41.

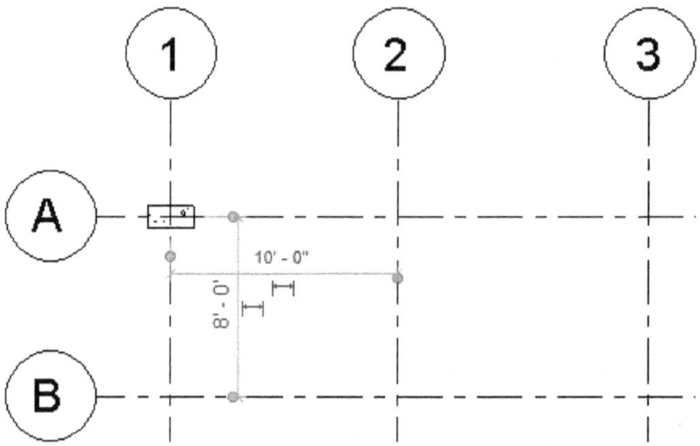

Figure 4-41 *A column added at the intersection of grid 1 and grid A*

10. Repeat steps 5 to 9 to insert other columns with the same type and alignment at different grid intersections, refer to Figure 4-42. Do not exit the **Column** tool even though you have inserted all columns at different grid intersections.

Note

While inserting columns at different grid intersections, you can zoom out the current view and then zoom into the area where you need to insert columns.

11. After inserting all the columns, choose the **At Grids** tool from the **Multiple** panel of the **Modify | Place Structural Column** tab; the **Modify | Place Structural Column > At Grid Intersection** tab is displayed.

12. Place the cursor on grid B. Click when the grid is highlighted; the grid B is selected.

13. Press and hold the CTRL key and click on grid 4, grid 5, and grid G.1 to select them; columns are inserted at the grid intersections of the selected grids, as shown in Figure 4-43

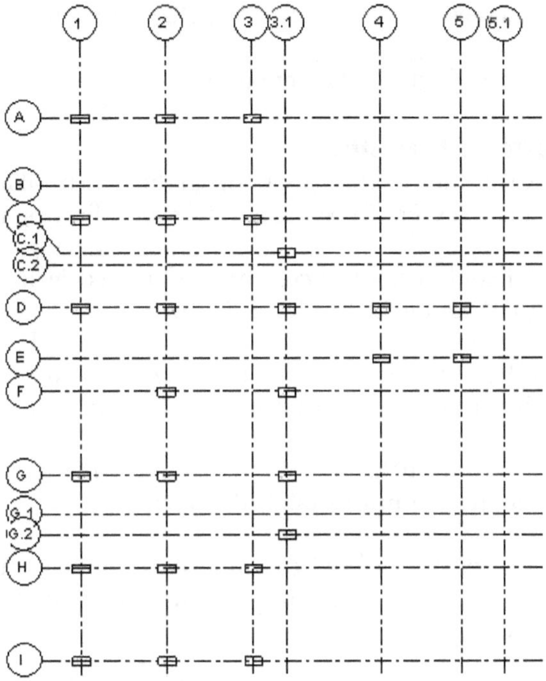

Figure 4-42 Columns of the same type and alignment inserted in the grid

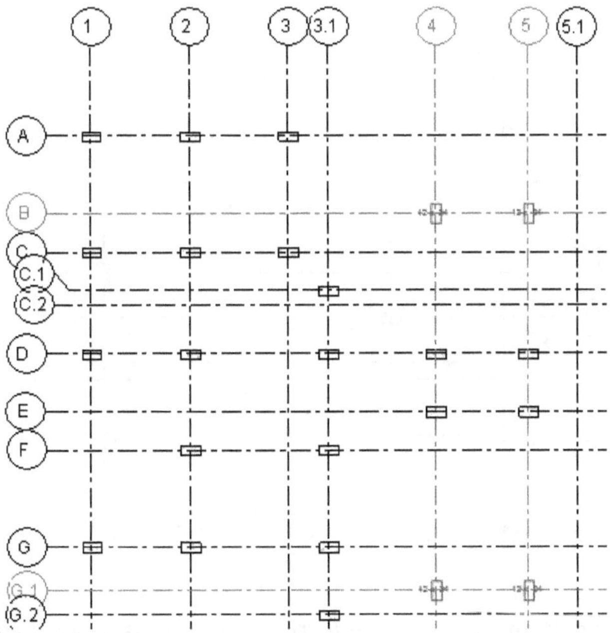

Figure 4-43 Columns added at the intersections of the highlighted grids

14. Now, choose the **Finish** button from the **Multiple** panel in the **Modify | Place Structural Column > At Grid Intersection** tab; the **Modify | Place Structural Column** tab is displayed.

15. Choose the **Modify** button from the **Select** panel.

Modifying the Height of Columns

In this section, to adjust the thickness of the slab (floor) in the structural model, you will modify the height of the inserted columns (except for stairs and lifts) by adjusting their top offset distances.

1. To modify the height of columns, select columns from the drawing window. To do so, place the cursor between grid A and grid 1, refer to Figure 4-44.

2. Next, press and hold the left mouse button and then drag the cursor along the selection window, downward toward the right until it is placed at the location shown in Figure 4-45.

3. Release the mouse button and click; all the columns inside the selection window are selected and the **Modify | Structural Columns** tab is displayed.

4. In the **Properties** palette of the selected columns, click in the value field corresponding to the **Top Offset** instance parameter. Clear the existing value and then enter **-1'** (in Imperial) and **-304.8mm** (in Metric).

5. Next, choose the **Apply** button in the **Properties** palette and then the **Modify** button in the **Select** panel of the **Modify | Structural Columns** tab.

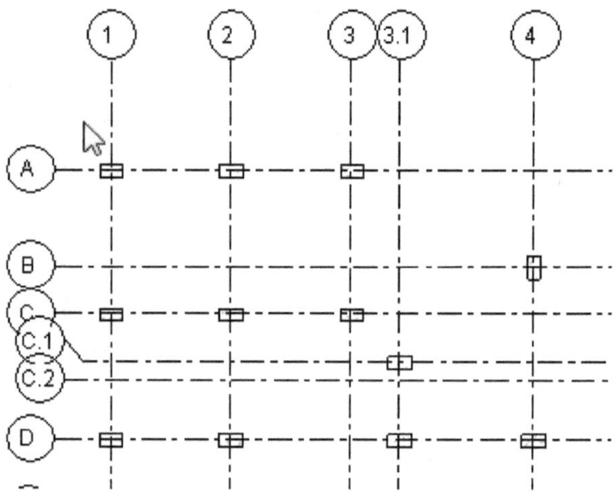

***Figure 4-44** Specifying the first point of the selection window*

6. Now, select all the columns using the selection window, as done earlier.

7. In the **Properties** palette of the selected columns, select the **Entry** option for the **Base Level** parameter and the **Second Floor** option for the **Top Level** parameter if not selected.

8. Choose the **Apply** button; the height of the columns is set between the **Entry** and **Second** floor levels.

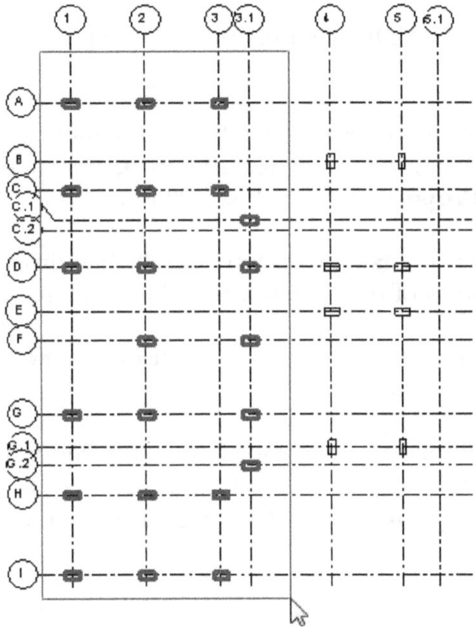

Figure 4-45 *Specifying the other point of the selection window*

Adding Structural Walls

In this section, you will create a structural wall type and add it to the project view.

1. Choose the **Wall: Structural** tool from **Structure >Structure > Wall** drop-down; the **Modify | Place Structural Wall** tab is displayed.
2. In the **Properties** palette, select the desired option from the **Type Selector** drop-down list.
   For Imperial **Exterior - 10" Concrete**
   For Metric **Exterior - 225mm Concrete**
   Now, choose the **Edit Type** button; the **Type Properties** dialog box is displayed
3. Choose the **Duplicate** button in this dialog box; the **Name** dialog box is displayed.
4. Type the specified values in the **Name** edit box of the **Name** dialog box.
   For Imperial **Core-Shaft - 12" Concrete**
   For Metric **Core-Shaft - 300mm Concrete**
   Next, choose **OK**; the **Name** dialog box is closed.
5. In the **Type Properties** dialog box, choose the **Edit** button displayed in the **Value** column of the **Structure** parameter; the **Edit Assembly** dialog box is displayed.

6. In the **Layers** area of the **Edit Assembly** dialog box, click in the edit box in the **Thickness** column corresponding to the **Structure [1]** function and enter **12"**(for Imperial) and **300mm** (for Metric). Next, press ENTER to replace the existing value.

7. Choose the **OK** button; the **Edit Assembly** dialog box closes and the **Type Properties** dialog box is displayed.

8. In the **Type Properties** dialog box, select the **Core-shaft** option from the drop-down list corresponding to the **Function** parameter from the **Construction** area and then choose the **OK** button; the **Type Properties** dialog box is closed.

9. In the **Options Bar**, select the **Height** option from the first drop-down list; a warning message box appears at the bottom of the screen. Close the message box.

10. Select the **Second Floor** option from the second drop-down list in the **Options Bar** and then clear the **Chain** check box.

11. Zoom in the drawing and move the cursor toward the column at the intersection of grid 4 and grid B. Place the cursor on the right face of the column. Click when the right face is highlighted and the **Midpoint** snap is displayed, as shown in Figure 4-46.

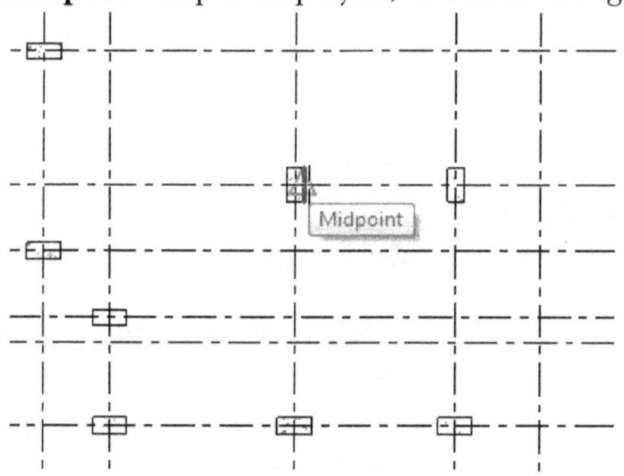

Figure 4-46 *Specifying the start point at the intersection of grid 4 and grid B*

12. Now, move the cursor horizontally toward the right and place it at the midpoint of the left face of the column located at the intersection of grid 5 and grid B, as shown in Figure 4-47. Click when the **Midpoint and Horizontal** snap appears.

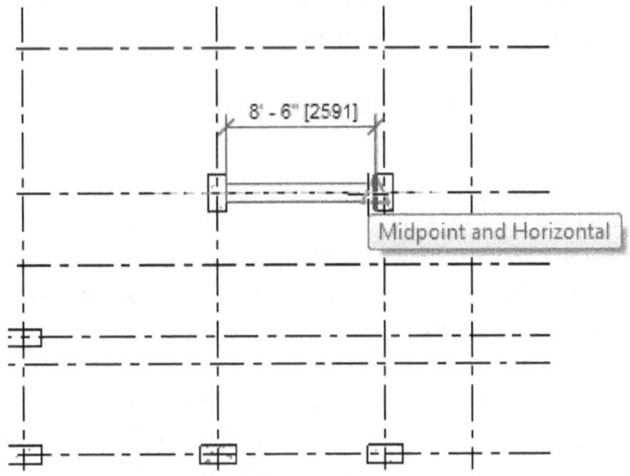

Figure 4-47 *Specifying the endpoint at the intersection of grid 5 and grid B*

13. Move the cursor and place it at the lower face of the column, located at the intersection of grid 5 and grid B. Click when the **Midpoint** snap appears, as shown in Figure 4-48.

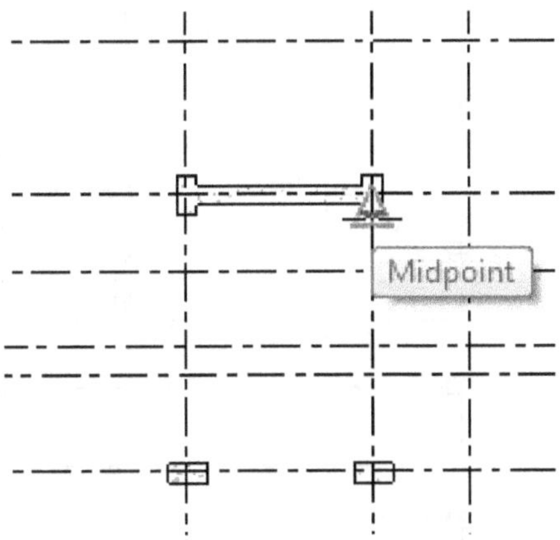

Figure 4-48 *Specifying the start point of the other wall at the midpoint of the lower face of the column at the intersection of grid 5 and grid B*

14. Move the cursor vertically down and place it at the upper face of the column at the intersection of grid 5 and grid D. Click when the **Midpoint and Vertical** snap appears.

15. Next, place the cursor at the lower face of the column, at the intersection of grid 5 and grid D. Click when the lower face is highlighted and the **Midpoint** snap appears.

16. Move the cursor vertically downward and place it at the upper face of the column located at the intersection of grid 5 and grid E. Click when the **Midpoint and Vertical** snap appears at the intersection point.

17. Repeat the method of adding structural walls discussed in Steps 13 to 16 to add the remaining structural walls to the project, as shown in Figure 4-49.

18. Choose the **Modify** button from the **Select** panel to exit the **Wall: Structural** tool.

Saving the Project

In this section of the tutorial, you will save the project file using the **Save As** tool.

1. To save the project, choose **Save As > Project** from the **Application Menu**. On doing so, the **Save As** dialog box is displayed.

2. In this dialog box, browse to the *C:\rst_2017* location and create a folder with the name *c04_rst_2017_tut*. Next, open the *c04_rst_2017_tut* folder and enter **c04_Commercial-Complex_tut1** (for Imperial) and **M_c04_Commercial-Complex_tut1** (for Metric) in the **File name** edit box.

3. Choose the **Save** button; the **Save As** dialog box closes and the project file is saved.

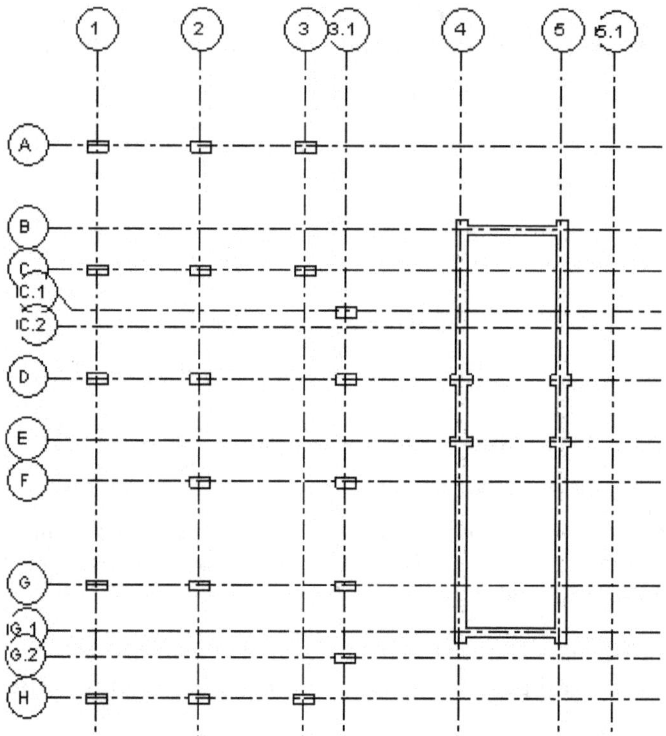

Figure 4-49 *The structural walls added*

Closing the Project

1. To close the project, choose the **Close** option from **Application Menu**.

Tutorial 2 Industrial Complex

In this tutorial, you will add structural columns to the *Industrial-Complex* project created in Tutorial 2 of Chapter 3. Use the project parameters and specifications that are given next:

(Expected time: 30 min)

1. Structural column:

   For Imperial Type: **Concrete-Rectangular-Column**
   Dimension: **24" x 24"**
   Type: **W-Wide Flange-Column**
   Dimension: **W10x 33**

   For Metric Type: **M_Conrete-Rectangular-Column**
   Dimension: **600 x 600mm**
   Type: **M_W-Wide Flange-Column**
   Dimension: **W840 x 251**

2. File name to be assigned:

   For Imperial *c04_Industrial-Complex_tut2.rvt.*
   For Metric *M_c04_Industrial-Complex_tut2.rvt.*

The following steps are required to complete this tutorial:

a. Start a Revit session and open Tutorial 2 of Chapter 3.
b. Create a structural column type.
c. Add columns, refer to Figures 4-50 to 4-51.
d. Modify the height of structural columns by using the **Base Level** parameter.
e. Save the project using the **Save As** tool.
f. Close the project using the **Close** tool.

Opening the Project

In this section, you will open the project file created in Tutorial 2 of Chapter 3.

1. To open the project file, choose **Open > Project** from the **Application Menu**; the **Open** dialog box is displayed

2. In this dialog box, browse to the *C:\rst_2017\c03_rst_2017_tut* location and then choose the *c03_Industrial-Complex_tut2.rvt* (for Imperial) and *M_c03_Industrial-Complex_tut2.rvt* (for Metric) project file. You can also download this file from *http://www.cadcim.com*. The path of the file is as follows: *Textbooks > Civil/GIS > Revit Structure > Exploring Autodesk Revit 2017 for Structure*.

3. Next, choose the **Open** button from the **Open** dialog box; the selected project file opens in the drawing window in the **Entry** structural plan view.

Creating a Structural Column Type

In this section, you will create a concrete rectangular column type and add it to the structural model.

1. To create a structural column type, ensure that the **Entry** structural plan view is opened as the current view and then choose the **Column** tool from the **Structure** panel of the **Structure** tab; the **Modify | Place Structural Column** tab is displayed.

2. In the **Properties** palette, select the specified option from the **Type Selector** drop-down list.
   For Imperial **Concrete-Rectangular-Column: 24 x 30**
   For Metric **M_Concrete-Rectangular-Column: 600 x 750mm**

3. Choose the **Edit Type** button; the **Type Properties** dialog box is displayed.

4. In this dialog box, choose the **Duplicate** button; the **Name** dialog box is displayed.

5. In the **Name** dialog box, enter the specified name in the **Name** edit box.
   For Imperial **24 x 24**
   For Metric **600 x 600mm**
   Next, choose **OK**; the **Name** dialog box closes.

Note

*In case you do not find the above mentioned column in the **Type Selector** drop-down list, you can load the families from the **Load Family** dialog box.*

6. In the **Type Properties** dialog box, click in the **Value** field corresponding to the **h** parameter; the existing **2'6"** (for Imperial) and **750.0** (for Metric) value is highlighted.

7. Replace this value with **2' 0"** (for Imperial) and **600.0** (for Metric) and choose **Apply** and then the **OK** button; the **Type Properties** dialog box closes and the preview of the new column type is displayed with the cursor in the drawing area.

Adding Structural Columns

In this section, you will add new and existing structural column types to the project.

1. In the **Options Bar**, select the **Height** option from the first drop-down list on left and ensure that the **Second Floor** option is selected in the other drop-down list.

2. Choose the **At Grids** tool from the **Multiple** panel of the **Modify | Place Structural Column** tab; the **Modify Placement Structural Column > At Grid Intersection** tab is displayed.

3. Place the cursor over grid C and click when it is highlighted; grid C is selected.

4. Next, press and hold the CTRL key and click over grid A, grid B, grid 7, grid 8, and grid 9 to select them; columns are inserted at the intersections of the selected grids, as shown in Figure 4-50.

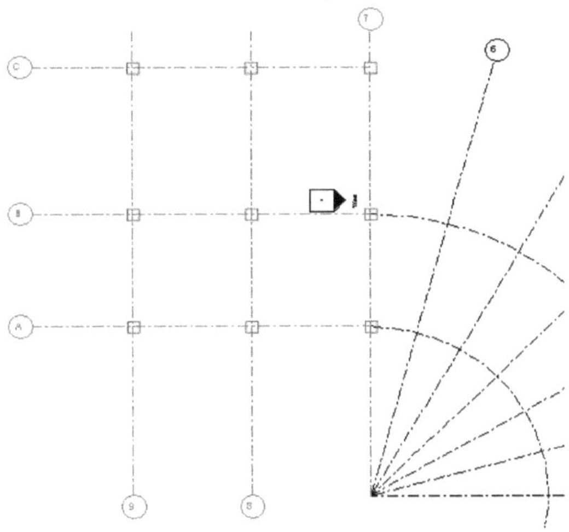

Figure 4-50 *Columns added at the intersection of the selected grids*

5. Choose the **Finish** button from the **Multiple** panel; the **Modify | Place Structural Column** tab is displayed.

6. In the **Properties** palette, select the specified option from the **Type Selector** drop-down list.

   For Imperial **W-Wide-Flange-Column W10x33**
   For Metric **W-Wide-Flange-Column W840x251**

7. Choose the **At Grids** tool from the **Multiple** panel in the **Modify | Place Structural Column** tab; the **Modify | Place Structural Column > At Grid Intersection** tab is displayed.

8 Move the cursor and place it on grid A' and click when it is highlighted; grid A' is selected.

9. Press and hold the CTRL key and click over the grid B', grid 1, grid 2, grid 3, grid 4, grid 5, and grid 6 to select them; columns are inserted at the intersections of the selected grids, as shown in Figure 4-51.

10. Choose the **Finish** button from the **Multiple** panel; the **Modify | Place Structural Column** tab is displayed.

11. Choose the **Modify** button from the **Select** panel.

    Notice that a column is added at the intersection of all grid lines.

12. Select the column and press DELETE to remove it from the model.

Note

*Some of the structural columns added may not be aligned as required in the project. These columns can be aligned by using the **Rotate** tool, which will be discussed later in Tutorial 2 of Chapter 6.*

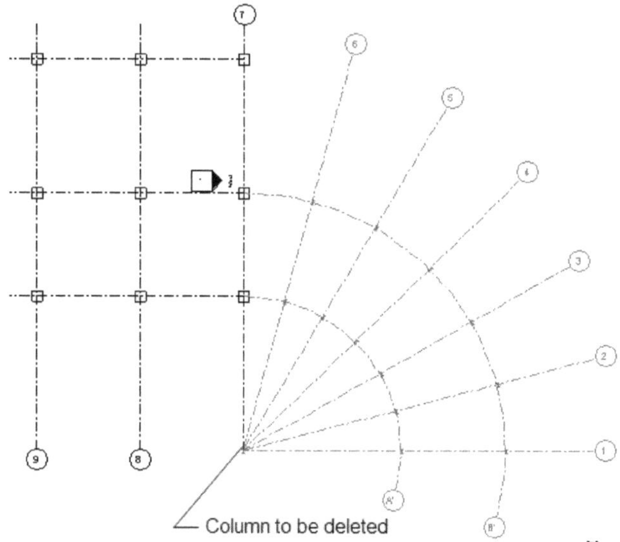

Figure 4-51 The ***W-Wide-Flange-Column W10x33*** *type added to the selected grids*

Modifying the Height of Columns

In this section, you will modify the height of the inserted rectangular concrete columns by adjusting their **Base Level** parameter.

1. To modify the height of the concrete columns, you need to select all the **Concrete-Rectangular-Column: 24x24** type (of Imperial) and **M_Concrete-Rectangular-Column: 600x600mm** type (of Metric) concrete columns from the drawing. To select the desired columns, click on an individual **Concrete-Rectangular-Column: 24x24** (of Imperial) and **Concrete-Rectangular-Column: 600x600mm** (of Metric) column type. Then, press and hold the CTRL key and click on other columns for multiple selection.

2. On selecting all the rectangular concrete columns, the **Modify | Structural Columns** tab is displayed. In the **Properties** palette of the selected columns, click in the value field corresponding to the **Base Level** instance parameter and then select the **T.O.F.** level from the drop-down list displayed.

3. Choose the **Apply** button in the **Properties** palette and then choose the **Modify** button from the **Select** panel in the **Modify | Structural Columns** tab.

Saving the Project

In this section of the tutorial, you will save the project file using the **Save As** tool.

1. To save the project, choose **Save As > Project** from the **Application Menu**; the **Save As** dialog box is displayed.

2. In this dialog box, browse to the *C:\rst_2017\c04_rst_2017_tut* location and enter **c04_Industrial-Complex_tut2** (for Imperial) and **M_c04_Industrial-Complex_tut2** (for Metric) in the **File name** edit box.

3. Now, choose the **Save** button; the **Save As** dialog box closes and the project file is saved.

Closing the Project

1. To close the project, choose the **Close** option from the **Application Menu**.

   The file is closed and this completes Tutorial 2 of Chapter 4.

Self-Evaluation Test

Answer the following questions and then compare them to those given at the end of this chapter:

1. In Revit, you can add three types of columns based on material: Concrete, Steel, and __________.

2. When you start a new project, by default you can use two families for inserting columns: **W-Wide-Flange Column** and __________.

3. The __________ walls play an active role in transferring load from one structural element to the other.

4. In the **Properties** palette for a structural wall, the __________ parameter specifies the level that constrains the base of a wall.

5. The __________ parameter is used to specify the location line of a structural wall that is placed in the model.

6. The __________ instance parameter for a column specifies the plane of the column used for analysis and design.

7 Invoke the __________ sketching tool to sketch straight walls in the drawing.

8. The __________ columns are load-bearing members that are placed in a structural project to add ambience to the structure.

9. In Revit, you cannot create openings in structural columns. (T/F)

10. You can add slanted structural columns to a structural model. (T/F)

Review Questions

Answer the following questions:

1. You can use the __________ tool to create openings in structural columns.

2. The __________ parameter can be used to specify whether to keep rigid link between the analytical model of a beam and column.

3. The __________ parameter specifies the information regarding the phase in which a wall is created.

4. To constrain the **Top** of the wall, you can use the **Top Constraint** and __________ parameters from the **Properties** palette.

5. The __________ tool can be used to create a wall from existing model lines, edge, or grid lines.

6. The __________ tool from the **Modify | Place Structural Column** tab is used to display tags with columns in a drawing.

7. The __________ tool located in the **Draw** panel of the **Modify | Place Structural Wall** tab is used to sketch a circular wall profile in a drawing.

8. The __________ parameter can be used to apply a material to a column.

9. In Revit, the faces of a massing object can be used to create walls. (T/F)

10. In Revit, you can add a structural column to an architectural column. (T/F)

EXERCISES

Exercise 1 — Academic Institution

In this exercise, you will add structural columns with tags and structural walls to the exercise file created in Exercise 1 of Chapter 3, refer to Figure 4-52. Use the following project parameters and specifications: **(Expected time: 1 hr 15 min)**

1. Structural column:

   For Imperial — Type: **Concrete-Rectangular-Column**
   Dimension: **18" x 24"**
   For Metric — Type: **M_Concrete-Rectangular-Column**
   Dimension: **450 x 600mm**

2. Structural wall:

   For Imperial — Type: **Exterior-10" Concrete**.
   For Metric — Type: **Exterior-250mm Concrete**.

3. Project file to be used:

   For Imperial — *c03_Academic-Institution_exer1.rvt.*
   For Metric — *M_c03_Academic-Institution_exer1.rvt*

4. Base and Top constraints for walls and columns are **Entry** and **Second Floor** levels.

5. Columns and walls are to be added in the **Entry** structural plan view.

6. File name to be assigned:

   For Imperial — *c04_Academic-Institution_exer1.rvt.*
   For Metric — *M_c04_Academic-Institution_exer1.rvt*

7. For locations and placement of columns and walls, refer to Figure 4-52.

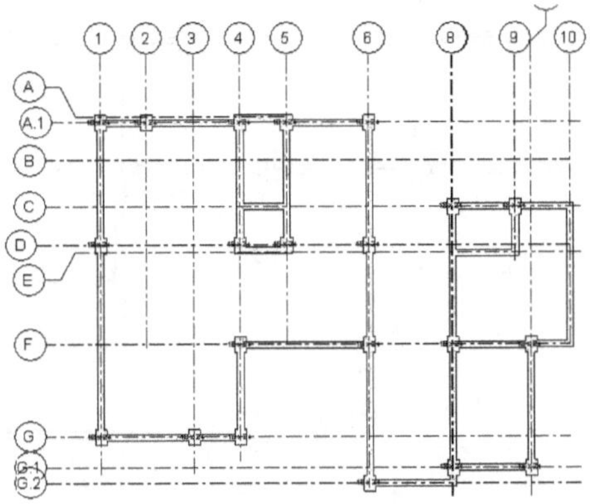

Figure 4-52 *The walls and columns added to the Academic-Institution project*

Exercise 2 Factory Shed

In this exercise, you will add structural columns and walls to the project file created in Exercise 2 of Chapter 3. Refer to Figure 4-53 for grid dimensions and names. Use the following project parameters and specifications: **(Expected time: 45 min)**

1. Project file to be used:
   For Imperial *c03_Factory-Shed_exer2.rvt.*
   For Metric *M_c03_Factory-Shed_exer2.rvt.*

2. Structural column:
   For Imperial Type: **Concrete-Rectangular-Concrete**
   Dimension: **18" x 24"**
   For Metric Type: **M_Concrete-Rectangular-Concrete**
   Dimension: **450 x 600mm**
3. Structural wall:
   For Imperial Type: **Exterior-18" Concrete** (Create a new type)
   For Metric Type: **Exterior-450mm Concrete** (Create a new type)

4. Project file to be used:
   For Imperial *c03_Factory-Shed_exer2.rvt*
   For Metric *M_c03_Factory-Shed_exer2.rvt*

5. Columns and walls need to be added in the **Plinth** structural plan view.

6. File name to be assigned:
   For Imperial *c04_Factory-Shed_exer2.rvt*
   For Metric *M_c04_Factory-Shed_exer2.rvt*

7. Base and Top constraints for walls and columns are the **Plinth** and **Second Floor** levels.

8. For the location and placement of columns and walls, refer to Figure 4-53.

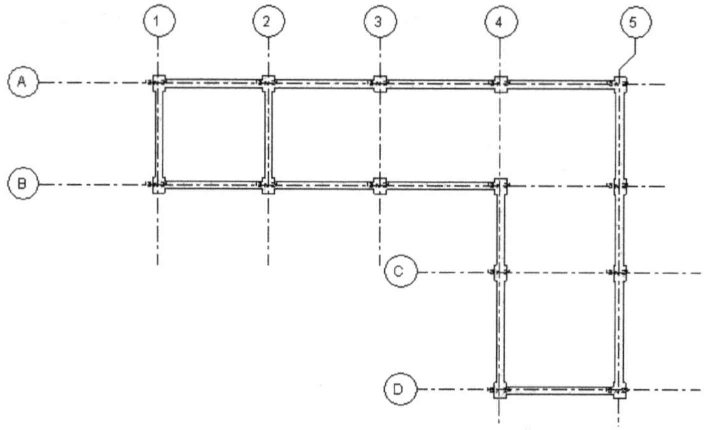

Figure 4-53 *The walls and columns added to the Factory-Shed project*

Exercise 3 Residential Building

Download the *c04_residential_bldg_exer3* (for Imperial) and *M_c04_residential_bldg_exer3* (for Metric) file from *http://www.cadcim.com*. The path of the file is as follows: *Textbooks > Civil/GIS > Revit Structure > Exploring Autodesk Revit 2017 for Structure*. In this exercise, you will add structural columns and walls to the *Residential_Building* project. Refer to Figure 4-54 for grid dimensions, column locations and placement of the structural walls. Use the following project parameters and specifications: **(Expected time: 45 min)**

1. Structural column:
   - For Imperial
     - Type: **Concrete-Rectangular-Column**
     - Dimension: **18" x 24"**
     - Type: **Concrete-Rectangular-Column**
     - Dimension: **24" x 24"**
   - For Metric
     - Type: **M_Concrete-Rectangular-Column**
     - Dimension: **450 x 600mm**
     - Type: **M_Concrete-Rectangular-Column**
     - Dimension: **600 x 600mm**
2. Structural wall:
   - For Imperial: Type: **Retaining -18" Concrete** (Create a new type)
   - For Metric: Type: **Retaining -450mm Concrete** (Create a new type)
3. Project file to be used:
   - For Imperial: *c04_residential_bldg_exer3.rvt.*
   - For Metric: *M_c04_residential_bldg_exer3.rvt.*
4. Top and Base constraints for columns at intersections (B,1.1), (C,1.1), (3,3.1) and (4,3.1) are the **FL** and **Subgrade** levels.
5. Top and Base constraints for other columns are TL and Subgrade.

6. Top offset for columns at intersections (B,1.1), (C,1.1), (3,3.1) and (4,3.1) are:
   For Imperial -2'6".
   For Metric -762mm.

7. Top offset for columns is 2'6" (for Imperial) and 762mm (for Metric).

8. File name to be assigned:
   For Imperial *c04_residential_bldg_exer3a.rvt.*
   For Metric *M_c04_residential_bldg_exer3a.rvt.*

9. For the location and placement of columns and walls, refer to Figure 4-54.

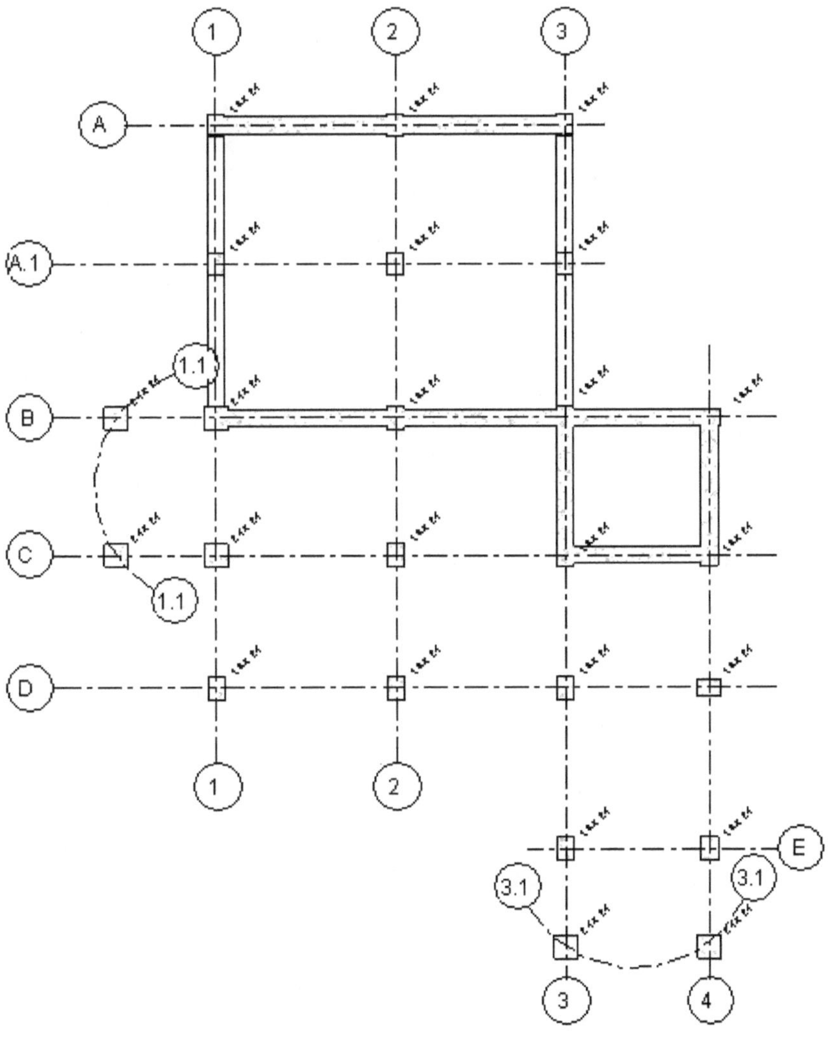

Figure 4-54 *The walls and columns added to the Residential_Building project*

Answers to Self-Evaluation Test

1. Wood, **2. Concrete-Rectangular-Column**, **3.** Load-Bearing, **4. Base Constraint**, **5. Location Line**, **6. Top Vertical Project**, **7. Line**, **8.** Architectural, **9.** F, **10.**T

Chapter 5

Foundations, Beams, Floors, and Open Web Joists

Learning Objectives

After completing this chapter, you will be able to:

- *Create foundations*
- *Add structural floors*
- *Modify structural floors*
- *Add beams*
- *Modify beams*
- *Understand open web joists*

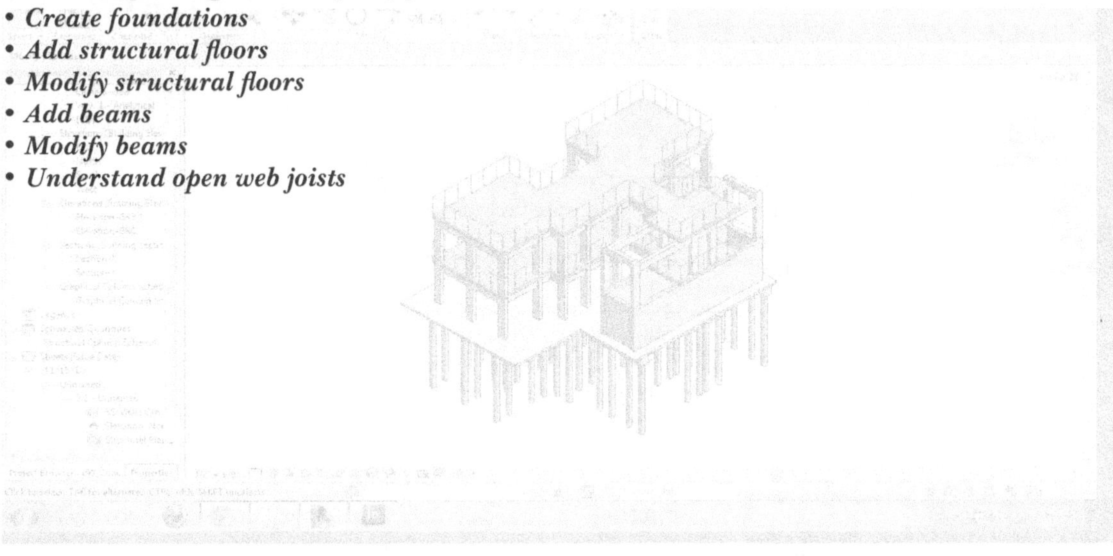

In the previous chapter, you learned to add different structural components such as walls, columns, and so on. In this chapter, you will learn about different types of foundations. You will also understand how to add and modify beams and structural floor and later on add them in your project. Moreover, you will learn to add slab edge to floor and also about the open web steel joists.

FOUNDATIONS

Foundation is one of the important components in a structural model. The function of foundation in any of the construction project is to provide support to the entire building. Foundation acts as an interface between building structure and earth strata for transferring load. For a structural model, you can either have a shallow or deep foundation depending on the bearing capacity of the soil on which the building structure is to be erected. Shallow foundation can be further subcategorized as strip footing and raft footing while deep foundation as pile foundation. Depending on soil bearing capacity (SBC) and structure we can provide either of them.

In Autodesk Revit , you can add foundations of different types such as Isolated, Pile, Wall, and Slab. These types represent different physical foundations that are applicable to sites such as isolated footing (for columns), retaining wall, stepped footing, bearing footings and raft foundation. In this chapter, you will learn how to add both the shallow and deep foundations.

Foundation Wall

After adding structural walls, you can add a foundation wall to the existing structural model. This wall serves as a basement wall or a shear wall that goes down below the ground level to a level from where the top of the foundation starts. To add a foundation wall, first you need to create the required levels to constrain the ends of the wall. For example, if you want to create a basement wall for an existing structural model, you need to have two levels, one level to constrain the top of the basement wall and the other level to constrain the bottom of the basement wall. The top of the basement wall may be at a level above or at ground level whereas the bottom of this wall will be at a level that coincides with the top of the footing. Figure 5-1(a) and Figure 5-1(b) display a basement wall with its top constrained to the **Entry Level** level and its bottom constrained to the **T.O. Footing** level in the imperial and metric form respectively.

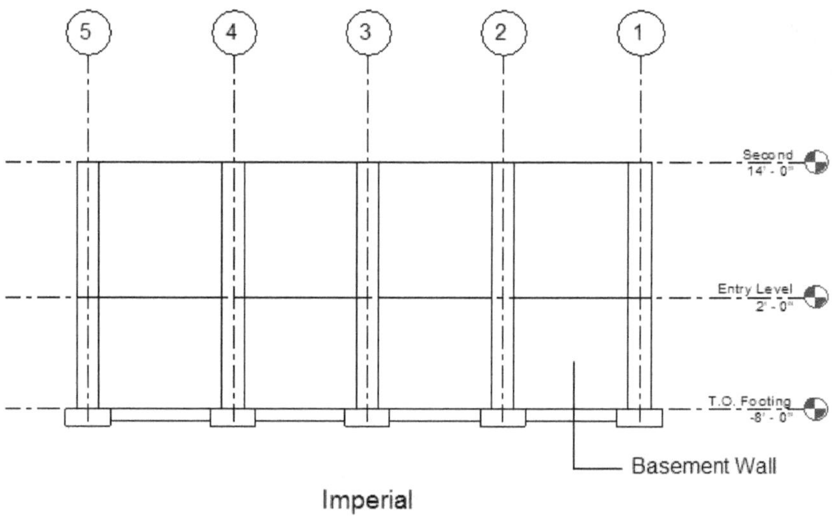

Figure 5-1(a) *The basement wall with the top and bottom constraints imperial*

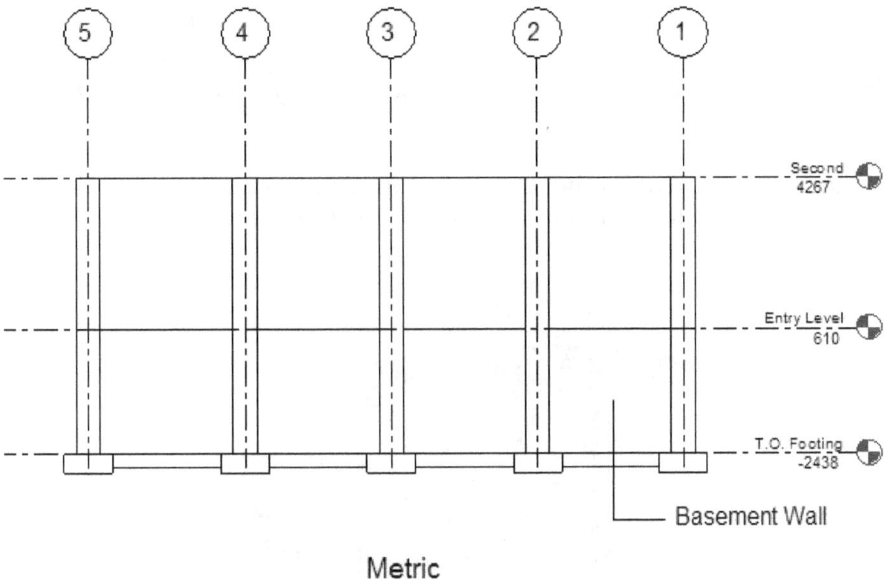

Figure 5-1(b) *The basement wall with the top and bottom constraints*

To add a foundation wall to a structural model, change the display of the current view to a level such that it displays the top of the wall. For example in Figure 5-1, the level that represents the top of the wall is **Entry Level**. Now, after displaying the current view, invoke the **Wall: Structural** tool from **Structure > Structure > Wall** drop-down; the **Modify | Place Structural Wall** tab will be displayed. In the **Type Selector** drop-down list, select the **Foundation-12" Concrete** option for Imperial unit system or select the **Foundation-300mmConcrete** for Metric unit system, refer to Figure 5-2. After selecting the desired type for the foundation wall, notice that in the **Options Bar**, the **Depth** option is selected by default from the first drop-down list from the left. To assign a level that will constrain the depth to the foundation wall, you can select a level from the drop-down list displayed next to it, refer to Figure 5-3. The selected level (for example the **T.O. Footing** level) will have a lower elevation value as compared to the current level. Alternatively, you can assign an unconnected depth to the foundation wall by selecting the **Unconnected** option in the second drop-down list from the left in the **Options Bar** and then specifying a value in the edit box displayed next to it. The unconnected depth will be measured in reference to the elevation of the current view.

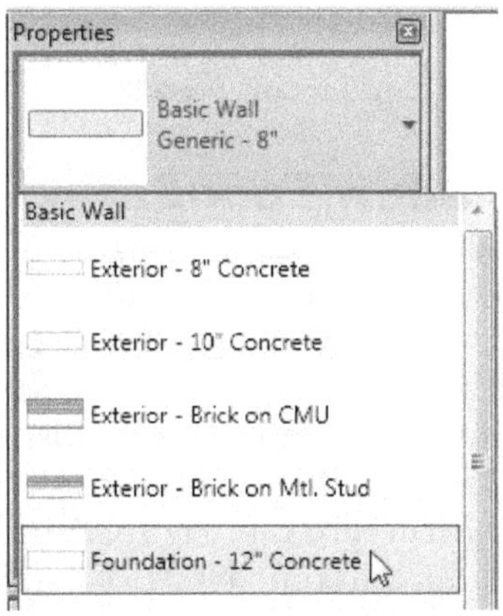

Figure 5-2 *Selecting the* ***Foundation - 12" Concrete*** *option from the* ***Type Selector*** *drop-down list*

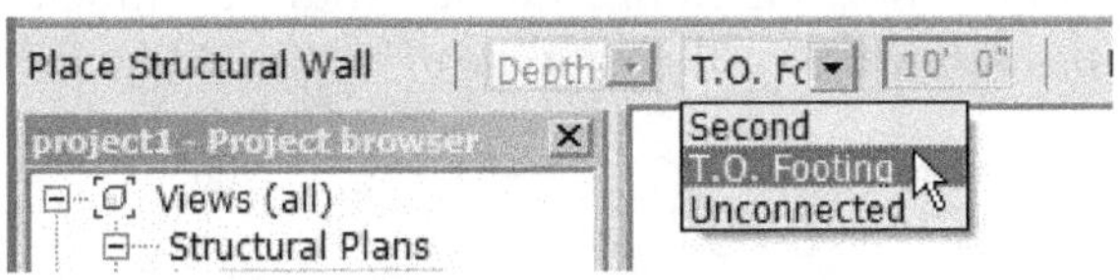

Figure 5-3 *Selecting the* ***T.O. Footing*** *option from the drop-down list in the* ***Options Bar***

Note

You cannot enter negative values or 0 in the edit box to specify the unconnected depth of the foundation wall. If you do so, a warning message box will appear asking you to enter a positive value less than 30000' in Imperial system and for Metric system enter a positive value less than 9144000 mm.

Now, after setting the constraint for the base of the foundation wall, you can click in the drawing area and start creating the foundation wall. Figure 5-4 displays a foundation wall added to a structural model.

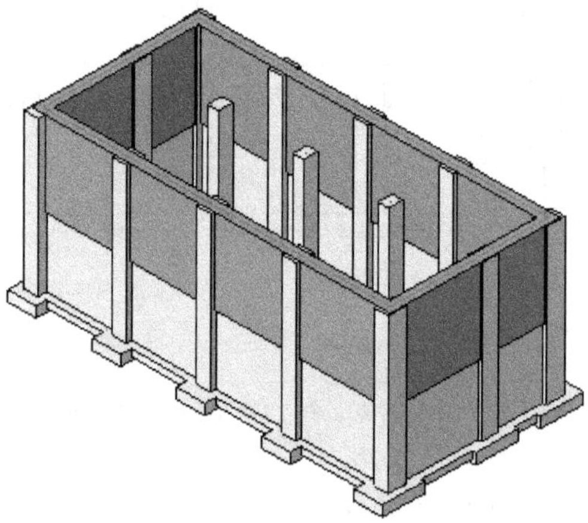

Figure 5-4 *The foundation wall added to a structural model*

Shallow Foundations

Shallow foundations are used to transfer load from a building to the earth strata near the ground surface. Shallow foundations include strip footing foundations, spread footing foundations, mat-slab, and slab-on-grade foundations. In Revit , you can add a shallow foundation as a wall foundation (strip footing foundation), isolated foundation (spread footing foundation) for columns, or foundation slabs. To add these foundations, you need to invoke various tools such as **Isolated**, **Wall**, and **Foundation Slab** from the **Foundation** panel in the **Structure** tab. Figure 5-5 shows various tools that can be invoked to add different types of foundations to a structural model. In the **Foundation** panel, you can invoke the **Wall** tool to add a wall foundation in the existing structural model. You can use this tool to add two types of wall foundations: Bearing footing foundation and Retaining footing foundation. Each of these foundations has separate element properties and placement methods. These properties and methods will be discussed later in this chapter. You can use the **Isolated** tool to add the spread or isolated foundations to the columns of the structural model. The element properties and the placement method for this type of foundation will also be discussed later. Note that the wall and the isolated footings are dependent on the host element to which they are attached. Figure 5-6 shows the associativity of a structural wall with a wall foundation.

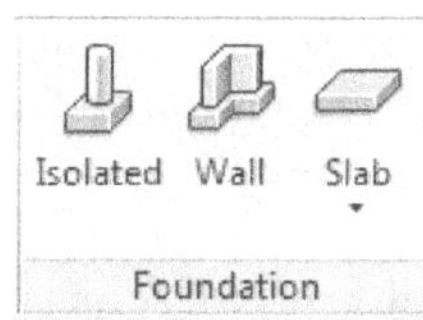

Figure 5-5 *Various tools to add foundation*

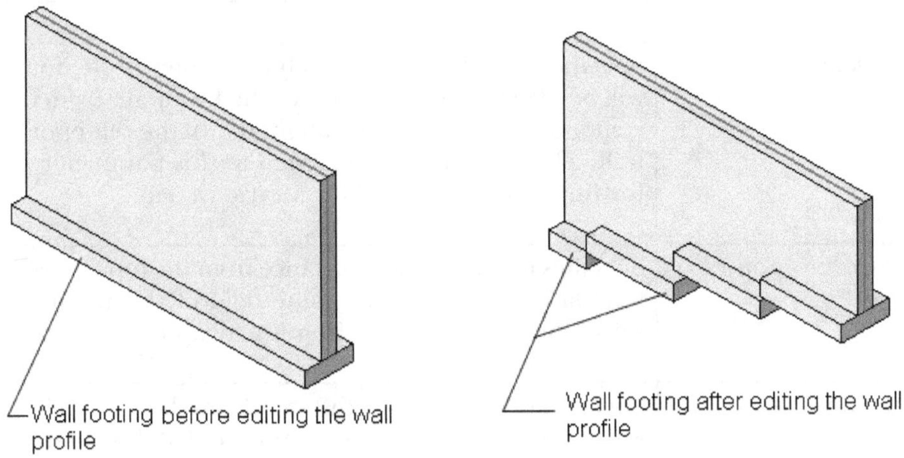

***Figure 5-6** The wall foundation before and after editing the wall profile*

The **Slab** tool in the **Foundation** panel can be used to add a foundation slab to the structural model. Unlike the isolated and wall foundations, the slab foundations are not attached to any host element. Slab foundations are created independent of other structural elements present in a structural model. The placement and the properties of this type of foundation will be discussed later in this chapter.

Wall Footing Foundation

A wall foundation is a footing that is attached to a wall. In actual practice, strip footing is a good example of this kind of foundation. The advantage of using Revit over other similar software to place wall footing is that they are associated with the host wall. It means the foundation wall will remain attached to its host wall even if the host wall is moved, or its shape or type is changed. In Revit , you can add two types of wall foundations: bearing foundation and retaining foundation. These foundations differ from each other in terms of their functions and geometries. The method of adding and placing these foundations is discussed in the next sections.

Tip

The footings are attached to columns or walls. This means that a relationship is established between the footing and its host element (wall or column). As a result, whenever there is a change in the shape or position of the wall or the column, the footing also changes accordingly.

Adding Bearing Foundations

In a structural model, you can add a bearing shallow foundation to an existing structural wall by invoking the **Wall** tool from the **Foundation** panel of the **Structure** tab. On invoking this tool, the **Modify | Place Wall Foundation** tab will be displayed. Next, select the **Bearing Footing - 36" x 12"** option (for Imperial) or **Bearing Footing - 900 x 300** (for Metric) from the **Type Selector** drop-down list in the **Properties** palette. Now, you can change the instance parameters such as the eccentricity, cover lengths of the rebar (at top, bottom, or from other faces of the foundation), and analytical settings of the selected bearing footing type.

To change the instance parameter, you can use various options in the **Properties** palette. The different instance parameters available in this palette are discussed next.

Parameter	Description
Eccentricity	Specifies the distance at which the center of the foundation will be offset from the center of the host wall. By default, the center of the foundation is aligned with the center of the host wall. The default value assigned to this parameter is "**0' 0"**" for Imperial system or **0** for Metric system.
Rebar Cover - Top Face	Specifies the rebar cover distance from the top face of the wall foundation. Click on the value field of this parameter and select the desired option from the drop-down list displayed.
Rebar Cover - Bottom Face	Specifies the rebar cover distance from the bottom face of the wall foundation. Click on the value field of this parameter and select the desired option from the drop-down list displayed.
Rebar Cover - Other Face	Specifies the rebar cover distance from the other face of the wall foundation. Click on the value field of this parameter and select the desired option from the drop-down list displayed.
Enable Analytical Model	Specifies whether to enable analytical model of the wall foundation. Select the check box in its value field to enable the analytical model.

After assigning the instance parameters to the bearing foundation from the **Properties** palette, you can assign its type parameters in the **Type Properties** dialog box. To invoke this dialog box, choose the **Edit Type** button from the **Properties** palette or choose the **Type Properties** tool from the **Properties** panel of the **Modify | Place Wall Foundation** tab. Using this dialog box, you can access the type properties of the wall foundation and change them based on your project requirement. Various type properties in the **Type Properties** dialog box are discussed next.

Parameter	Description
Structural Material	Specifies the material assigned to the concrete wall foundation. To change the material of the current foundation type, choose the Browse button displayed in the **Value** field of this parameter; the **Material Browser** dialog box will be displayed. In this dialog box, you can specify settings to change the appearance and physical properties of the material to be assigned.
Structural Usage	Specifies the type of wall usage for the foundation wall. For a bearing foundation wall, the value assigned to this parameter is **Bearing**.
Default End Extension Length	Specifies the distance of the foundation wall that will extend beyond the end of the wall.

Width	Specifies the width of the bearing wall foundation.
Foundation Thickness	Specifies the thickness of the bearing wall foundation.
Do Not Break At Inserts	Specifies whether the wall foundation will remain continuous beneath the opening of the structural wall when doors and windows are added. Select the check box displayed in the **Value** field of this parameter to enable the wall foundation to continue beneath the opening of the structural wall, refer to Figure 5-7.

In a structural project, you may need to add different types of bearing wall foundations based on your requirement. To add a different type of bearing wall foundation, first you need to create it. To create a new type, invoke the **Type Properties** dialog box from the **Properties** palette of the bearing foundation wall. In the **Type Properties** dialog box, choose the **Duplicate** button; the **Name** dialog box will be displayed. Replace the existing name in this dialog box with a suitable name for the new foundation type and then choose **OK**; the **Name** dialog box will close. In the **Type Properties** dialog box, you will notice that the name of the bearing foundation type in the **Type** drop-down list has been changed to the new name that you have entered in the **Name** dialog box. Next, change the desired type parameter from its corresponding **Value** fields and choose the **Apply** button; the modified parameters will be applied to the new type. To close the **Type Properties** dialog box, choose the **OK** button.

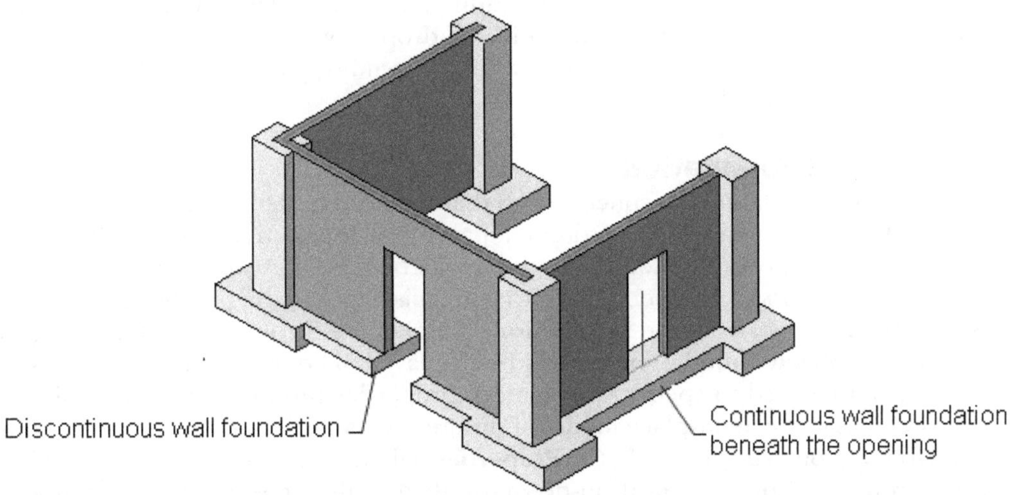

Figure 5-7 *Foundations with the **Do not Break At Inserts** parameter enabled and disabled*

Note

*You can assign a new bearing foundation type to an existing bearing foundation. To do so, select the desired wall foundation from the structural model; the **Modify / Structural Foundations** tab will be displayed. In the **Properties** palette, select the newly created type from the **Type Selector** drop-down list. On doing so, the selected wall foundation will inherit the properties of the newly created type and change according to the parameters assigned to the new property.*

After specifying the instance and type parameters, you can add the selected bearing foundation type to a structural wall. To do so, click on an existing structural wall in the model; the bearing foundation will be added to the entire length of the selected wall, refer to Figure 5-7. Note that you can also select multiple walls while adding wall foundations. To do so, choose the **Select Multiple** button from the **Multiple** panel of the **Modify | Place Wall Foundation** tab and then make a window selection or press the CTRL key for selecting multiple structural walls to host wall foundations. You will notice that on choosing the **Select Multiple** button; the other two buttons in this panel, **Finish** and **Cancel**, will be activated. Choose the **Finish** button to finish the multiple selection process or choose the **Cancel** button to terminate the multiple selection process.

Tip

*The added foundation will be visible only in the plan view that has view depth to display the bottom of the wall. To display the added foundation in any other plan view, make the plan view active and then choose the **Edit** button in the value field of the **View Range** parameter in the **Properties** palette. On doing so, the **View Range** dialog box will be displayed. In this dialog box, select the **Unlimited** option from the **Level** drop-down list in the **View Depth** area and then choose **OK** to close the dialog box. On doing so, you will notice that the added foundation is displayed along its host wall.*

Adding Retaining Wall Foundations

You can add the retaining footing foundation to a structural model in a similar way as you did in the Bearing footing foundation. To add a retaining wall foundation to a structural wall, invoke the **Wall** tool from the **Foundation** panel of the **Structure** tab and then select the retaining wall type from the **Type Selector** drop-down list in the **Properties** palette. After selecting the desired option from the drop-down list, you can add the foundation to an existing structural wall.

Isolated Footing Foundation

An isolated footing foundation is a concrete slab that is placed under a column subjected to the application of load. The isolated foundation is a part of the foundation family in Revit . Unlike the wall footing foundation, these foundations may not be hosted. Also, these foundation types can be inserted into a model independently. To add an isolated footing to the model, you need to invoke the **Isolated** tool from the **Foundations** panel of the **Structure** tab. On doing so, the **Modify | Place Isolated Foundation** tab will be displayed. The options in this tab, as shown in Figure 5-8, can be used to specify the element properties (instance and type), alignment, locations, as well as to create in-place model of different geometries or load existing family for the isolated foundation to be placed. In the **Properties** palette, you can select an existing isolated footing type from the **Type Selector** drop-down list. By default, the **Footing-Rectangular 72" x 48" x 18"** option (in Imperial) or **M_Footing-Rectangular 1800 x 1200 x 450 mm** (in Metric) is selected in this drop-down list.

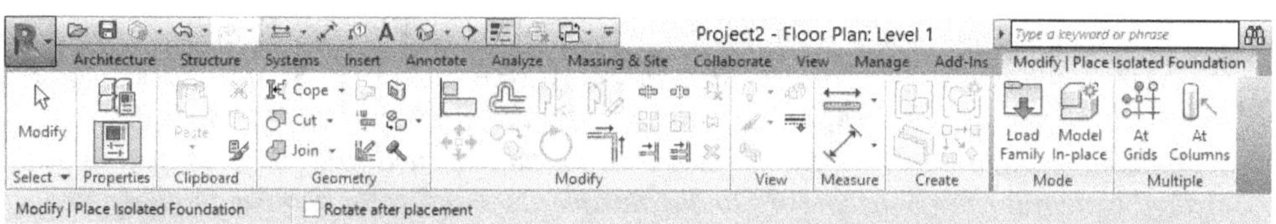

***Figure 5-8** Various options displayed in the **Modify / Place Isolated Foundation** tab*

You can also add other footing type to the **Type Selector** drop-down list. To do so, you need to load families from the existing foundation library. To load a family, choose the **Load Family** tool from the **Mode** panel of the **Modify | Place Isolated Foundation** tab; the **Load Family** dialog box will be displayed. In this dialog box, browse to **US Imperial** or **US Metric** and then choose the **Structural Foundations** folder; the files of foundation families will be displayed, as shown in Figure 5-9.

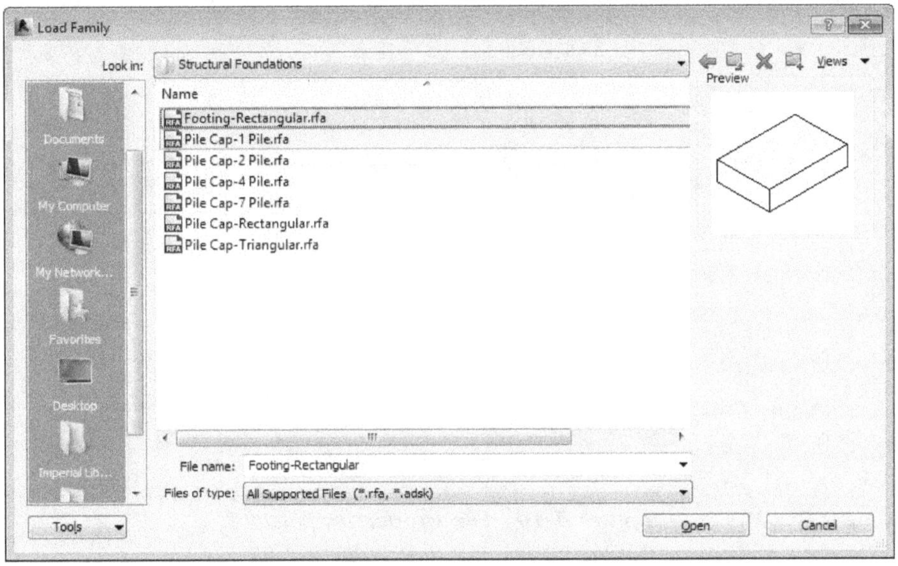

Figure 5-9 *Various families in the* ***Load Family*** *dialog box*

You can select any of the families from this list. To add multiple families, select multiple families from the dialog box and then choose the **Open** button; the **Load Family** dialog box will close and the selected families will be loaded into the current drawing file. You can select the desired foundation family type from the **Type Selector** drop-down list.

After selecting the desired type from this list, you can modify the existing element properties (instance and type) using various options in the **Properties** palette, as shown in Figure 5-10.

In the **Properties** palette, the **Structural Material** parameter can be used to specify the constructional material associated with the isolated footing foundation to be added in the project. To specify a different constructional material, click on the value field corresponding to the **Structural Material** parameter and then choose the **Browse** button that will be displayed in it; the **Material Browser** dialog box will be displayed. You can use this dialog box to specify different materials for the graphic, render, and physical appearances of the foundation. After specifying the desired materials, choose the **OK** button to close the **Material Browser** dialog box.

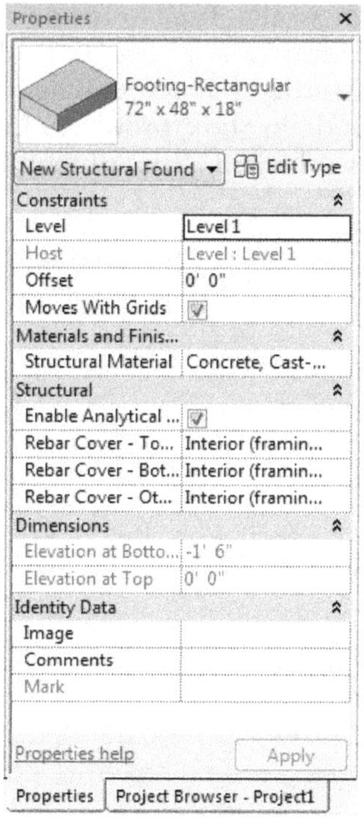

***Figure 5-10** The **Properties** palette displaying the instance properties for the isolated footing*

In the **Properties** palette of the isolated footing foundation, the parameters listed under the **Structural** head are used to set the rebar cover distance of the footing foundation. The parameters listed under the **Structural** head of this palette are **Rebar Cover - Top Face**, **Rebar Cover - Bottom Face**, and **Rebar Cover - Other Faces**. These parameters can be modified by clicking on their respective value fields and selecting the desired options from the drop-down lists displayed. To change the analytical properties, select **New Analytical Isolated Foundation** option from the drop-down list located on the left of the **Edit Type** button in the **Properties** palette. Various parameters related to analytical properties of foundation are displayed under **Identity Data**, **Analytical Model** and **Analytical Properties** heads. The **Analyze As** parameter can be used to specify whether the concerned isolated footing foundation should be analyzed as a Foundation or not. By default, the **Foundation** option is specified in the value field of this parameter. To change the value in the value field, click on it and then select the **Not for analysis** option from the drop-down list displayed.

Slab Foundation

Ribbon: Structure > Foundation > Slab drop-down > Structural Foundation: Slab

Slab foundations are used to transmit the load of the superstructure elements (wall, columns, and other structural elements) to the earth. Slab foundations can be constructed near the ground surface or at the bottom of the basements of a structure as in high-rise buildings. These foundations can be several meters thick with extensive reinforcement to ensure that the load is uniformly transferred to the earth strata.

To add a slab foundation to a structural model, invoke the **Structural Foundation: Slab** tool from the **Structure >Foundation > Slab** drop-down; the **Modify | Create Floor Boundary** tab will be displayed. You can use this tab to invoke the sketching tools to draw the boundary of the foundation slab as well as edit the sketched boundary. You can also use the **Modify | Create Floor Boundary** tab to modify the instance and type properties of the foundation slab to be created.

The **Draw** panel of the **Modify | Create Floor Boundary** tab contains the tools that can be used to sketch the boundary and the slope arrows for the foundation slab. The slope arrows define the sloping direction of the foundation slab. Before you sketch the boundary line of the slab foundation, you need to specify the instance and type parameters required for it. Various options required for specifying the instance and type properties of the foundation slab are discussed next.

Defining Instance Properties

To specify the instance properties for the foundation slab, you can use various options in its **Properties** palette, as shown in Figure 5-11. The instance parameters in the **Properties** palette are categorized under seven heads: **Constraints, Structural, Dimensions, Identity Data, Phasing, Analytical Properties, Analytical Alignment** and **Analytical Model**, refer to Figure 5-11. The heads and the options in them are discussed next.

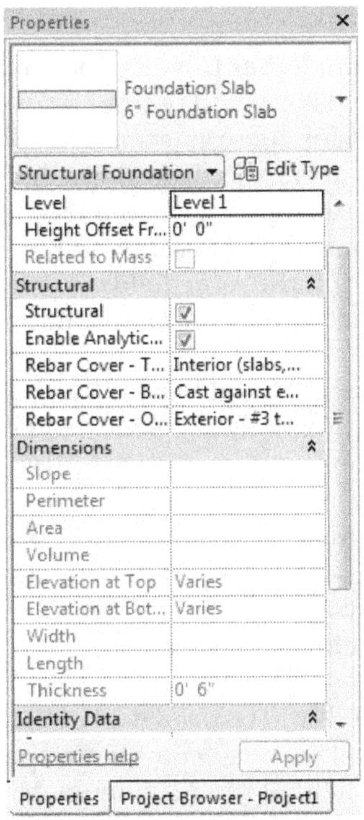

***Figure 5-11** The **Properties** palette displaying the instance properties for the slab foundation*

Constraints: The **Constraints** head contains three parameters: **Level, Height Offset From Level**, and **Related to Mass**. To specify a desired level to constrain the foundation slab, click on the value field corresponding to the **Level** parameter; a drop-down list will be displayed. Select the desired option from the drop-down list to constrain the foundation slab. After specifying the desired level for the foundation slab, you can specify the offset distance of the foundation slab from the constraint level. The offset distance will be measured from the top surface of the slab to the constraint level. To assign an offset distance to the foundation slab, click on the value field corresponding to the **Height Offset From Level** parameter and enter the desired distance value in this field. The **Related to Mass** parameter is a read-only parameter and shows whether or not a foundation slab has been created from a mass object.

Structural: In the **Properties** palette for a foundation slab instance, you can specify the structural parameters under the **Structural** head. This head contains five parameters: **Structural**, **Rebar Cover - Top Face**, **Rebar Cover - Bottom Face**, **Rebar Cover - Other Faces**, and **Enable Analytical Model**. The **Structural** parameter is used to specify whether or not the foundation slab will have an analytical model that can be exported to another analytical software for structural analysis. To specify the **Structural** parameter, you can use the check box displayed in the value field corresponding to it. By default, this check box is selected, which indicates that the foundation slab will have an analytical model and can be exported to an analytical software for analysis. You can clear this check box if you do not want it to be used for analysis.

The **Rebar Cover - Top Face** and **Rebar Cover - Bottom Face** parameters are used to specify the rebar cover distances from the top and bottom faces of the foundation slab, respectively. To specify appropriate values for these parameters, click on their respective value fields and select the desired option from the drop-down lists displayed. Apart from specifying the rebar cover distance from the top and bottom faces of the foundation slab, you can also specify the rebar cover distance in the foundation slab in reference to the face of the adjacent foundation slab. To do so, click on the value field corresponding to the **Rebar Cover - Other Faces** parameter and select the desired option from the drop-down list displayed.

Analytical Model: The **Analytical Model** head will be displayed in the **Properties** palette for the analytical model of the foundation slab. This head contains the **Analyze As** parameter. By default, the **Foundation** option is specified in this parameter. To modify the default value, click on the value field corresponding to this parameter; a drop-down list will be displayed. You can select the **Slab on Grade** option from this list. Note that if you specify the **Foundation** option for the **Analyze As** parameter, the foundation slab will bear the load of other elements connected to it. If you specify the **Slab On Grade** option, the foundation slab will only bear its own weight.

Analytical Alignment: This head will be displayed in the **Properties** palette for the analytical model of the foundation slab. This head contain two parameters: **Alignment Method** and the **z Projection**. In the value field of the **Alignment Method** parameter, the **Auto - Detect** option is selected by default. As a result, Revit will automatically detect the projection plane of the foundation slab. Also, in the **Properties** palette, you will notice that **z Projection** parameter is disabled. To enable the **z Projection** parameter, select the **Projection** option from the drop-down list corresponding to the **Alignment Method** parameter. To assign a value to the **z Projection** parameter, click on its value field, a drop-down list will be displayed. From the drop-down list, you can select any of the options to define the projection plane of the slab.

Analytical Properties: This head will be displayed in the **Properties** palette for the analytical model of the foundation slab. This head contains four parameters: **Family Type**, **Physical Properties Asset**, **Area**, and **Perimeter**. These parameters are read-only parameters. In the value field of the **Area** and **Perimeter** parameters, you can view the area and perimeter of the slab. The **Family Type** parameter in the **Properties** palette displays the family of slab.

Dimensions: The parameters under the **Dimensions** head display the values associated with the slope, perimeter, area, volume, width, length, elevation at bottom, and thickness of an existing selected slab. All parameters displayed under this head are read-only parameters.

Identity Data: The **Identity Data** head in the **Properties** palette contains three parameters: **Image**, **Comments**, and **Mark**. In the **Image** parameter, you can place the image by choosing the **Browse** button from the corresponding value field. You can use the other two parameters to write a general comment and to specify the mark for the foundation slab, respectively. Note that the value specified for the **Mark** parameter should be unique for each foundation slab element.

Phasing: The parameters under the **Phasing** head for the foundation slab are used to specify the creation and demolition phases of the foundation slab. Phasing allows us to complete our project on time.

Defining Type Properties

After specifying the desired instance parameters of the foundation slab from the **Properties** palette, choose the **Edit Type** button to invoke the **Type Properties** dialog box, as shown in Figure 5-12. In addition, you can use the options in this dialog box to modify the type properties of the foundation slab to be created. You can use this dialog box to add a new family type of a foundation slab and change its parameters as per your requirement. To do so, choose the **Duplicate** button; the **Name** dialog box will be displayed. In this dialog box, enter a name of the new type in the **Name** edit box and then choose the **OK** button; the **Name** dialog box will be closed and new type will be added. You can now use the various options in the **Type Properties** dialog box to assign different properties to the new foundation slab type. Apart from the new family type that has been added, you can select any of the options from the **Type** drop-down list to specify the foundation type to be assigned to the foundation slab. The options that you can select from the drop-down list for Imperial unit system are **6" Foundation Slab**, **8" Foundation Slab**, **10" Foundation Slab**, and **12" Foundation Slab** and the options for the Metric unit system are **150mm Foundation Slab**, **200mm Foundation Slab**, **250mm Foundation Slab, and 300mm Foundation Slab**. After selecting the desired type, you can edit its type properties such as its thickness, structural details. Now, choose the **OK** button to return to the **Type Properties** dialog box. Notice that the option selected in the **Type** drop-down list in the **Type Properties** dialog box displays the new name that you have entered in the **Name** dialog box. Also, the type parameters are displayed with their values in a tabular format in the **Type Properties** dialog box. The type parameters are displayed under five heads: **Construction**, **Graphics**, **Material and Finishes**, **Identity Data**, and **Analytical Properties**. These heads are discussed next.

Note

The new family type created will be project specific.

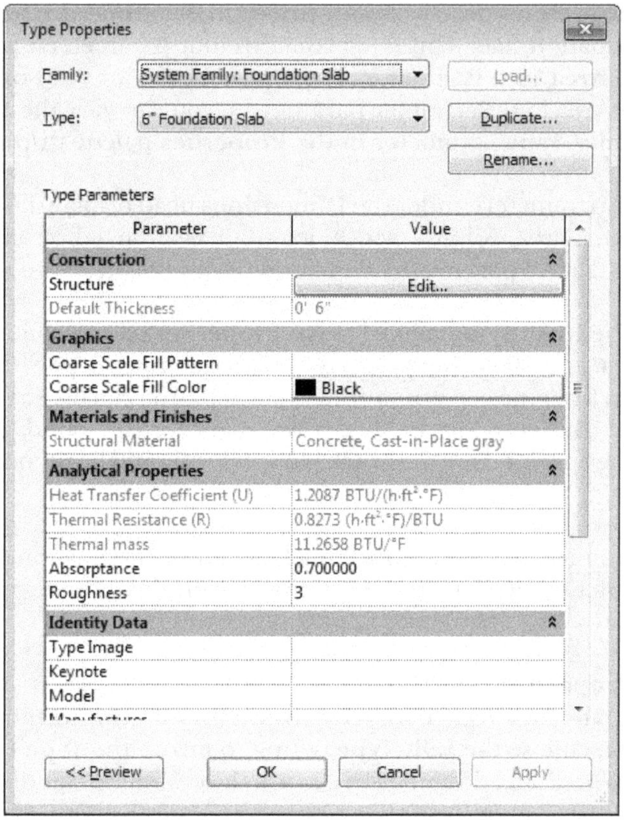

***Figure 5-12** The **Type Properties** dialog box of the foundation slab*

Construction: The **Construction** head contains two parameters: **Structure** and **Default Thickness**. The **Default Thickness** parameter remains inactive. This parameter displays the thickness of the foundation slab that can be defined in the **Edit Assembly** dialog box and is discussed next.

Note

*The value displayed for the **Default Thickness** parameter indicates the thickness of a foundation slab type which is determined by the cumulative thickness of its layers.*

To change the value of the **Structure** parameter, choose the **Edit** button displayed in the **Value** column corresponding to it; the **Edit Assembly** dialog box will be displayed, as shown in Figure 5-13.

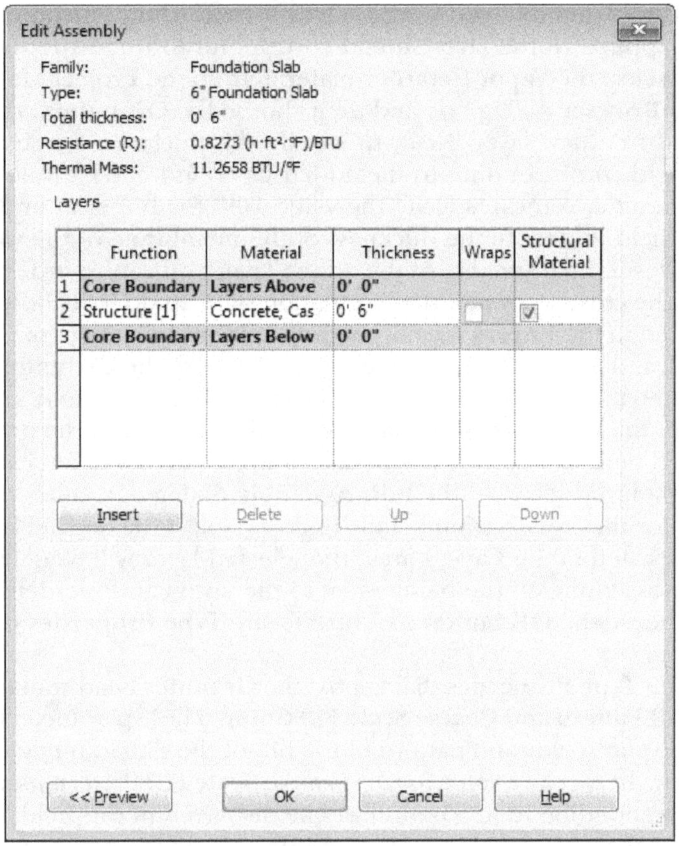

***Figure 5-13** The **Edit Assembly** dialog box for the slab foundation*

In this dialog box, the **Layers** area displays the layers that constitute the structural detail of the foundation slab. The structural detail of the foundation slab is displayed in tabular format. The table contains five columns: **Function**, **Material**, **Thickness**, **Wraps**, and **Structural Material**. Each row of the table represents the layer in the cross-section of the foundation slab. Note that the top and bottom rows in the table denote the outermost layers between which the other layers are wrapped. Also, these layers are used as references while adding a new layer or deleting an existing layer. To add a layer to the existing cross-section of the foundation slab, choose the **Insert** button from the **Layers** area of the **Edit Assembly** dialog box; a new row denoting a new layer (**Structure[1]**, default name) will be added in the table.

The **Function** field is used to specify the exact function of a layer added. To specify the function of a layer, click on the **Function** field of the new layer; a drop-down list will be displayed. You can select an option from the drop-down list to specify the exact function of the layer added. For example, if you have added a layer in the foundation slab for a vapor retarder, you can select the **Thermal/Air Layer [3]** option from the drop-down list displayed. To assign a material to the layer, click on the **Material** field of the desired layer; a browse button will be displayed. Choose that browse button; the **Material Browser** dialog box and the **Material Editor** pane will be displayed. You can use this dialog box to specify the shading, rendering, identity, and physical parameters of the material to be assigned to the desired layer. For example, if you assign the vapor retarding function for a new layer, the material to be applied to the layer should correspond to it. To specify a material for the vapor retarder, select the **Vapor Retarder** material from the **Project Materials: <All>** area in the **Material Browser** dialog box and then choose the **OK** button; the selected material will be assigned to a new layer. Next, to specify the thickness for the layer added, click on the **Value** field corresponding to the added layer and enter a value in the **Thickness** column. For example, you can specify the value **1/8"** for Imperial or **3 mm** for Metric in the **Thickness** field to denote the thickness of the membrane of the vapor retarder. After specifying the desired parameters of the added layer you can re-order the location of the added layer in the cross-section of the foundation slab. To do so, choose the **Up** and **Down** buttons displayed in the **Layers** area of the **Edit Assembly** dialog box. To move a layer up in the order, select the row corresponding to it and choose the **Up** button; the selected layer will move one step up. To see the change in the cross-section due to re-ordering, choose the **Preview** button; the **Edit Assembly** dialog box will expand and the preview of the section will be displayed in a preview box on the left of the dialog box. You can choose the **Down** button to move a layer down. In the **Edit Assembly** dialog box, a layer can be deleted if it is not required for the cross-section detail. To do so, select the desired layer and choose the **Delete** button located in the **Layers** area; the selected layer will be deleted from the list in the table. After assigning all the parameters to the layers and reordering them as per the requirement, choose the **OK** button to return to the **Type Properties** dialog box.

Graphics: In the **Type Properties** dialog box, the **Graphics** head contains two parameters: **Coarse Scale Fill Pattern** and **Coarse Scale Fill Color**. These parameters are used to control the graphical display (color and pattern of the fill) of the slab foundation cross-section in a coarse-scale view. To specify a value for the **Coarse Scale Fill Pattern** parameter, click on the **Value** field corresponding to it; a button will be displayed in this field. Choose the button; the **Fill Patterns** dialog box will be displayed. In this dialog box, you can select different patterns for the slab foundation cross-section, create a new pattern, or edit an existing pattern. The list box in the left of this dialog box displays the list of available patterns that can be applied for the display of the foundation slab cross-section in a coarse-scale view. The **Aluminum** pattern is selected by default.

To select a different pattern, click on any of the available patterns; the selected pattern will get highlighted. Alternatively, you can create a new pattern as per your requirement. To do so, choose the **New** button from the **Fill Patterns** dialog box; the **New Pattern** dialog box will be displayed. In this dialog box, you can define a simple pattern by using lines, or import a customized pattern file (*.pat* format) and apply it to the slab foundation cross-section. After specifying the options in the **New Pattern** dialog box, choose the **OK** button from it; a new pattern will be displayed with its name and preview in the list box.

You can also edit the patterns displayed in the list box of the **Fill Patterns** dialog box. To do so, select any of the patterns from the list box and choose the **Edit** button; the **Modify Pattern Properties** dialog box will be displayed. You can modify the properties of the selected pattern using the options in this dialog box. After editing the selected pattern, choose the **OK** button to apply the modification and return to the **Fill Patterns** dialog box. In this dialog box, choose the **Delete** button to delete the unwanted fill patterns from the list. Next, in the **Fill Patterns** dialog box, you can either choose the **No Pattern** button, if the cross-section of the slab foundation does not need a fill pattern in its display, or choose the **OK** button to apply the selected pattern. When you choose either of the options, the **Fill Pattern** dialog box will close and you will return to the **Type Properties** dialog box.

After a filled pattern has been assigned to the cross-section of the foundation, you can assign a different color to it. By default, the pattern is displayed in black. To assign a different color to the fill pattern, choose the **Black** button displayed in the **Value** field corresponding to the **Coarse Scale Fill Color** parameter; the **Color** dialog box will be displayed. In this dialog box, you can select a color from the color swatch and then choose the **OK** button. On doing so, the **Color** dialog box will close and the selected color will be assigned to the **Coarse Scale Fill Color** parameter.

Identity Data: The **Identity Data** head of the **Type Properties** dialog box contains the following parameters: **Type Image**, **Keynote**, **Model**, **Manufacturer**, **Type Comments**, **URL**, **Description**, **Assembly Description**, **Assembly Code**, **Type Mark**, and **Cost**. The **Keynote** parameter indicates the keynote assigned to the slab foundation. To add or edit the value of this parameter, click on the **Value** field corresponding to this parameter; a browse button will be displayed. Choose the browse button; the **Keynotes** dialog box will be displayed. In this dialog box, select the desired keynote and then choose the **OK** button; the selected keynote will be assigned to the **Keynote** parameter.

The **Model** parameter is used to assign an internal code or number that the manufacturer of the foundation slab will use. The **Manufacturer** parameter is used to specify the name of the manufacturer of the foundation slab. The **Type Comments** parameter is used to specify a general comment about the type of the foundation slab. The value assigned to this parameter can be used in the schedule for the foundation slab.

The **URL** parameter can be used to specify a link to a web page that contains information about the foundation slab type. Using the **Type Properties** dialog box, you can specify the desired information regarding the foundation slab type. To do so, you can use the **Description** parameter. The **Assembly Code** parameter is used to specify a uniformat assembly code for the foundation slab. The value in this parameter can be used to mark or refer the foundation slab in the construction phase of the building structure associated with it. To specify a value for the **Assembly Code** parameter, click on the **Value** field corresponding to this parameter; a button will be displayed. Choose this button; the **Choose Assembly Code** dialog box will be displayed. In this dialog box, ensure that the **Structural Foundations** option is selected in the **Show classifications for** drop-down list. This dialog box displays two columns: **Uniformat Classification** and **Revit Category**. The **Uniformat Classification** column contains the list of classifications for the desired foundation type and the **Revit Category** column shows the category to which the classification belongs. Select the desired assembly code from the list and then choose the **OK** button; the **Choose Assembly Code** dialog box will close and the selected code will be assigned to the **Assembly Code** parameter in the **Type Properties** dialog box. The **Assembly Description** parameter in the **Type Properties** dialog box is a non-editable parameter, which takes value from the **Assembly Code** parameter assigned to the desired foundation type.

The **Type Mark** parameter is used to specify a unique value for a specific slab. The value used in this parameter is used to demarcate slab foundations. The **Cost** parameter is used to specify the cost of manufacturing the desired foundation slab. After assigning the type properties from the **Type Properties** dialog box, choose the **OK** button; the **Type Properties** dialog box will close and the **Properties** palette for the foundation slab instance will be displayed.

After specifying the instance and type properties for the foundation slab, you will use the tools in the **Draw** panel of the **Modify | Create Floor Boundary** tab to sketch the boundary and define the slope arrow of the foundation slab. Various options and tools to draw the boundary line and to define the slope arrow for the foundation slab are discussed next.

Sketching the Boundary Line

To sketch the boundary line and the slope arrow for the foundation slab, you can use different tools from the **Draw** panel of the **Modify | Create Floor Boundary** tab. In this tab, you can use the **Boundary Line** and **Slope Arrow** tools to draw the boundary line and the slope arrow for the foundation slab. It is recommended that you draw the boundary of the foundation slab before drawing the slope arrow for it.

To draw the boundary of the foundation slab, invoke the **Boundary Line** tool from the **Draw** panel; various tools to sketch the boundary line of the foundation slab will be displayed in a list box next to it, refer to Figure 5-14. In the list box, the **Pick Walls** tool is invoked, by default.

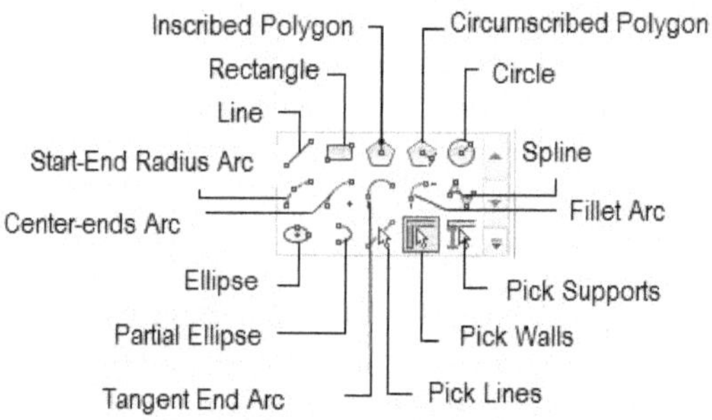

***Figure 5-14** Various sketching tools for sketching the boundary line of the foundation slab*

The **Pick Walls** tool is used to define the boundary of the foundation slab along the perimeter of the existing structural walls. On invoking this tool, the **Options Bar** will display the **Offset** edit box and the **Extend into wall (to core)** check box. In the **Offset** edit box, you can enter a value to offset the boundary line from a selected wall. The **Extend into wall (to core)** check box is selected by default. As a result, the offset distance entered in the **Offset** edit box will be measured from the core boundary of the wall. If you clear the check box, the offset distance will be measured from the outermost face of the wall. After specifying the options in the **Options Bar**, click on the wall in the drawing to make the reference for the boundary line. On clicking the wall, a pink line with flip arrow will be displayed along the perimeter of the wall, refer to Figure 5-15. You can click on the flip arrow to flip the boundary line to either side of the selected wall. You can select the other walls to finish the definition of the boundary of the foundation slab. Note that the boundary that you define should form a closed loop. After defining the boundary, choose the **Finish Edit Mode** button from the **Mode** panel of the **Modify | Create Floor Boundary** tab; the **Revit** message box will be displayed. Choose the **Yes** button from this message box; the foundation slab will be created at the perimeter of the picked walls and the walls will be attached to the foundation slab. If you choose the **No** button from the **Revit** message box, the foundation slab will be created but it will not be attached to the walls that are picked to define its boundary.

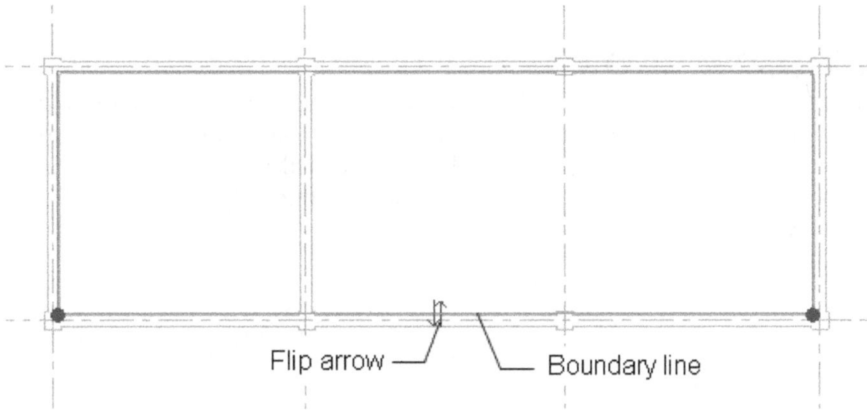

Figure 5-15 *The boundary line displayed along the wall perimeter*

Similar to the **Pick Walls** tool, you can use the **Pick Supports** tool from the list box in the **Draw** panel to sketch the boundary of the foundation slab. The **Pick Supports** tool can be used when you require to create a foundation slab above an existing structural wall or beam. The structural wall or the beam will be the supporting element for the foundation slab. After invoking the **Pick Supports** tool from the list box in the **Draw** panel, pick the existing structural walls or beams from the drawing to define the boundary of the foundation slab. Note that the boundary that you define by picking the support should be a closed loop. On invoking the **Pick Supports** tool, various options related to the position of the boundary line will be displayed in **Options Bar**. The options in the **Options Bar** are the same as those discussed for the **Pick Walls** tool.

The **Pick Lines** tool is used to sketch the boundary of the foundation slab by picking the edges of beams or structural walls or model lines present in a structural model. On invoking this tool, the **Options Bar** will display various options to position the boundary lines. In the **Options Bar**, you can enter a suitable value in the **Offset** edit box to specify the offset distance of the boundary lines from the picked edges. The **Lock** check box is cleared by default. You can select this check box to lock the position of the boundary lines defined for the foundation slab. After selecting the desired options from **Options Bar**, click on the desired wall edges, beam edges, or lines to define the boundary of the foundation slab.

From the list box in the **Draw** panel of the **Modify | Create Floor Boundary** tab, you can use other tools such as **Line**, **Rectangle**, **Circle**, and so on to sketch the boundary of the foundation.

Note

*The options in the **Options Bar** will differ depending on the sketching tool invoked.*

After sketching the boundary of the foundation slab, you can use the **Slope Arrow** and **Span Direction** tools to specify the slope of the foundation slab and the direction of the metal decks in the foundation slab, respectively. To finish the creation of the foundation slab, choose the **Finish Edit Mode** button from the **Mode** panel of the **Modify | Create Floor Boundary** tab.

STRUCTURAL FLOORS

The structural floor in a structural model represents a floor deck or a floor slab. In Autodesk Revit, you can specify various structural parameters such as thickness, structural composition, rebar sizes, and more to the added structural floor. You can also add slope and create opening in the added structural floor. Figure 5-16 shows a structural floor in a model. In the next sections, you will learn how to add structural floors.

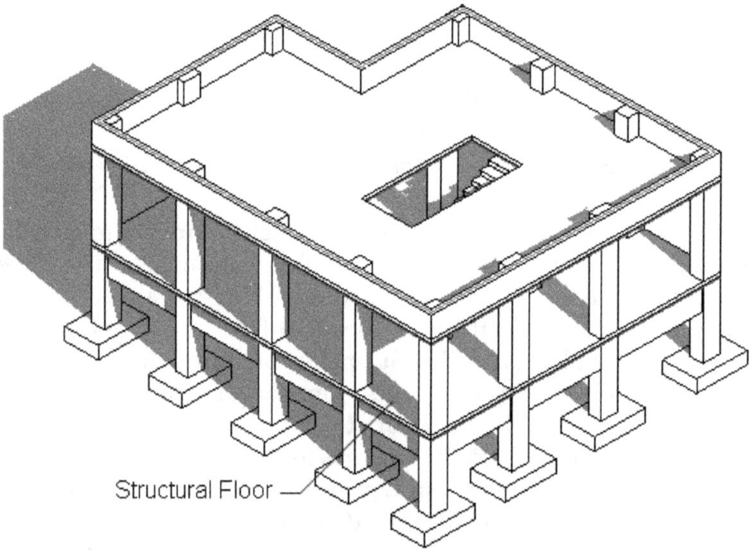

***Figure 5-16** Structural model displaying a structural floor deck*

Adding Structural Floors

Ribbon:	Structure > Floor drop-down > Floor: Structural

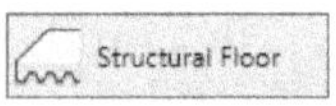

To add a structural floor to a structural floor plan, choose the **Floor: Structural** tool from **Structure > Floor** drop-down; the **Modify | Create Floor Boundary** tab will be displayed, as shown in Figure 5-17.

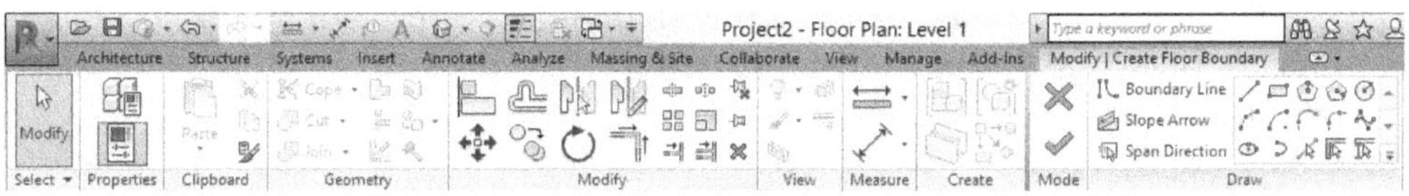

***Figure 5-17** Various options in the **Modify / Create Floor Boundary** tab*

You can use the options in this tab to draw, modify, and assign various properties to the structural floor. In the **Draw** panel of this tab, there are three tools: **Boundary Line**, **Slope Arrow**, and **Span Direction**. The **Boundary Line** tool is chosen by default. You can use this tool to create the floor boundary or the periphery of the structural floor of your project. To draw the floor boundary, you can choose a sketching tool from the list box located next to the **Boundary Line** option. In the list box, the **Pick Walls** tool is chosen by default. As a result, you can pick structural walls from your structural model to define the floor boundary of the structural floor. You can use the other sketching tools such as **Pick Supports**, **Pick Lines**, **Rectangle**, **Circle**, and so on from the list box to create the floor boundary. While sketching the floor boundary, you can use various options displayed in the **Options Bar** to set the floor boundary based on the project requirement. The options in the **Options Bar** change according to the sketching tool selected from the list box next to the **Boundary Line** option. In the **Offset** edit box of the **Options Bar**, you can enter a value to specify the offset distance of the edges of the floor boundary from the sketched location. This option is useful for creating cantilever slabs. In the **Options Bar**, select the **Radius** check box; the edit box located on its right will become active. Enter a value in the edit box to specify the radius of the curved boundary that you will sketch using any of the curve sketching tools such as **Circle**, **Center-ends Arc**, and so on. Notice that when you use tools such as the **Line** tools and the arc tools, the **Chain** check box is selected in the **Options Bar**. As a result, you can create the floor boundary continuously.

After creating the floor boundary, you can choose the **Slope Arrow** tool in the **Draw** panel to assign slope to the structural floor in the project. On choosing the **Slope Arrow** tool, the list box on its right will display two sketching tools: **Line** and **Pick Lines**. You can use these tools to sketch the slope arrow and define its direction for the structural floor. In the list box, the **Line** tool is chosen by default. As a result, you can sketch the slope arrow as a line. To sketch the slope arrow, you need to specify the start point and the end point of the arrow, refer to Figure 5-18. To specify the start point of the slope arrow, move the cursor and place it inside the floor boundary by clicking at a point inside the floor boundary. On doing so, a blue dashed line with an arrow head will be displayed. Next, move the cursor toward the direction of the slope of the floor and then click at an appropriate location to define the endpoint of the slope arrow. As you click, the **Properties** palette will display various properties of the slope arrow. In the **Properties** palette, click on the value field corresponding to the **Specify** parameter to specify the method for defining the slope of the structural floor. On doing so, a drop-down list will be displayed with two options: **Height at Tail** and **Slope**. In this drop-down list, the **Height at Tail** option is selected by default. As a result, you can define the levels and the offset distances of the endpoints of the slope arrow. To specify the level of the start point of the slope arrow, click on the value field corresponding to the **Level at Tail** parameter in the **Properties** palette, and then select an option from the drop-down list displayed. To specify the offset distance of the start point from the specified level, click on the value field corresponding to the **Height Offset at Tail** parameter in the **Properties** palette and enter a value in the value field. Similarly, you can use the **Level at Head** and **Height Offset at Head** parameters to specify the desired values in their corresponding fields to assign a level and offset distance of the endpoint of the slope arrow.

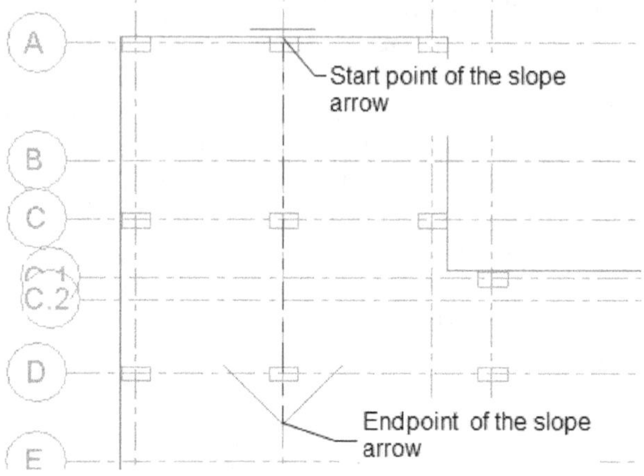

Figure 5-18 *Specifying the start and end points of slope arrow*

Alternatively, if you select the **Slope** option from the **Specify** drop-down list in the **Properties** palette, you can define the slope of the slope arrow by using the **Level at Tail** and **Height Offset at Tail** parameters.

Note

In a structural floor, the start and end points of the slope arrow define the start and end, respectively, of a slope.

After specifying the slope arrow of the structural floor, you can specify the boundary line of the structural floor corresponding to which the metal decks in the floor structure will run parallel. In the structural floor boundary, you will notice two parallel lines at one of its edge. When you sketch the floor boundary, by default these lines are displayed on the first edge you sketch. These lines represent the direction of the run of metal decks in the structural floor. To change the direction of these parallel lines, you can choose the **Span Direction** tool from the **Draw** panel in the **Modify | Create Floor Boundary** tab. When you choose the **Span Direction** tool, the list box next to it displays the **Line** and **Pick Lines** tools. You can use any of these tools to define the boundary line of the structural floor to which metal decks in the floor structure will run parallel.

After defining the boundary line, slope arrow, and span direction of the structural floor, you can specify its instance and type properties by using the options in the **Properties** palette. The instance and type properties for the structural floor are discussed in the following sections:

Specifying the Instance and Type Properties of a Structural Floor

You can change the instance and type properties of a structural floor while or after adding them to a structural model. To change the instance properties, use the various parameters displayed in the **Properties** palette, as shown in Figure 5-19. Some of the instance parameters in the **Properties** palette are discussed in the following table:

Parameter	Description
Level	Specifies a level to which a structural floor is constrained. Click in its value field to select an option from the drop-down list displayed.
Height Offset From Level	Specifies the distance of the top surface of the structural floor relative to the level selected from the **Level** parameter.
Structural	The check box in its value field is selected by default.

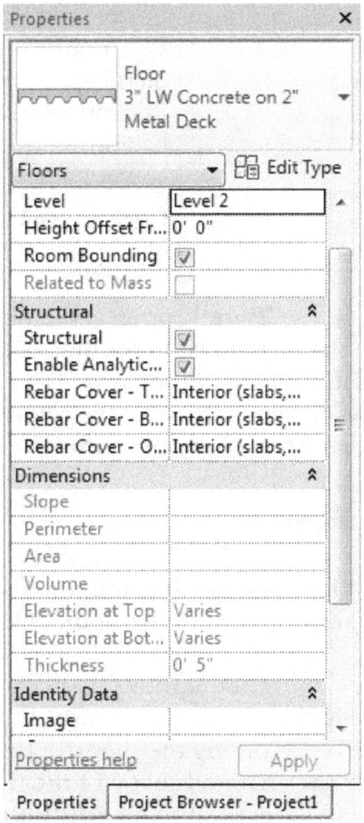

Figure 5-19 *The* ***Properties*** *palette displaying the instance properties of a structural floor*

In the **Properties** palette, choose the **Edit Type** button to invoke the **Type Properties** dialog box. In this dialog box, you can select an option in the **Type** drop-down list and change the type parameters of the concerned structural floor from the **Type Parameters** area. By default, the **3" LW Concrete on 2" Metal Deck** option in Imperial unit or **Floor: Generic 300mm** in Metric unit is selected in this list. In the **Type Properties** dialog box, you can choose the **Duplicate** button to create a new type by modifying various parameters. On choosing the **Duplicate** button, the **Name** dialog box will be displayed. In the **Name** edit box of this dialog box, enter a name for the new type of structural floor and then choose the **OK** button; the **Name** dialog box will be closed. Note that the **Type** drop-down list displays the name of the type created. In the **Type Parameters** area, you can change the type parameters displayed under various heads. Some of the type parameters in the **Type Properties** dialog box are described in the table given next.

Parameter	Description
Structure	Specifies the structural composition of the structural floor. Choose the **Edit** button in the **Value** field to display the **Edit Assembly** dialog box. You can use this dialog box to specify the composition of a compound floor.
Default Thickness	Is a read-only parameter. The **Value** field of the parameter displays the value of the thickness of a floor type, which is determined by adding the thickness of its layers.
Function	Specifies whether a floor is interior or exterior. Click in its **Value** field and select any of the two options: **Interior** and **Exterior**.

After specifying various options in the **Type Properties** dialog box, choose the **OK** button to close this dialog box. Next, to exit the **Structural Floor** tool, choose the **Finish Edit Mode** button from the **Mode** panel of the **Modify | Create Floor Boundary** tab; the **Modify | Floors** tab will be displayed. You can use various options in this tab to change the shape properties and type of the created structural floor. Some of the options in this tab are discussed in the next section.

Modifying Structural Floors

To modify a structural floor in a structural model, select it; the **Modify | Floors** tab will be displayed. You can use various options in this tab to edit the shape and boundary of a floor, edit the boundary of the floor, add reinforcement to the floor, and modify geometry.

To edit the shape of a floor, you can use the options displayed in the **Shape Editing** panel of the **Modify | Floors** tab. The tools in this panel are: **Modify Sub Elements**, **Add Points**, **Add Split Line**, and **Pick Supports**. After changing the shape of the selected structural floor, choose the **Reset Shape** button to restore the changed shape of the floor to its original shape. You can edit the sketch of the floor boundary by choosing the **Edit Boundary** tool from the **Mode** panel of the **Modify| Floors** tab. On choosing the **Edit Boundary** tool; the **Modify | Floors > Edit Boundary** tab will be displayed and the selected structural floor will change to edit mode and its boundaries will be highlighted in pink. Now, you can use various editing tools from this tab to change the geometry of the floor boundary. Next, choose the **Finish Edit Mode** button from the **Mode** panel to exit the editing mode and return to the options displayed in the **Modify | Floors** tab.

To modify the floor type and change the instance and type properties of the selected structural floor, use the **Type Selector** drop-down list and various options displayed in the **Properties** palette.

Adding Openings to a Structural Floor

You can create an opening in a structural floor to create stairs, ramps, and shafts for allowing access between levels. In Autodesk Revit , you can create openings in the structural floors by using the options displayed in the **Opening** panel of the **Structure** tab of the ribbon. In the **Opening** panel, you can use the **Shaft**, **By Face**, and **Vertical** tools to create the openings in the structural floors. In the next section, you will learn how to create openings by using the **By Face** tool.

Creating Openings Using the By Face Tool

To create an opening on a face of a structural floor, choose the **By Face** tool from the **Opening** panel of the **Structure** tab; the cursor will change to a cross mark and you will be prompted at the **Status Bar** to select a planar face of the floor surface. Move the cross mark toward the edge (any) of the floor surface and click when the floor surface corresponding to this edge gets highlighted. On clicking, the **Modify | Create Opening Boundary** tab will be displayed. Use the sketching tools in the **Draw** panel of this tab and then choose the **Finish Edit Mod**e button from the **Mode** panel; the **Modify | Floor opening cut** tab will be displayed. You can choose the **Edit Boundary** tool from the **Mode** panel of this tab to edit the boundary of the opening created. Figure 5-20 shows an opening created in a structural floor.

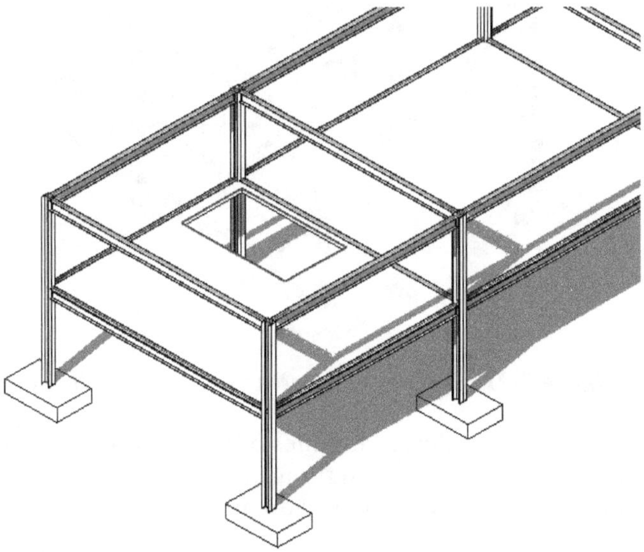

Figure 5-20 *Opening created in a structural floor*

ADDING A SLAB EDGE TO A STRUCTURAL FLOOR

You can also add a slab edge to your structural floor. To do so, choose the **Floor: Slab Edge** tool from the **Structure** panel of the **Structure** tab; the **Modify | Place Slab Edge** tab will be displayed and you will be prompted at the **Status Bar** to click on the floor edge to add the slab edge. Move the cursor at the desired edge of the floor; the edge will get highlighted. Click at the highlighted edge; slab edge under the floor will be created. After clicking on the edge, two flip arrows will be displayed at the selected edge. You can flip the direction of the slab edge using these arrows. In the **Properties** palette you can modify the type and instance properties of the created slab. To modify the type properties, choose the **Edit Type** button from the **Properties** palette; the **Type Properties** dialog box will be displayed. In this dialog box, you can specify the desired values in the **Value** fields of the parameters either by entering a new value or by selecting an option from the drop-down list in the **Type Properties** dialog box. Similarly, you can modify the instance properties of a slab by entering a new value or by selecting an option from the drop-down list displayed in the value field of the corresponding parameter.

Note

If you place a slab segment at the adjacent side of the existing slab edge, both the existing and the new segments will mitter and form a single continuous slab.

BEAMS

Beams are structural elements that can be used to resist load. Beams generally carry vertical forces. They can also carry horizontal forces as that of wind and earthquake. Figure 5-21 shows steel beams between vertical steel columns.

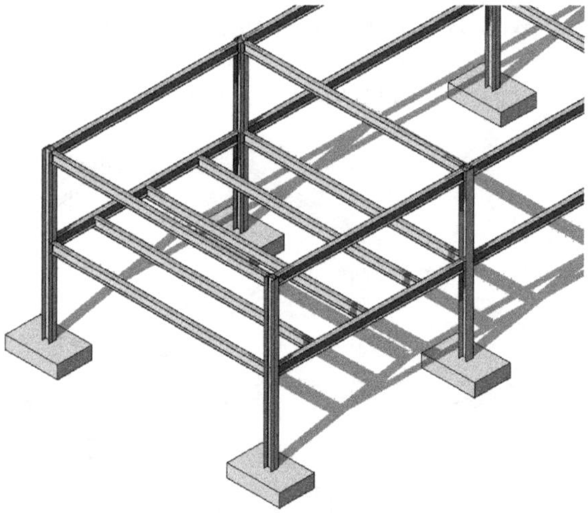

Figure 5-21 *Steel beams between vertical steel columns*

In Autodesk Revit , you can add different families of beams to your structural model. You can assign different properties to a beam by assigning various instance and type parameters to it. In the following sections, you will learn how to add and modify beams.

Adding Beams

Ribbon: Structure > Structure > Beam

In Autodesk Revit , you can add beams to a structural plan view. To add beams to a project, choose the **Beam** tool from the **Structure** panel of the **Structure** tab; the **Modify |Place Beam** tab will be displayed. In this tab, you can choose the **Tag on Placement** button to make the beam tags visible while adding the beams to the project. In the **Multiple** panel of the **Modify |Place Beam** tab, you can choose the **On Grids** option to place the beam between the intersection points of grid lines in the project view.

Note

*The **Tag on Placement** tool will be displayed and activated when you are in 2D view.*

In the **Draw** panel of the **Modify | Place Beam** tab, you can use various sketching tools available in the list box to sketch a beam in your project. By default, the **Line** tool is chosen in this panel. To add more beam family types to your project, choose the **Load Family** tool from the **Mode** panel of the **Modify |Place Beam** tab; the **Load Family** dialog box will be displayed. In this dialog box, for Imperial system, browse to **US Imperial > Structural Framing** folder location and then choose any of the sub-folders such as **Concrete**, **Light Gauge Steel**, **Precast Concrete**, **Steel**, and **Wood** or browse to **US Metric > Structural Framing** folder location and then choose any of the sub-folders such as **Concrete**, **Light Gauge Steel**, **Precast Concrete**, **Steel**, and **Wood** for Metric system. Next, select the family files from the chosen folder to load the desired beam types to the project view. After selecting the desired families, choose the **Open** button to close the **Load Family** dialog box and load the selected families into your project.

Note

After placing beams, the Flip Structural Framing ends arrow will be displayed which allow you to flip the start and end points of the beam.

In the **Properties** palette of beam, you can select a beam type from the **Type Selector** drop-down list. By default, the **W- Wide Flange : 12X26** option in Imperial unit system or **UB- Universal Beam 305x165x40UB** in Metric unit system is selected in this drop-down list. In the **Properties** palette, you can also specify various instance and type parameters for a beam. To specify type parameters for a beam, choose the **Edit Type** button from the **Properties** palette; the **Type Properties** dialog will be displayed. In this dialog box, you can modify various type parameters of the beam. Next, in the project view, specify the start point and endpoint of the beam to add it. In the following sections, various instance and type parameters of beams are discussed.

Instance Parameters of Beam

In the **Properties** palette of a beam, you can specify different instance parameters that may be different for different families. Some of the instance parameters are discussed in the table next.

Parameter	**Description**
Reference Level	Specifies the level that constrains the beam.
z Justification	Specifies the z justification (vertical) for the beam cross-section. You can select the **Center**, **Top**, **Bottom**, or **Origin** option from the drop-down list displayed in its value field.
Structural Usage	Specifies the structural usage of a beam. You can select the **Girder**, **Horizontal Bracing**, **Joist**, **Other**, or **Purlin** option from the drop-down list in the value field.

Apart from the parameters discussed in the previous table, there are other parameters available in the **Properties** palette of the beam. These parameters may differ depending upon the beam family type selected from the **Type Selector** drop-down list.

Type Parameters of the Beam

To specify type parameters, you need to invoke the **Type Properties** dialog box. To do so, choose the **Edit Type** button from the **Properties** palette of the beam. Note that the type parameters are different for the steel and concrete members. These parameters are discussed next.

Parameter	Description
b	Specifies the width of a concrete member.
h	Specifies the depth of a concrete member.
Section Area	Specifies the section area of a steel member.
Nominal Weight	Specifies the nominal weight of a steel member.
Flange Thickness	Specifies the width of flange width.

You can specify more type parameters in the **Type Properties** dialog box by clicking in their corresponding **Value** fields.

Modifying Beams

After adding beams in a project, you can modify them. To do so, select the beam that you want to modify; the **Modify | Structural Framing** tab will be displayed. In this tab, you can use various options to modify the beam. You can also use the grip editing tools to increase the length of the selected beam. You can change the instance and type properties of a selected beam by using the **Properties** palette and the **Type Properties** dialog box.

Adding a Structural Beam System

In a project, you may require to place similar parallel beams in a specified area of a structure. In such a situation, you can use quickly the **Beam System** tool. This tool allows you to create a single structural framing element containing a series of parallel placed beams. You can use this tool to create a structural beam system either by picking the structural supporting elements such as beams and structural walls, or by sketching a border.

To create a structural beam system by picking a structural support, sketch the supporting elements that you require for the beam system. Next, invoke the **Beam System** tool from the **Structure** panel of the **Structure** tab; the **Modify | Place Structural Beam System** tab will be displayed, as shown in Figure 5-22.

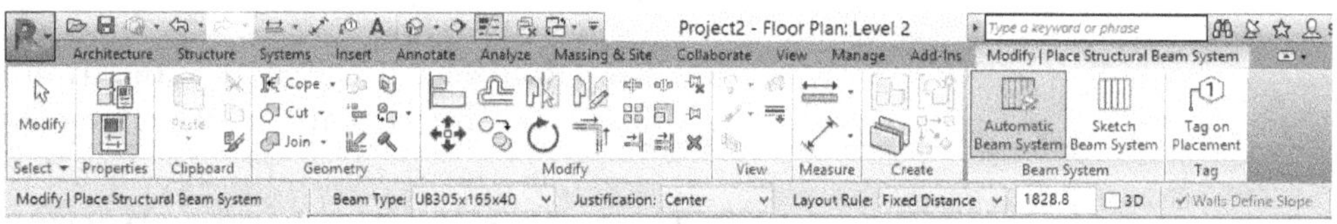

Figure 5-22 *Various options in the* ***Modify / Place Structural Beam System*** *tab*

In the **Beam System** panel of this tab, choose the **Sketch Beam System** tool; the **Modify | Create Beam System Boundary** tab will be displayed. In the **Draw** panel of this tab, ensure that the **Boundary Line** tool is chosen and then choose the **Pick Supports** tool displayed in the list box next to the **Boundary Line** tool. On doing so, you will be prompted in the **Status Bar** to pick structural walls or beams to create the boundary lines for the beam system. In the drawing area, click on a wall or a beam that you want to define as the starting edge of the beam system. Note that the starting edge of the beam system defines the alignment of the beams that will be placed. For example, on clicking a wall, a magenta line along with a symbol of parallelism will be displayed at the center of the wall cross-section, refer to Figure 5-23. Next, click on the other walls that you want to include as a support for the system. Note that the sketched boundary for the beam system should be a closed boundary. After selecting the walls as supports for the beam system, you can specify various instance parameters for the beam system such as **Layout Rule**, **Fixed Spacing**, **Beam Type**, **Justification**, and more in the **Properties** palette.

Figure 5-23 *Beams placed inside the boundary of a beam system*

After specifying the various instance parameters, choose the **Finish Edit Mode** button from the **Mode** panel of the **Modify | Create Beam System Boundary** tab; the **Modify | Structural Beam Systems** will be displayed. Also, beams will be placed inside the defined boundary, refer to Figure 5-24. You can use various tools in the **Modify | Create Beam System Boundary** tab to modify the created beam system, if required. You can also use the options in the **Options Bar**, displayed below the tab to change the size of the beams and edit the layout rules. Next, press ESC to exit the **Beam System** tool.

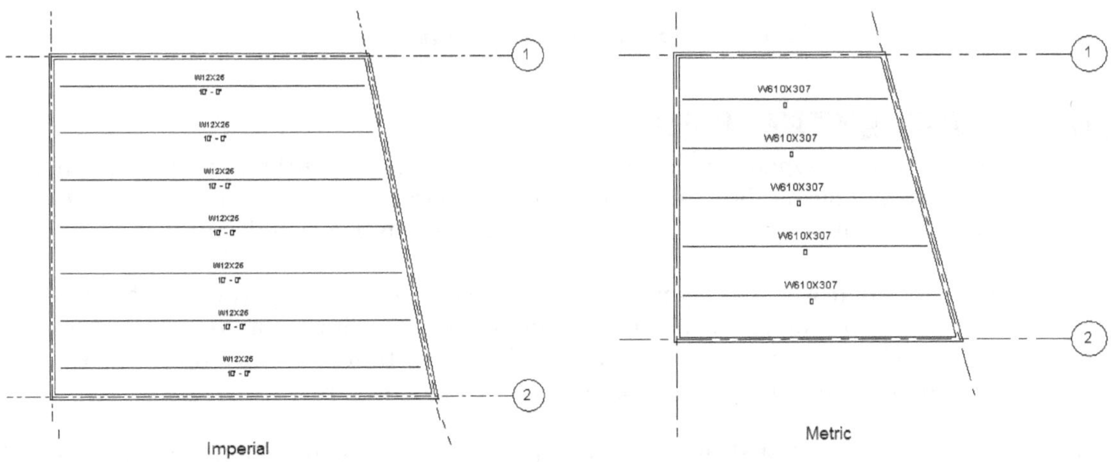

Figure 5-24 *Beams placed inside the boundary of a beam system*

Alternatively, you can sketch the boundary of the beam system by choosing any of the sketching tools from the **Draw** panel of the **Modify | Create Beam System Boundary** tab. In the next section, you will learn to add a beam system automatically in a closed loop of supporting elements.

Adding a Structural Beam System Automatically

You can add a beam system automatically inside a boundary enclosed by structural walls or beams. To do so, choose the **Beam System** tool from the **Structure** panel of the **Structure** tab; the **Modify | Create Beam System Boundary** tab will be displayed. In the **Beam System** panel of this tab, choose the **Automatic Beam System** tool, if not chosen by default, and then move the cursor to the structural members to which you want to add the beam system; the beam system will be highlighted in blue, as shown in Figure 5-25. Now click on the desired member; the beam system will be added inside the closed loop of supporting elements (walls or beams) already drawn.

Note

The beam system will be aligned parallel to the structural member selected by you.

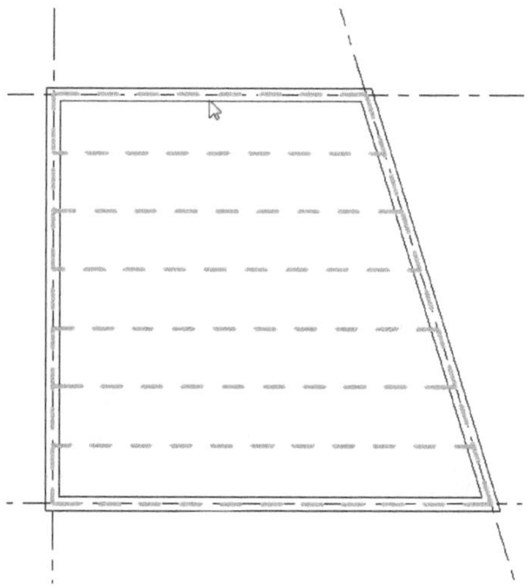

Figure 5-25 *The highlighted beam system*

OPEN WEB STEEL JOISTS

In a structural project, the Open web steel joists or bar joists are very efficient structural members. The main purpose of placing the Open web steel joists is to provide support to roof and floor deck and transfer the imposed load to the beams and columns associated with it.

In a project, the Open web steel joists are not considered as structural steel as they are manufactured as proprietary structural steel members by various manufacturers that are recognized by Steel Joist Institute (SJI). The Steel Joist Institute specifies the standards for manufacturing of joists as well as designing and construction of joists.

The Open Web Steel Joist (OWSJ) is a lightweight steel truss that consist of parallel chords and a triangulated web system. The truss is designed to span between the bearings points.

The Open Web Steel Joists system have various types: K Series Joists, LH and BLH Series Joists, and Joist Girders. In Autodesk Revit , you can add these joists by using the **Load Family** dialog box and then using the **Beam** tool from the **Structure** tab. The families that you can use for adding an Open web joist in Imperial system and Metric system are: **LH-Series Bar Joist, SLH-Series Bar Joist**, **G Joist Girder**, and **K- Series Bar Joist**, **M_ G Joist Girder**, **M_K- Series Bar Joist**, and **M_SLH-Series Bar Joist**.

Adding K Series Joist System

In a structural project, the K Series joist system is most commonly used to support roofs and floor decks. In this system, the depth can range from 8" to 30" (2'6") for Imperial system or for Metric system depth range from 200 to 700 mm. This system can be used economically to span up to 60'-0"(or 18288 mm). To add a K Series joist system, choose the **Load Family** tool from the **Load from Library** panel of the **Insert** tab; the **Load Family** dialog box will be displayed. In this dialog box, for Imperial system, browse to **US Imperial > Structural Framing > Steel** then select the **K-Series Bar Joist-Angle Web** file or for Metric system browse to **US Metric > Structural Framing > Steel** and then select the **M_K-Series Bar Joist-Angle Web** file from the list box. Next, choose the **Open** button; the **Specify Types** dialog box will be displayed. In this dialog box, select the desired family type from the **Family** area. In the **Types** area, select the type of sizes that are required and then choose the **OK** button; the selected file will be loaded in the current project. Next, select the structural plan view from the **Project Browser** and then choose the **Beam** tool from the **Structure** panel of the **Structure** tab; the **Modify | Place Beam** contextual tab will be displayed. In the **Properties** palette, you can select the **K-Series Bar Joist-Angle Web** type for Imperial or **M_K-Series Bar Joist-Angle Web** type for Metric from the **Type Selector** drop-down list. In the **Properties** palette, you can modify various instance properties of the Joist system. The instance properties of the joist system are same as that of beams. The instance properties of the beams has already been discussed in the previous sections of this chapter. To modify the type properties of the K-Series joist system, choose the **Edit Type** button from the **Properties** palette; the **Type Properties** dialog box will be displayed. In this dialog box, you can edit various type parameters such as **Web Thickness**, **Seat Length**, **Seat Depth**, **Chord Leg**, **Chord Thickness**, and **Depth**. The graphical representation of the parameters for the **K-Series Bar Joist -Angle Web 8K1** type is shown in Figure 5-26.

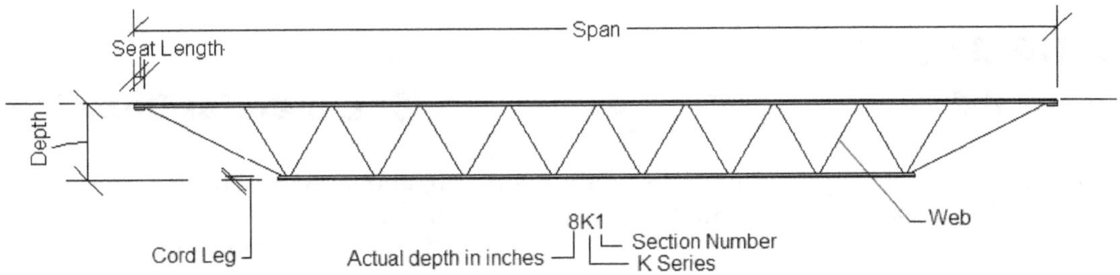

Figure 5-26 *The graphical representation of a K type joist system*

After editing the type properties in the **Type Properties** dialog box, choose the **OK** button to close it. Now, in the **Options Bar**, select an option from the **Placement Plane** drop-down list to specify a level in which the joist system will be placed. In the **Options Bar**, you can select an option from the **Structural Usage** drop-down list to specify the structural usage of the joist system. Also, you can select the **3D Snapping** and **Chain** check boxes in the **Options Bar** to activate snapping in 3D environment and the chain method, respectively. After specifying various options in the **Options Bar**, click at the desired point in the drawing area to specify the first bearing point to start creating the joist system; the joist system will be created at the specified point. Next, click on the other point to specify the other bearing of the joist system. On doing so, the joist system will be created between the two bearing points. Choose the **Modify** button from the **Modify | Structural Framing** tab to exit the joist creation tool. To view the joist system in 3D view, choose the **Default 3D View** tool from the **View > Create > 3D View** drop-down. Figure 5-27 displays the 3D view of a joist system.

Note
The Steel joists are usually fastened to their supporting members by field welding.

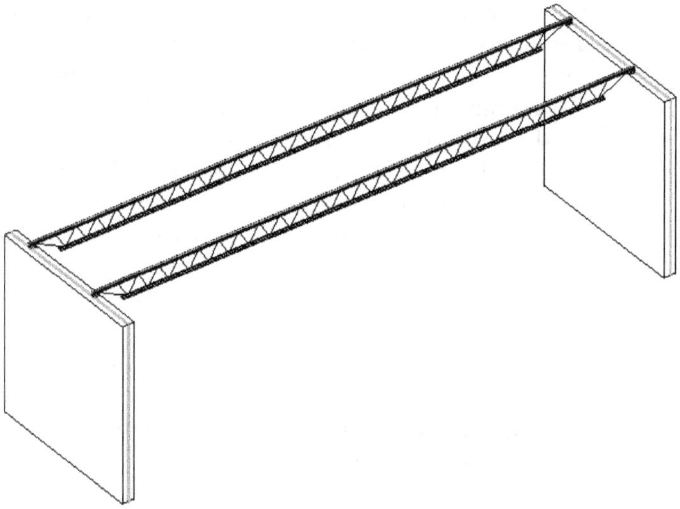

Figure 5-27 *The 3-D view of a K-joist system supported between the walls*

TUTORIALS

Tutorial 1 Commercial Complex

In this tutorial, you will add foundations, foundation slab, and structural framing members to the *Commercial Complex* project created in Tutorial 1 of Chapter 4.

(Expected time: 1 hr 45 min)

Use the following project parameters and specifications:

1. Foundations:

For Imperial Types: **Pile Cap-2 : Pile 31"x61"x35"**
Pile Cap-4 : Pile 72"x72"x35"
Wall Foundation : Bearing Footing -36"x12"

For Metric Types: **M_Pile Cap-2 : Pile 800 x 1500 x 900mm**
M_Pile Cap-4 : Pile 1800 x 1800 x 900mm
Wall Foundation : Bearing Footing - 900 x 300mm

2. Foundation Slab:
   For Imperial Type: **Foundation Slab : 12" Foundation Slab**
   For Metric Type: **Foundation Slab : 300mm Foundation Slab**
3. Beams:
   For Imperial Type: **Concrete-Rectangular Beam : 12 x 24**
   For Metric Type: **M_Concrete-Rectangular Beam: 300 x 600mm**

4. Structural Floor:
   For Imperial Type: **Generic 12"**
   For Metric Type: **Generic 300mm**

5. File name to be assigned:
   For Imperial *c05_Commercial-Complex_tut1.rvt.*
   For Metric *M_c05_Commercial-Complex_tut1.rvt.*

The following steps are required to complete this tutorial:

a. Open the *c04_Commercial-Complex_tut1.rvt* project file.
b. Load the isolated foundation families.
c. Create new types for isolated foundations.
d. Add isolated foundation.
e. Add foundation wall.
f. Add foundation slab.
g. Add beams.
h. Modify beams.
i. Add structural floor.
j. Save the project using the **Save As** tool.
k. Close the project using the **Close** tool.

Opening the Existing Project

In this section, you will open the project file created in Tutorial 1 of Chapter 4.

1. To open the project file, choose **Open > Project** from the **Application Menu**; the **Open** dialog box is displayed.

2. In this dialog box, browse to the *C:\rst_2017\c04_rst_2017_tut* location and choose the project file *c04_Commercial-Complex_tut1.rvt* for Imperial unit system and *M_c04_Commercial-Complex_tut1.rvt* for Metric unit system. You can also download this file from *http://www.cadcim.com*. The path of the file is as follows: *Textbooks > Civil/GIS > Revit Structure > Exploring Autodesk Revit 2017 for Structure*.

3. Choose the **Open** button from the **Open** dialog box; the selected project file opens in the drawing window in the **Entry** structural plan view.

Loading Isolated Foundation Families

In this section, you will load the isolated foundation families to the current project.

1. Ensure that the **Entry** structural plan view is the current view in the drawing area. Choose the **Isolated** tool from the **Foundation** panel of the **Structure** tab; the **Modify | Place Isolated Foundation** tab is displayed.

2. In this tab, choose the **Load Family** tool from the **Mode** panel; the **Load Family** dialog box is displayed.

3. In the **Load Family** dialog box, browse to folder location.

   For Imperial **US Imperial > Structural Foundations**
   For Metric **US Metric > Structural Foundations**

4. Press and hold the CTRL key and then click on the *Pile Cap-2 Pile.rfa* and *Pile Cap-4 Pile.rfa* files displayed in the **Structural Foundations** folder for Imperial or select *M_Pile Cap-2 Pile.rfa* and *M_Pile Cap-4 Pile* for Metric system; the files are highlighted indicating they have been selected.

Note

If the required family files are not available in the library, you can download them from the website: www.cadcim.com.

5. Choose the **Open** button in the **Load Family** dialog box; the **Shared Family Already Exists** window is displayed, as shown in Figure 5-28.

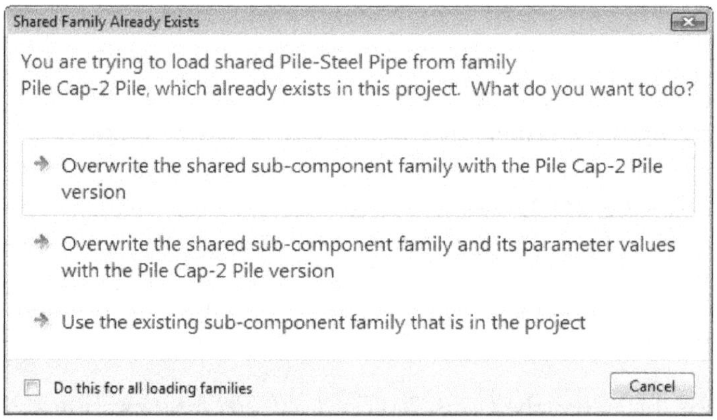

***Figure 5-28** The **Shared Family Already Exists** window*

6. Choose the **Overwrite the shared sub-component family with the Pile Cap-2 Pile version** option from the window; the window closes and the highlighted family files are loaded in the project.

Creating New Types for the Isolated Foundation

In this section, you will create two new types of the isolated foundation.

1. In the **Properties** palette, ensure that the **Pile Cap-2 : Pile 31"x71"x35"** option for Imperial unit system or **M_Pile Cap-2 : Pile 800 x 1800 x 900mm** option for Metric unit system is selected from the **Type Selector** drop-down list.

2. Choose the **Edit Type** button from the **Properties** palette; the **Type Properties** dialog box is displayed.

3. Choose the **Duplicate** button in the **Type Properties** dialog box; the **Name** dialog box is displayed.

4. In the **Name** dialog box, enter the specified name in the **Name** edit box-

   For Imperial **31"x61"x35"**

   For Metric **800 x1500 x 900**

   Next, choose the **OK** button; the **Name** dialog box is closed and the **Type Properties** dialog box is displayed.

5. In the **Type Properties** dialog box, select the desired option, if not selected by default, from the drop-down list in the **Value** field of the **Pile Type<Structural Foundations>** parameter.

   For Imperial **Pile-Steel Pipe : 16" Diameter**

   For Metric **M_Pile-Steel Pipe : 400mm Diameter**

6. Click in the **Value** field of the **Width** parameter and replace the existing value with **5' 1"** for Imperial unit system or **1500mm** for Metric unit system.

7. Click in the **Value** field of the **Clearance** parameter and replace the existing value with **1'** for Imperial or **300mm** for Metric.

8. In the **Type Properties** dialog box, select the **Pile Cap-4 Pile** option from the **Family** drop-down list and then select the desired option from the **Type** drop-down list, if not selected by default.

   For Imperial **79"x 79" x35"**

   For Metric **2000 x 2000 x 900mm**

9. Choose the **Duplicate** button; the **Name** dialog box is displayed.

10. In the **Name** edit box of the **Name** dialog box, enter the specified value-

    For Imperial **72"x 72" x35"**

    For Metric **1800 x1800 x900mm**

    Next, choose the **OK** button. the **Name** dialog box is closed and the **Type Properties** dialog box is displayed.

11. In the **Type Properties** dialog box, select the desired option, if not selected by default, from the drop-down list in the **Value** field of the **Pile Type<Structural Foundations>** parameter.

    For Imperial **Pile-Steel Pipe : 16" Diameter**

    For Metric **M_Pile-Steel Pipe : 400mm Diameter**

12. Click in the **Value** field corresponding to the **Width** and **Length** parameters and enter **6'** in Imperial or **1800mm** in Metric unit systems in each of them.

13. Click in the **Value** field corresponding to the **Clearance** parameter and enter **1' 3"** for Imperial or enter **380mm** for Metric to replace the existing value.

14. Choose the **OK** button; the **Type Properties** dialog box is closed.

Adding Isolated Foundations

In this section, you will add the isolated foundation types that you have created in the project.

1. In the **Properties** palette, select the option from the **Type Selector** drop-down list.
   For Imperial **Pile Cap-4 Pile: 72"x72"x35"**
   For Metric **M_Pile Cap-4 Pile: 1800 x1800 x900mm**

2. Choose the **At Columns** tool from the **Multiple** panel of the **Modify | Place Isolated Foundation** tab.

3. Move the cursor and place it over the column located at the intersection of grid C and grid 2.

4. Click when the column is highlighted; a foundation with four piles and a pile cap is added at the center of the highlighted column.

5. Press and hold the CTRL key and repeat steps 3 and 4 to add foundation to all the internal columns. Refer to Figure 5-29 for the location of foundations.

6. Release the CTRL key and choose the **Finish** button from the **Multiple** panel of the **Modify |Place Isolated Foundations > At Structural Column** tab; the 4 pile foundation is added to the selected columns. Ignore if any warning message is displayed. Next, press ESC.

7. Choose the **Isolated** tool from the **Foundation** panel of the **Structure** tab to add the foundations to the remaining columns.

8. In the **Properties** palette, select the option from the **Type Selector** drop-down list.
   For Imperial **Pile Cap -2 Pile : 31"x61"x35"**
   For Metric **M_Pile Cap -2 Pile : 800 x1500 x900mm**

9. Choose the **At Columns** tool from the **Multiple** panel of the **Modify | Place Isolated Foundation** tab.

10. Next, move the cursor and place it over the column located at the intersection of grid 1 and grid A; the column at the intersection of the grids is highlighted.

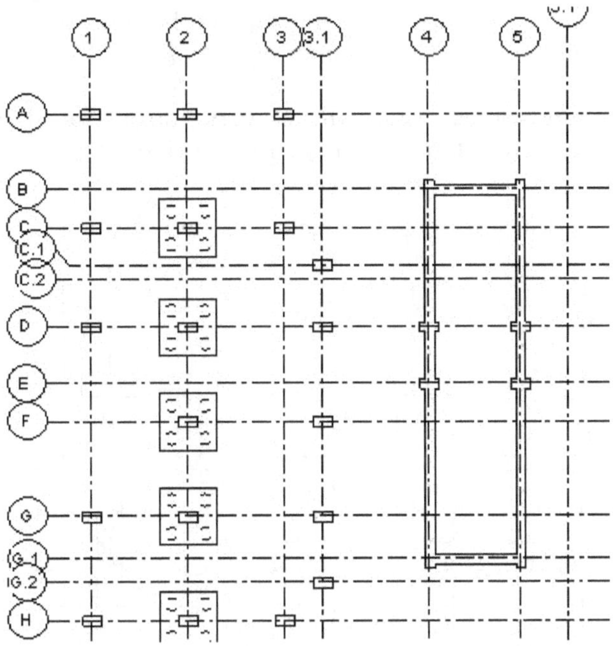

Figure 5-29 *The 4-pile foundations added to all internal columns*

11. Click on the highlighted column; the foundation is added to the highlighted column.

12. Next press and hold the CTRL key and repeat steps 10 and 11 to add the foundation type to all the columns that do not have a foundation attached to them (except the columns at the intersection of grid C.1 and grid 3.1 and at the intersection of grid G.2 and grid 3.1), refer to Figure 5-30.

13. Release the CTRL key, and choose the **Finish** button from the **Multiple** panel of the **Modify | Place Isolated Foundations > At Structural Column** tab. Ignore, if any warning message is displayed. Next, press ESC.

14. Next, choose the **Isolated** tool from the **Foundation** panel of the **Structure** tab; the **Modify | Place Isolated Foundation** tab is displayed.

15. In the **Properties** palette, ensure that the option is selected in the **Type Selector** drop-down list.

    For Imperial **Pile Cap -2 Pile : 31"x61"x35"**
    For Metric **M_Pile Cap -2 Pile : 800 x1500 x900mm**

16. Now place the cursor over the column at the intersection of grid C.1 and grid 3.1; both the grids at the intersection get highlighted.

17. Next, click on the highlighted grids; the selected foundation type is added to the column placed at the intersection of grid C.1 and grid 3.1. Ignore if any warning message is displayed.

18. Repeat steps 16 and 17 to add the foundation type to the column at the intersection of grid G.2 and grid 3.1. Ignore, if any warning message is displayed.

19. After adding the foundations, choose the **Modify** button from the **Select** panel of the **Modify | Place Isolated Foundation** tab to exit the **Isolated** tool.

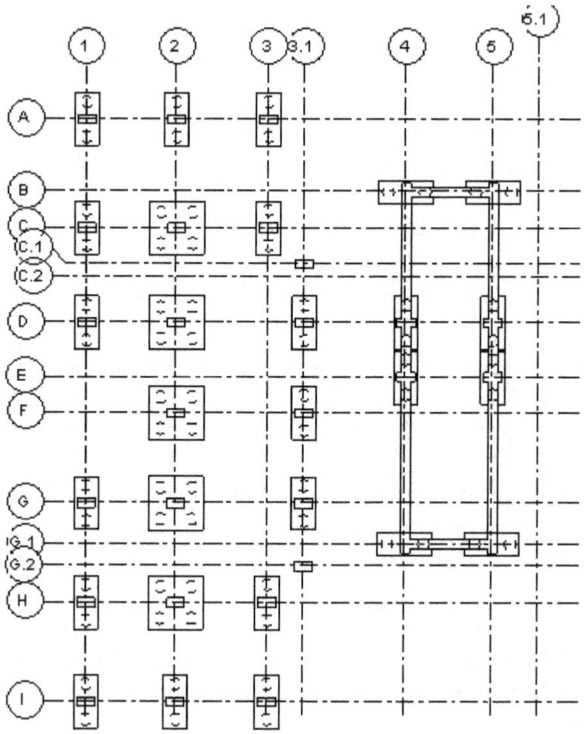

Figure 5-30 *The 2-pile foundations added to the columns*

Adding the Foundation Wall

In this section, you will add the foundation wall to the structural walls in the project.

1. Choose the **Wall** tool from the **Foundation** panel of the **Structure** tab; the **Modify | Place Wall Foundation** tab is displayed.

2. In the **Properties** palette, ensure that the **Wall Foundation : Bearing Footing -36" x12"** option in Imperial unit system or **Wall Foundation : Bearing Footing -900 x300mm** in Metric unit system is selected from the **Type Selector** drop-down list.

3. Click on the structural wall located at grid B; the foundation wall of the selected type is added throughout the length of the wall, as shown in Figure 5-31.

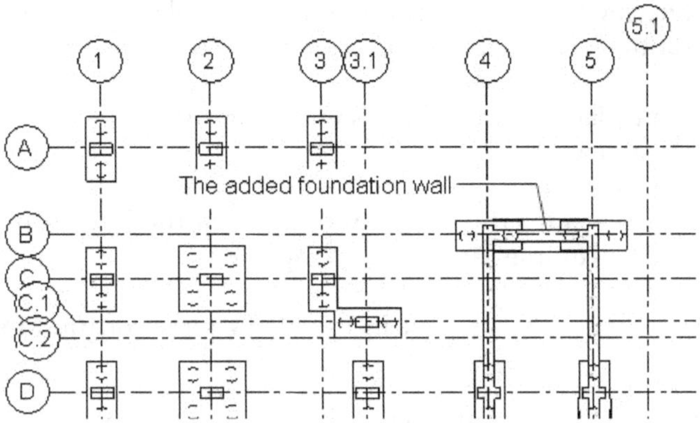

***Figure 5-31** The foundation wall added to the structural wall located at grid B*

4. Now, click on the remaining structural walls in the model; the foundation walls are added to all the structural walls in the project, as shown in Figure 5-32.

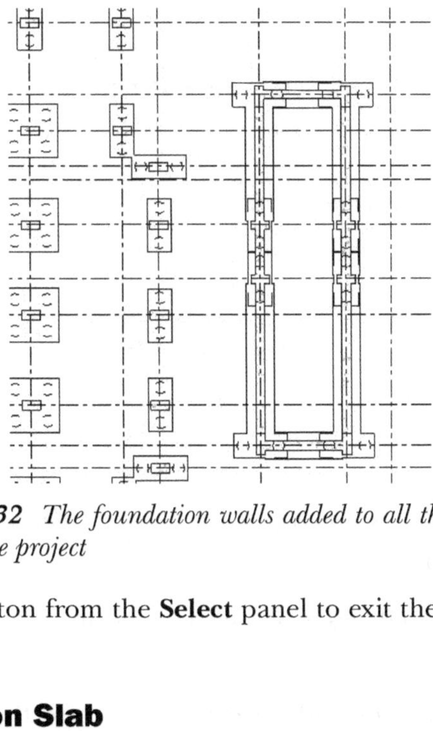

***Figure 5-32** The foundation walls added to all the structural walls in the project*

5. Choose the **Modify** button from the **Select** panel to exit the **Wall** tool in the **Foundation** panel.

Adding the Foundation Slab

In this section, you will add foundation slab to the project.

1. Invoke the **Structural Foundation: Slab** tool from **Structure > Foundation > Slab** drop-down; the **Modify | Create Floor Boundary** tab is displayed.

2. In the **Draw** panel of the **Modify | Create Floor Boundary** tab, ensure that the **Boundary Line** tool is chosen, and then choose the **Rectangle** tool from the list box displayed next to it.

3. In the **Options Bar**, click in the **Offset** edit box and replace the existing value with **6'** (for Imperial) or **1800mm** (for Metric). Next, press ENTER.

4. Move the cursor and place it at the intersection of grid 1 and grid A and click when the **Midpoint** snap is displayed.

5. Move the cursor down toward the right and place it at the intersection of grid I and grid 3.1, and click when the **Intersection** snap is displayed.

6. Invoke the **Line** tool from the list box in the **Draw** panel of the **Modify | Create Floor Boundary** tab.

7. In the **Options Bar**, enter **3'** in Imperial unit or **900 mm** in Metric unit in the **Offset** edit box to replace the existing value.

8. Move the cursor and place it at the intersection of grid B and the right edge of the rectangle created previously, refer to Figure 5-33.

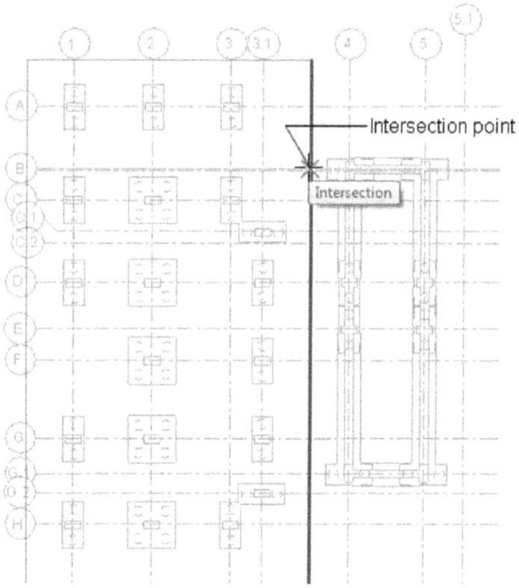

Figure 5-33 *Intersection of grid B and the right edge of the rectangle created previously*

9. Click when grid B and the right edge of the rectangle are highlighted and the **Intersection** snap is displayed.

10. Move the cursor toward the right and place it at the intersection of grid B and grid 5. Click when the **Midpoint and Horizontal** snap is displayed.

11. Move the cursor vertically down and place it at the intersection of grid G.1 and grid 5. Click when the **Midpoint and Vertical** snap is displayed.

12. Move the cursor horizontally left and place it at the intersection of grid G.1 and the right edge of the rectangle created.

13. Click when the **Intersection and Horizontal** snap is displayed. Next, press ESC.

14. In the **Modify** panel of the **Modify | Create Floor Boundary** tab, choose the **Split Element** tool; the cursor changes into a pen symbol.

15. Place the cursor on the point marked **A**, refer to Figure 5-34, and click when the split symbol appears. On doing so, a blue dot is displayed at the point you clicked. Similarly, click at the point marked **B**, refer to Figure 5-34.

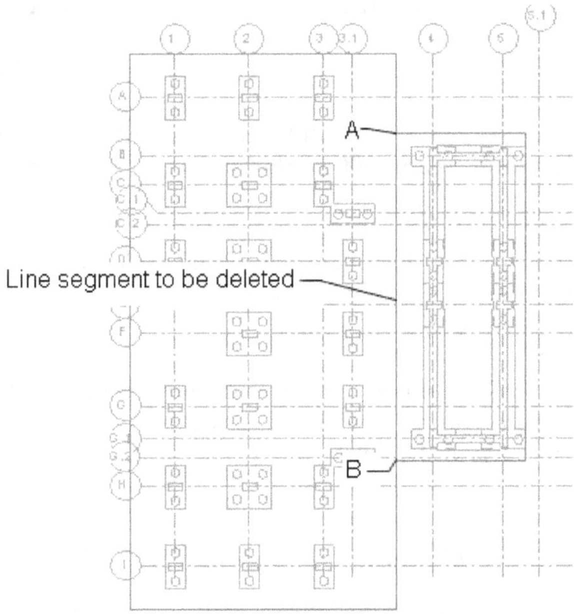

Figure 5-34 *Reference points for splitting the slab boundary and the line segment to be deleted*

16. Press ESC twice and select the split line between the points marked **A** and **B**, refer to Figure 5-34.

17. Press DELETE to remove the split line from the slab boundary.

18. Next, choose the **Finish Edit Mode** button from the **Mode** panel of the **Modify | Create Floor Boundary** tab to finish the creation of the boundary of the foundation slab. Note that a warning message may be displayed at the bottom of the screen. Choose **OK** in the warning message box to ignore it.

19. In the **Properties** palette, select the **Foundation Slab : 12" Foundation Slab** option in Imperial unit system or **Foundation Slab : 150 mm Foundation Slab** option for Metric unit system from the **Type Selector** drop-down list. Ignore any warning message, if displayed.

20. Choose the **Modify** button from the **Select** panel of the **Modify | Structural Foundations** tab to create the foundation slab.

Adding Beams

In this section, you will add beams to the **Second Floor** structural plan view of the project.

1. To change the current view to the **Second Floor** structural plan view, double-click on the **Second Floor** sub-node under the **Structural Plans** head in the **Project Browser**.

2. Now, choose the **Beam** tool from the **Structure** panel of the **Structure** tab.

3. In the **Properties** palette, select the **Concrete-Rectangular Beam : 12x24** option for Imperial or **M_Concrete-Rectangular Beam: 300 x 600mm** option for metric from the **Type Selector** drop-down list.

4. In the **Options Bar,** ensure that the **Chain** check box is cleared. Now, move the cursor and place it at the intersection of grid 1 and grid A. Click at the intersection when the **Endpoint** snap is displayed.

5. Move the cursor to the right and place at the endpoint of column. Click at the intersection when the **Endpoint and Horizontal** snap is displayed for the column placed at the intersection of grid A and grid 2. Notice that a beam is added between the intersections of grids.

6. Next, click when the **Endpoint** snap is displayed for the column placed at the intersection of grid 1 and grid A.

7. Move the cursor horizontally right and place it at the start point of the column placed at intersection point of grid A and grid 3. Click at the intersection when the **Endpoint and Horizontal** snap is displayed.

8. Repeat the procedure followed in steps 6 and 7 to add the remaining beams to the structural plan view. Refer to Figure 5-35 for the placement of beams. Note that while adding beams, a warning message may be displayed at the bottom of the screen. Choose the **OK** button to ignore the warning message box and close it.

> **Note**
> *While referring to Figure 5-35 for the placement of beams, notice the tags on the beams to distinguish between beams and structural walls.*

9. Choose the **Modify** button from the **Select** panel of the **Modify |Place Beam** tab to exit the **Beam** tool.

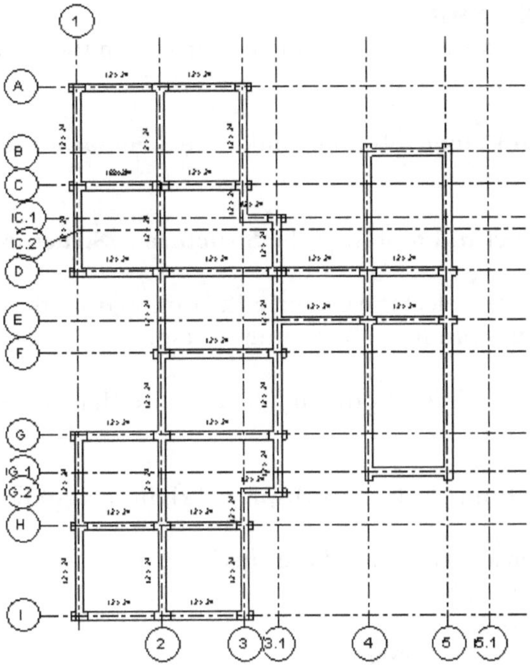

Figure 5-35 *The beams and other structural elements in the* ***Second Floor*** *structural plan view*

Modifying Beams

In this section, you will select all the instances of the beams in the project view and modify their instance properties.

1. Click on the beam placed at grid A between grid 1 and grid 2, and then right-click; a shortcut menu is displayed.

2. Choose **Select All Instances > Visible in View** from the shortcut menu, as shown in Figure 5-36; all beams in the project view area are selected.

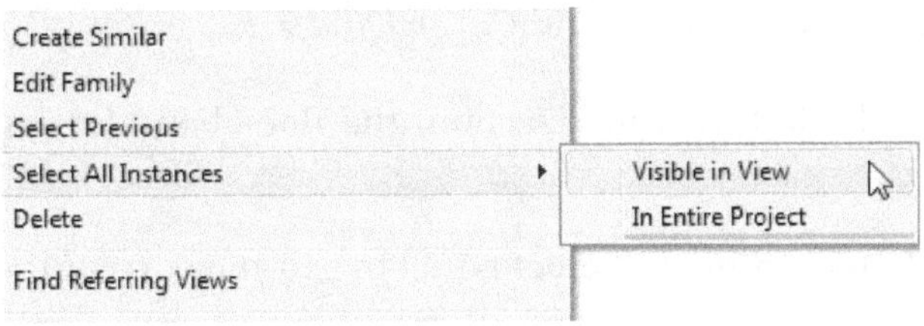

Figure 5-36 *Choosing* ***Select All Instances > Visible in View*** *from the shortcut menu*

3. In the **Properties** palette, specify the instance parameters as:

For Imperial	Start Level Offset = **-1'**	End Level Offset = **-1'**
For Metric	Start Level Offset = **-300mm**	End Level Offset = **-300mm**

4. Next, choose **Apply** in the **Properties** palette and then press ESC.

Adding a Structural Floor

In this section, you will add a structural floor to the project in the **Second Floor** structural plan view.

1. Ensure that the **Second Floor** sub-node under the **Structural Plans** node is highlighted in the **Project Browser**.

2. Invoke the **Floor: Structural** tool from the **Structure > Structure > Floor** drop-down.

3. In the **Properties** palette, select the **Generic 12"** option from the **Type Selector** drop-down list or **Generic 300mm** option in Metric unit system.

4. In the **Modify | Create Floor Boundary** tab, ensure that the **Boundary Line** option is chosen in the **Draw** panel.

5. Choose the **Line** tool from the list box displayed right next to the **Boundary Line** tool.

6. In the **Options Bar**, ensure that the **Chain** check box is selected and then in the **Offset** edit box, replace the default value.

   For Imperial **1'**
   For Metric **300 mm**

7. Move the cursor and place it at the intersection of grid A and grid 1 and click when the **Endpoint** snap is displayed.

8. Move the cursor toward the right and place it at the intersection of grid A and grid 3. Click when the **Endpoint and Horizontal** snap is displayed.

9. Click on the other grid intersections in the following order:
   (C.1, 3); (C.1, 3.1); (D, 3.1); (D, 4); (C.1, 4); (C.1, 5); (F, 5); (F, 4); (E, 4); (E, 3.1); (G.2, 3.1); (G.2, 3); (I, 3); (I, 1); (G,1); (G, 2); (F, 2); (D, 2); (D, 1); and (A, 1)

   Note
   The grid intersection points are denoted by their intersecting grid lines. For example, (C.1, 3.1) represents the intersection point of grid C.1 and grid 3.1

   After you click on the last grid intersection, the floor boundary is created, as shown in Figure 5-37.

10. Choose the **Modify** tool from the **Select** panel to exit the current tool.

11. Choose the **Trim/Extend to Corner** tool and then select the recently created line and the horizontal line starting from grid 1; the two lines get joined.

12. Next, choose the **Finish Edit Mode** button from the **Mode** panel of the **Modify | Create Floor Boundary** tab; the **Revit** message box is displayed.

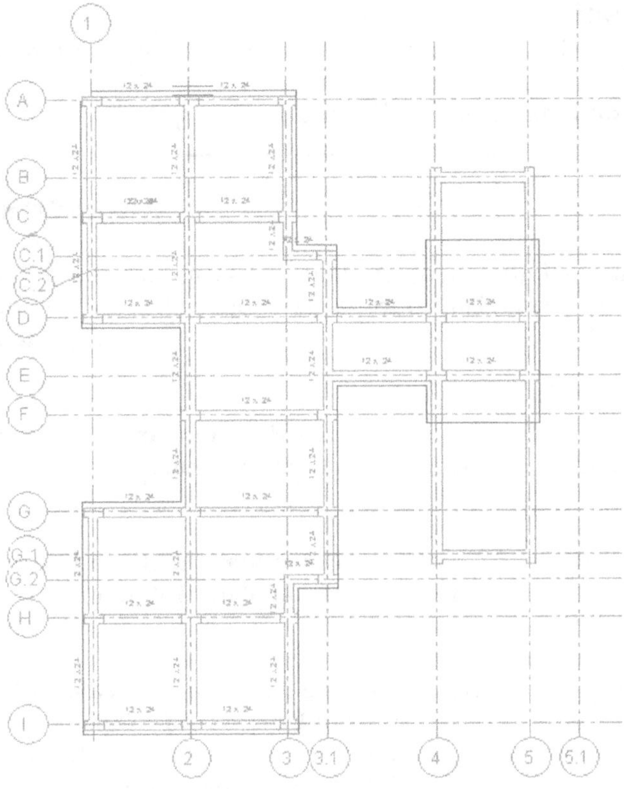

Figure 5-37 *Structural plan displaying the floor boundary*

13. Choose **Yes** from the message box; the **Autodesk Revit 2017** warning box is displayed. Choose **OK** from the warning box; a structural floor is created and highlighted in blue. Ignore any other warning messages, if displayed.

14. Choose the **Modify** button from the **Select** panel of the **Modify | Floors** tab to complete the floor creation.

Saving the Project

In this section, you will save the project file using the **Save As** tool.

1. To save the project, choose the **Save As > Project** option from the **Application Menu**; the **Save As** dialog box is displayed.

2. In this dialog box, browse to the *C:\rst_2017* location and create a folder with name *c05_rst_2017_tut*. Open the *c05_rst_2017 _tut* folder and enter **c05_Commercial-Complex_tut1** for Imperial and **M_c05_Commercial-Complex_tut1** for Metric in the **File name** edit box.

3. Choose the **Save** button; the **Save As** dialog box closes and the project file is saved.

Closing the Project

1. To close the project, choose the **Close** option from the **Application Menu**.

   The file is closed and this completes Tutorial 1 of Chapter 5.

Tutorial 2 Industrial Complex

In this tutorial, you will add foundation, foundation slab, and structural framing members to the Industrial Complex project created in Tutorial 2 of Chapter 4. Use the following project parameters and specifications: **(Expected time: 1 hr 45 min)**

1. Foundation:
   For Imperial Type: **Footing - Rectangular : 72"x48"x18"**
   For Metric Type: **M_Footing - Rectangular : 1800 x1200 x450mm**

2. Foundation Slab:
   For Imperial Type: **Foundation Slab : 6" Foundation Slab**
   For Metric Type: **Foundation Slab : 150mm Foundation Slab**

3. Beams:
   For Imperial Type: **Concrete-Rectangular Beam : 12x24**
   W-Wide Flange : W14X30
   For Metric Type: **M_Concrete-Rectangular Beam : 300 x 600mm**
   M_W-Wide Flange : W760X350
4. Structural Floor:
   For Imperial Type: **LW Concrete on Metal Deck**
   1 1/2" Metal Roof Deck
   For Metric Type: **160mm Concrete with 50mm Metal Deck**
   75mm Metal Roof Deck

5. File name to be assigned:
   For Imperial *c05_ Industrial-Complex_tut2.rvt*.
   For Metric *M_c05_ Industrial-Complex_tut2.rvt*.

   The following steps are required to complete this tutorial:

a. Open the project file.
   For Imperial *c04_Industrial-Complex_tut2.rvt*
   For Metric *M_c04_Industrial-Complex_tut2.rvt*
b. Add foundations.
c. Add foundation slab.
d. Add beams.
e. Add structural floor.
f. Save the project using the **Save As** tool.
g. Close the project using the **Close** tool.

Opening the Existing Project

In this section, you will open the project file created in Tutorial 2 of Chapter 4.

1. To open the project file, choose **Open > Project** from the **Application Menu**; the **Open** dialog box is displayed.

2. In this dialog box, browse to the *C:\rst_2017\c04_rst_2017_tut* location and then choose the *c04_Industrial-Complex_tut2.rvt* for Imperial or *M_c04_Industrial-Complex_tut2.rvt* for Metric project file. You can also download this file from *http://www.cadcim.com*. The path of the file is as follows: *Textbooks > Civil/GIS > Revit Structure > Exploring Autodesk Revit 2017 for Structure* .

3. Choose the **Open** button from the **Open** dialog box; the selected project file opens in the drawing window in the **Entry** structural plan view.

Adding Isolated Footing Foundations

In this section, you will add the isolated footing foundations to columns.

1. In the **Entry** structural plan view, choose the **Isolated** tool from the **Foundation** panel of the **Structure** tab; the **Modify | Place Isolated Foundation** tab is displayed.

2. In the **Properties** palette, select the specified option from the **Type Selector** drop-down list if it is not selected by default.
   For Imperial **Footing - Rectangular : 72"x48"x18"**
   For Metric **M_Footing - Rectangular : 1800 x1200 x450mm**

3. Choose the **At Grids** tool from the **Multiple** panel of the **Modify | Place Isolated Foundation** tab .

4. Next, move the cursor and place it over grid C and click when the grid is highlighted.

5. Press and hold the CTRL key, and select the grids marked 1, 2, 3, 4, 5, 6, 7, 8, 9, A', B', A, and B.

6. Release the CTRL key and choose the **Finish** button from the **Multiple** panel of the **Modify | Place Isolated Foundation > At Grid Intersection** tab; the isolated foundations are added to the project view, refer to Figure 5-38. Ignore and close the warning message box if displayed.

7. Press ESC to finish the insertion of the isolated foundations and then exit from the **Isolated** tool.

8. Next, you need to delete the excess foundation added to the project view. To do so, move the cursor and place it toward the isolated foundation added at the intersection of the radial grids. For location of the excess foundation, refer to Figure 5-38. Click when the foundation is highlighted.

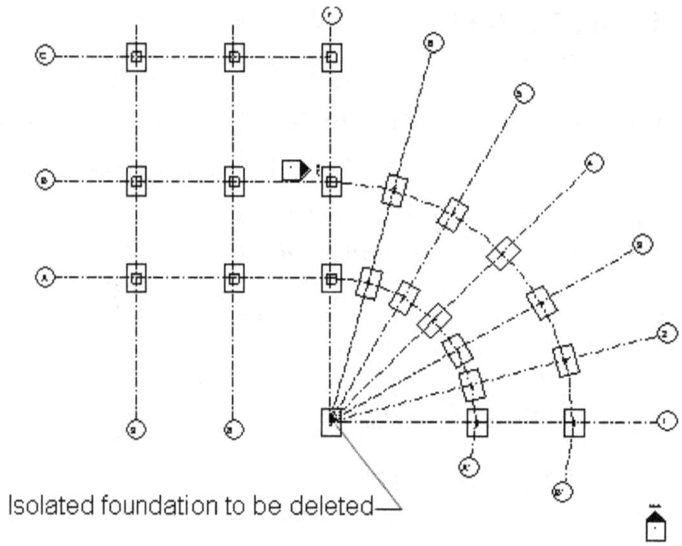

Figure 5-38 Added isolated foundations

9. Press DELETE; the isolated foundation is deleted from the project.

Adding a Foundation Slab

In this section, you will add a foundation slab in the **Foundation Level** structural plan.

1. Double-click on the **Foundation Level** node under the **Structural Plans** head in the **Project Browser** and then invoke the **Structural Foundation: Slab** tool from **Structure > Foundation > Slab** drop-down.

2. In the **Draw** panel of the **Modify | Create Floor Boundary** tab, ensure that the **Boundary Line** tool is chosen and then invoke the **Rectangle** tool from the list box displayed next to it.

3. In the **Options Bar**, click in the **Offset** edit box. Replace the existing value in this edit box with **1'** for Imperial or **300mm** for Metric then press ENTER.

4. Move the cursor and place it at the intersection of grid C and grid 9. Click when the **Intersection** snap is displayed at the grid intersection.

Note

*If the **Intersection** snap is not displayed, you can right-click and choose **Snap Overrides > Intersection** from the shortcut menu displayed.*

5. Move down the cursor toward the right and place it at the grid intersection of grid A and grid 7. Click when the **Endpoint** snap is displayed at the intersection.

6. Choose the **Finish Edit Mode** button from the **Mode** panel of the **Modify | Create Floor Boundary** tab; the floor boundary is highlighted in blue.

7. In the **Properties** palette, select the option from the **Type Selector** drop-down list if not selected by default.

   For Imperial **Foundation Slab : 6" Foundation Slab**
   For Metric **Foundation Slab : 150mm Foundation Slab**

8. Choose the **Modify** button from the **Select** panel of the **Modify| Structural Foundations** tab to exit the **Foundation Slab** tool.

Adding Beams

In this section, you will add concrete and flange beams to the project view.

1. Double-click on the **Second Floor** node under the **Structural Plans** head in the **Project Browse**r; the **Second Floor** structural plan view is displayed.

2. Invoke the **Beam** tool from the **Structure** panel of the **Structure** tab.

3. In the **Properties** palette, select the option from the **Type Selector** drop-down list.

   For Imperial **Concrete- Rectangular Beam : 12 x 24**
   For Metric **M_Concrete-Rectangular Beam : 300 x 600mm**

4. Choose the **On Grids** tool from the **Multiple** panel of the **Modify | Place Beam** tab. Also, ensure that the **Tag on Placement** tool is selected in the **Tag** panel of this tab.

5. Select grid C by clicking on it; the beams of the selected type are added between the intersections, as shown in Figure 5-39.

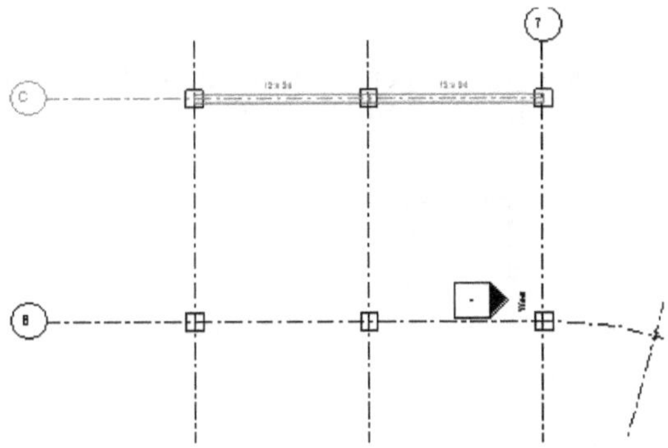

***Figure 5-39** Beams added at grid C*

6. Press and hold the CTRL key and select the grids A, B, 7, 8, and 9; the beams are added between the intersections, as shown in Figure 5-40.

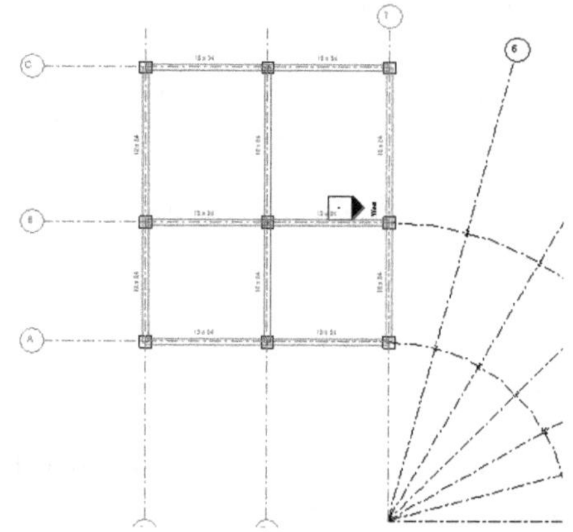

Figure 5-40 Rectangular concrete beams added to the project

7. Choose the **Finish** button from the **Multiple** panel of the **Modify | Place Beam > On Grid Lines** tab.

8. In the **Properties** palette, select option from the **Type Selector** drop-down list.
   For Imperial **W-Wide Flange : W14X30**
   For Metric **M_W-Wide Flange : W760X350**

9. Choose the **On Grids** tool from the **Multiple** panel of the **Modify | Place Beam** tab.

10. Select grid 6; a beam of the selected type is added between the intersections of grid 6 with other grids.

11. Press and hold the CTRL key and select the grids 5, 4, 3, 2, 1, A', and B'; the flange beams are added between grids, as shown in Figure 5-41.

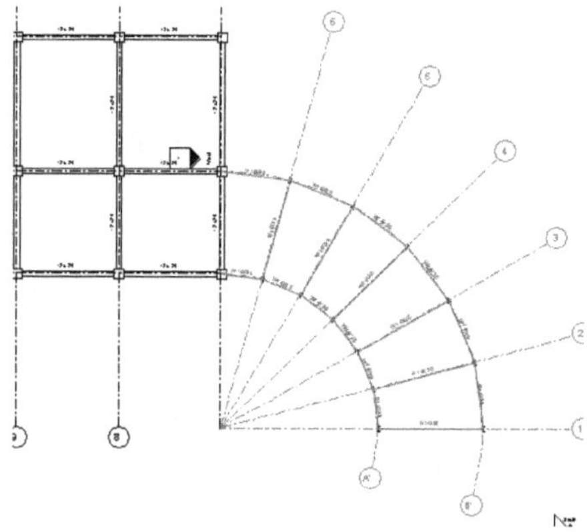

Figure 5-41 Flange beams added to the project view

12. Choose the **Finish** button from the **Multiple** panel of the **Modify | Place Beam > On Grid Lines** tab. Press ESC to exit the **Beam** tool.

Adding the Structural Floors

In this section, you will add the structural floors to the project view.

1. Invoke the **Floor: Structural** tool from **Structure > Structure > Floor** drop-down; the **Modify | Create Floor Boundary** tab is displayed.

2. In the **Properties** palette, click in the value field corresponding to the **Height Offset From Level** parameter and enter **19/4"** for Imperial or enter **120mm** for Metric and then press ENTER. Next, choose the **Edit Type** button; the **Type Properties** dialog box is displayed.

3. In the **Type Properties** dialog box, select required option from the **Type** drop-down list.
   For Imperial **LW Concrete on Metal Deck**
   For Metric **160mm Concrete with 50mm Metal Deck**

4. Choose the **OK** button to close the **Type Properties** dialog box.

5. In the **Draw** panel of the **Modify | Create Floor Boundary** tab, choose the **Boundary Line** tool, if not chosen.

6. Invoke the **Rectangle** tool from the list box displayed next to the **Boundary Line** tool.

7. In the **Options Bar**, replace the existing value in the **Offset** edit box by entering 1 and press ENTER.

8. Move the cursor and place it at the intersection of grid 9 and grid C. Click when the **Endpoint** snap is displayed.

9 Next, move the cursor downward toward the right and place it at the intersection of grid A and grid 8.

10. Click when the **Endpoint** snap appears at the intersection.

11. Choose the **Finish Edit Mode** button from the **Mode** panel of the **Modify | Create Floor Boundary** tab; the floor boundary is created, as shown in Figure 5-42 and is highlighted in blue color. Press ESC to exit the **Structural Floor** tool.

12. Invoke the **Floor: Structural** tool from **Structure > Structure > Floor** drop-down; the **Modify | Create Floor Boundary** tab is displayed.

13. In the **Draw** panel of the **Modify | Create Floor Boundary** tab, choose the **Boundary Line** tool if not chosen by default.

14. Choose the **Pick Supports** tool from the list box displayed next to the **Boundary Line** tool chosen.

15. Place the cursor on the beam at grid 7 between grids A and B, as shown in Figure 5-43; the beam gets highlighted. Click on the highlighted beam to add the first edge of the structural floor.

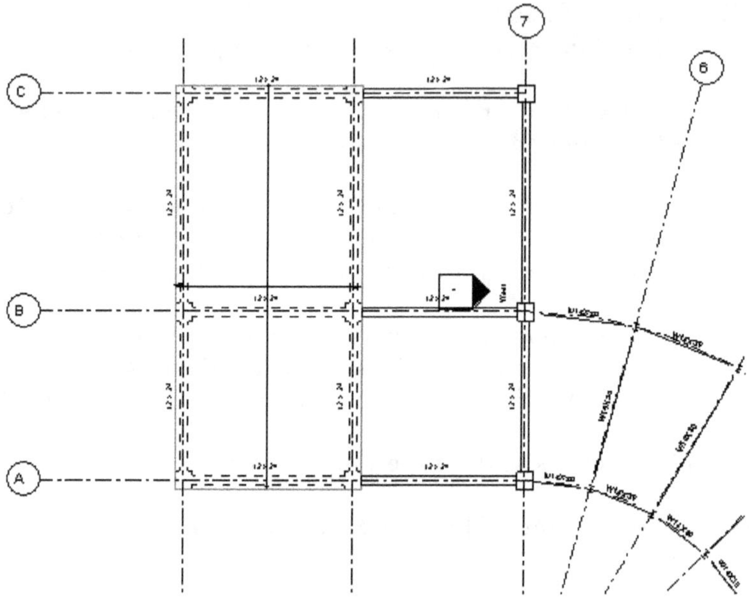

Figure 5-42 *The added structural floor*

16. Next, choose the **Pick Lines** tool from the list box in the **Draw** panel of the **Modify | Create Floor Boundary** tab.

17. Move the cursor and place it over the beam located at grid B' between grids 6 and 7; the beam gets highlighted. Click on the highlighted beam.

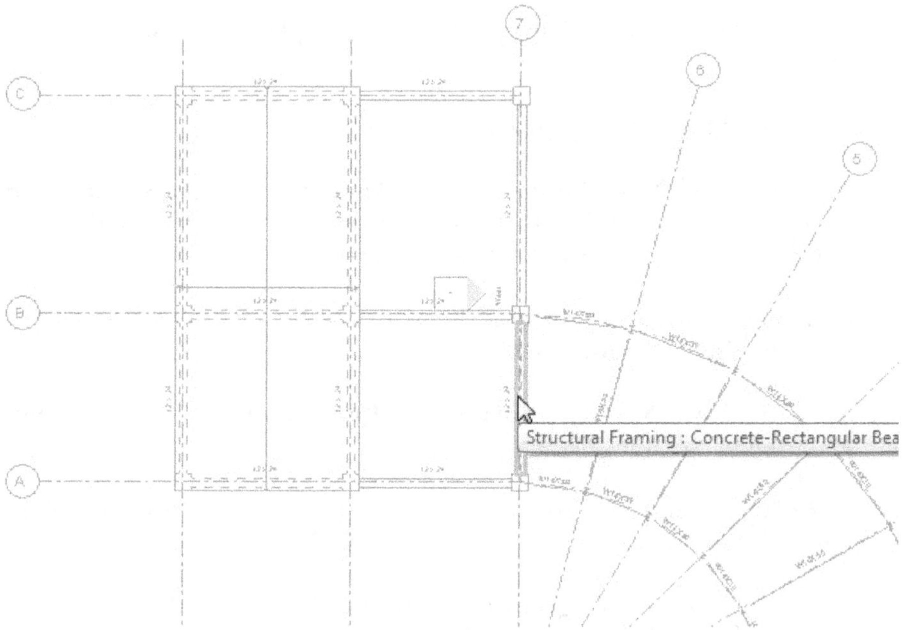

Figure 5-43 *The selected beam highlighted*

18. Next, one by one, click on all the outer flange beams placed on grids B', 1, and A', to define the floor boundary. After you have clicked on all the outer flange beams, a floor boundary is created, as shown in Figure 5-44.

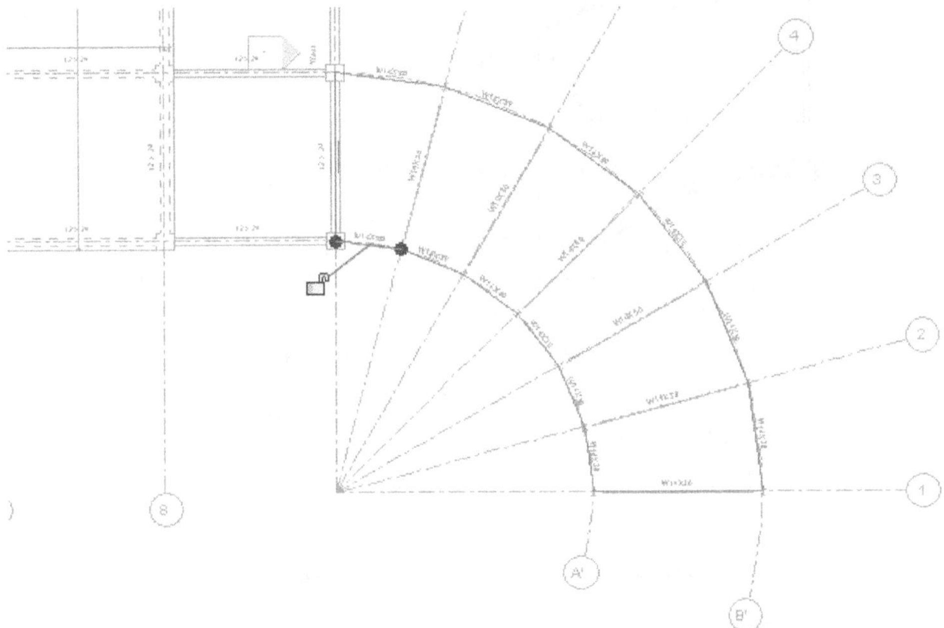

Figure 5-44 *The structural floor boundary over the flanges*

19. Choose the **Finish Edit Mode** button from the **Modify | Create Floor Boundary** tab; floor will be created, as shown in Figure 5-45.

20. In the **Properties** palette, select the option from the **Type Selector** drop-down list.
    For Imperial **Floor : 1 1/2" Metal Roof Deck**
    For Metric **Floor : 75mm Metal Roof Deck**

21. Press ESC to exit the **Structural Floor** tool.

Saving the Project

In this section, you will save the project file using the **Save As** tool.

1. To save the project, choose the **Save As > Project** option from the **Application Menu**; the **Save As** dialog box is displayed.

2. In this dialog box, browse to the *C:\rst_2017\c05_rst_2017_tut* location and enter **c05_Industrial-Complex_tut2** for Imperial or enter **M_c05_Industrial-Complex_tut2** for Metric in the **File name** edit box.

3. Choose the **Save** button; the **Save As** dialog box closes and the project file is saved.

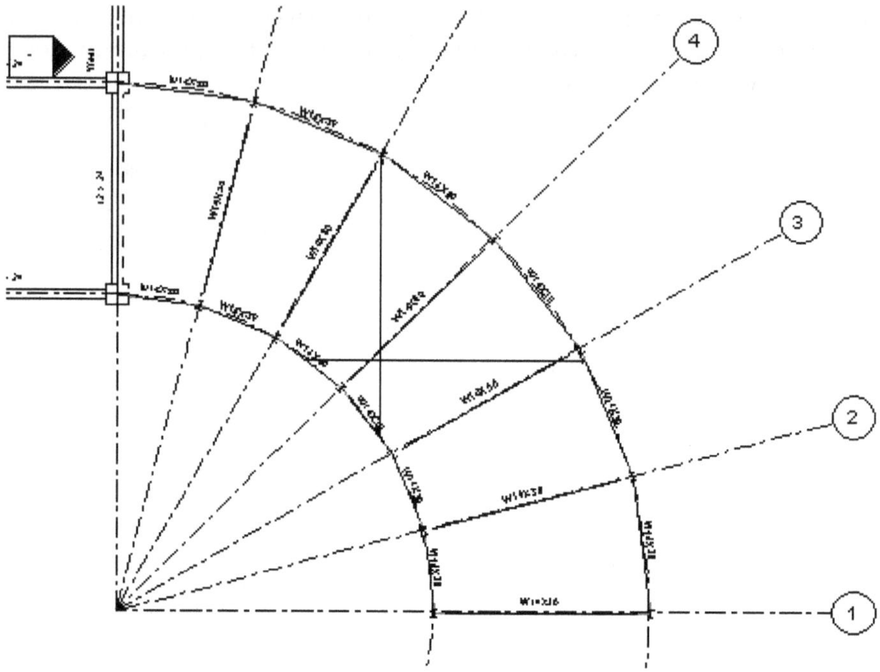

Figure 5-45 *The structural floor created*

Closing the Project

1. To close the project, choose the **Close** option from the **Application Menu**.

   The file is closed and this completes Tutorial 2 of Chapter 5.

Self-Evaluation Test

Answer the following questions and then compare them to those given at the end of this chapter:

1. __________ acts as an interface to transfer the load of a building structure to the earth strata.
2. __________ foundation is a footing that is attached to a wall.
3. The __________ instance parameter is used to specify the type of wall usage for the foundation.
4. __________ instance parameter is used to specify the distance of the top surface of structural floor from a level.
5. The __________ parameter is a read-only instance parameter which shows whether or not a foundation slab has been created from a mass object.

6. You can specify the distance at which the center of a foundation will be offset from the center of the host wall by using the __________ parameter.

7. An Isolated Foundation is a concrete slab placed under a column. (T/F)

8. A structural floor represents a floor deck or a floor slab. (T/F)

9. You can associate a material with an isolated footing foundation. (T/F)

10. Slab foundation does not transmit the load of a superstructure to the earth. (T/F)

Review Questions

Answer the following questions:

1. You can use the __________ parameter to specify whether a wall foundation remains continuous beneath the opening of a structural wall or not.

2. You can create an opening on the face of a structure by using the __________ tool from the **Structure** tab of the ribbon.

3. You can use the __________ tool to define a slope for a foundation slab.

4. In the **Properties** palette of a foundation slab, the instance parameters under the __________ head can be used to specify the creation and demolition phase.

5. In the **Properties** palette of a slab foundation, the **Structural Analysis** head contains the __________ parameter.

6. The __________ tool in the **Foundation** panel can be used to add a foundation slab to a structural model.

7. The **Structural** parameter specifies whether or not a foundation slab will have an analytical model. (T/F)

8. An isolated footing foundation is placed under a column. (T/F)

9. Shallow foundations transfer load from a building to the earth strata near the ground surface. (T/F)

10. In Autodesk Revit , you can add a bearing foundation to a structural model. (T/F)

EXCERCISES

Exercise 1 Academic Institution

In this exercise, you will add foundations, beams, and structural floor to the project file created for Exercise 1 of Chapter 4, refer to Figures 5-46 and 5-47. Use the following project parameters and specifications: **(Expected time: 1 hr 15 min)**

1. Foundations: (To be added in the **Entry** structural plan view)
   For Imperial Type: **Wall Foundation Bearing Footing : 36"x12"**
   Footing Rectangular : 72"x48"x18"
   For Metric Type: **Wall Foundation Bearing Footing : 900 x 300**
   M_Footing - Rectangular : 1800 x1200 x450mm

2. Beams: (To be added in the **Second Floor** structural plan view)
   For Imperial Type: **Concrete-Rectangular Beam : 12x24**
   For Metric Type: **M_Concrete-Rectangular Beam : 300 x600**

3. Structural Floor:
   For Imperial Type: **3" LW Concrete on 2" Metal Deck**
   1 1/2" Metal Roof Deck
   For Metric Type: **160mm LW Concrete on 50mm Metal Deck**
   75mm Metal Roof Deck

4. Project file to be used:
   For Imperial *c04_Academic-Institution_exer1.rvt.*
   For Metric *M_c04_Academic-Institution_exer1.rvt.*

5. File name to be assigned:
   For Imperial *c05_Academic-Institution_exer1.rvt.*
   For Metric *M_c05_Academic-Institution_exer1.rvt.*

6. For location and placement of beams, foundations and other structural elements, refer to Figure 5-46 and Figure 5-47.

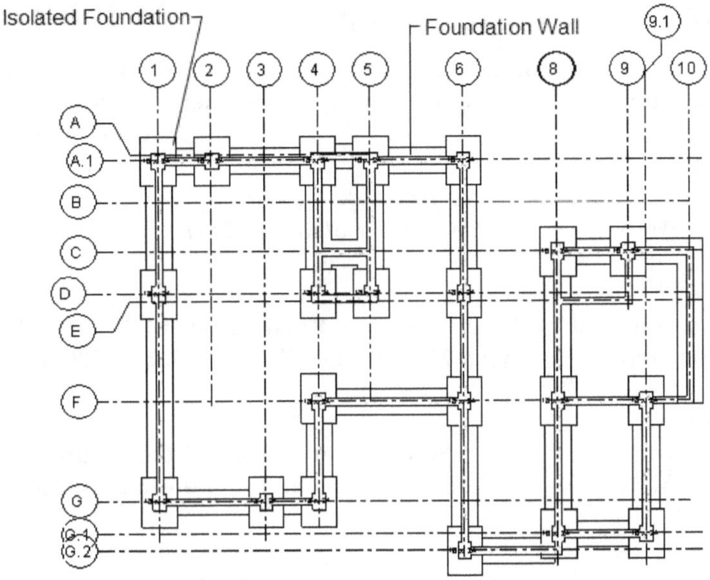

Figure 5-46 *Structural plan view at the **Entry** level*

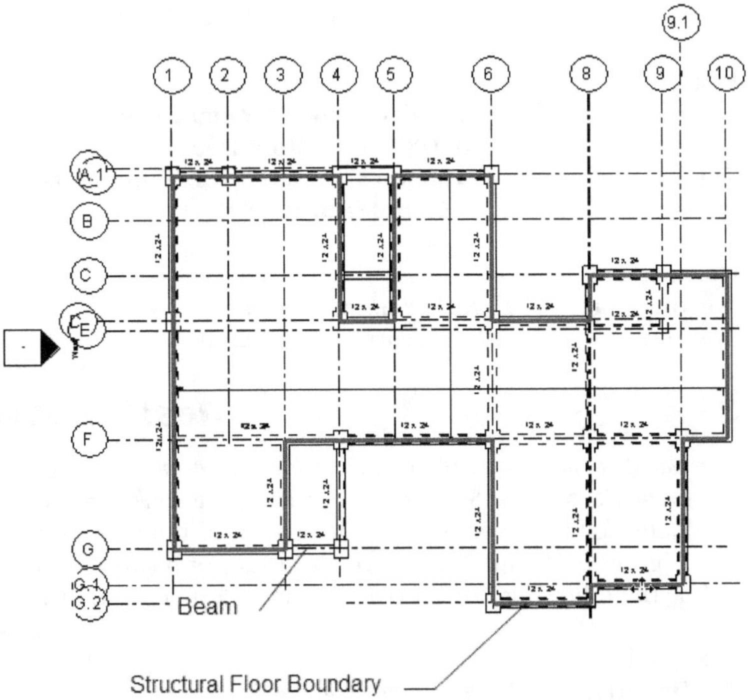

Figure 5-47 *Structural plan view at the **Second Floor** level*

Exercise 2 Factory Shed

In this exercise, you will add foundations, beams, and structural floor to the project file created in Exercise 2 of Chapter 4. Use the following project parameters and specifications:

(Expected time: 1 hr 45 min)

1. Project file to be used:
   For Imperial *c04_Factory-Shed_exer2.rvt.*
   For Metric *M_c04_Factory-Shed_exer2.rvt.*

2. Foundations:
   For Imperial Type: **Footing - Rectangular : 72"x48"x18"**
   For Metric Type: **M_Footing - Rectangular : 1800 x1200 x450mm**

3. Foundation Slab:
   For Imperial Type: **Foundation Slab : 12" Foundation Slab**
   For Metric Type: **Foundation Slab : 300mm Foundation Slab**

4. Beams:
   For Imperial Type: **Concrete-Rectangular Beam : 12x24**
   W-Wide Flange : W14X30
   For Metric Type: **M_Concrete-Rectangular Beam : 300 x 600mm**
   W-Wide Flange : W760X350

5. Structural Floor:
   For Imperial Type: **LW Concrete on Metal Deck**
   1 1/2" Metal Roof Deck
   For Metric Type: **160mm Concrete with 50mm Metal Deck**
   75mm Metal Roof Deck

6. File name to be assigned:
   For Imperial *c05_ Factory-Shed_exer2.rvt.*
   For Metric *M_c05_ Factory-Shed_exer2.rvt.*

Exercise 3 Residential Builiding

Download the *c05_residential_bldg_exer3* file from *http://www.cadcim.com*. The path of the file is as follows: *Textbooks > Civil/GIS > Revit Structure > Exploring Autodesk Revit 2017 for Structure*. In this exercise, you will add foundations, beams, and structural floor to the residential building project file. Refer to Figures 5-48 and 5-49 for the locations of beams, floors, and foundations in the project file. Use the following project parameters and specifications:

(Expected time: 1 hr)

1. Project file to be used:
   For Imperial *c05_residential_bldg_exer3.rvt.*
   For Metric *M_c05_residential_bldg_exer3.rvt.*

2. Foundations:
   For Imperial Type: **Footing - Rectangular : 72"x48"x 20"**
   For Metric Type: **M_Footing - Rectangular : 1800 x1200 x450mm**

3. Foundation Slab:

For Imperial	Type: **Foundation Slab : 6" Foundation Slab**
For Metric	Type: **Foundation Slab : 150mm Foundation Slab**

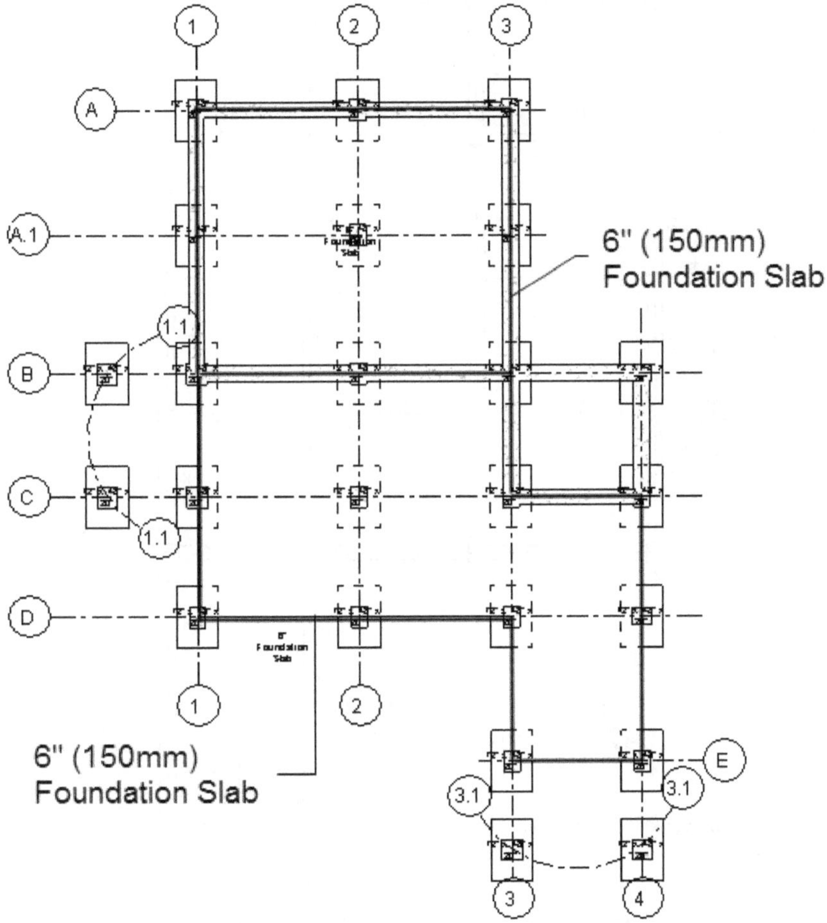

Figure 5-48 *Structural plan view at the **Entry** level*

4. Beams:

For Imperial	Type: **Concrete-Rectangular Beam : 16x32**
For Metric	Type: **M_Concrete-Rectangular Beam : 400x800mm**

5. Structural Floor:

For Imperial	Type: **6" Concrete Floor**
For Metric	Type: **Insitu Concrete 225mm**

6. File name to be assigned:

For Imperial	*c05_ residential_bldg_exer3a.rvt.*
For Metric	*M_c05_ residential_bldg_exer3a.rvt.*

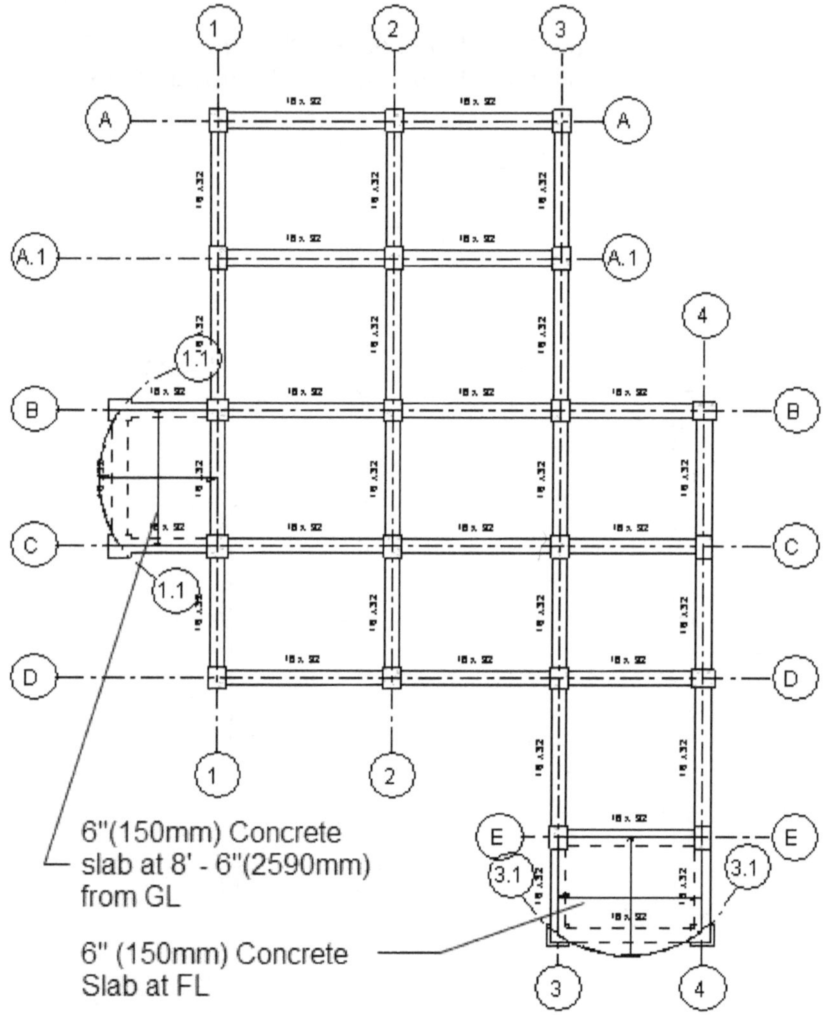

***Figure 5-49** Structural plan view at the **FL** level*

Answers to Self-Evaluation Test

1. Foundation, **2.** Wall, **3. Structural Usage**, **4. Height Offset From Level**, **5. Related to Mass**, **6. Eccentricity**, **7.** T, **8.** T, **9.** T, **10.** F

Chapter 6

Editing Tools

Learning Objectives

After completing this chapter, you will be able to:

- *Create a selection set of elements*
- *Move and copy elements*
- *Use the Cut and Paste tools*
- *Understand the use of the Rotate and Mirror tools*
- *Create an array of elements using the Array tool*
- *Group elements*
- *Use the Match, Align, Trim, Offset, and Delete tools*

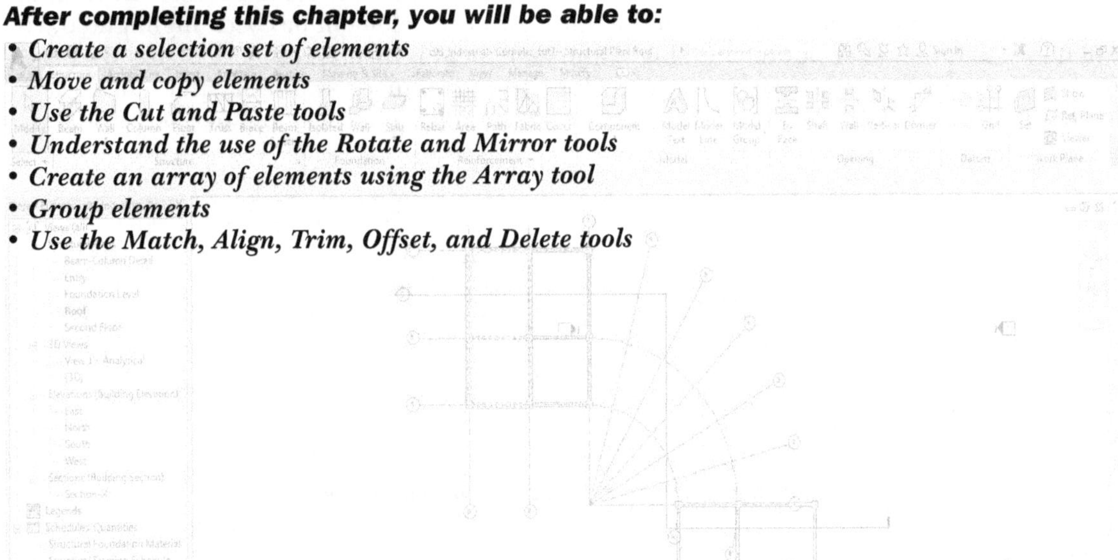

In the previous chapters, you have learned to create structural elements such as structural walls, columns, and framing elements. In this chapter, you will learn about the editing process and the tools that are used to modify elements based on the design requirement. Some of the editing processes such as modifying type and instance properties, flipping orientation, and many more have already been discussed in the previous chapters. Also, you will learn to select and edit elements by using the object controls and editing tools such as **Move**, **Copy**, **Trim/Extend**, **Delete**, **Rotate**, **Mirror**, **Array**, **Align**, **Match**, and so on.

SELECTING ELEMENTS

Selecting elements in a drawing is an important and regular task needed for editing elements. Therefore, it is important to learn the basic and advanced methods of selection. Learning these methods will help you speed up the selection process as well as reduce the errors that may occur due to the wrong selection of elements.

In Revit Structure, you need to select the element or group of elements for editing. There are several methods that can be used to select the elements from a drawing. The method used to select an element may differ depending on the nature of selection. For instance, you may need to select a single element, multiple elements of different categories, multiple elements of same categories, and so on. These methods are discussed in detail next.

Selecting a Single Element

You can select a single element from a drawing by using the automatic highlighting feature. To select an element using this feature, move the cursor near or on the element to be selected; the outline of the desired element will be highlighted. After the element has been highlighted, you will notice that the description of the highlighted element is displayed in the **Status Bar**. Also, note that a similar description may appear along with the cursor as tooltip after a brief delay, as shown in Figure 6-1. Once the element has been highlighted, click on it to select it.

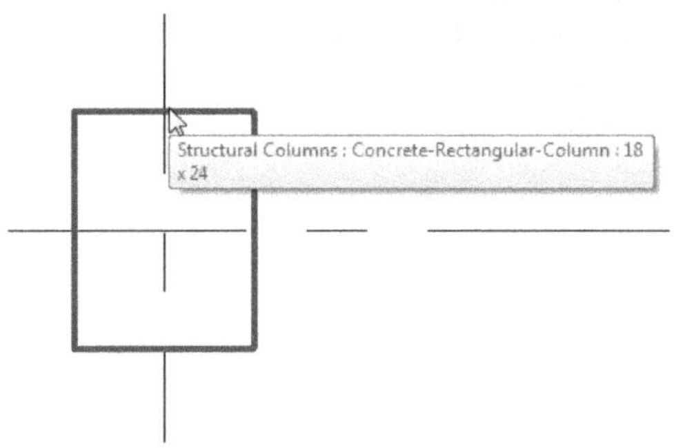

Figure 6-1 *Highlighted wall outline with the tooltip*

Note

In a project, if an element is selected in one view, it will remain selected in other views.

When an element is selected, it becomes transparent and its color changes to blue, and also the temporary dimensions are displayed. You can modify the temporary dimensions to adjust the position of the element with respect to the adjacent elements. Additionally, when you select an element, a contextual tab consisting of editing tools is also displayed. You will also notice the selection count on the **Status Bar**. Selection count displays the number of elements selected.

In Revit Structure, you can also select multiple elements. Various methods to select multiple elements from a drawing are discussed next.

Selecting Multiple Elements

You can select multiple elements from a drawing in a number of ways, such as using the CTRL key, the TAB key, window selection method, and so on. The selection method used to select multiple elements depends on the user as well as the project requirement. Various methods for selecting multiple elements are discussed next.

Selecting Elements Using the CTRL Key

You can select multiple elements from a drawing by using the CTRL key. This method is used when you need to select multiple elements from the present selection set. To select multiple elements from a drawing, first select any one of the elements from the drawing area and then press and hold the CTRL key; a '+' symbol along with the cursor will appear. Next, move the cursor to the next element. When this element has been highlighted, click on it; the desired element will be added to the current selection set. If you need to add more elements to the current selection set, keep the CTRL key pressed and click on the elements to be added to the current selection set. However, if you have released the CTRL key and want to add another element to the current selection set, move the cursor to the desired element until the element has been highlighted. After the element has been highlighted, press and hold the CTRL key and then click on the desired element; the desired element will be added to the current selection set.

Selecting Structural Elements Using a Selection Box

You can select multiple elements of same or different categories simultaneously, by drawing a selection box around them. To draw a selection box, place the cursor at one corner of the region where the elements to be selected are placed and then drag it diagonally toward the other side to form a rectangular boundary. When you drag the cursor from left to right, a rectangle is formed in continuous lines and the elements that are completely inside it are selected, refer to Figure 6-2. Alternatively, when you drag the cursor from right to left, a dashed rectangle appears and all items that are partially or fully inside it are selected.

Selecting Walls or Lines Using the TAB Key

You can use the TAB key to select walls, lines, or both the walls and lines that are joined together and form a chain. To select elements by using this method, first highlight any of the elements in the selection chain and then press the TAB key; all the elements that are joined to the highlighted element will be highlighted for selection. Click to select the elements. Figure 6-3 shows a wall chain selected by using the TAB key.

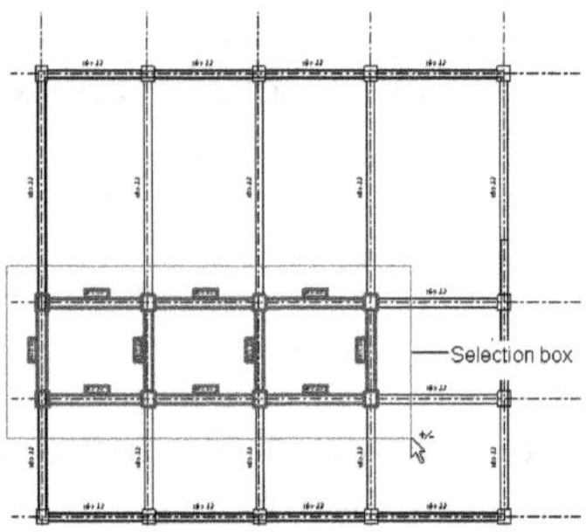

***Figure 6-2** The selection box*

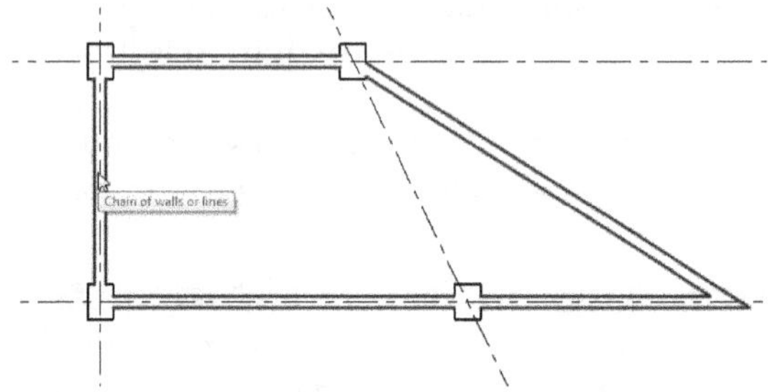

***Figure 6-3** The chain of walls selected*

Note

The method of selection of elements using the TAB key has certain limitations. This method cannot be used if the walls are not properly joined to each other.

You can also use the TAB key to select multiple wall or line elements joined at a common endpoint. To select the elements using this method, first place the cursor on or near any wall or line; the wall or line will be highlighted, as shown in Figure 6-4. Next, press the TAB key; the next wall or line that is closest to the cursor will be highlighted, refer to Figure 6-5. In this figure, the cursor is positioned on the right side of the lower left wall and therefore is closest to the wall connected on the right. As a result, the wall connected on the right side also gets highlighted, refer to Figure 6-5. Now, you can click to select both the highlighted walls that are connected to a common endpoint. However, if you want to select all the walls connected at the common endpoint, press the TAB key; all the walls at the common endpoint will be highlighted, refer to Figure 6-6. Next, click to select the highlighted walls.

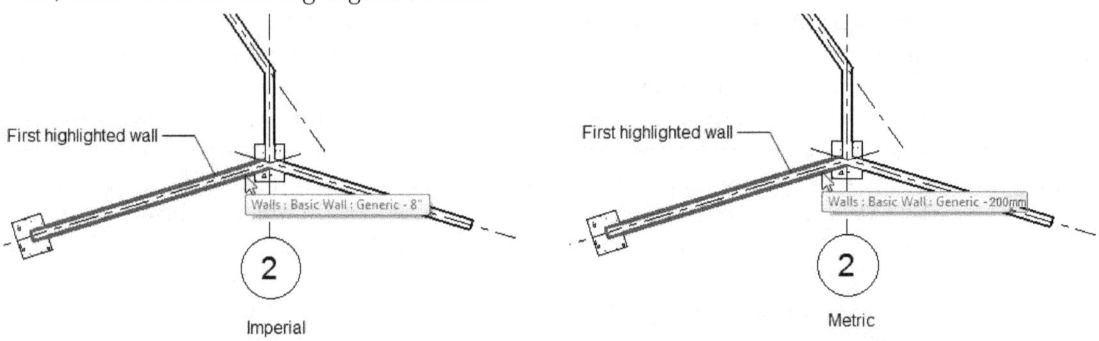

***Figure 6-4** The first wall highlighted*

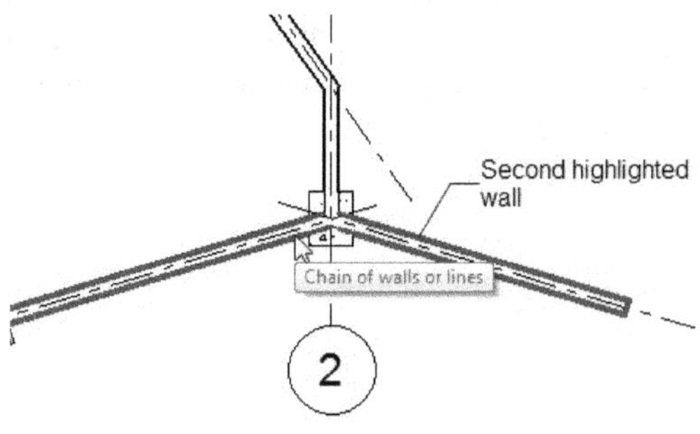

***Figure 6-5** The second wall highlighted*

Selecting Elements Using the Select All Instances Option

In Revit Structure, you can select the elements of similar instances from a drawing. For example, if you need to select all instances of columns of the **Concrete-Rectangular-Column-24x30** type and change them to a different type, you can use the **Select All Instances** option. To use this option, first select any of the instances of the **Concrete-Rectangular-Column-24x30** type from the drawing area, refer to Figure 6-7. Next, right-click to display a shortcut menu. Choose **Select All Instances > Visible in View** from the shortcut menu; all elements of similar instances will be selected, refer to Figure 6-8. Alternatively, to select all instances from the drawing, select an element and then type **SA**. You can also select all instances of the selected element in the entire structural model. To do so, right-click and then choose **Select All Instance > In Entire Project** from the shortcut menu displayed.

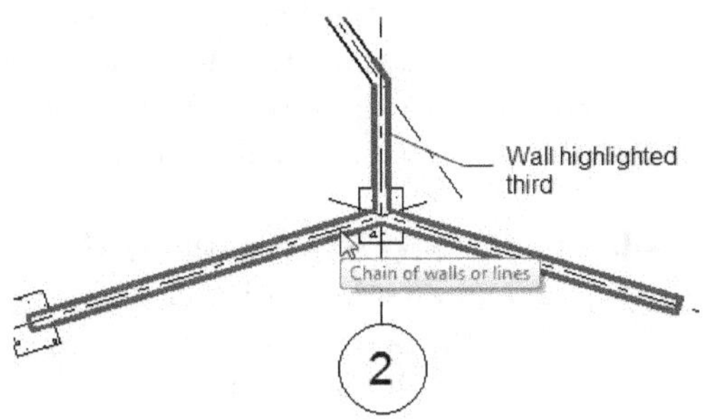

Figure 6-6 *The third wall highlighted*

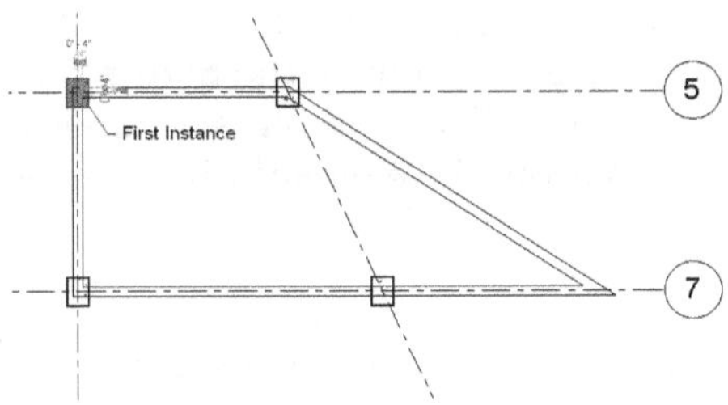

Figure 6-7 *Selecting the first instance from the drawing area*

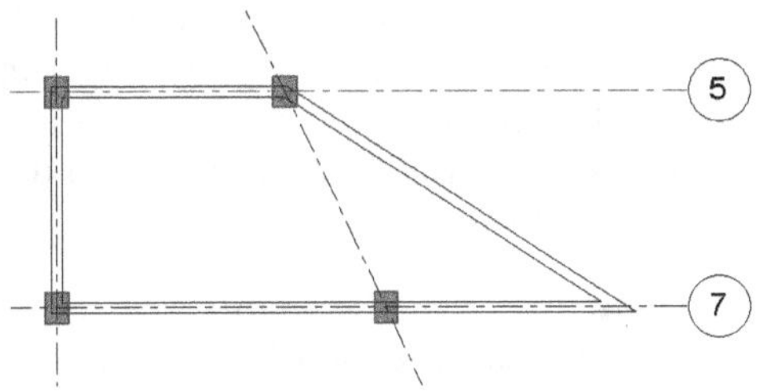

Figure 6-8 *All instances of the element selected*

Selecting Elements Using the Advanced Selection Tools

In Autodesk Revit Structure, you can use the advanced selection tools for the selection and modeling process. These tools are available in the **Status Bar**. You can also invoke these tools from the **Modify** panel drop-down. Some of the advanced selection tools are discussed next.

Select Links Tool

In a project, you may require to select an entire linked model or an individual element inside the linked model. To do so, choose the **Select Links** tool from the **Status Bar** and then select the element or the model.

Tip

To select an individual element in a model, place the cursor on it and then press the Tab key; the element gets highlighted. Click on the highlighted element to select it.

Select Underlay Element Tool

You can use the **Select Underlay Element** tool to select elements which are not part of the current plan view. To do so, choose this tool from the **Status Bar**. Next place the cursor on the underlaid element; the element gets highlighted. Next, click to select the highlighted element. You can disable this tool by choosing it again.

Select Pinned Element Tool

You can enable or disable the selection of the pinned elements in a model by using the **Select Pinned Elements** tool. To enable the selection of the pinned elements, choose this tool from the **Status Bar**.

Select Element by Face Tool

You can choose this tool to enable the selection of an element by face. To do so, choose the **Select Element by Face** tool from the **Status Bar** and then click at any point on the face of the element(s) such as wall or floor; the element will be selected. You can disable this tool by choosing it again.

Drag Elements on Selection Tool

You can choose the **Drag Elements on Selection** tool to drag an element without selecting it first. To do so, choose the **Drag Elements on Selection** tool from the **Status Bar** and then hover the cursor over the desired element such as wall or floor without selecting it; the element gets highlighted. Now, press and drag the mouse button, the element will be moved to the desired location.

Note

*It is recommended to disable the **Drag Elements on Selection** tool to avoid accidental displacement of the elements.*

Restoring a Selection

You can restore a selection which got cleared intentionally or unintentionally, using several methods. One of the options for restoring the previous selection is to choose the **Modify** tool and then right-click in the drawing area to display a shortcut menu. Next, choose the **Select Previous** option from the shortcut menu to restore the previous selection. Alternatively, press CTRL+Left Arrow to restore the previous selection.

Selecting Elements Using the Filter Tool

In Revit Structure, you can use the **Filter** tool to refine the selection process by removing the elements of specific categories from the current selection set. For example, suppose a current selection set includes walls, structural foundations, structural columns, and beams, and you need to remove the elements of structural columns categories from the selection. You can do so by using the **Filter** tool. On using this tool, the walls, structural foundation and beams will remain selected in the existing selection set. Figures 6-9 and 6-10 show the elements of different categories selected before and after using the **Filter** tool. Before using this tool, you need to select the elements from the drawing area. Next, invoke the **Filter** tool from the **Filter** panel of the **Multi-Select** tab; the **Filter** dialog box will be displayed along with a list of categories of the currently selected elements, refer to Figure 6-11. Alternatively, click on the filter icon displayed at the extreme right corner of the **Status Bar** to display the **Filter** dialog box.

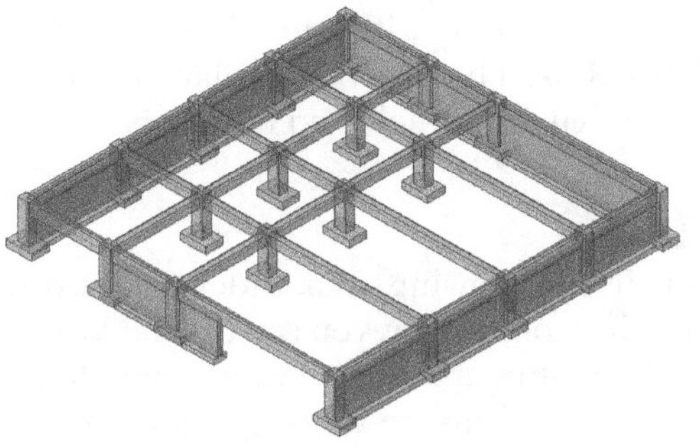

Figure 6-9 *Selected elements before using the **Filter** tool*

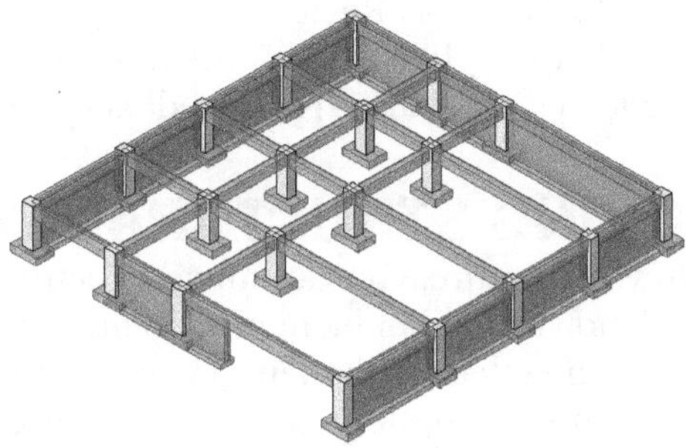

Figure 6-10 *Selected elements after using the **Filter** tool*

The **Category** column in the **Filter** dialog box displays all the element categories present in the selection set. The **Count** column displays the number of selected elements in each category. At the bottom of the dialog box, the **Total Elements** area displays the total number count of the elements currently selected.

In the **Filter** dialog box, by default the check boxes corresponding to the listed categories are selected. To exclude an element from the category, you can clear the check box corresponding to it. You can use the **Check All** or **Check None** button to select or clear all the check boxes in the element categories, respectively.

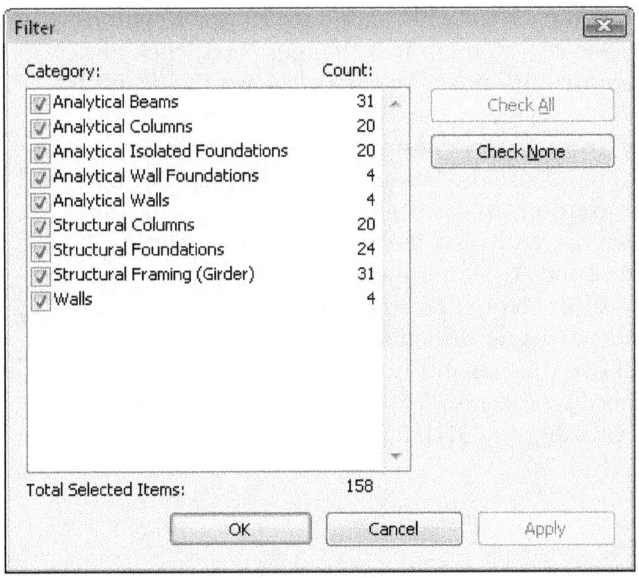

Figure 6-11 *The **Filter** dialog box*

Note

*The **Total Selected Items** at the bottom of the **Filter** dialog box displays the count of the total number of elements selected. This count keeps on changing as you select or clear the check boxes.*

The Filter Icon

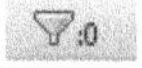

In Autodesk Revit Structure, you can see a filter icon along with a number located at the lower right corner of the interface on the **Status Bar**. The number displayed next to the **Filter** icon gives you a visual feedback of the number of currently selected elements.

MOVING STRUCTURAL ELEMENTS

In Revit Structure, you can move structural elements in the drawing area, independently or in association with other elements. To move a component independently, you can use the **Move** tool or drag the selected element. However, to move an element in association with other elements, you can use the options such as changing temporary dimensions, re-hosting elements, and so on. You can move a hosted component from one position to another. The methods used for moving an element are discussed next.

Moving Elements by Changing Temporary Dimensions

You can move an element by changing its temporary dimensions. The temporary dimensions for an element are displayed at the time you create or select the element. These dimensions are displayed in reference to a nearby element. To move an element, select it; its distance with respect to the nearby element will be displayed in the temporary dimensions. Click on the temporary dimension and enter a new value in the edit box to move the element to the desired location.

Moving Elements by Dragging

You can also move elements by dragging them to the desired location in the drawing area. To drag an element, select the element and place the cursor over it; the cursor will change into a move symbol and the color of the selected element will change. Next, click and hold the left mouse button and move the cursor toward the desired location. On doing so, an image of the selected element will move along with the cursor, as shown in Figure 6-12. When the element reaches desired location, release the left mouse button; the element will move to the new location. In case, you move a structural wall, the adjoining walls will extend with the moved wall. Figure 6-13 shows the adjustment of the adjoining walls when the selected wall is dragged to a new location. Note that some elements such as walls can be moved only in a particular direction. However, you can press the SHIFT key to remove this constraint. The rest of them can be moved in all directions. You can restrict the direction of movement of these elements by pressing the SHIFT key.

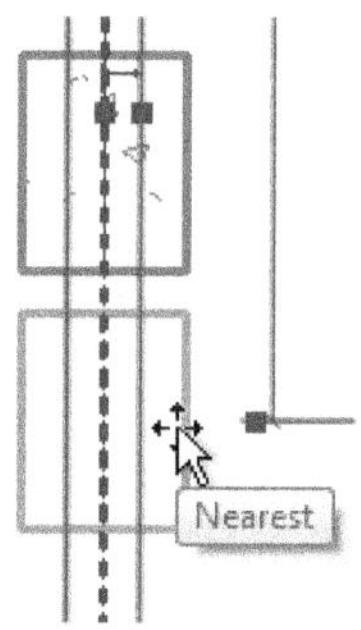

***Figure 6-12** Image of the selected element displayed along with the cursor symbol*

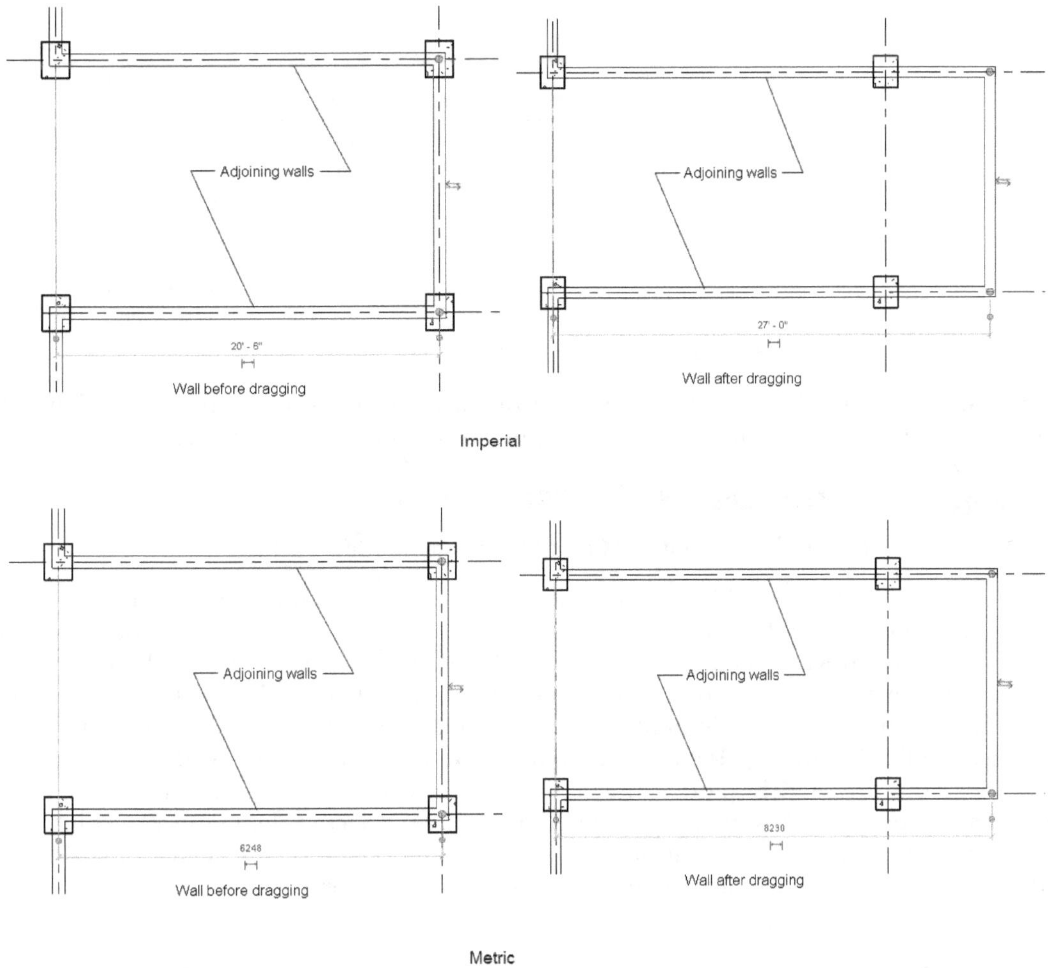

Figure 6-13 The selected wall before and after dragging

Dragging End-joined Components

Revit Structure provides you with the option to drag the end-joined components by using drag controls. You can drag and move more than two components simultaneously, provided they are joined. Note that this feature can be applied to the line based components only such as walls, beams, or braces.

To drag the joined elements, select one of the members of the joined elements from the drawing and right-click; a shortcut menu will be displayed. Choose **Select Joined Elements** from the shortcut menu; all the joined members will be highlighted and the circular drag controls will be displayed, refer to Figure 6-14. Next, move the cursor to the common joint; the **Drag Wall End** tooltip will be displayed. Drag the joint to a new location, as shown in Figure 6-14. You will notice that all walls joined to that common joint have been dragged.

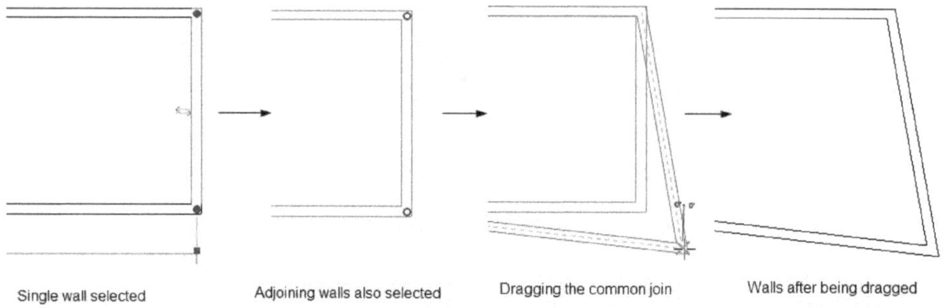

Figure 6-14 *The joined walls being dragged*

Doors and windows can be moved along their host walls. It can be done by picking and dragging them. When you move doors and windows, an opening is automatically created at the new location.

Moving Elements Using the Move Tool

Ribbon: Modify | (Elements or Components) > Modify > Move

You can also use the **Move** tool to move elements. The usage of the **Move** tool is similar to that of the tools used for dragging with the only difference that in case of the **Move** tool, the preview of elements is not displayed. This tool is more useful for moving the elements as it increases the accuracy of the position of the structural model by specifying the exact distance for moving the elements. To invoke the **Move** tool, select the element(s) from the drawing area; the **Modify | (Elements or Components)** tab will be displayed. Next, choose the **Move** tool from the **Modify** panel. Alternatively, to invoke the **Move** tool, select the desired element and then type **MV**. On invoking the **Move** tool, the selected elements will be highlighted inside a dashed rectangle. The **Options Bar** shows three check boxes: **Constrain**, **Disjoin**, and **Multiple**. These check boxes are discussed in the table given next.

Note

*The **Move** tool, like other editing tools such as **Copy**, **Rotate**, **Mirror**, **Array**, and **Scale**, is available only after a selection has been made. You can access these tools from the contextual tabs such as **Modify / Structural Columns**, **Modify / Structural Framing**. The name of the contextual tab will be based on the type of element or component selected.*

Name of the Check box	Description
Constrain	If this check box is selected, the movement of the selected element along with its collinear or perpendicular vectors will be restricted.
Disjoin	If this check box is selected, you can break an element from its associated elements and move it to a new location. For example, you can move a wall without moving the joined wall. You can also move the elements from their current host element to another host element. For example, you can easily move a door or window from one wall to another. The selection of this check box can be effectively used if the **Constrain** check box is cleared.
Multiple	This check box is disabled by default.

After invoking the **Move** tool, click in the drawing area to specify the start point. Next, move the cursor to the desired location and click to specify the endpoint; the element will move to a new location. Alternatively, you can move the element by specifying distance. To do so, invoke the **Move** tool and specify the start point for the movement in the drawing area. Next, move the cursor away from the specified start point and toward the desired direction of movement; the image of the selected element will be displayed. Also, you will notice that a temporary dimension along with the image is displayed. Type a value to specify the distance; you will notice that the typed value is displayed in the edit box. To confirm the specified distance, press ENTER; the selected element will move by the specified distance.

COPYING STRUCTURAL ELEMENTS

In Autodesk Revit, you can copy structural elements to the drawing area by using the CTRL key, the **Copy** tool, the clipboard to copy-and-paste elements method, and the **Create Similar** tool. The methods of copying structural elements are discussed next.

Copying Elements Using the CTRL Key

Using the CTRL key for copying elements is the most commonly used method. To copy an element using this method, select the desired element and move the cursor over it; the move symbol will replace the existing cursor. Next, press and hold the CTRL key and the left mouse button. Then, drag the cursor away from the selected element or component at a desired direction; the image of the selected element will be highlighted and displayed. Also, the temporary dimensions with respect to the nearby elements will be displayed, refer to Figure 6-15. Now, you can release the CTRL key and the left mouse button, and then click any temporary dimension to specify the position of the copied element. On clicking the temporary dimension, an edit box will be displayed. You can enter a value in the edit box to fix the position of the copied element or component in reference to the nearby element. After entering a value in the edit box, press the ESC key or double-click in the drawing area to finish copying elements or components.

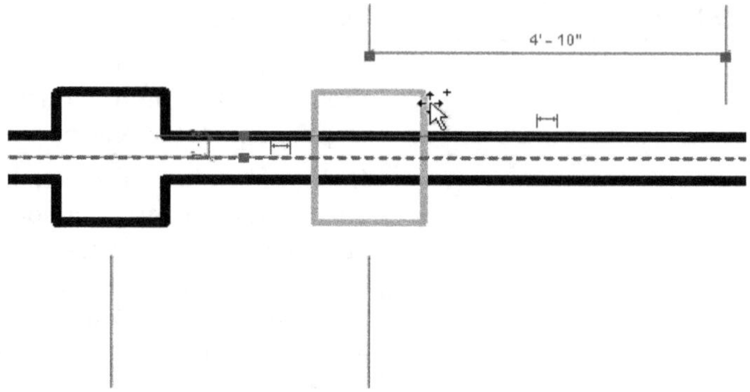

Imperial

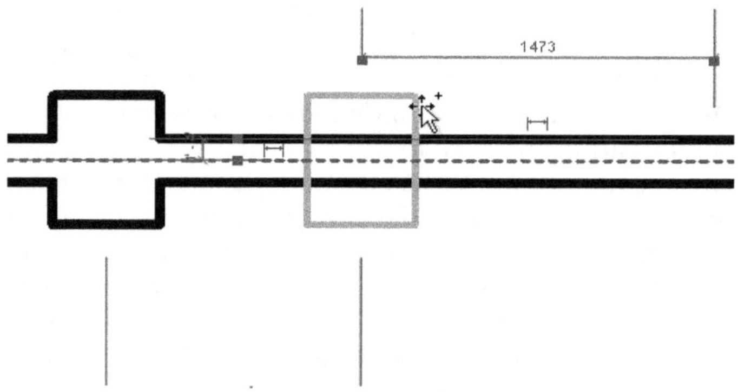

Metric

Figure 6-15 *The pasted elements highlighted and displayed along with their temporary dimension*

Copying Elements Using the Copy Tool

Ribbon: Modify | (Elements or Components) > Modify > Copy

The usage of the **Copy** tool is similar to that of the **Move** tool. You can use the **Copy** tool to create a single or multiple copies of element(s). To use the **Copy** tool, select the element(s) that you want to copy from the drawing area; the **Modify | (Elements or Components)** contextual tab will be displayed. In this tab, invoke the **Copy** tool from the **Modify** panel; the selected element will be highlighted inside a rectangular dotted box. Next, click in the drawing area, preferably inside the rectangular dotted box to specify the start point of displacement for the element(s) to be copied. Next, move the cursor along the direction you want to copy the element(s). On doing so, the image of the rectangular dotted box will be displayed. Also, a temporary dimension in reference to the start point will be displayed. Now, you can either click at the desired location in the drawing area or use the temporary dimensions to specify the end position of the copied element(s) in reference to the selected element(s). After specifying the end position of the dotted rectangular box, you will notice that the copy of the selected element has replaced the dotted rectangular box and the temporary dimensions to align the copied element (s) will also be displayed at the same time. You can use the temporary dimensions of the copied elements to fix its horizontal and vertical distances. Next, press ESC to exit the selection mode. You can also use the **Copy** tool to copy element(s) multiple times. The method for copying elements multiple times is discussed next.

Creating Multiple Copies

You can create several copies of an element using the **Copy** tool. To do so, select the elements that you want to copy and then invoke the **Copy** tool. Select the **Multiple** check box in the **Options Bar**. Next, click in the drawing window to specify the start point. Move the cursor and click at the location where you want to create a copy. Each successive click will create a copy of the original element, refer to Figure 6-16. Alternatively, when the temporary dimensions are displayed, type the distance between the currently selected elements and the next element to create copies at a specified distance. The multiple copy option remains active until you choose some other tool or press the ESC key twice.

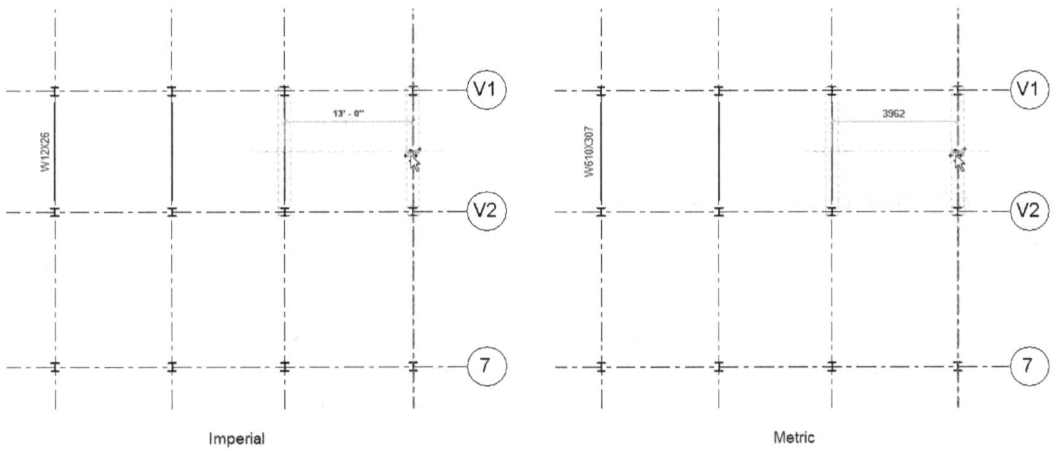

***Figure 6-16** Element copied multiple times*

Using the Create Similar Tool

Ribbon: Modify | (Elements or Components) > Create > Create Similar

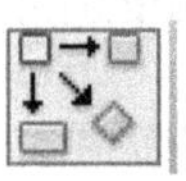

In a project, you can create and place similar type of elements at the required space. For this, you need to use the **Create Similar** tool. This tool will be available only after a selection has been made. Before using the **Create Similar** tool, select the element whose similar types you want to create and place in the drawing. Next, choose the **Create Similar** tool from the **Create** panel; the creation tool, similar to that of the element you have selected, will be invoked. Next, you can create the element of similar instance parameter as that of the selected element in the drawing area. For example, if you select a structural framing member from the drawing and then invoke the **Create Similar** tool; the **Modify | Place Beam** tab will be displayed. By using the options in this tab, you can place a structural framing member of the same instance that you have selected from the drawing area. Alternatively, to invoke the **Create Similar** tool, select the element and then right-click; a shortcut menu will be displayed. Next, choose the **Create Similar** option from the shortcut menu; the **Create Similar** tool will be invoked.

Copying Elements Using the Clipboard

In Revit Structure, you can use the functionality of cut, copy, and paste to speed up work and reduce the time required for creating new elements. These functions are generally called Clipboard operations and can be used to cut or copy one or more elements from the drawing area to the clipboard and then paste them in the drawing area of the same project or in another project. This method of copying elements is different from that of using the **Copy** tool. These tools and their usage are discussed next.

Cutting Elements from a Project

Ribbon:	Modify | (Elements or Components) > Clipboard > Cut to Clipboard

The **Cut to Clipboard** tool is used to remove one or more elements from a project. The selected elements are copied on the clipboard where they remain until another selection has been pasted. When you cut and paste new elements to the clipboard, the existing selection is overwritten by a new one. You can then paste it in the same or another project. The **Cut to Clipboard** tool will be available only after a selection has been made. Select the elements or components from the drawing using any of the selection tools described earlier in this chapter and then invoke the **Cut to Clipboard** tool from the **Clipboard** panel of the contextual tab. Alternatively, press the CTRL+X keys. The selected elements will be removed from the project and pasted on the clipboard.

Copying Elements to the Clipboard

Ribbon:	Modify | (Elements or Components) > Clipboard > Copy to Clipboard

You can create a copy of an element on the clipboard and retain the original ones in the current project. To copy an element, select the desired elements or components and then invoke the **Copy to Clipboard** tool from the **Clipboard** panel; copy of the selected elements will be created on the clipboard. Alternatively, to copy the desired elements or components to the clipboard, select them and then press the CTRL+C keys.

Pasting Elements from the Clipboard

In Revit Structure, once elements are copied, you can paste them in the drawing area of the current project or in another project by using any of the pasting options available in the **Paste** drop-down located in the **Clipboard** panel of the **Modify | (Elements or Components)** tab. To display the options used for pasting elements, click on the **Paste** drop-down; a list of pasting tools will be displayed, as shown in Figure 6-17. The tools displayed in the drop-down are **Paste from Clipboard**, **Aligned to Selected Levels**, **Aligned to Selected Views**, **Aligned to Current View**, **Aligned to Same Place**, and **Aligned to Picked Level**. The different tools in the drop-down are discussed next.

Using the Paste from Clipboard Tool

The **Paste from Clipboard** tool is used to paste one or more elements from the clipboard to the current or another project. This tool will be available only after the element(s) has been cut. You can invoke the **Paste from Clipboard** tool from **Modify | (Elements or Components) > Clipboard > Paste** drop-down. On doing so, a preview box representing the pasted selection will be displayed with a temporary dimension and the original selection, refer to Figure 6-18. Notice that the original selection is displayed inside a rectangular dashed box. You can use the temporary dimension to fix the distance of the pasted selection from the original selection, or click on the desired location in the drawing area to place the pasted selection. lternatively, you can press the CTRL+V keys to paste the copied element(s) with respect to the clipboard to the drawing area of the current project or another project. Once these elements have been pasted, they remain selected as a group in a dashed rectangular box and the **Modify | Model Groups** tab is displayed. In this tab, there are other editing tools such as **Move**, **Copy**, **Rotate**, **Align**, and so on to modify the selected elements. In the **Options Bar**, you can choose the **Activate Dimensions** button to view the dimensions of the pasted elements from the nearby element(s). After pasting, choose the **Finish** tool from the **Edit Pasted** panel to return to the **Modify** tab (default mode). To quit the pasting operation even when the group of elements is selected, choose the **Cancel** button in the **Modify | Model Groups** tab. On doing so, the recently pasted group of elements will be removed from the project and you will exit the **Paste** tool.

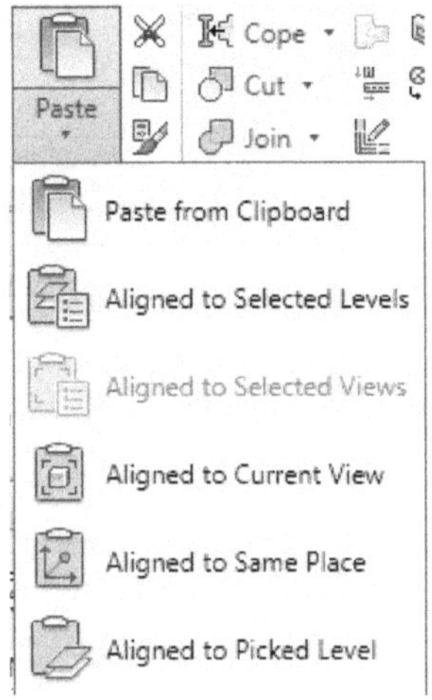

***Figure 6-17** Various pasting options*

You can also edit the elements that you have pasted in the drawing area by changing the position of the individual element or changing the type properties of the elements while the **Modify | Model Groups** tab is active. To do so, choose the **Edit Pasted Elements** tool from the **Edit Pasted** panel in the **Modify | Model Groups** tab; the **Modify | Multi-Select** or **Modify | (Elements or Components)** tab will be displayed and the elements, other than the pasted elements, will be inactive and appear fade. You can select an element from the pasted element and reposition them by using the tools available in the **Modify | Multi-Select** or **Modify | (Elements or Components)** tab. As you select an element from the model group, the corresponding contextual tab will be displayed. For example, if you select a column, the **Modify | Structural Columns** tab will be displayed. You can use various editing tools such as **Move** and **Copy** from the **Modify** panel of the contextual tab to reposition or realign the selected element. You can also rehost the element using various tools in the **Modify | Structural Columns** tab.

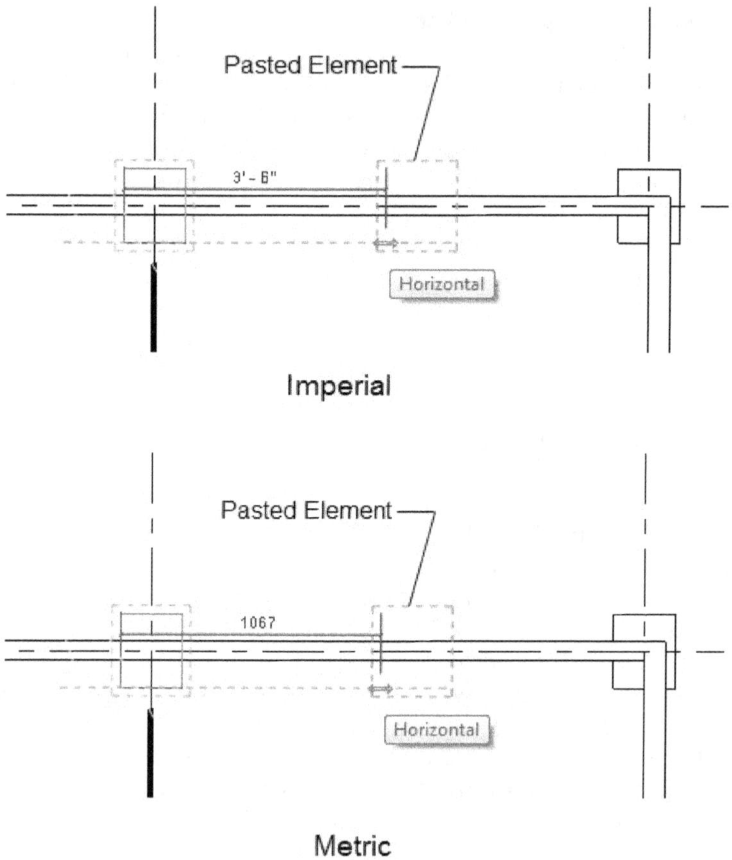

Figure 6-18 *Pasted element shown in a preview box*

Using the other Pasting Tools

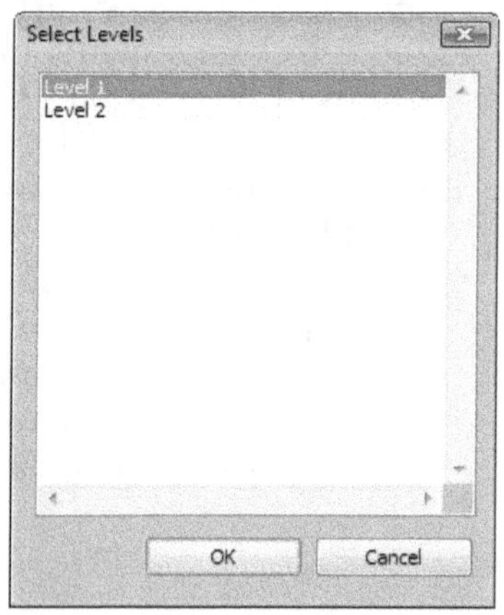

Figure 6-19 *The* ***Select Levels*** *dialog box*

After you cut or copy element(s) for pasting them to the clip board, you can paste the copied or cut element(s) in various ways such as to one level, to many levels, in a specific view, in the current view, at same place over the original element, or in an elevation or sectional view. To align the pasted elements, you need to click on the **Paste** drop-down from the **Clipboard** panel of the **Modify | <Elements or Components>** tab. On doing so, a drop-down list will be displayed with the following tools: **Aligned to Selected Levels**, **Aligned to Selected Views**, **Aligned to Current View**, **Aligned to Same Place**, and **Aligned to Picked Level**. By default, the **Aligned to Select Views** tool is inactive.

You can use the **Aligned to Selected Levels** tool to paste the copied elements to one or more levels. On choosing this tool from the drop-down list, the **Select Levels** dialog box will be displayed, as shown in Figure 6-19. In this dialog box, the names of the levels present in the structural model are displayed in a list box. You can select the individual level by clicking on it. You can use the CTRL key to select multiple elements. After selecting levels by their names, choose the **OK** button; the **Select Levels** dialog box will close and the selected elements will be pasted on the selected levels.

You can invoke the **Aligned to Selected Views** tool to paste view-specific elements such as dimensions, tags, and texts to a similar type of view. The **Aligned to Current View** tool is used to paste the elements that are cut or copied from a different view and has to be pasted in the current view. Note that you can paste elements at the same place where you have copied them. To do so, you can use the **Aligned to Same Place** tool. The **Aligned to Picked Level** tool can be used to paste elements in an elevation or section view. Note that while using this tool, you need to copy and paste elements only in elevation or section view.

ROTATING STRUCTURAL ELEMENTS

Ribbon: Modify | (Elements or Components) > Modify > Rotate

You can rotate elements in a project by using the **Rotate** tool. This tool is used to rotate the selected element(s) about a specified axis or point. You can rotate the selected elements about an axis perpendicular to the plan, elevation, or section view. In 3D view, elements rotate about an axis perpendicular to the current work plane. This tool will be available only after the elements have been selected.

To invoke the **Rotate** tool, select elements or components from the drawing area and then choose the **Rotate** tool from the **Modify** panel of the contextual tab displayed. Alternatively, you can type **RO**. On invoking the **Rotate** tool, a rotation symbol will be displayed at the center of the selected element(s). It represents the perpendicular axis or the center about which the elements will rotate. In the drawing area, you can drag the rotation symbol or use the **Place** button from the **Options Bar** to specify a different location as the center of rotation. If you want to place the center of rotation to its default location, choose the **Default** button from the **Options Bar**. Alternatively, you can snap the rotation symbol and drag it to the desired center of rotation.

Once the center of rotation has been specified, a ray moving along the direction of the cursor is generated from the center. Click at the desired start point of the rotation to display a line indicating its direction. You can use various object snap options to specify the start point of the rotation. Once the start point of the rotation has been specified, another line will rotate with the cursor to specify the endpoint. Also, you will notice that a preview box, indicating the selected elements, is rotating along with the selected elements. The element rotates to a new direction. When you move the cursor to the required angle and click to specify the second point of rotation. On rotating the entities, you will notice that a temporary angular dimension is displayed. You can enter the exact value of the angle of rotation to rotate the elements to the specified angle. When you invoke the **Rotate** tool, three options become available in the **Options Bar**, refer to Figure 6-20.

Figure 6-20** Different rotation options in the **Options Bar

Before rotating the entities, you can select the **Disjoin** check box to break the association between the selected elements and the other elements associated with the selected elements. For example, you can independently rotate a wall that is joined to other walls. The selection of the **Copy** check box enables you to create a rotated copy of the selected element without changing its original location. The **Angle** edit box can be used to rotate elements by a specific angle. You can enter the value of the angle of rotation in this edit box. On entering a value, the selected elements will rotate about the specified center point with specified angle. For example, in the **Structural Plan** view, to rotate a horizontally aligned wall about its intersection point with the vertically aligned wall, by 15 degrees, click on the horizontal wall to select it and then invoke the **Rotate** tool from the **Modify** panel of the **Modify | Walls** tab. On doing so, the rotation symbol will be placed at the midpoint of the wall and a ray will be generated from it, as shown in Figure 6-21. Drag this symbol to the intersection of the two walls, as shown in Figure 6-22. The rotation symbol moves to the new location and the ray will be generated from this point. Next, move the cursor vertically upward such that it snaps to the vertical wall. To specify the start point of the rotation, click when the snap symbol is displayed. To create an angle of 15 degrees, move the cursor clockwise and click when the temporary dimension shows the desired angle value. As a result, the vertical wall will rotate about the specified center point. Alternatively, enter **15** for the new angle.

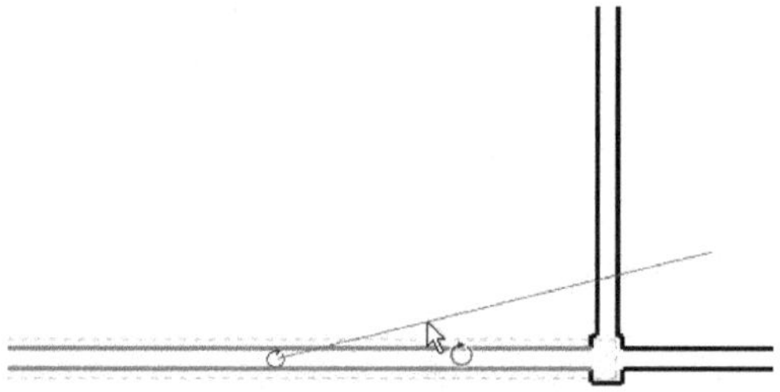

Figure 6-21 *Ray generated at the midpoint of the wall*

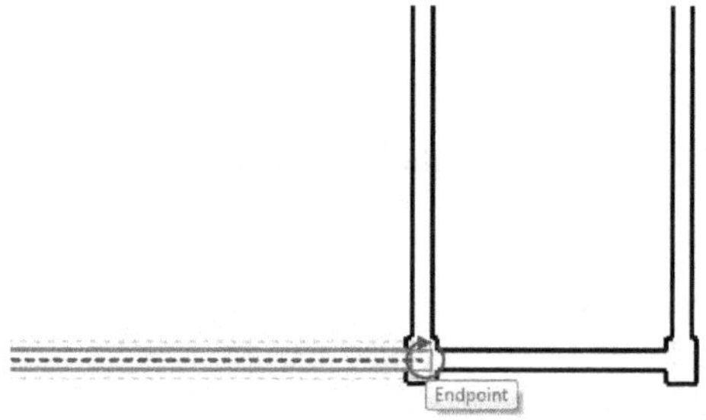

Figure 6-22 *Rotation symbol dragged to the intersection of the wall*

MIRRORING STRUCTURAL ELEMENTS

In Revit Structure, you can mirror elements and components to create design symmetry in a structural model. You can mirror elements using any of the following two tools: **Mirror - Pick Axis** and **Mirror - Draw Axis**. These tools can be chosen from the **Modify** panel of the contextual tab. To mirror elements or components in the structural model, first you need to select them from the drawing area; the contextual tab for the selected elements or components will be displayed. In the **Modify** panel of the contextual tab, choose any of these two tools: **Mirror - Draw Axis** and **Mirror - Pick Axis**. The selection of the tool depends on the project requirement. Note that when you invoke any of the tools, the **Options Bar** displays the **Copy** option. By default, this option is selected, and therefore enables you to create a mirrored copy of the elements. Clear the **Copy** check box if you do not require a copy. The two tools for mirroring elements or components are discussed next.

Mirror - Pick Axis Tool

Ribbon: Modify | (Elements or Components) > Modify > Mirror - Pick Axis

The **Mirror - Pick Axis** tool can be used to mirror elements by selecting an element from the drawing to define the mirror axis. To mirror elements, select them from the drawing and then invoke the **Mirror - Pick Axis** tool from the **Modify** panel; the mirror symbol will appear with the cursor in the drawing area. Alternatively, you can type **MM** to invoke the **Mirror - Pick Axis** tool. Next, move the cursor over the element that will define the mirror axis and click when the appropriate object snap appears. Figure 6-23 shows beam, column, and wall elements selected for mirroring and the grid line tagged **D** as the mirror axis. Figure 6-24 shows the elements created after mirroring.

Note

*Using the **Mirror-Pick Axis** tool, you can only select a line or a reference plane that the cursor can snap to. You cannot mirror elements about an imaginary axis created in the open space.*

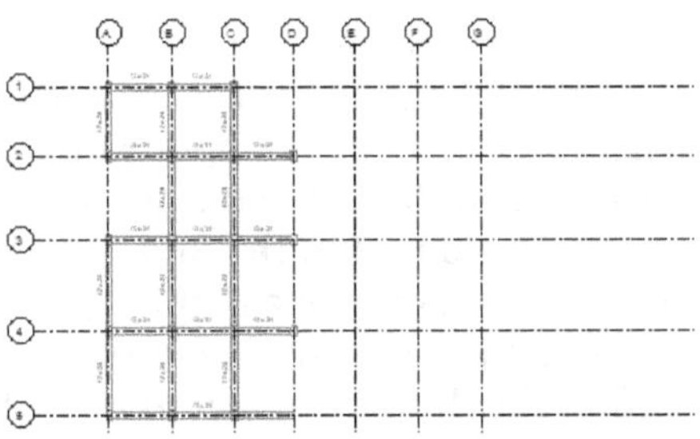

__Figure 6-23__ Beams, columns, and walls selected for mirroring

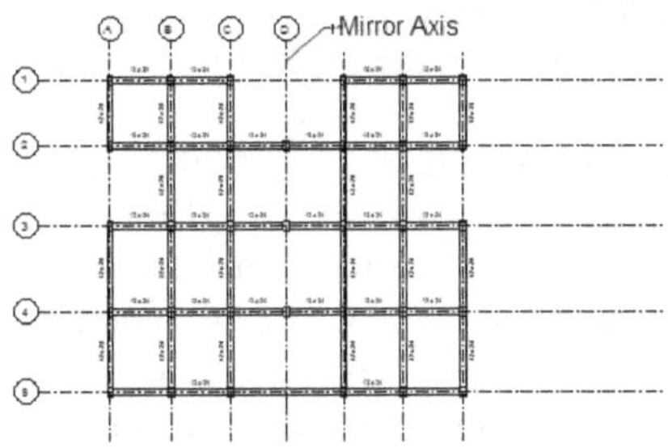

__Figure 6-24__ Beams, columns, and walls mirrored on either side of the grid line D

Mirror - Draw Axis Tool

Ribbon: Modify | (Elements or Components) > Modify > Mirror - Draw Axis

The **Mirror - Draw Axis** tool can be used to mirror an element or a component by creating a mirror line with the help of two points to define the mirror axis. This tool is represented by a pencil with a mirror symbol. To mirror elements, select the elements from the drawing and then choose the **Mirror - Draw Axis** tool from the **Modify** panel. You can also invoke this tool by typing **DM**. After invoking the tool, click in the drawing area to specify the start point of the mirror line; a ray will emerge from the specified point. Move the cursor to the desired location and click to specify the second point; the selected elements will be mirrored about the line that represents the mirror axis.

ARRAYING STRUCTURAL ELEMENTS

Ribbon: Modify | (Elements or Components) > Modify > Array

The **Array** tool, which is used to create the copies of a selected element in a rectangular or radial pattern, will be available only after elements have been selected for an array. You can use any of the methods discussed earlier in this chapter to select an element. Revit Structure, provides you the option of arraying multiple elements that are associated as a group. To invoke the **Array** tool, select the element(s) that you want to array and then choose the **Array** tool from the **Modify** panel of the contextual tab. Alternatively, you can type **AR** to invoke this tool. On invoking the **Array** tool, you will notice that various array options are displayed in the **Options Bar**. In the **Options Bar**, the **Group and Associate** check box is selected by default. As a result, the elements are copied as a group rather than individual elements. You can clear this check box so that the arrayed elements are not grouped. In the **Options Bar**, the **Linear** and **Radial** tools can be used to create two different types of arrays. The tools for creating different types of arrays are discussed next.

Linear Tool

The **Linear** tool can be used to create an array of selected elements in a rectangular or linear pattern. When you invoke the **Array** tool, the **Linear** tool is also invoked by default in the **Options Bar**. As a result, various options which are available in the **Options Bar** to create an array are displayed according to the **Linear** tool. In the **Number** edit box, enter the number of copies you want to create. In the **Move To** area of the **Options Bar**, the **2nd** radio button is selected by default. As a result, you can click the cursor in the drawing area to specify the distance through which the array will be created. Alternatively, select the **Last** radio button in the **Move to** area to specify the entire distance of the array. Click in the drawing area to specify the entire distance of the array. In the **Options Bar**, choose the **Activate Dimension** button to view various dimensions of elements. You can specify various options in the **Options Bar** before or during the creation of an array. After selecting the **2nd** radio button (if not selected by default) in the **Move To** area, click at any point in the drawing window to specify the start point of the array distance. As you move the cursor in the desired direction, a temporary dimension is displayed. You can either enter the value of the distance in that direction or simply move the cursor to the appropriate location and click. Revit Structure creates an array of the selected elements by calculating the distance and direction between the two specified points. After the second point is specified, you are prompted to enter the number of items for the array in an edit box. The number of items specified in the **Options Bar** will be displayed as the default value. You have the option of overriding it and specifying a new number. On specifying the number of items for the array, an array of the selected items will be created with the specified distance between each group of items. Figure 6-25 shows the selected elements and the distance specified between each element. The preview of the array to be created is shown in Figure 6-26.

You can select the **Constrain** check box in the **Options Bar** to restrict the movement of the cursor in orthogonal direction only. This helps in creating a horizontal or vertical linear array. You can select the **Last** radio button to create an array of elements between two points. Figure 6-27 shows the selection of the start point for specifying the array distance and the selected window to be arrayed. Move the cursor to the total distance that needs to be divided, as shown in Figure 6-28.

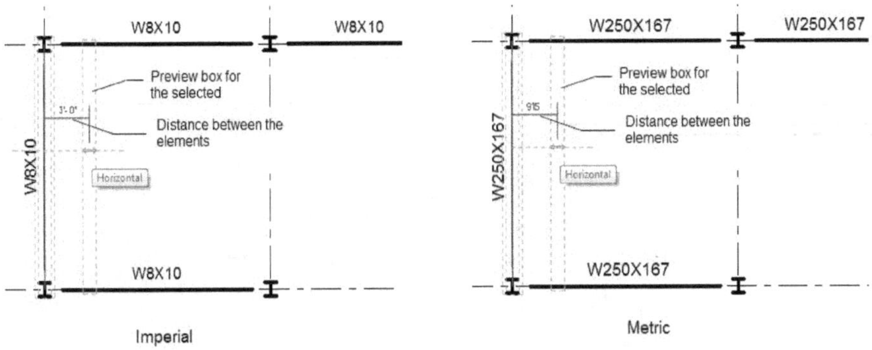

Figure 6-25 *Specifying the distance between elements after selecting the **2nd** radio button from the **Options Bar***

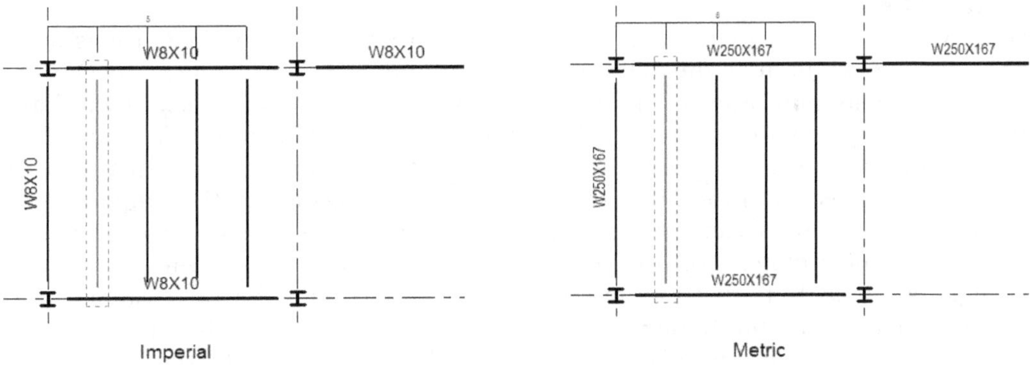

Figure 6-26 *The preview of the array to be created*

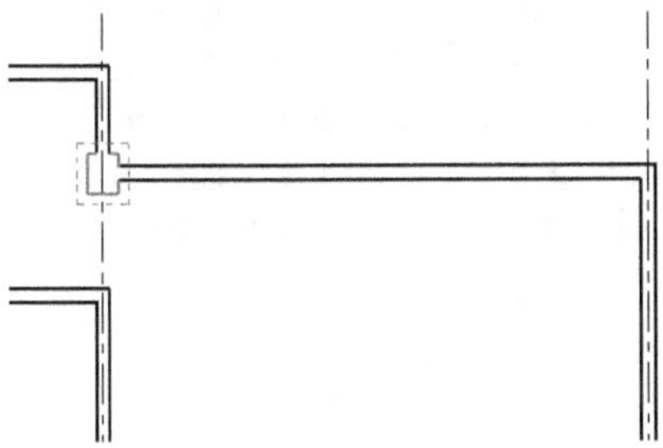

Figure 6-27 *Selection of the start point for the array*

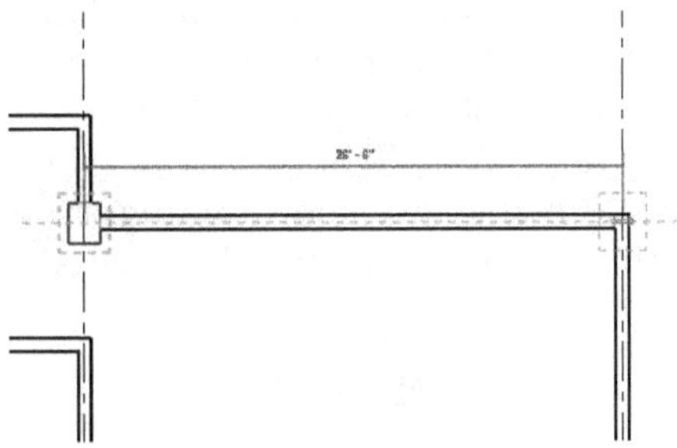

Figure 6-28 *Specifying the endpoint to mark the total distance of the array*

As you click to specify the endpoint of the array distance, an edit box appears displaying the number of items in the array. You can specify the number in it, as shown in Figure 6-29, and press ENTER to complete the procedure. An array of the selected column created on the wall is shown in Figure 6-30.

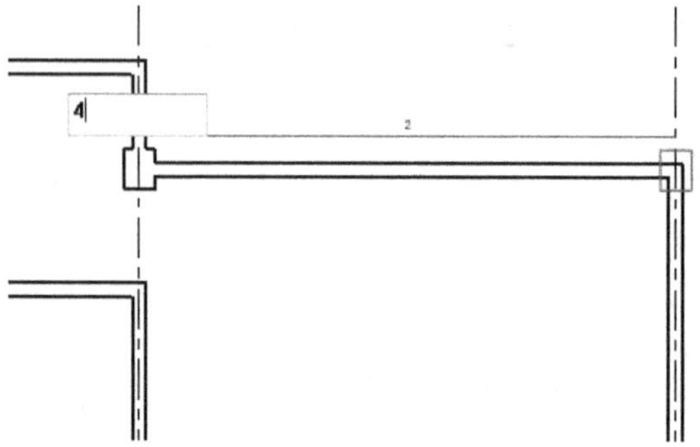

Figure 6-29 Specifying the array count in the edit box

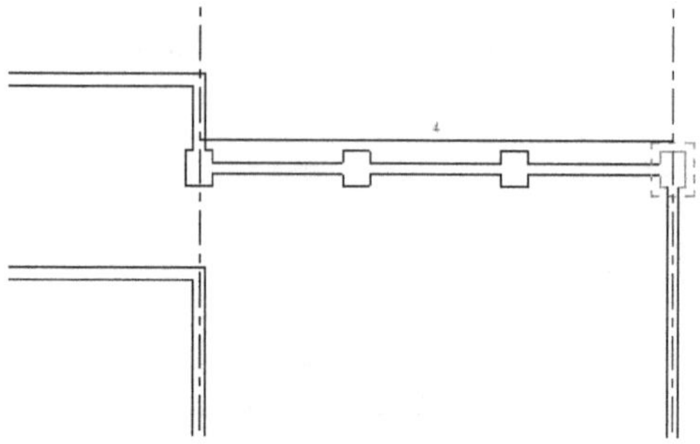

Figure 6-30 The preview of the array to be columns

Radial Tool

You can create radial arrays using the **Radial** tool available in the **Options Bar**. When you invoke this tool, the following setting options will be available in the **Options Bar**: the **Group and Associate** button, the **Number** edit box, and the **2nd** and **Last** radio buttons. These options are the same as those discussed for the **Linear** tool. The **Angle** edit box in the **Options Bar** is used to specify the angle about which the array is to be created. If the **2nd** radio button is selected in the **Move To** area, the angle specified in the **Angle** edit box will become the angle between each item, whereas if the **Last** radio button is selected from the **Move To** area, the number of items will be created by equally dividing the specified angle entered in the **Angle** edit box. Note that the procedure for creating a radial array is similar to that of rotating elements using the **Rotate** tool, which has been described earlier in this chapter. For arraying the selected elements radially, invoke the **Radial** tool from the **Options Bar**; a rotational symbol is displayed at the center of the selection. You can choose the **Place** button in the **Options Bar** or drag the rotation symbol to a location that will be used as the center of the array. Figure 6-31 shows a semi-circular wall in which the rectangular column is selected for an array. The center of the semi-circular wall is specified as the center of the array. Next, click to specify the start point of the angle of the array. Select the **Last** radio button from the **Move To** area to specify the total angle of the array. Now, move the cursor radially and click to specify the end point of the angle, as shown in Figure 6-32; an edit box will be displayed, prompting you to enter the number of items to be created between the specified angles, as shown in Figure 6-33.

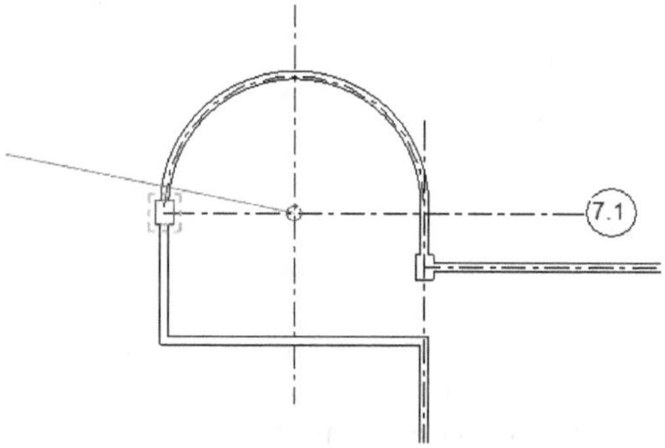

Figure 6-31 Selecting the rectangular column to array radially over the semi circular wall

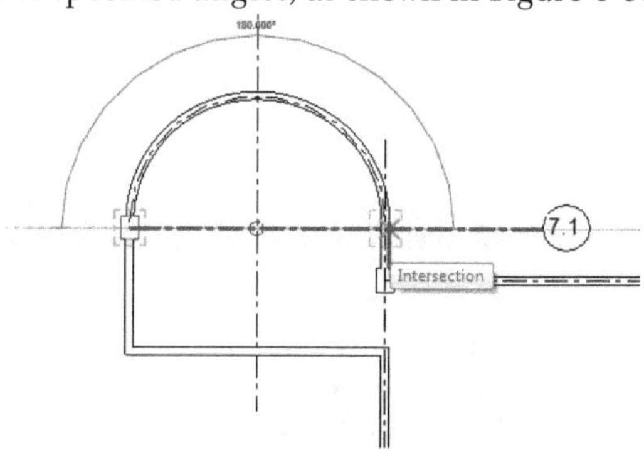

Figure 6-32 Specifying the endpoint to define the end angle of the array

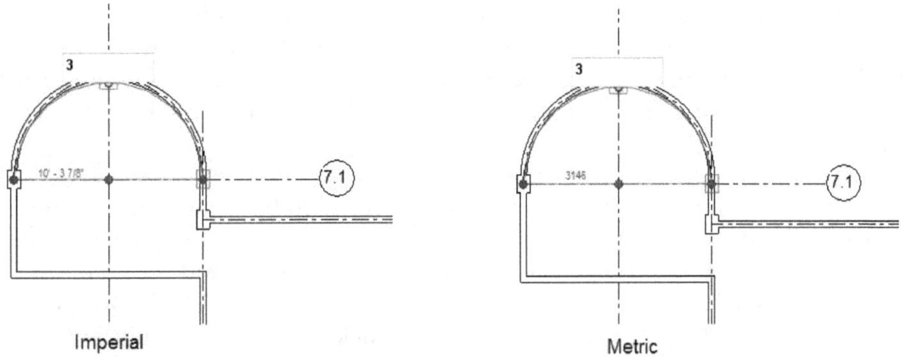

Figure 6-33 *Specifying the array count for the radial array*

After specifying the number, press ENTER to complete the procedure. Figure 6-34 shows the resultant radial array of the rectangular columns. You will notice that the number of items is displayed in blue. You can click on it and enter a new value, if required. The two controls at the end of the arc can be used to resize it. The middle control can be used to modify the radius of the radial array, while the central control lets you drag it.

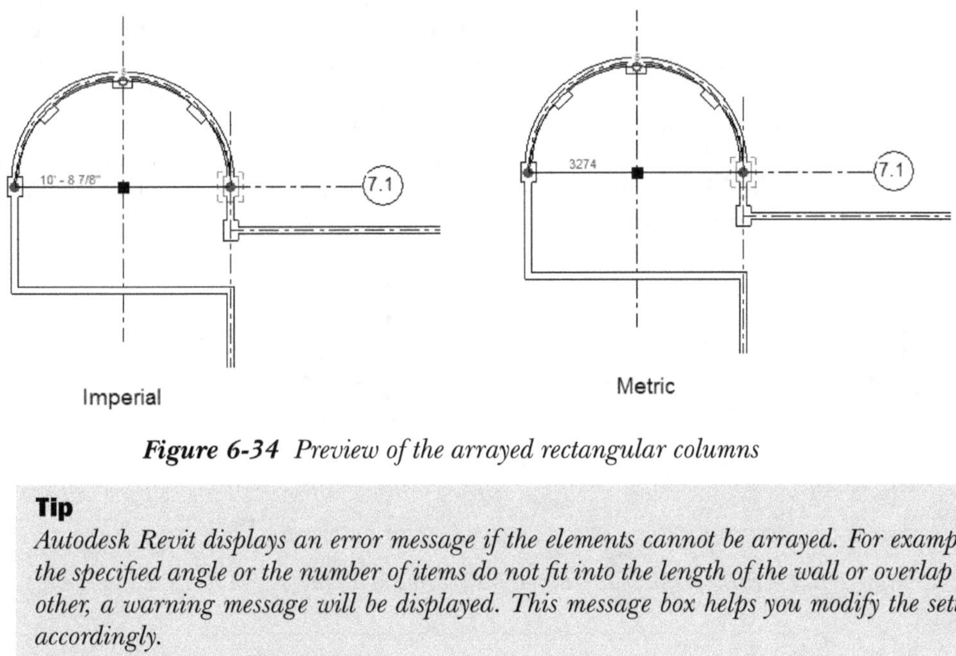

Figure 6-34 *Preview of the arrayed rectangular columns*

Tip

Autodesk Revit displays an error message if the elements cannot be arrayed. For example, if the specified angle or the number of items do not fit into the length of the wall or overlap each other, a warning message will be displayed. This message box helps you modify the settings accordingly.

RESIZING ELEMENTS

Ribbon: Modify |(Elements or Components) > Modify > Scale

In your project, you may need to resize or scale elements when required. To resize the elements in the project, you can use the **Scale** tool. This tool will help to modify the scale of a single element or a group of elements. This tool is only available for walls, lines, images, *.dwg* and *.dxf* imports, reference planes, and position of dimensions. To scale an element, select it from the drawing and then invoke the **Scale** tool from the **Modify** panel of the **Modify | (Elements or Components)** tab. On doing so, the **Options Bar** will display two radio buttons: **Graphical** and **Numerical**. These radio buttons are selected to specify a method for scaling.

In the Options Bar, the Graphical radio button is selected by default. As a result, you can resize elements graphically. After selecting this radio button, you need to specify three points in the drawing area. The first point is to specify the origin of scaling and the other two points define the scale vectors. Once the origin is specified, move the cursor and click to specify the length of the first vector. Then, move the cursor again and click to define the length of the second vector. Revit Structure will calculate the proportion of the two vectors and resize the selected element in the same proportion using the specified origin. The temporary dimensions displayed can also be used to resize the element to the desired scaling. For example, if the length of the first vector is 10' for Imperial or 3048mm for Metric and the second vector is 15' for Imperial or 4572mm for Metric, the selected element is resized 1.5 times. Figure 6-35 shows a 10' for Imperial or 3048mm for Metric long wall for which the left endpoint has been selected as the origin and the length of its final vector is 7'6" in Imperial or 2286mm in Metric. When you specify the length of the second vector as 15' in Imperial or 4572mm in Metric, as shown in Figure 6-36, Revit Structure calculates the proportion of the two vectors as 2 in this case and scales the wall length to 20' in Imperial or 6096 in Metric. When the numeric value of the scaling is known, you can select the Numerical radio button from the Options Bar. After selecting the Numerical radio button, specify a numeric value in the Scale edit box. Next, you need to specify the origin of the scaling by clicking at the desired location in the drawing area. Revit Structure instantly will resize the element by the specified scaling proportion. For example, when you resize a 50' long wall in Imperial or 15240 mm long wall in Metric by a numerical factor of 2.5, its length is resized to 125' for Imperial or 38100mm for Metric systems.

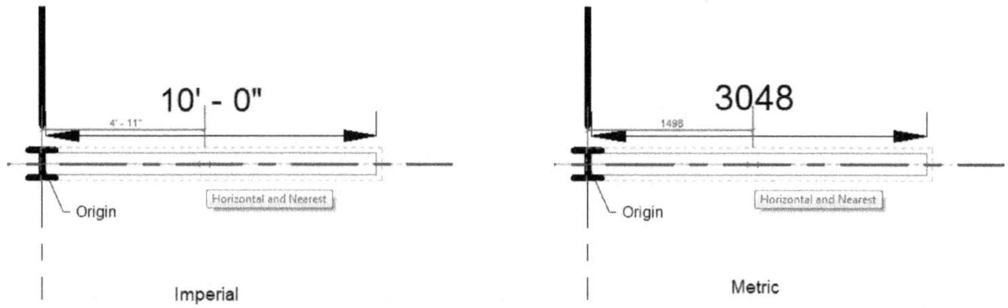

Figure 6-35 *Structural wall selected for scaling with the origin at its left endpoint*

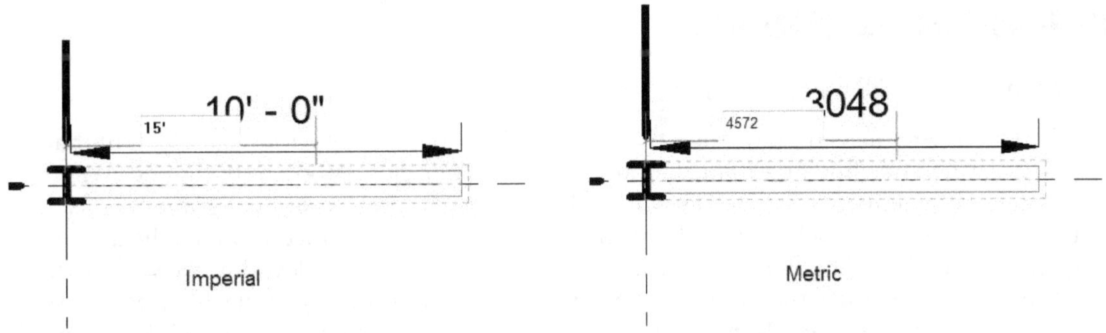

Figure 6-36 Specifying the length of the second vector

PINNING ELEMENTS

Ribbon: Modify | (Elements or Components) > Modify > Pin / Unpin

Elements can be pinned to their position by using the **Pin** tool. To invoke the **Pin** tool, make a selection and choose this tool from the **Modify** panel. Alternatively, you can type **PN** to invoke the **Pin** tool. Once pinned, a graphical pushpin symbol will appear with the element, as shown in Figure 6-37, indicating that it is pinned and cannot be moved from its position. You can also pin the position with respect to other elements and groups. For example, you can pin the position of columns with the intersecting grid lines. Once pinned, column and grid lines cannot be moved without first unpinning their position. You can use the **Unpin** tool to unpin the position constraint for the pinned elements. Note that the **Unpin** tool will replace the **Pin** tool after an element has been pinned. You can click on the pin symbol to pin or unpin an element.

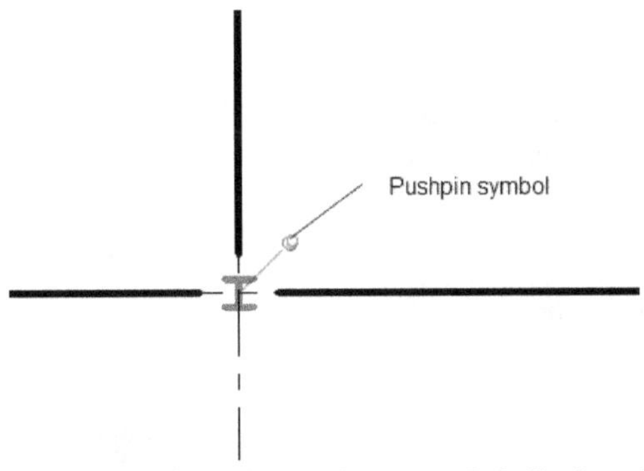

Figure 6-37 The pushpin symbol displayed with the pinned column

ALIGNING ELEMENTS

Ribbon: Modify > Modify > Align

The **Align** tool is used to align elements collinearly. You can choose an element that acts as a reference line or a plane along which the other elements can be aligned. When you align elements, a padlock key will be displayed, representing the alignment constraint. Using this key, you can lock or unlock the alignment of the walls. Once you lock the alignment, walls move together. The **Align** tool can be used only in plan or elevation views that can be invoked from the **Modify** panel of the **Modify** tab. On doing so, a symbol will appear with the cursor. Now in the **Options Bar**, you can select the **Multiple Alignment** check box from the **Options Bar** to align a number of elements in a common line. While aligning walls, you can choose to align wall centerlines or wall faces by selecting an appropriate option from the **Prefer** drop-down list in the **Options Bar**.

After specifying the options from the **Options Bar**, you will be prompted to select the reference line or point for alignment. The reference line can be a grid line or can be referenced from the center or any face of an element. Move the cursor to the desired reference; the reference will be highlighted. Click on it; a dashed line will appear along the selected reference. This dashed line represents the reference line for alignment. Next, move the cursor over the element that you want to align; the reference of the element will be highlighted. For example, if you move the cursor over a rectangular concrete column, any of the central axes or edges may get highlighted. Click on the reference of the element when it gets highlighted; the selected element will be aligned with the reference line specified. After aligning the desired element, you will notice a padlock displayed along the reference line. You can click the padlock key to lock the alignment of these two walls. To align them to the third wall, select them as the reference line, as shown in Figures 6-38 through 6-41. These figures show the alignment of a column to the center of the structural wall.

Note

The pin symbol becomes invisible after invoking other tools. In such a case, the element or group must be selected to display the pin symbol.

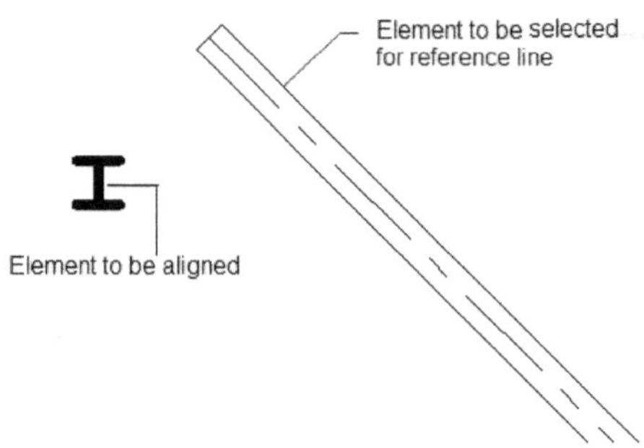

Figure 6-38 *Element to be selected for reference line and the element to be aligned*

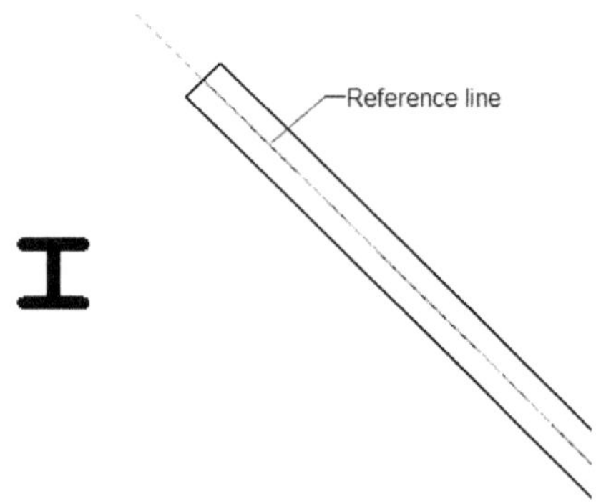

Figure 6-39 *The reference line passing through the center of the selected structural wall*

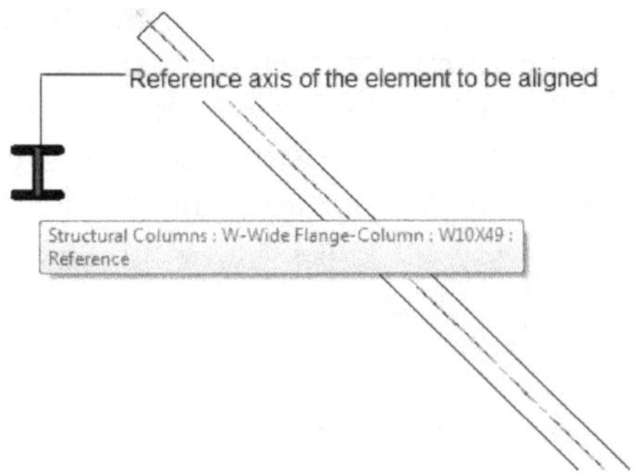

Figure 6-40 Vertical axis of the structural column selected for alignment

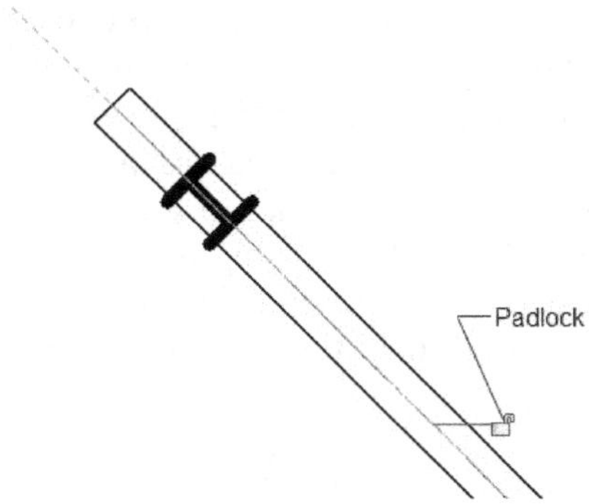

Figure 6-41 The aligned column with the structural wall with padlock displayed

MATCHING TYPE PROPERTIES

Ribbon: Modify > Clipboard > Match Type Properties

In Revit Structure, you can use the **Match Type Properties** tool to copy the properties of a source element and transfer them to another element belonging to the same family. For example, you can select two walls and then match second wall's properties with the first wall.

Invoke the **Match Type Properties** tool from the **Clipboard** panel of the **Modify** tab. On doing so, you will notice that as you move the cursor in the drawing window, it changes to a paint brush. To select an element to match, move the paint brush near the element and click. The empty paint brush will change into a filled one indicating that the element type has been selected. To match another element of the same family, move the dropper over it and click; the dropper will become empty again indicating that the element type has been changed. In case of walls, the matching wall type does not change the instance parameters. Revit Structure provides option for matching multiple family elements to a selected element of the same family.

TRIMMING AND EXTENDING ELEMENTS

In Revit Structure, you can use various tools to trim or extend one or more elements to a boundary defined by the element of the same type. You can also use the tools to extend or trim (if the elements intersect) the non-parallel elements to create a corner. These tools can be used to create corners, trim or extend a reference line or a wall to another reference line or wall. You can use these tools with walls, lines, beams, or braces. In Autodesk Revit Structure, you can trim and extend elements by using three tools: **Trim/Extend to Corner**, **Trim/Extend Single Element**, and **Trim/Extend Multiple Elements**. The three tools are discussed next.

Trim/Extend to Corner Tool

Ribbon: Modify > Modify > Trim/Extend to Corner

The **Trim/Extend to Corner** tool is used to extend a wall or a beam to another wall or beam, or trim any extended portion at the apparent intersection to form a corner. For example, to extend the vertical structural wall to the horizontal structural wall such that the left portion of the horizontal structural wall is retained, use the **Trim/Extend to Corner** tool, refer to Figure 6-42. You can invoke the **Trim/Extend to Corner** tool from the **Modify** panel of the **Modify** tab. Next, select the horizontal structural wall which is to be trimmed or extended. If you want to retain its left portion then click anywhere on the structural wall on the left region of the apparent intersection, as shown in Figure 6-43. To select the next structural wall to be trimmed or extended, move the cursor near the vertical wall. As you move the cursor, you will notice that a dashed line is displayed indicating that the corner will be formed, as shown in Figure 6-44. When you click on the vertical structural wall, it extends and the horizontal wall is trimmed to form the desired corner, as shown in Figure 6-45.

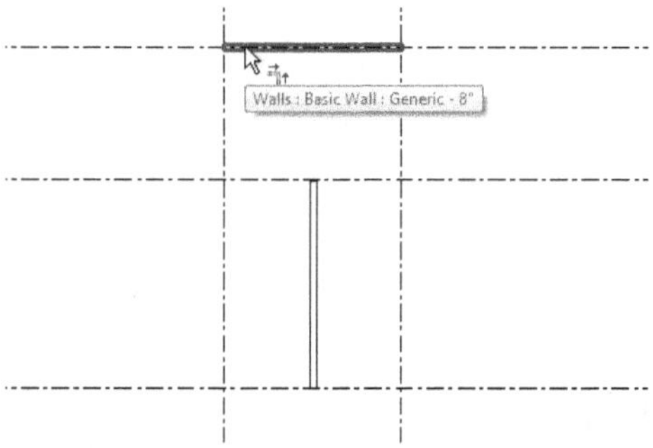

Figure 6-42 *Horizontal wall to be trimmed with the extended vertical wall*

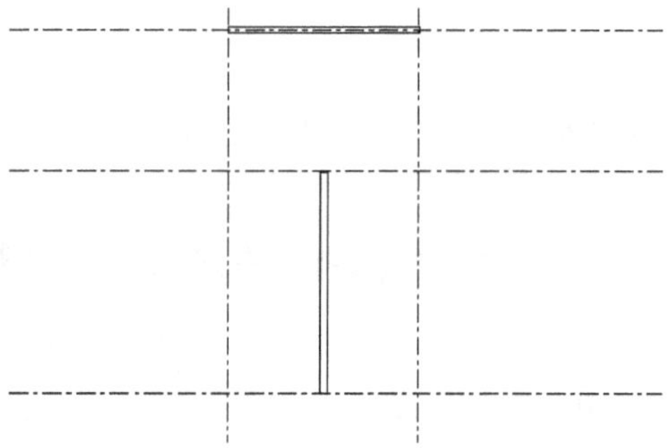

Figure 6-43 *Selecting the first wall to trim/extend on the side to be retained*

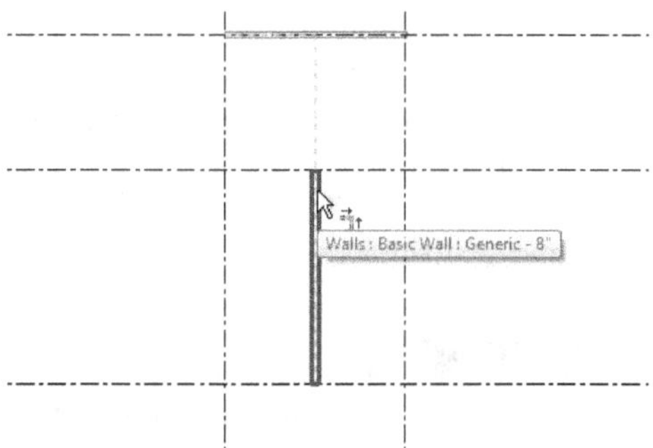

Figure 6-44 *Selecting the second wall to be trimmed/extended*

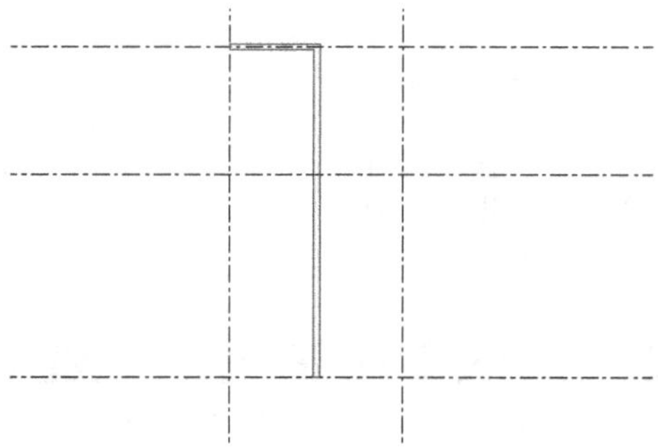

Figure 6-45 *Trimmed and extended walls*

Trim/Extend Single Element Tool

Ribbon: Modify > Trim/Extend Single Element

The **Trim/Extend Single Element** tool enables you to extend or trim an element by defining a reference line or boundary. Using this tool, you can trim a wall or extend a wall to another wall without creating a corner at the intersection. You can also select the desired side of the wall to be retained. For example, you can use this tool to trim the vertical wall at its intersection with the horizontal wall so that the lower portion of the vertical wall is retained. To do so, invoke the **Trim/Extend Single Element** tool from the **Modify** tab, and then to retain the lower portion of the horizontal wall, move the cursor near it until the lower edge of the horizontal wall is highlighted, as shown in Figure 6-46. Next, click on the highlighted edge to select the boundary. Move the cursor to the lower side of the vertical wall and click anywhere on this wall to retain the lower portion. On doing so, the vertical wall will be trimmed at the specified boundary, as shown in Figure 6-47.

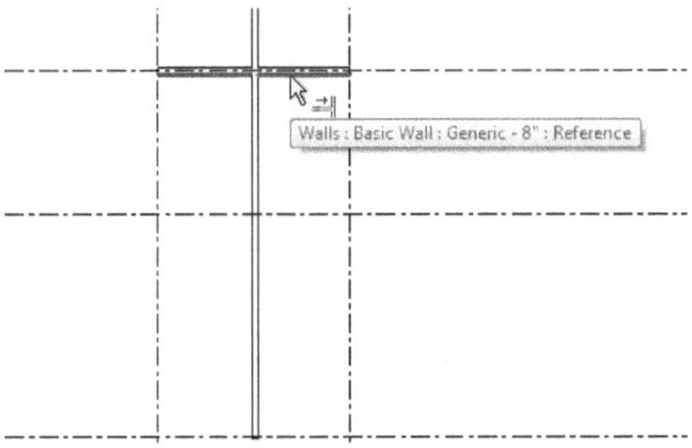

Figure 6-46 *Selecting the edge of the wall to specify the boundary*

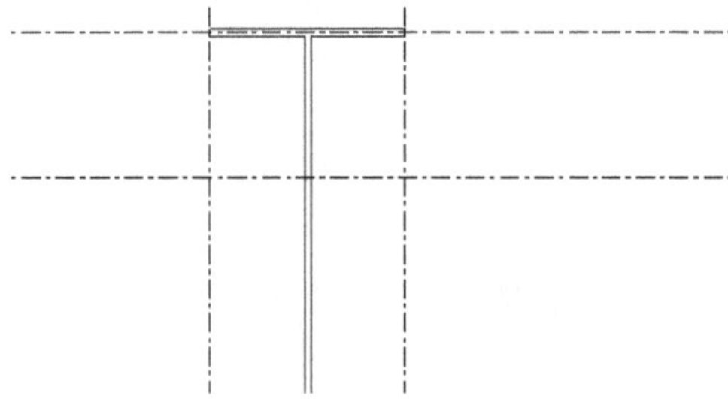

Figure 6-47 *Trimmed wall*

Trim/Extend Multiple Elements Tool

Ribbon:	Modify > Trim/Extend Multiple Elements

The **Trim/Extend Multiple Elements** tool is used to extend or trim a number of elements by using a common boundary or edge. The common boundary or edge can be defined by clicking on an element. After specifying the boundary, select the elements to be extended or trimmed; Revit Structure will extend or trim them at the defined boundary. For example, you may want to extend and trim multiple vertical and inclined walls by using the horizontal wall as the boundary defining element such that the lower portions of the walls are retained. For this, invoke the **Trim/Extend Multiple Elements** tool from the **Modify** tab; you will notice that a symbol appears along with the cursor. Next, move the cursor and place it on the lower edge of the wall and click when the edge is highlighted. On doing so, the lower edge of the horizontal wall will be defined as the cutting boundary and will be highlighted, as shown in Figure 6-48. Next, move the cursor and place it on the lower portion of the middle vertical wall; it will be highlighted, as shown in Figure 6-49. Next, click to trim the wall at the specified boundary. Similarly, you can click on the inclined wall at its lower portion to trim it. Without exiting the **Trim** tool, move the cursor near the vertical wall which you want to extend, and click when the dashed line appears, as shown in Figure 6-50. The wall will extend to the horizontal wall, as shown in Figure 6-51. In this way, you can extend or trim multiple building elements to a specified boundary. Alternatively, you can trim walls by drawing a selection set and selecting the length of the elements to be trimmed.

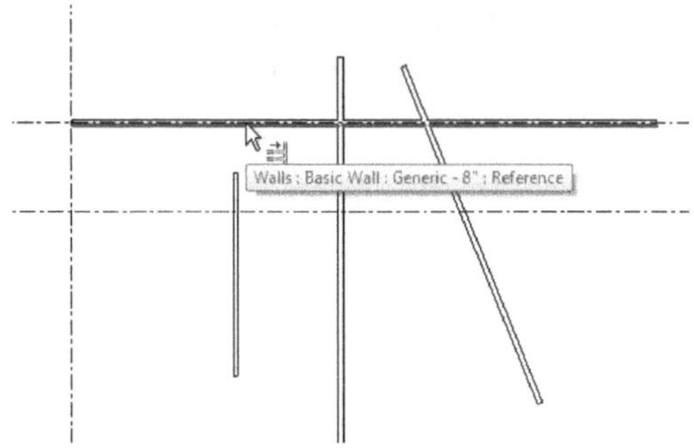

Figure 6-48 Selecting the lower edge of the horizontal wall

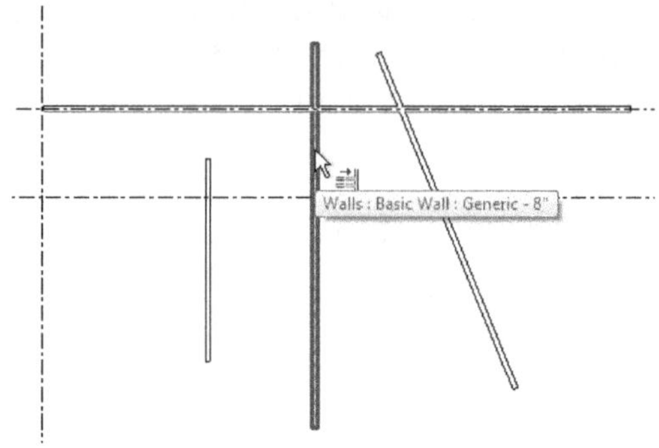

Figure 6-49 Placing the cursor on the lower portion of the middle wall

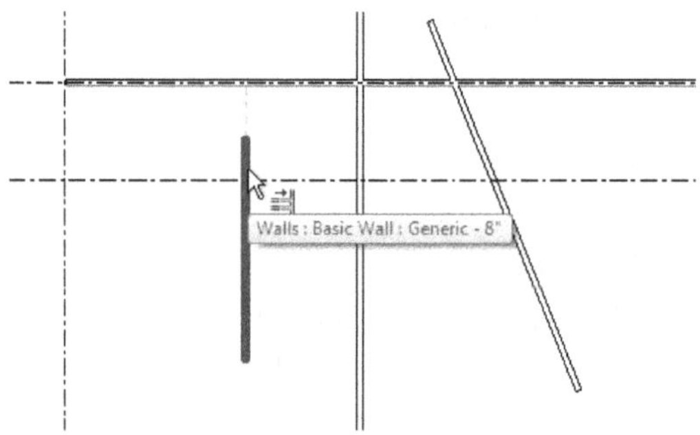

Figure 6-50 Selecting the wall to extend

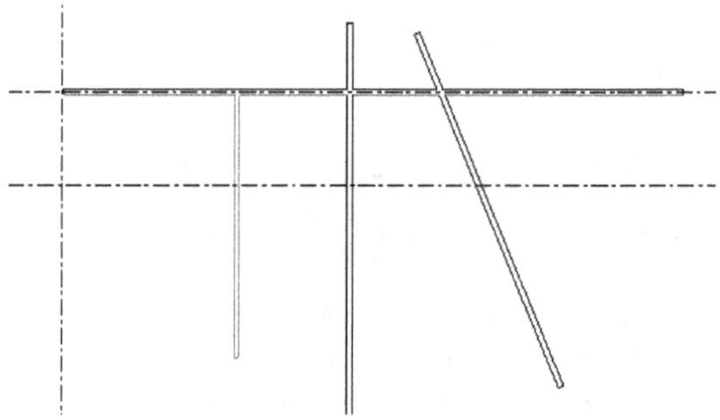

Figure 6-51 The selected wall extended

OFFSETTING ELEMENTS

Ribbon: Modify > Modify > Offset

In Autodesk Revit, you can use the **Offset** tool to offset elements based on your project requirement. This tool is used to create the copy of a selected element at a specified distance from the original element. You can select a single element or multiple elements. The selection can be made even after invoking this tool. By default, Revit Structure, creates the copy of a selected element while creating an offset. To invoke the **Offset** tool, choose it from the **Modify** panel of the **Modify** tab. Alternatively, you can type **OF** to invoke this tool. On invoking the **Offset** tool, you will notice that the **Copy** check box is displayed in the **Options Bar**. By default, this check box is selected. As a result, you can create a copy of the element at the offset distance. Clear the **Copy** check box if you do not need a copy of the element. The **Options Bar** also displays two radio buttons: **Graphical** and **Numerical**. The use of these radio buttons is discussed next.

Tip

While selecting walls or lines for extending and trimming, ensure that the portion of the element to be retained is clicked. In trimming, the portion of the element on the other side of the intersection is removed.

The **Graphical** radio button can be selected to specify the offset distance graphically. After selecting elements and then this radio button, click in the drawing area to specify the start point of the offset distance. On doing so, a preview image of the selected element will be displayed. Next, click again to specify the second point; the selected element will be copied at the distance and direction defined by the two specified points. As you move the cursor to specify the second point, the temporary dimension will be displayed. To specify the exact offset distance, move the cursor in the appropriate direction and enter the offset value in the edit box, as shown in Figure 6-52. Figure 6-53 shows the offset element that is created. The **Numerical** radio button of the **Offset** tool provides you the flexibility to create an offset by specifying the offset distance in the **Offset** edit box of the **Options Bar**. Note that the value you specify in the edit box must be a positive number. After specifying the offset distance, if you move the cursor near the element, a preview image profile of the selected element at the specified offset distance will be displayed in the form of a dashed line. At this stage, you can change the offset distance. You will notice that as you move the cursor over the element, the preview image flips the sides. To select a chain of elements, press the TAB key when the preview image is displayed; all the connected elements will be selected and the preview will change accordingly. Figure 6-54 shows the connected wall profile with its preview offset image. When you see the preview image on the desired side, click to offset the walls; the offset will be created at the distance specified in the **Offset** edit box in the **Options Bar**. Figure 6-55 shows the wall profile created by offsetting it.

Note

You can offset elements only in the same work plane. You cannot offset the elements created as in-place families.

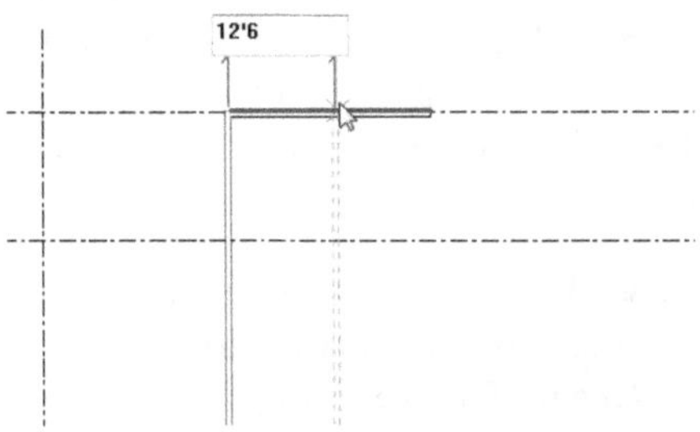

Figure 6-52 *Specifying the offset distance on selecting the* ***Graphical*** *radio button*

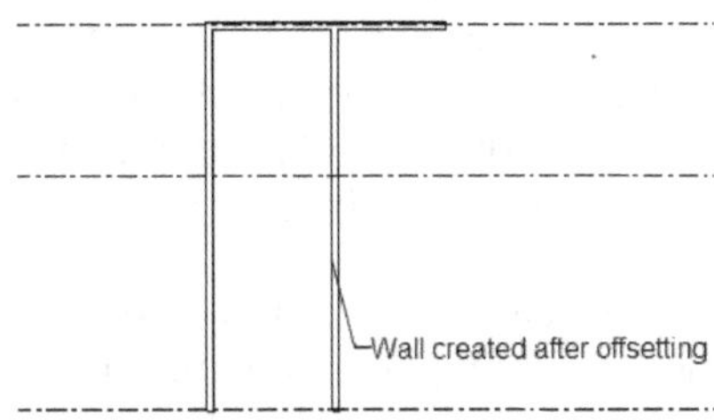

Figure 6-53 *Offset wall created at the specified distance on selecting the* ***Graphical*** *radio button*

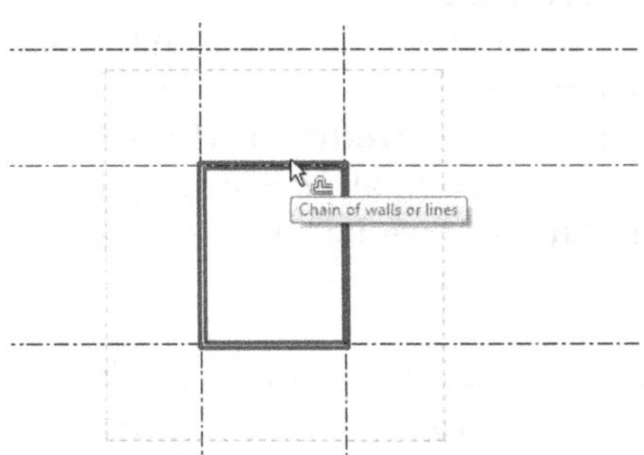

Figure 6-54 *Previewing the offset and selecting a chain of walls using the TAB key*

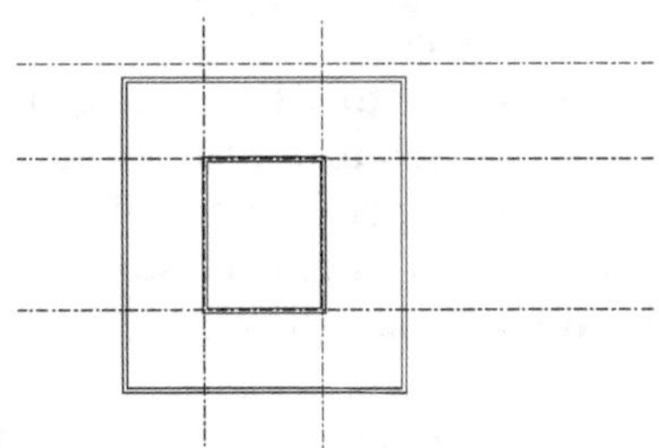

Figure 6-55 *Walls created at the specified offset distance*

DELETING ELEMENTS

Ribbon: Modify | (Elements or Components) > Modify > Delete

You can use the **Delete** tool to remove the selected elements from a project. On doing so, the deleted elements will permanently be removed and will not be pasted on the clipboard. The tool is available only after the elements have been selected. To invoke the **Delete** tool, select the element that you want to delete and then choose it from the **Modify** panel of the contextual tab. Alternatively, you can press the DELETE key or enter **DE** to invoke the **Delete** tool.

GROUPING ELEMENTS

Ribbon: Structure > Model > Model Group drop-down > Create Group

Grouping elements helps you to represent a number of elements as a single entity. Once grouped, these elements can be modified individually. For example you can move or copy the grouped elements as a single unit. Each instance of the group acts as a single unit and any modification done in the group is reflected in all its instances.

For example, you may create a group with a collection of beams and then create its multiple instances. When you modify the beam type in one group, the change will reflect in all the instances of that group. In a project, you can create three types of groups for the elements in the model. These types are: Model Groups, Detail Groups, and Attached Detail Groups. You can create groups either by selecting the elements first and then grouping them, or by creating the groups first and then adding elements to the group. To create a group, invoke the **Create Group** tool from the **Model** panel; the **Create Group** dialog box will be displayed. In this dialog box, enter the name of the group in the **Name** edit box.

Next, in the **Create Group** area, you can select the **Model** or **Detail** radio button. You can select the **Model** radio button to create a model group or select the **Detail** radio button to create a detail group. Now, choose the **OK** button to close the **Create Group** dialog box. However, if you invoke the **Create Group** tool after selecting the elements, you would need to name the group first and then you can add elements in the group. In the next sections, you will learn various methods and techniques to create, modify, and place groups in a project.

Creating Groups by Selecting Elements

In a project, you can create a group by selecting only the model elements from the project views or using the **Group Editor**. To create groups of elements, select the elements to group and then choose the **Create Group** tool from the **Create** panel of the **Modify | Multi-Select** tab; the **Create Model Group** dialog box will be displayed. Alternatively, after selecting the elements to be grouped, choose the **Create Group** tool from **Structure > Model > Model Group** drop-down; the **Create Model Group** dialog box will be displayed.

In the **Create Model Group** dialog box, specify the name of the group in the **Name** edit box and choose the **OK** button; the name of the created group will be automatically displayed under the **Group** heading in the **Project Browser**. The default name of a model group is **Group 1** and the number in it increments as more groups are created. You can rename a group by selecting and right-clicking on its name and choosing **Rename** from the shortcut menu.

Note

While creating a group, if you have selected any one element type, the corresponding ***Modify | (Elements / Components)*** *tab will be displayed instead of the* ***Multi-Select*** *tab.*

Creating Groups Using the Group Editor

This option allows you to create groups prior to the selection of the elements. You can create the **Model** group or the **Annotation** group based on the requirement and later on, select the elements from the project views and add them to the group. To do so, choose the **Create Group** tool from **Structure > Model > Model Group** drop-down; the **Create Group** dialog box will be displayed. In this dialog box, specify the name of the group in the **Name** edit box. The **Group Type** area in this dialog box displays two radio buttons, **Model** and **Detail**. Specify the type of group by selecting the **Model** or **Detail** radio button and then choose the **OK** button; the screen will turn into the edit mode with a yellow background and the **Edit Group** panel will be displayed in the **Structure** tab. Figure 6-56 shows the options in the **Edit Group** panel.

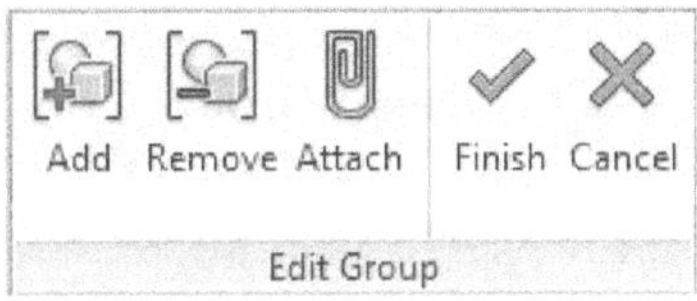

Figure 6-56 *The options in the* ***Edit Group*** *panel*

In the **Edit Group** panel, you can choose the **Add** button to add elements into a group. On choosing the **Add** button, the add (+) symbol will appear with the cursor. You can select the elements from the project view to group them. If there are no elements in the project view, you can create the elements using different tools from the **Structure** panel in the **Structure** tab. The elements will be automatically grouped. You can remove the elements from the existing group, attach details, and even view properties of the group using the respective buttons from the toolbar. Choose the **Finish** button after adding elements to the group to finish editing.

Creating a Detail Group

Revit Structure, allows you to group all detail lines, fill regions, and the detail components separately. To group detail components, select them in the drawing and choose the **Create Group** tool from **Annotate > Detail > Detail Group** drop-down; the **Create Detail Group** dialog box will be displayed. Specify the name of the detail group in the **Name** edit box and choose the **OK** button; the name of the detail group will be displayed in **Groups > Detail** head in the **Project Browser**.

You can create a new group by duplicating an existing group. You can also modify the group without affecting the parent group. To create a duplicate group, select the group to be duplicated in the **Project Browser** and right-click; a shortcut menu will be displayed. Choose **Duplicate** from the shortcut menu; a new group will be added in the **Project browser** under the parent group. You can rename the group, if required.

Alternatively, to create a duplicate group, select the group of which the duplicate is to be created from the drawing area; the **Modify | Detail Groups** tab is displayed. Choose the **Type Properties** tool from the **Properties** panel of this tab; the **Type Properties** dialog box is displayed. Choose the **Duplicate** button from the dialog box; the **Name** dialog box will be displayed. Name the group in the **Name** edit box of this dialog box and choose the **OK** button; the **Name** dialog box will be closed. Now, choose the **OK** button in the **Type Properties** dialog box; the dialog box will be closed and the group will be added to the **Project Browser** with a new name.

Placing Groups

Once a group is created, you can easily place its instances in the project by dragging the group from the **Project Browser**. Select the required **Detail** or **Model** group from the **Project Browser**, drag and drop it at the desired location in the drawing area, and then click to place it. Alternatively, to place a model group, choose the **Place Model Group** tool from **Structure > Model > Model Group** drop-down; the **Modify | Place Group** tab will be displayed. Select the required group from the **Type Selector** drop-down list and click in the drawing area to place the model group.

Alternatively, to place a detail group, choose the **Place Detail Group** tool from **Annotate > Detail > Detail Group** drop-down; the **Modify | Place Group** tab will be displayed. Now, select the desired detail group from the **Type Selector** drop-down list and click in the drawing area to place the detail group.

Modifying Groups

You can modify a group by ungrouping it. To modify a group, select it from the drawing; the **Modify | Model Groups** tab will be displayed. The **Group** panel of this tab displays the **Edit Group**, **Ungroup**, and **Link** buttons. On choosing the **Ungroup** button, the elements in the selected group will be disassociated and ungrouped. Using the **Edit Group** button, you can add or remove elements from the group and also attach details to it. When you choose this button, the screen will enter into the edit mode displaying the **Edit Group** panel, refer to Figure 6-56. The **Add** and **Remove** buttons can be chosen to add and remove elements from the group, respectively. After editing, choose the **Finish** button to complete the modifications. The **Cancel** button provides the option of discarding the modifications and returning to the **Modify | Model Groups** tab. You can use different editing tools on the grouped elements such as **Move**, **Copy**, **Mirror**, **Cut**, **Paste**, **Rotate**, and so on.

Excluding Elements from a Group

You can exclude elements from a group if some elements of the group overlap with the elements of the other group while placing them together. To exclude an element from a group, place the cursor over the element that you want to exclude. Next, press TAB to highlight it, and click; a blue color icon will be displayed. When you place the cursor on the icon, a tooltip will be displayed prompting you to click on the icon to exclude the selected element. Click on the icon and press ESC; the element will be excluded from the group, and will not be displayed in the drawing. Alternatively, to exclude an element from the group select the element in the drawing as explained above and right-click; a shortcut menu will be displayed. Choose **Exclude** from the shortcut menu; the element will disappear from the drawing and will be excluded from the group. The elements that are excluded will not be visible in the drawing and will not be included in the schedule. To restore the excluded elements in the group, place the cursor over the desired group and right-click; a shortcut menu will be displayed. Choose **Restore All Excluded** from the shortcut menu; the elements will be restored in the group and will be displayed in the drawing.

Saving and Loading Groups

To save groups, choose **Save As > Library > Group** from **Application Menu**; the **Save Group** dialog box will be displayed. Select the group to be saved from the **Group To Save** drop-down list and then save it using either the same name as the group name or enter a new name in the **File name** edit box. The group(s) will be saved with the assigned name with *.rvt* extension in case of a project file, or *.rfa* extension in case of a family file.

EDITING ELEMENTS

In Autodesk Revit, you can use editing tools to modify the elements so as to enhance their properties. Some of the editing tools are discussed in detail next.

Applying and Removing Coping

Coping is a process which is used to attach steel members of a model such as beams and columns together. This is basically applied to those beams which are intersecting each other, as shown in Figure 6-57. The **Apply Coping** tool is used to attach the beams to the girders on which they rest. You can apply and then remove copings between the structure steel members.

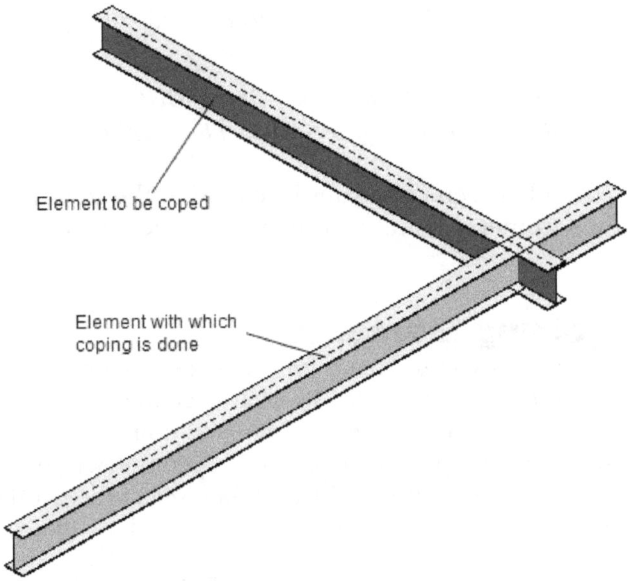

Figure 6-57 *Two beams crossing each other*

Choose the **Apply Coping** tool from the **Cope** drop-down of the **Geometry** panel; the cursor will change to cross hairs and will prompt you to select the element to be coped, refer to Figure 6-57. Select the first element and then next element with which coping is to be done. The two beam members will get coped, as shown in Figure 6-58.

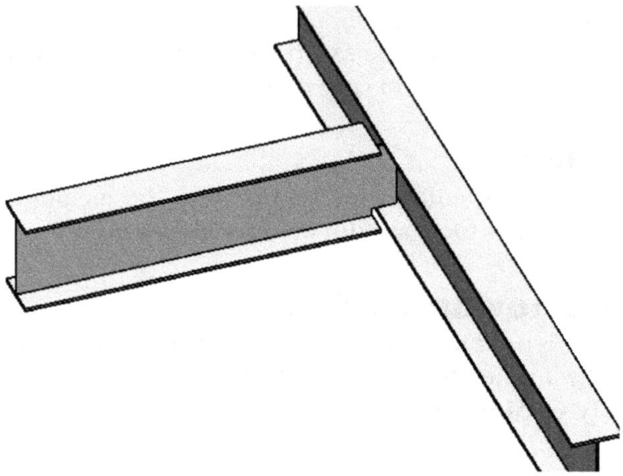

Figure 6-58 *The beams after applying coping*

Note

*Once two elements cope with each other, the **Coping Distance** parameter will be displayed in the **Properties** palette. You can change the value of the **Coping Distance** parameter as desired.*

You can remove the coping between the beams by choosing the **Remove Coping** tool from the **Cope** drop-down of the **Geometry** panel and then select the two beams from which coping is to be removed.

Joining and Unjoining Elements

The **Join Geometry** tool allows you to create joins between two or more host elements which share common face such as columns and walls. To join the geometries, choose the **Join Geometry** tool from the **Join** drop-down of the **Geometry** panel. On doing so, the join geometry tool gets attached with the cursor and prompts you to select the geometry to be joined. First, select the geometry to be joined, as shown in Figure 6-59, and then select the geometry with which it is to be joined, as shown in Figure 6-60. The geometries get joined, as shown in Figure 6-61.

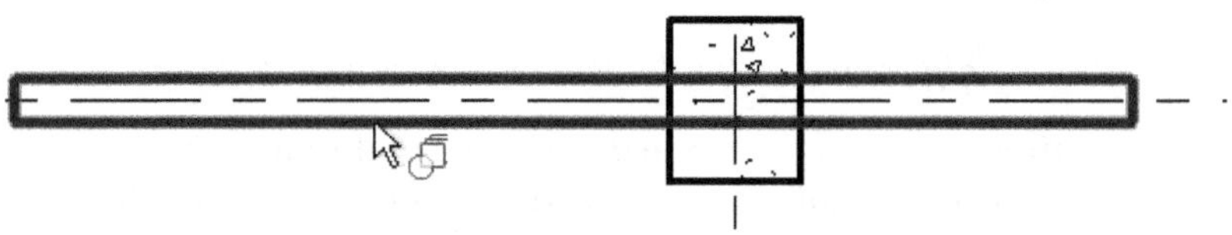

Figure 6-59 *Select first entity to be joined*

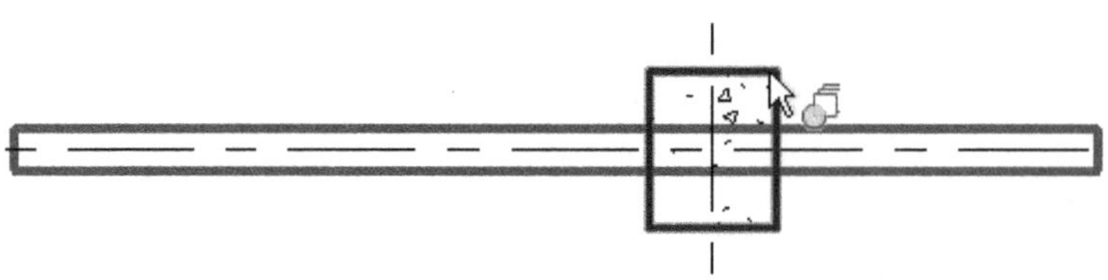

Figure 6-60 *Select second entity to be joined*

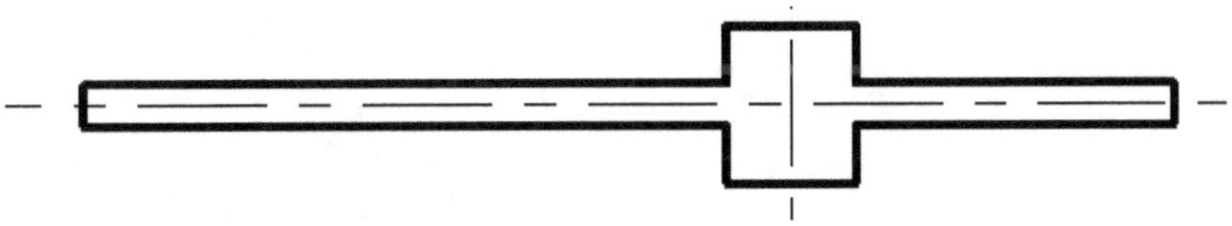

Figure 6-61 *First entity gets joined with the second*

Similarly, you can unjoin the geometries from the drawing. To unjoin the geometries, choose the **Unjoin Geometry** tool from the **Join** drop-down of the **Geometry** panel. On doing so, the tool gets attached with the cursor and prompts you to select the geometries to be removed. Select both geometries one by one and the joins will be removed.

SPLITTING FACE

You can use this the **Split Face** tool to divide the face of an element such as wall or an architectural column. This tool allows you to divide the face into different regions for application of different materials. To apply the split on the face, choose the **Split Face** tool from the **Geometry** panel of the **Modify** tab. On doing so, a cube gets attached with the cursor and will prompt you to select a wall of whose face is to be splitted. Select a wall to split the face, as shown in Figure 6-62; the **Modify | Split Face > Create Boundary** contextual tab will be displayed.

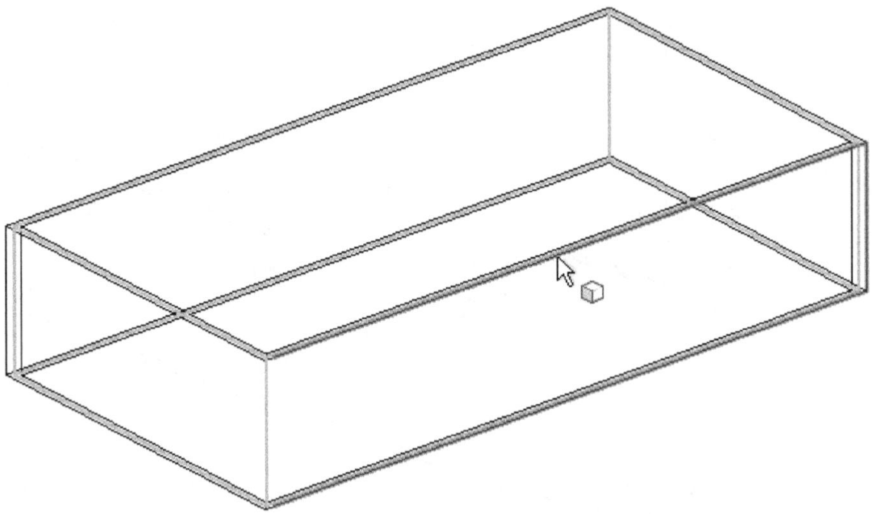

Figure 6-62 *Selecting the face to be split*

In this tab, you can select the required tool to sketch the geometry from the **Draw** panel, as shown in Figure 6-63 and then choose the **Finish Edit Mode** button; the face get splits from the selected face, as shown in Figure 6-64. Now, you can apply separate material to the splitted face.

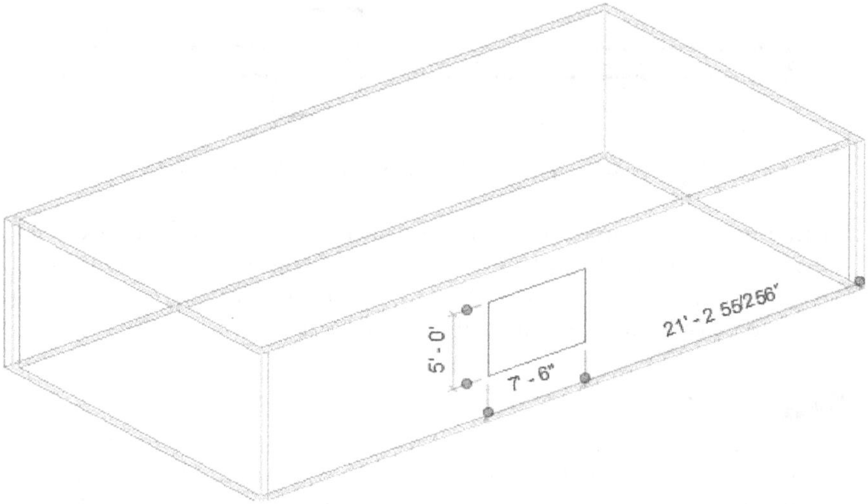

Figure 6-63 *The geometry created on the selected face*

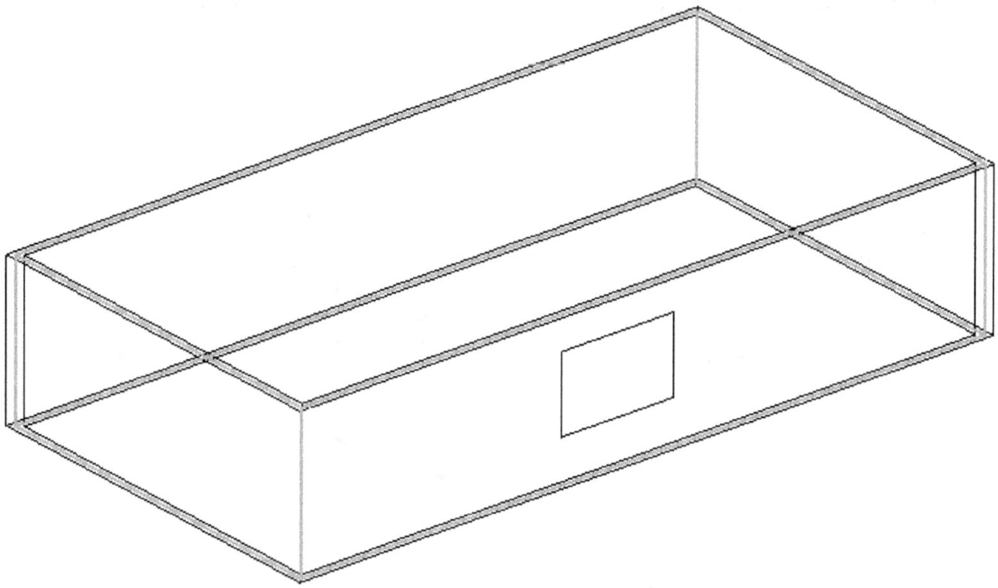

Figure 6-64 *The created geometry gets split from the selected face*

TUTORIALS

Tutorial 1 Commercial Complex

In this tutorial, you will rotate some of the structural foundations and columns of the *Commercial Complex* project created in Tutorial 1 of Chapter 5. Further, you will copy beams, columns and floor in the **Second Floor** structural plan view and paste it on the **Roof** level, using the **Filter** tool. Use the following project parameters and specifications:

(Expected time: 45 min)

1. Rotation Angle: **90°**
2. Elements to be copied to level: columns, beams, and structural floor
3. File name to be assigned:
   For Imperial *c06_Commercial-Complex_tut1.rvt*
   For Metric *M_c06_Commercial-Complex_tut1.rvt*

The following steps are required to complete this tutorial:

a. Open Tutorial 1 of Chapter 5.
b. Rotate structural foundations. (at the **Entry** level).
c. Rotate columns (at the **Entry** level).
d. Copy and paste columns, floors, and beams.
e. Save the project using the **Save As** tool.
f. Close the project using the **Close** tool.

Opening the Existing Project

In this section, you will open the project file created in the Tutorial 1 of Chapter 5.

1. To open the project file, choose **Open > Project** from the **Application Menu**; the **Open** dialog box is displayed.

2. In this dialog box, browse to *C:\rst_2017\c05_rst_2017_tut* location and then choose the *c05_Commercial-Complex_tut1.rvt* for Imperial *or M_c05_Commercial-Complex_tut1.rvt* for Metric project file. You can also download this file from *http://www.cadcim.com*. The path of the file is as follows: *Textbooks > Civil/GIS > Revit Structure > Exploring Autodesk Revit 2017 for Structure*.

3. Choose the **Open** button from the **Open** dialog box; the selected project file opens in the drawing window in the **Second Floor** structural plan view.

Rotating the Structural Foundations

In this section, you will select some of the structural foundations in the **Entry** project view and rotate it.

1. Double-click on the **Entry** node in the **Structural Plans** head in the **Project Browser** to open the corresponding structural floor plan view.

2. Place the cursor on the structural foundation located at the intersection of grid 3.1 and grid

C.1; the structural foundation is highlighted, as shown in Figure 6-65.

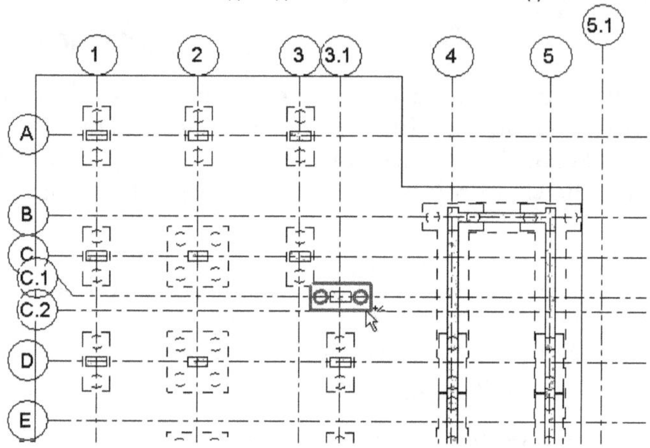

Figure 6-65 The structural foundation highlighted

3. Click on the highlighted foundation to select it.

4. Choose the **Rotate** tool from the **Modify** panel of the **Modify | Structural Foundations** tab; a dashed preview box is displayed around the selected foundation.

5. In the **Options Bar**, click on the **Angle** edit box and enter **90** and then press ENTER; the selected structural foundation is rotated by 90 degrees about its center, refer to Figure 6-66.

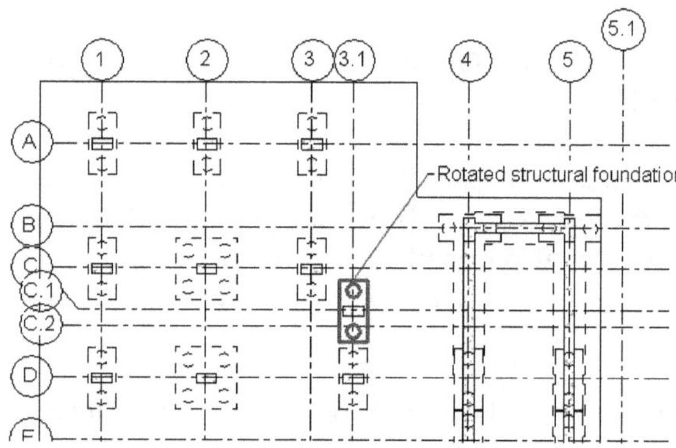

Figure 6-66 The structural foundation rotated

6. Place the cursor on the structural foundation located at the intersection of grids 3.1 and G.2. Click to select the highlighted structural foundation.

7. Repeat the procedure in steps 4 and 5 to rotate the selected structural foundation.

8. Choose the **Modify** tool from the **Select** panel to exit the selection.

Rotating Columns

In this section, you will rotate some of the structural columns in the structural model.

1. Place the cursor over the column located at the intersection of grid 4 and grid D; the column is highlighted, as shown in Figure 6-67. Click on the highlighted column to select it.

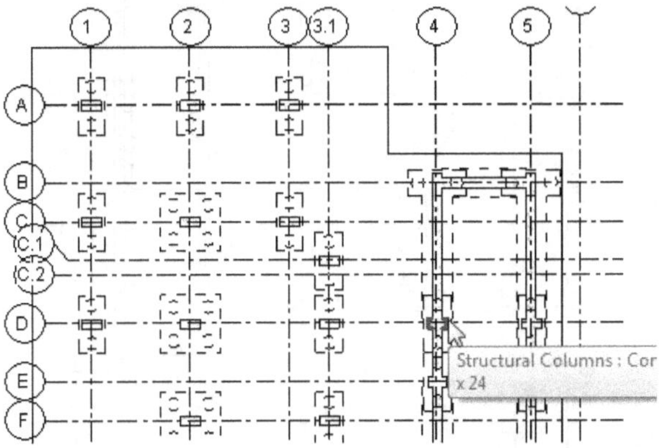

Figure 6-67 The structural column highlighted

2. Choose the **Rotate** tool from the **Modify** panel of the **Modify | Structural Columns** tab; a dashed preview box is displayed around the structural column.

3. In the **Options Bar**, click in the **Angle** edit box and enter **90**. Press ENTER; the selected column is rotated 90 degrees about its center.

4. Next, repeat the procedure in steps 1 to 3 to rotate the columns located at the following grid intersections: (4, E); (5, D); and (5, E). After rotating the desired columns, the structural plan view at the **Entry** level appear, as shown in Figure 6-68.

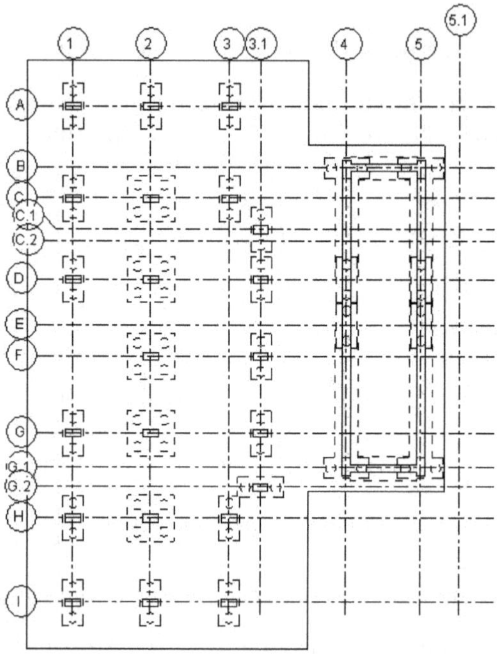

*Figure 6-68 The structural plan view at the **Entry** level*

Copying and Pasting Structural Columns, Floors, and Beams

In this section, you will select structural columns and beams in the **Second Floor** structural plan view and then copy and paste them at the **Roof** level. To select the beams and columns, you will use the **Filter** tool.

1. Double-click on **Second Floor** in the **Structural Plans** head in the **Project Browser** to open the corresponding structural floor plan view.
2. Place the cursor between grids A and 1, as shown in Figure 6-69.
3. Now, press and hold the left mouse button and drag the cursor diagonally down and place it between grid I and grid 5.1, as shown in Figure 6-70. On dragging the cursor down, a selection window emerges and all the elements including grids enclosed inside the selection box are highlighted.
4. Release the left mouse button; all the elements in the selection box are selected.
5. Choose the **Filter** tool from the **Section** panel of the **Modify | Multi-Select** tab; the **Filter** dialog box is displayed.
6. In the **Filter** dialog box, choose the **Check None** button; all the check boxes in the **Category** column are cleared.

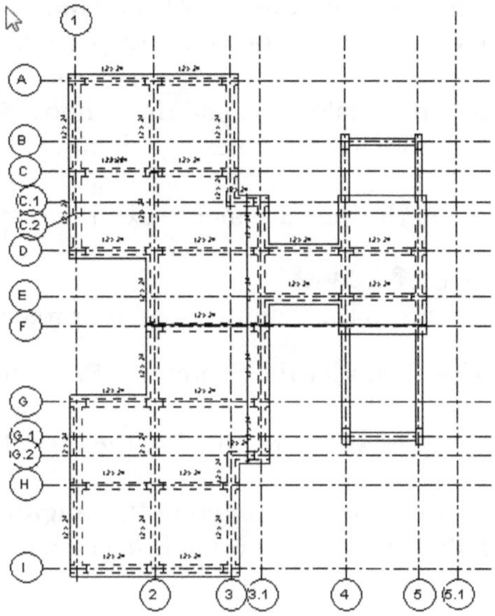

Figure 6-69 *Placing the cursor to select the start point of the selection window*

7. Now, select the following check boxes in the **Category** column: **Floors**, **Structural Columns**, **Structural Framing (Girder)**, and **Structural Framing (Other)**.
8. Choose **Apply** and then **OK**; the **Filter** dialog box closes.
9. Invoke the **Copy to Clipboard** tool from the **Clipboard** panel of the **Modify | Multi-Select** tab; the selected elements are copied to the clipboard.

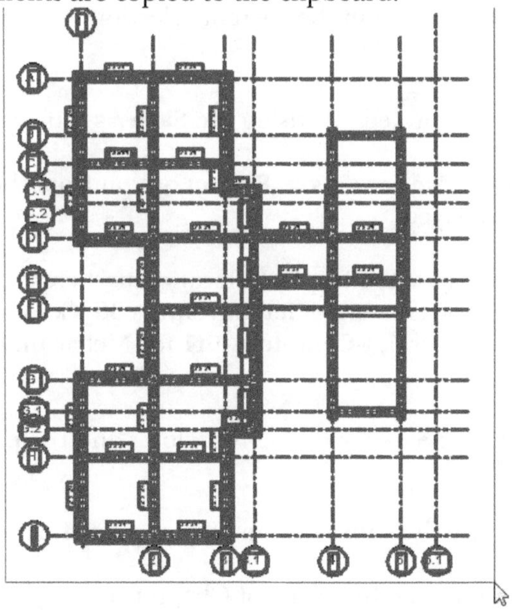

Figure 6-70 *Placing the cursor to select the endpoint of the selection window*

10. Next, choose the **Aligned to Selected Levels** tool from **Modify | Multi-Select > Clipboard > Paste** drop-down; the **Select Levels** dialog box is displayed.

11. Click on the **Roof** option in this dialog box and then choose **OK**; the selected elements are copied to the **Roof** level. Ignore warning message, if any.

12. Choose the **Modify** tool from the **Select** panel to exit the selection.

Adding Beams to the Roof Level

In this section, you will add beams to the columns that are projected to the **Roof** level.

1. Double-click on the **Roof** node under the **Structural Plans** head in the **Project Browser**.

2. Invoke the **Beam** tool from the **Structure** panel in the **Structure** tab.

3. In the **Properties** palette, ensure that the **Concrete-Rectangular Beam: 12x24** (**M_Concrete-Rectangular Beam: 300x600** for metric) option is selected in the **Type Selector** drop-down list.

4. Now, using the **Beam** tool that is already invoked, add beams between the following grid intersections:
   (G.1,4) and (E,4); (E,4) and (D,4); (D,4) and (B,4); (B,4) and (B, 5); (B,5) and (D, 5); (D,5) and (E, 5); (E,5) and (G.1, 5); and (G.1,5) and (G.1, 4)

5. Next, choose the **Modify** tool from the **Select** panel and then select the beams created in the previous step by using the CTRL key, and in the **Properties** palette replace the existing values in the value fields of the **Start Level Offset** and **End Level Offset** parameters with **-1'** for Imperial **(-300mm** for Metric), respectievly.

6. Choose the **Modify** tool to exit from the current selection.

Saving the Project

In this section, you will save the project file using the **Save As** tool.

1. To save the project, choose the **Save As > Project** option from the **Application Menu**; the **Save As** dialog box is displayed.

2. In this dialog box, browse to the *C:\rst_2017* location and create a folder with name **c06_rst_2017_tut**. Next, open the created folder and enter **c06_Commercial-Complex_tut1** for Imperial and **M_c06_Commercial-Complex_tut1** for Metric unit system in the **File name** edit box.

3. Choose the **Save** button; the **Save As** dialog box closes and the project file is saved.

Closing the Project

1. To close the project, choose the **Close** option from the **Application Menu**.

The file is closed and this completes Tutorial 1 of Chapter 6.

Tutorial 2 Industrial Complex

In this tutorial, you will rotate some of the structural foundations and columns to the *Industrial-Complex* project created in Tutorial 2 of Chapter 5. Further, you will copy beams, columns and floor in the **Second Floor** structural plan view and paste it on the **Roof** level using the **Filter** tool. You will also mirror the structural elements at different levels. Use the following project parameters: **(Expected time: 45 min)**

1. Rotation Angle: **90°**
2. Elements to be copied to level: columns, beams, and structural Floor
3. File name to be assigned:
   For Imperial *c06_Industrial-Complex_tut2.rvt.*
   For Metric *M_c06_Industrial-Complex_tut2.rvt*

The following steps are required to complete this tutorial:

a. Open the project file.
   For Imperial *c05_Industrial-Complex_tut2.rvt*
   For Metric *M_c05_Industrial-Complex_tut2.rvt*
b. Copy and paste structural elements.
c. Rotate structural foundations and columns.
d. Delete structural columns and foundations.
e. Mirror structural elements at different levels.
f. Adjust the height of the column and the offset of the beams.
g. Save the project using the **Save As** tool.
h. Close the project using the **Close** tool.

Opening the Existing Project

In this section, you will open the project file created in Tutorial 2 of Chapter 5.

1. To open the project file, choose **Open > Project** from the **Application Menu**; the **Open** dialog box is displayed.
2. In this dialog box, browse to the *C:\rst_2017\c05_rst_2017_tut* location and choose the *c05_Industrial-Complex_tut2.rvt* for Imperial *or M_c05_Industrial-Complex_tut2.rvt* for Metric project file. You can also download this file from *http://www.cadcim.com*. The path of the file is as follows: *Textbooks > Civil/GIS > Revit Structure > Exploring Autodesk Revit 2017 for Structure.*
3. Choose the **Open** button from the **Open** dialog box; the selected project file opens in the drawing window in the **Second Floor** structural plan view.

Copying and Pasting the Structural Elements

In this section, you will select the columns, beams, and floor displayed in the **Second Floor** structural plan view, copy them to the clipboard, and then paste them to the **Roof** level.

1. Ensure that the **Second Floor** node is selected in the **Structural Plans** head in the **Project Browser**.

2. Place the cursor between grid 9 and grid C, refer to Figure 6-71. Now, press and hold the left mouse button and drag the cursor down diagonally, and place it in the area bounded by grid 6, grid 7 and grid A', refer to Figure 6-72. On dragging the cursor down, a selection window emerges from the start and all the elements that are enclosed completely inside the selection box get highlighted.

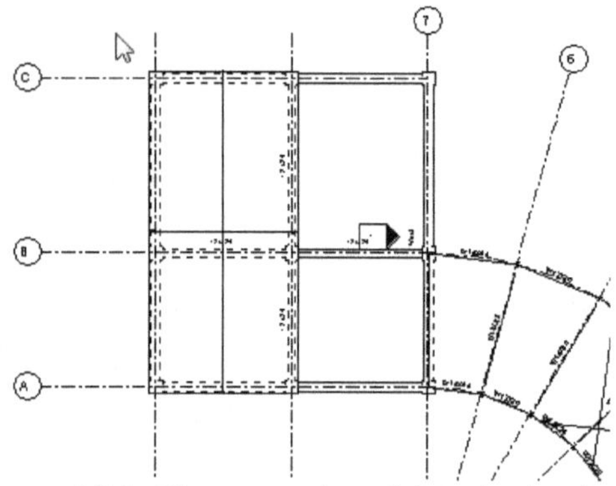

Figure 6-71 *The start point of the selection box*

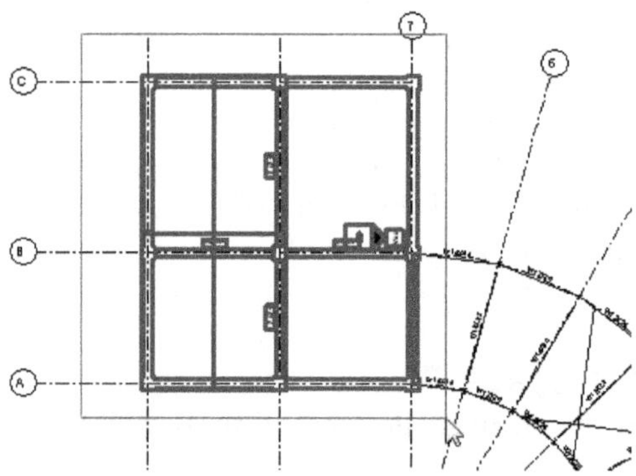

Figure 6-72 *Specifying the endpoint of the selection box*

3 Next, release the left mouse button to specify the endpoint of the selection box; all elements inside the selection box are selected.

4. Choose the **Filter** tool from the **selection** panel of the **Modify | Multi-Select** tab; the **Filter** dialog box is displayed, as shown in Figure 6-73.

5. In this dialog box, choose the **Check None** button, and then select the following check boxes: **Floors**, **Structural Columns**, and **Structural Framing (Girder)**, refer to Figure 6-73.

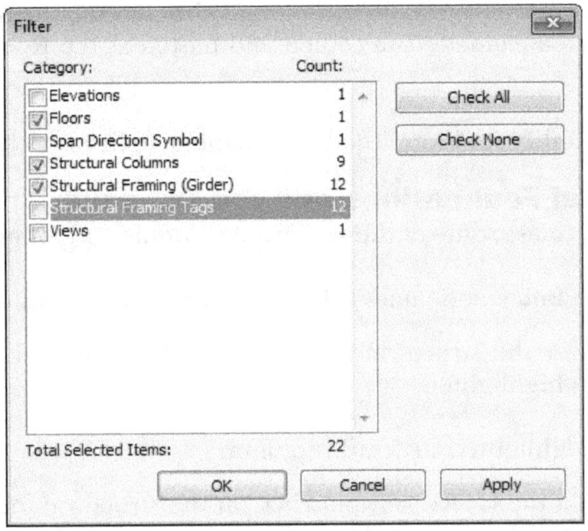

*Figure 6-73 The **Filter** dialog box displaying the selected check boxes*

6. In the **Filter** dialog box, choose **Apply** and then **OK** to apply the filter to the selection and close the **Filter** dialog box.

7. Invoke the **Copy to Clipboard** tool from the **Clipboard** panel of the **Modify | Multi-Select** tab.

8. Next, invoke the **Aligned to Selected Levels** tool from **Modify | Multi-Select > Clipboard > Paste** drop-down; the **Select Levels** dialog box is displayed, as shown in Figure 6-74.

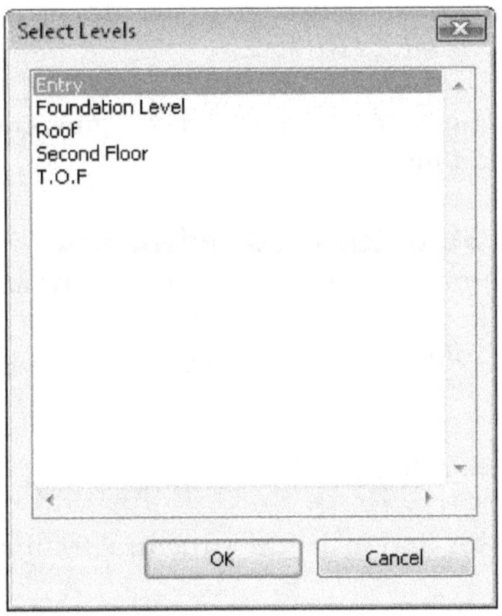

*Figure 6-74 The **Select Levels** dialog box*

9. In this dialog box, click on the **Roof** option and choose **OK**; the **Select Level** dialog box is closed. The selected elements are copied and pasted at the **Roof** level. Ignore warning message if any.

10. Next, choose the **Modify** tool from the **Select** panel to exit from the current selection.

Rotating Structural Foundations and the Columns

In this section, you will rotate some of the structural columns and foundations.

1. Double-click on the **Entry** node under the **Structural Plans** head in the **Project Browser**.

2. Place the cursor over the structural column located at the grid intersection (A',3); the structural column is highlighted.

3. Click to select the highlighted structural column.

4. Next, press and hold the CTRL key and click on the structural foundation located at the same grid intersection; the structural foundation is added to the selection.

5. Now, release the CTRL key and choose **Modify > Rotate** from the **Modify | Multi-Select** tab. Notice that a preview box in dashed line is displayed around the selected elements.

6. In the **Options Bar**, type **90** in the **Angle** edit box and press ENTER; the selected column and the structural foundation are rotated by 90 degrees.

7. Repeat steps 2 to 6 and rotate columns and foundations at the following grid intersections: (B',3); (B',2); and (A',2).

Note

While selecting the column and foundation, first you will select the column and then the structural foundation using the CTRL key.

After rotating the desired structural foundations and columns, choose the **Modify** tool from the **Select** panel to exit the selection mode.

Deleting Columns and Structural Foundations

In this section, you will delete some of the structural columns and foundations.

1. Place the cursor over the column at the grids intersection (A',1); the structural column is highlighted.

2. Click to select the structural column.

3. Press and hold the CTRL key and select the structural foundation located at the same grid intersection.

4. Next, click on the column and then click on the structural foundation located at the grid intersection (B',1).

5. Release the CTRL key and choose **Modify > Delete** from the **Modify | Multi-Select** tab; the selected elements are deleted from the project.

Mirroring the Structural Elements at Different Levels

In this section, you will mirror the left section of the structural model using grid 4 as mirror axis.

1. Double-click on the **Foundation Level** node under the **Structural Plans** head in the **Project Browser**.

2. Now, place the cursor between grid C and grid 9, as shown in Figure 6-75. Press and hold the left mouse button and drag the cursor downward and place it between grid 7 and grid A', as shown in Figure 6-76.

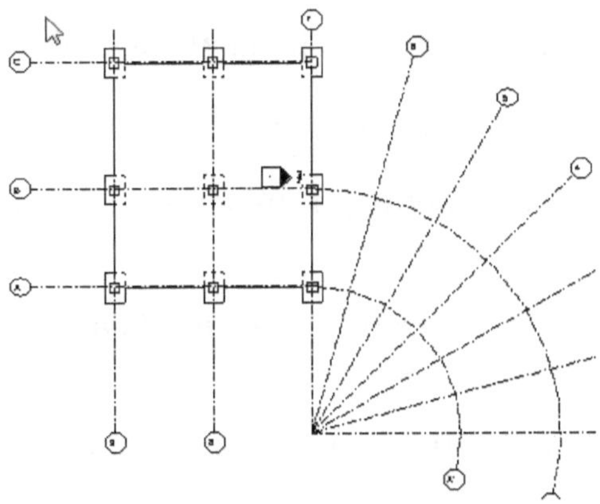

Figure 6-75 *The start point of the selection box*

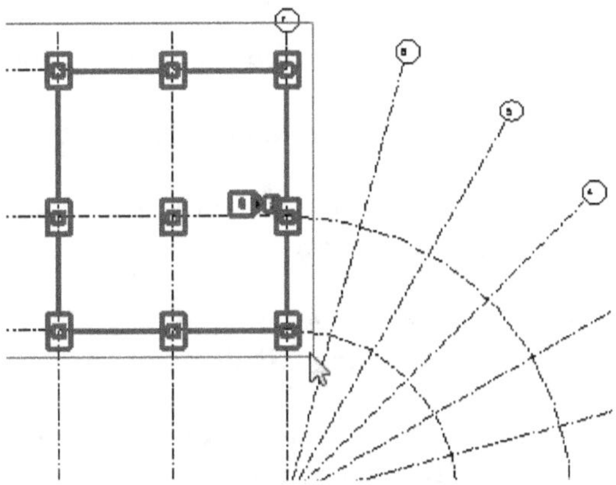

Figure 6-76 *The endpoint of the selection box*

3. Release the left mouse button; the elements within the selection box are selected.

4. Choose the **Filter** tool from the **Selection** panel of the **Modify| Multi-Select** tab; the **Filter** dialog box is displayed.

5. In the **Filter** dialog box, clear the **Foundation Span Direction Symbol** check box.

6. Choose the **Apply** and then the **OK** button; the filter is applied to the selection and the **Filter** dialog box is closed.

7. Now, choose the **Mirror - Pick Axis** tool from the **Modify** panel of the **Modify | Multi- Select** tab. Notice that the mirror symbol appears with cursor. Move the cursor, and place it over grid 4; grid 4 is highlighted.

8. Click at the highlighted grid; the selected elements are mirrored along grid 4. Ignore warning message if any.

9. Now, double-click on the **Second Floor** node under the **Structural Plans** head in the **Project browser**.

10. Next, create a selection box similar to that created using steps 2 and 3. refer to Figures 6-67 and 6-68.

11. Choose the **Filter** tool from the **Selection** panel of the **Modify | Multi-Select** tab; the **Filter** dialog box is displayed.

12. In the dialog box, choose the **Check None** button and then select the following check boxes: **Floors** and **Structural Framing (Girder)**.

13. Choose **Apply** and then **OK** from the **Filter** dialog box; the filter is applied to the selection and the **Filter** dialog box is closed.

14. Now, repeat steps 7 and 8 to mirror the selected elements.

15. Double-click on the **Roof** node under the **Structural Plans** head in the **Project Browser**.

16. Next, create a selection box similar to that created using steps 2 and 3.

17. Choose the **Filter** tool from the **Selection** panel of the **Modify |Multi-Select** tab; the **Filter** dialog box is displayed.

18. In the dialog box, ensure that the following check boxes: **Floors**, **Structural Columns**, and **Structural Framing (Girder)** are selected.

19. Choose **Apply** and then **OK** from the **Filter** dialog box; the filter is applied to the selection and the **Filter** dialog box is closed.

20. Now, repeat steps 7 and 8 to mirror the selected elements, as shown in Figure 6-77. After mirroring the selected elements, choose the **Modify** tool from the **Select** panel to exit the selection mode. Ignore, if a warning message is displayed.

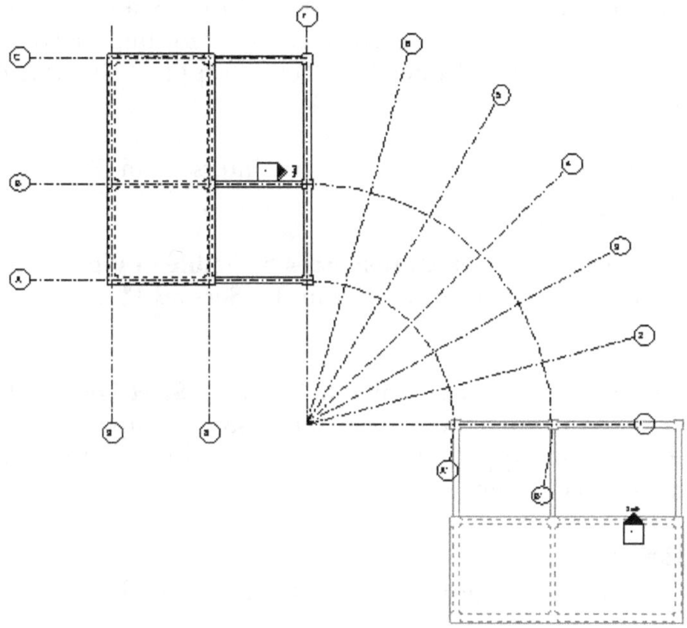

Figure 6-77 *The structural elements mirrored*

Adjusting the Height of Columns and the Offset of Beams

In this section, you will adjust the height of the columns and also to offset the beams.

1. Place the cursor over the column located at the grid-intersection (C,7). Click when it is highlighted.

2. Next, right-click and choose **Select All Instances > Visible in View** from the shortcut menu displayed; all the rectangular concrete columns displayed in the **Roof** structural plan view are selected.

3. In the **Properties** palette, click in the value field of the **Base Offset** parameter and replace the existing value **0** for Imperial unit system or Metric unit system. Choose the **Apply** button from the **Properties** palette to apply the modified parameter to the selected columns.

4. Double-click on the **Second Floor** node under the **Structural Plans** head in the **Project Browser**.

5. Move the cursor and place it over the column located at the grid intersection (B',6). Click when it is highlighted.

6. Right-click and choose **Select All Instances > Visible in View** from the displayed shortcut menu; all the w-flange columns displayed in the **Second Floor** structural plan view are selected.

7. In the **Properties** palette, click in the value field corresponding to the **Top Offset** parameter and delete the existing value. Next, enter **-5"** for Imperial or **-127mm** for Metric. Choose the **Apply** button in the **Properties** palette to apply the modified parameter to the selected columns.

8. Next, place the cursor on the beam located at grid intersection (B',5) and (A',5). Click when it is highlighted.

9. Right-click and then choose **Select All Instances > Visible in View** from the shortcut menu displayed; all the w-flange beams displayed in the **Second Floor** structural plan view are selected.

10. In the **Properties** palette, click in the value fields of the **Start Level Offset** and **End Level Offset** parameters and replace the existing value **-5"** in Imperial system or **-127mm** in Metric system. Choose the **Apply** button in the **Properties** palette to apply the modified parameter to the selected columns.

Saving the Project

In this section, you will save the project file using the **Save As** tool.

1. To save the project, choose **Save As > Project** option from the **Application Menu**; the **Save As** dialog box is displayed.

2. In this dialog box, browse to *C:\rst_2017\c06_rst_2017_tut* and enter **c06_Industrial-Complex_tut2** for Imperial or **c06_Industrial-Complex_tut2** for Metric in the **File name** edit box.

3. Now, choose the **Save** button; the **Save As** dialog box closes and the project file is saved.

Closing the Project

1. To close the project, choose the **Close** option from the **Application Menu**.

   The file is closed and this completes Tutorial 2 of Chapter 6.

Self-Evaluation Test

Answer the following questions and then compare them to those given at the end of this chapter:

1. You can use the __________ tool to select the elements of similar instances in a drawing.

2. The __________ tool is used to refine a selection process by removing the elements of specific categories from the current selection set.

3. The __________ tool is used to create and place the copies of same type of a selected element or a group at specified points.

4. The __________ tool is used to paste the copied elements to one or more levels.

5. The __________ tool enables you to extend or trim a number of elements using a common boundary or edge.

6. You can offset elements in your project by using the __________ tool.

7. The __________ tool is used to copy the properties of a source element and transfer them to another element belonging to the same family.

8. You can select multiple elements from the drawing area by using the CTRL key. (T/F)

9. In Autodesk Revit, you can trim and extend elements in your project by using a single tool. (T/F)

10. While rotating the elements using the **Rotate** tool, you cannot create a copy. (T/F)

Review Questions

Answer the following questions:

1. You can also use the __________ key to select multiple wall or line elements joined at a common endpoint.

2. The __________ tool is used to resize elements.

3. The __________ tool is used to lock a model in place.

4. For the current view of a structural model, you can select elements of similar instances by using the __________ tool.

5. You can use the __________ tool to mirror elements by sketching the mirror line in the drawing area.

6. The __________ tool is used to extend or trim an element by defining a reference line or boundary.

7. You can use the **Copy to Clipboard** tool to create a copy of the selected elements on the clipboard and retain the original ones in the current project. (T/F)

8. The **Match Type Properties** tool can be used to copy the properties of a source element and transfer them to another element belonging to a different family. (T/F)

9. The **Offset** tool is used to create a copy of an element at a specified distance in reference to the selected element. (T/F)

10. You can paste elements at the same place from where you have copied them. (T/F)

EXERCISES

Exercise 1 Academic Institution

In this exercise, you will mirror the structural model created in Exercise 1 of Chapter 5. You will define grid 10 as a mirror axis. Further, you will copy the structural floor, beams, and columns to the **Roof** level. Use the following project parameters and specifications:

(Expected time: 1 hr)

1. Use the **Second Floor** structural plan view to mirror the floor, and beams.

2. Use the **Entry** structural plan view to mirror the foundations.

3. File name to be assigned: *c06_Academic-Institution_exer1.rvt.*

4. Copy the structural floor, beams, and columns displayed in the **Second Floor** view and paste it at the **Roof** level.

5. For reference of the mirror line, refer to Figures 6-78 and 6-79.

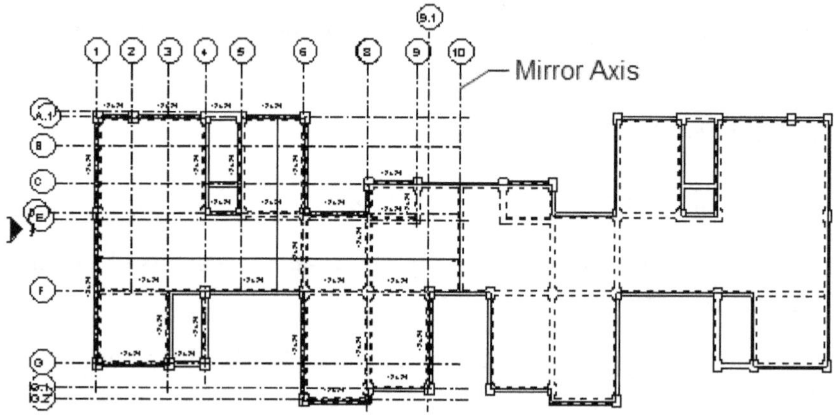

Figure 6-78 *The mirrored structural elements at the* ***Second Floor*** *level*

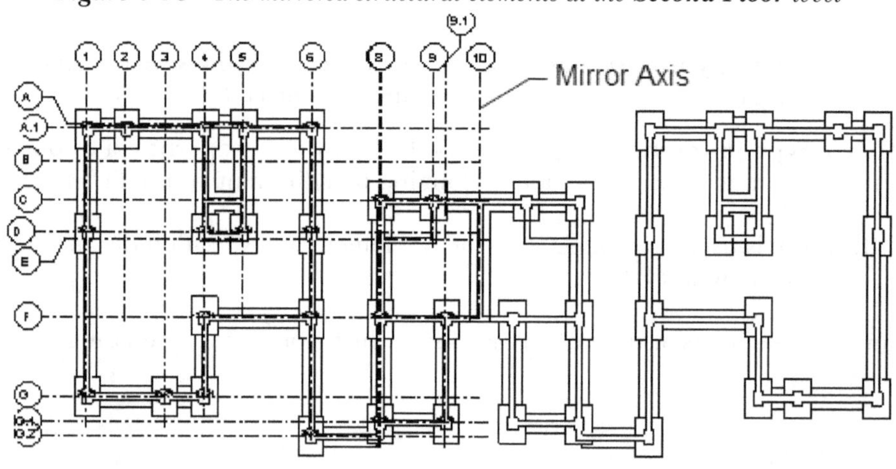

Figure 6-79 *The mirrored structural elements at the* ***Entry*** *level*

Exercise 2 Factory Shed

In this exercise you will mirror the structural model created in Exercise 2 of Chapter 5. You will define the grid 1 as mirror axis. Further, you will copy the structural floor, beams, and columns to the **Roof** level. Use the following project parameters:

(Expected time: 1 hr)

1. Use the **Second Floor** structural plan view to mirror floor and beams.
2. Use the **Plinth** structural plan view to mirror foundations.
3. File name to be assigned:
   For Imperial *c06_Factory-Shed_exer2.rvt.*
   For Metric *M_c06_Factory-Shed_exer2.rvt.*
4. For the reference of the mirror line, refer to the Figures 6-80 and 6-81.
5. Copy the structural floor, beams, and columns displayed in the **Second Floor** view and paste it at the **Roof** level.

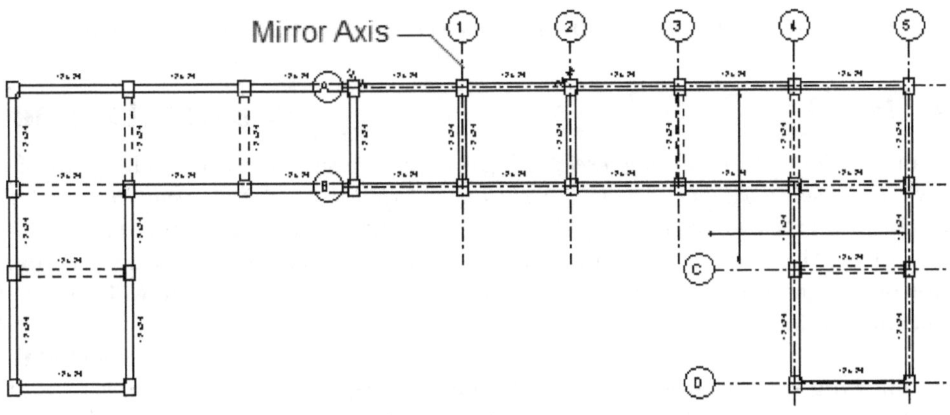

***Figure 6-80** The mirrored structural elements at the **Second Floor** structural plan view*

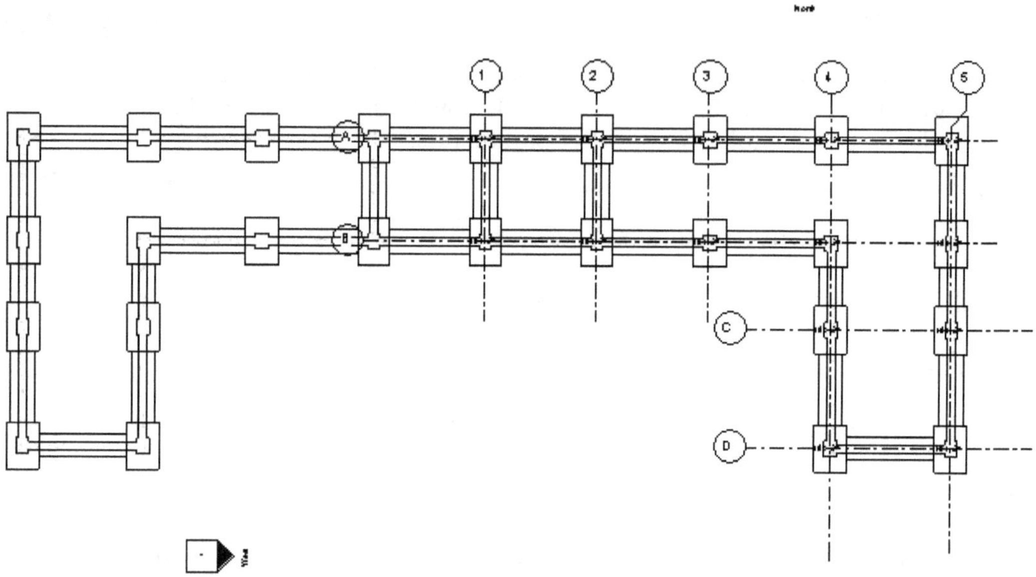

Figure 6-81 *The mirrored structural elements at the* ***Plinth*** *structural plan view*

Exercise 3 Residential Building

Download the *c06_residential_bldg_exer3* file from *http://www.cadcim.com*. The path of the file is as follows: *Textbooks > Civil/GIS > Revit Structure > Exploring Autodesk Revit 2017 for Structure*. In this exercise open the *c06_residential_bldg_exer3* file in Imperial unit system or *M_c06_residential_bldg_exer3* file in Metric unit system and copy the structural floors, columns, and beams displayed in the **Second** structural floor plan to Third and Roof levels. Further, you will mirror the entire model keeping grid D.1 as mirror axis

(Expected time: 1 hr)

Use the following specifications to complete the exercise:

1. In th**e Second** structural floor plan copy the structural floor, columns, and beams and paste them to **Third** and **Roof** level view.

2. Use the **North** elevation view to mirror the entire model. For reference of the mirror line refer to Figure 6-82.

3. File name to be assigned:

   For Imperial *c06_residential_bldg_exer3a.*
   For Metric *M_c06_residential_bldg_exer3a*

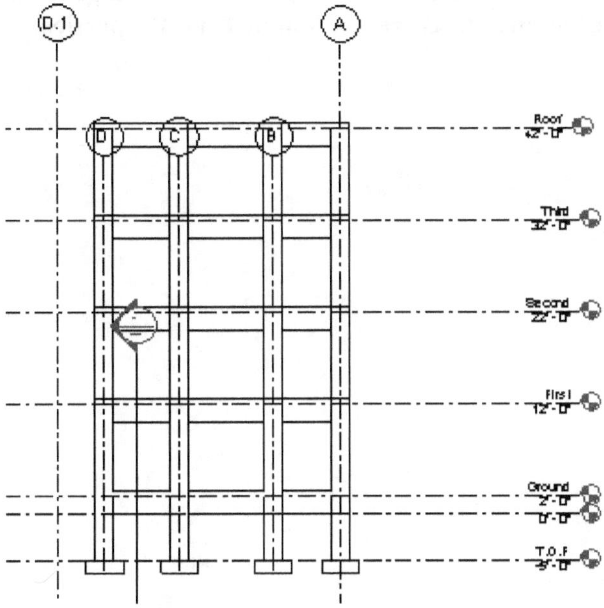

Figure 6-82 *The* **North** *elevation view of the Residential Building project displaying the grid D.1 as mirror line*

Answers to Self-Evaluation Test

1. Select All Instances, **2.** Filter, **3.** Paste from Clipboard, **4.** Aligned to Selected Levels, **5.** Trim/Extend Multiple Elements, **6.** Offset, **7.** Match Type Properties, **8.** T, **9.** T, **10.** F

Chapter 7

Documenting Models and Creating Families

Learning Objectives

After completing this chapter, you will be able to:

- *Add permanent dimensions*
- *Convert temporary dimensions into permanent dimensions*
- *Modify a dimension and its parameter*
- *Add tags to project*
- *Add Beam Annotations to a project*
- *Understand Family creations*

In the previous chapters, you learned how add structural grids, columns, framing members, and various other structural elements to a project. In this chapter, you will learn how to document a project and prepare it as a working drawing by using various tools and techniques in Revit. To document a model, you need to place dimensions to the structural plan view, add text notes and place tags to columns, beams, and structural elements wherever required. In this chapter, you will also learn about various dimensioning terms and tools, tools for placing text notes, and placing tags and beam annotations in your project.

DIMENSIONING

Dimensions play a crucial role in the presentation of a project. Although the structural model of a project can convey what the appearance or design of the model would be, yet to materialize it at the site, there must be information and statistics regarding each element involved in the model. Since the design of a project is used for the actual construction of the project, it is essential to describe each element of the project in terms of actual measurement parameters such as length, width, height, angle, radius, diameter, and so on. All these information can be added to a project by using dimensions. In most cases, information conveyed through dimensions is as important as the project view itself. The dimensions added in the project view ensure that the project drawings in the view are read and interpreted in an appropriate manner. Adding dimensions also help in avoiding the discrepancies that may creep in between the elements used for the generation of the building drawings. In a structural project, you can add dimensions based on the actual dimensions of the elements to be created. In other words, they are as truthful as the project elements themselves. The units used in the dimensions play an important role in describing the detailing that is required to complete a project. For example, the use of fractional inches in the dimensions indicates the amount of detailing required while generating a design. It also reflects the extent of detailing and the precision required to complete a project. Therefore, dimensions are used not only for specifying the sizes of elements, but also for instructing the persons involved in the project, such as cost estimators, project managers, site engineers, contractors, supervisors, and so on.

In Revit, you can create two types of dimensions for an element: temporary dimension and permanent dimension. Temporary dimensions appear while creating or selecting an element, but they do not appear in project views. On the other hand, permanent dimensions appear in the views in which they are created and describe the size or distance. In Revit, you can add the dimensions denoting the length of a wall, distances between rebars, distance between columns, arc length of a curved beam, angle between columns and braces, spot elevation of the top of a slab, and so on. While dimensioning a model, you can change various type parameters of the dimensions such as its size, color, and so on by using the **Type Properties** dialog box. In the next section, you will learn about various entities in a dimension, various types of dimensions, adding dimensions to a project, and modifying the added dimensions.

Types of Dimension

In Revit, you can use two types of dimension: temporary and permanent. By default, in the temporary and permanent dimensions, the units that are specified in the initial start-up of a project are used as the default units. Unlike temporary dimensions, the permanent dimensions are view-specific. It means if you change the view of a project, the permanent dimensions will not be visible.

Using Temporary Dimensions

Revit displays temporary dimensions around the component or element when you place or sketch it in a view. The dimension that appears dynamically while drawing or placing an element is called temporary dimension. This type of dimension is not view-specific and appears in any view while drawing or selecting an element. Temporary dimensions help you position elements at the desired location and references. While sketching the lines of desired length at the desired angle instantly, temporary dimensions can help you speed up your drafting work.

Temporary dimension appears only in three situations. First, when you draw an element; second, when you select an element; and third, when you select or place a component in a project.

When you place a component in a project, temporary dimension becomes a useful tool to guide the exact placing of the component with respect to a fixed element or a component. While placing a component, the temporary dimensions are displayed at the nearest perpendicular element or component with a predefined snap increment setting. This helps in placing a component in a drawing properly. Figure 7-1 shows the temporary dimensions displayed while adding a structural column in a project view.

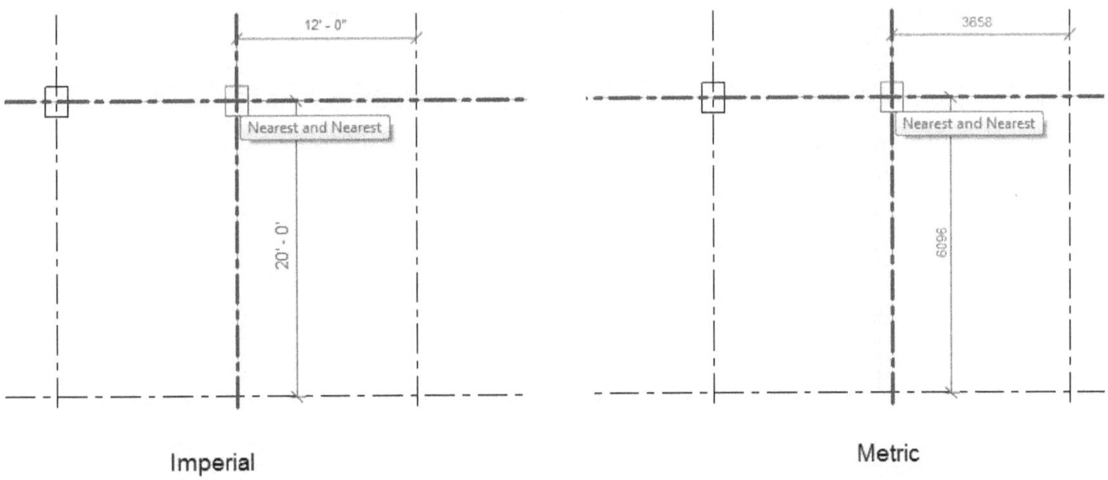

Figure 7-1 *Temporary dimensions displayed while placing a structural column*

While using a temporary dimension, you can set the point of reference of the element being placed. To change the setting of the temporary dimension, invoke the **Temporary Dimensions** tool from **Manage > Settings > Additional Settings** drop-down; the **Temporary Dimension Properties** dialog box will be displayed, as shown in Figure 7-2. This dialog box contains two areas: **Walls** and **Doors and Windows**.

In the **Walls** area of the **Temporary Dimension Properties** dialog box, you can select any one of the four radio buttons, namely **Centerlines**, **Center of Core**, **Faces**, and **Faces of Core** to set the reference point for displaying temporary dimensions of walls. To change the point of reference for placing doors and windows, you can select any of the two options, namely **Centerlines** and **Openings** from the **Doors and Windows** area of the **Temporary Dimension Properties** dialog box. These settings help you set the references of walls, doors, and windows for temporary dimensions while creating or placing elements or components in a model.

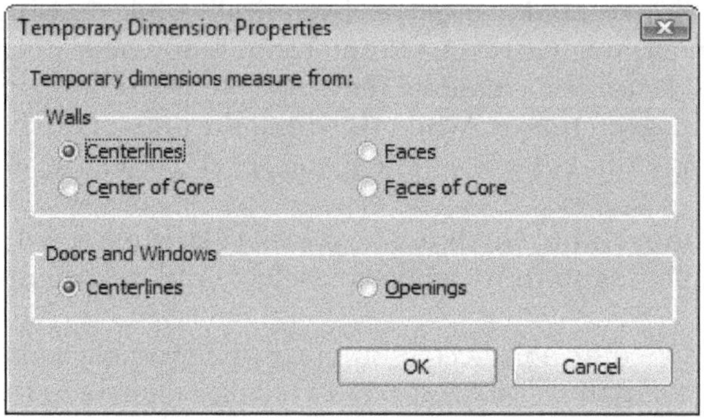

Figure 7-2 The options in the ***Temporary Dimension Properties*** *dialog box*

Entities in a Dimension

Before using the dimensioning tools, it is important to understand various entities in a dimension. Figure 7-3 shows the terms used for various entities in a dimension. In the next section, these entities are discussed briefly.

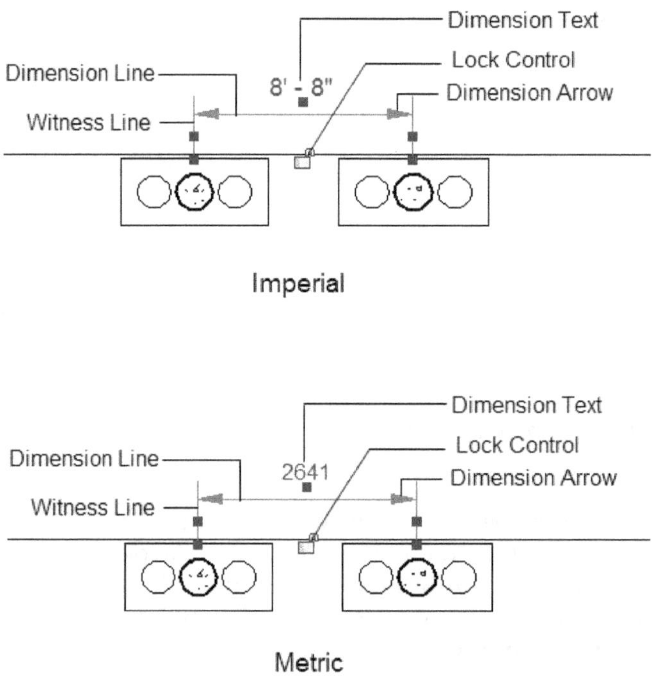

Figure 7-3 Various entities in a Permanent Dimension

Dimension Line

The dimension line in a dimension indicates the distance or angle being measured. By default, the dimension line has tick marks at both ends and the dimension text is placed along the dimension line. Note that in case of the angular or radial dimension, the dimension line will be an arc.

Dimension Text

The dimension text represents the actual measurement (dimension value) between the selected points. You cannot modify this value but can add prefixes or suffixes to it. Note that the dimension text value is automatically updated when the size of an element is modified.

Dimension Arrows

Dimension Arrows are added at the intersection of the dimension line and witness line. As drafting standards differ from company to company, Revit allows you to select tick marks from a range of in-built symbols.

Witness Lines

Witness lines are generated from a selected element and extend toward the dimension line. Generally, they are generated perpendicular to the dimension line. You can move the witness lines to a desired location by using the witness line drag controls. Alternatively, you can move the witness lines by first selecting the desired dimension and then moving the cursor and placing it above the witness line control; the color of the witness line control will change and at the same time, a tool tip displaying **Move Witness Line** text will also be displayed. Note that the tooltip will only be displayed if the setting in the **Options** dialog box (Refer to Chapter 2) permits so. Next, right-click to display a shortcut menu. In the shortcut menu, choose the **Move Witness Line** option; an additional witness line will appear. Move this witness line and click at the desired reference point for the new location; the selected dimension will change to a new value and the desired witness line will move to a new location.

You can add an extra witness line to the existing dimension. To do so, select the desired dimension and then right-click; a shortcut menu will be displayed. In this shortcut menu, choose the **Edit Witness Line** option; an additional witness line with a dimension line will emerge from any of the witness lines of the selected dimension. Next, click on the desired reference; a dimension will be added in extension to the selected dimension. Figures 7-4 and 7-5 show the addition of a witness line to an existing dimension.

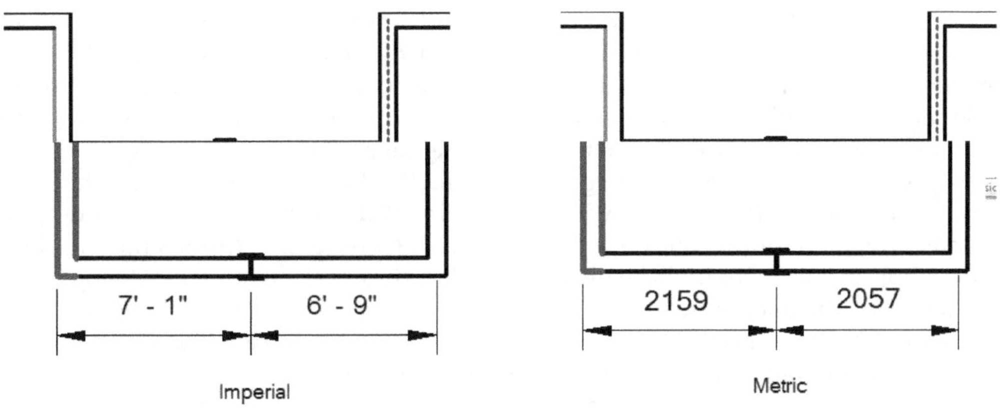

Figure 7-5 *Witness line added to the desired dimension*

Tip

*You can also edit witness lines by selecting the dimension and then choosing the **Edit Witness Lines** tool from the **Witness Lines** panel of the **Modify / Dimensions** contextual tab.*

Adding Permanent Dimensions

Permanent dimensions are added for a specific measurement. In Revit, you can access various dimension tools from the **Dimension** panel in the **Annotate** tab, as shown in Figure 7-6. You can choose appropriate dimension type and dimension tool to add dimensions to an element. The dimensioning tools are discussed next.

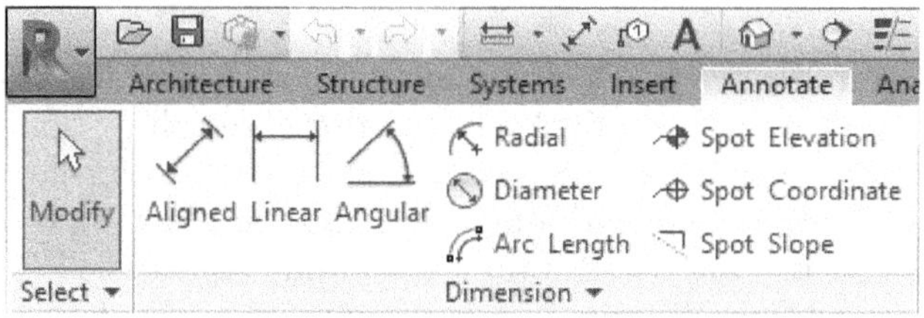

***Figure 7-6** Various dimensioning tools in the **Dimension** panel of the **Annotate** tab*

Aligned Tool

Ribbon: Annotate > Dimension > Aligned

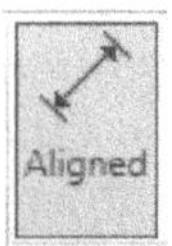

The **Aligned** tool is used to dimension two orthogonal references or points such as wall-ends. To do so, invoke this tool from the **Dimension** panel; the **Modify | Place Dimensions** tab will be displayed. To change the type of the aligned dimension, select an option from the **Type Selector** drop-down list in the **Properties** palette. Next, to set the snap point of the cursor for dimensioning, select an option from the first drop-down list on the left in **Options Bar**, refer to Figure 7-7. For example, if you select the **Wall centerlines** option from the drop-down list, the cursor if placed above the wall will snap to the centerline of the wall. Select the **Individual References** option from the **Pick** drop-down list. Place the cursor at a reference point on an element; the reference point will be highlighted. Next, left-click to specify the reference point and then place the cursor on the required location of the next reference point and click. As you move the cursor upward, a dimension line will appear. Move the cursor away from the component and left-click again; a permanent aligned dimension will appear, refer to Figure 7-8. The **Aligned** tool can also be used to dimension the center of the arc wall and other walls or lines. If you select **Entire Walls** from the **Pick**: drop-down list in the **Options Bar**, the **Options** button will be activated. Choose the **Options** button; the **Auto Dimension Options** dialog box will be displayed. Select the check boxes in the dialog box. Now, you can dimension the openings, doors, windows, and the intersecting walls simultaneously. You can also define the references for dimensioning by choosing the **Centers** and **Widths** radio buttons. Click on the wall that you want to dimension and then move the cursor away from the wall. Next, click in the drawing area; the walls, doors, and openings will be dimensioned separately based on the settings specified in the **Auto Dimension Options** dialog box.

***Figure 7-7** Various options in the **Options Bar** for the **Aligned** tool*

Linear Tool

Ribbon:	Annotate > Dimension > Linear

The **Linear** tool is used to dimension straight elements and distances. It measures the shortest distance between any two specified points. To measure a distance, invoke this tool from the **Dimension** panel. On doing so, the **Modify | Place Dimensions** tab will be displayed. From the **Properties** palette, click on the **Type Selector**

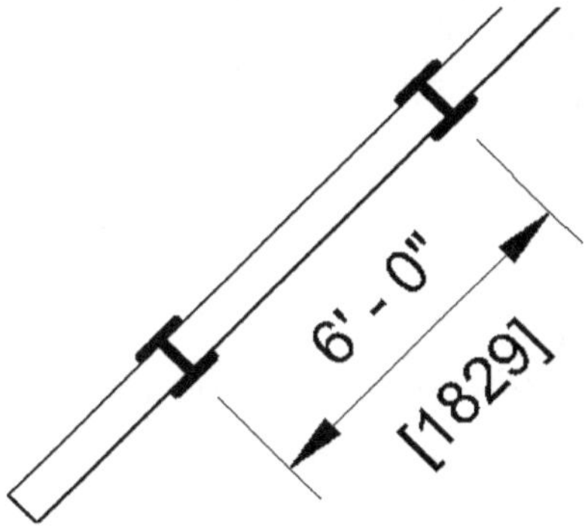

Figure 7-8 *Dimension created using the* ***Aligned*** *tool*

drop-down list and select an option to assign a type to the linear dimension that you want to create. After selecting this option, select the first point by clicking at the appropriate location. After selecting the first point, select the second point of reference for the dimension; the dimension line appears. You can now move the cursor to the desired location and click to place the dimension. The dimension thus created will display various parameters controls. You can invoke any other tool or press ESC to exit the **Linear** tool. The linear dimension will be created, as shown in Figure 7-9.

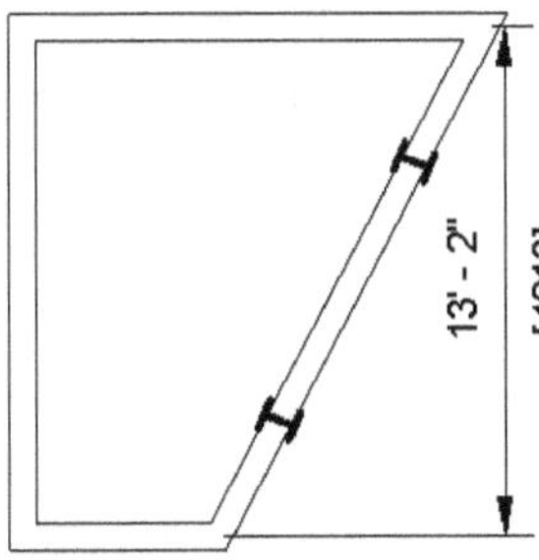

Figure 7-9 *Dimension created using the* ***Linear*** *tool*

Angular Tool

Ribbon:	Annotate > Dimension > Angular

The **Angular** tool is used to dimension an angle. It is also used to create a dimension arc (dimension line in the shape of an arc with arrows on both ends) to indicate the angle between two non-parallel elements, as shown in Figure 7-10. To dimension an angle, invoke the **Angular** tool from the **Dimension** panel; the **Modify | Place Dimension** tab will be displayed. From the **Properties** palette, click on the **Type Selector** drop-down list and select an option. Next, move the cursor, click on the first reference, and then click on the other reference.

On doing so, the angular dimension between the selected references will be displayed. Next, click on an appropriate place to locate the dimension displayed. Figure 7-10 shows an angular dimension between two structural walls.

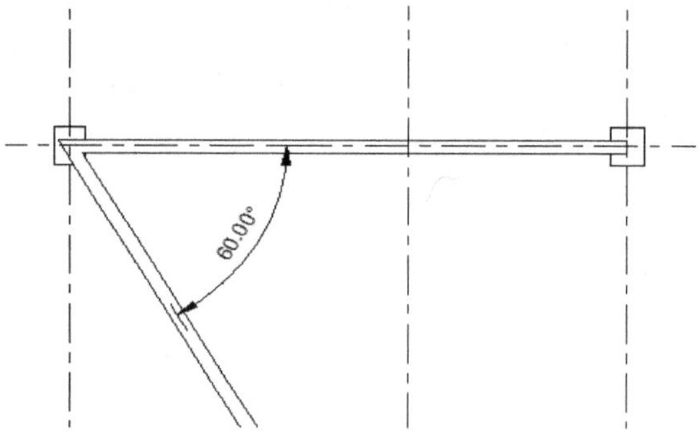

Figure 7-10 *Dimension created using the **Angular** tool*

Radial Tool

Ribbon:	Annotate > Dimension > Radial

The **Radial** tool is used to dimension the radius of a circular profile, as shown in Figure 7-11. To do so, invoke the **Radial** tool from the **Dimension** panel; the **Modify | Place Dimensions** tab will be displayed. In the **Properties** palette of this tab, click on the **Type Selector** drop-down list and select an option from the list of types displayed. After selecting an option, move the cursor near the profile and click when the appropriate snap option appears; a center mark along with the dimension will automatically be generated. Move the cursor and place the dimension.

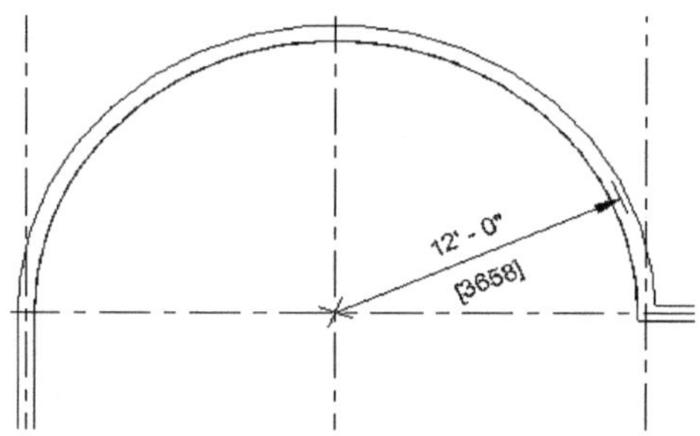

Figure 7-11 *Dimension created using the **Radial** tool*

Note

*In Revit, you can dimension the diameter of a circular profile. To do so, invoke the **Diameter** tool from the **Dimension** panel of the **Annotate** tab and place the dimension as required.*

Arc Length Tool

Ribbon:	Annotate > Dimension > Arc Length

The **Arc Length** tool is used to dimension a curved wall or a curved beam based on its overall length. To dimension a curved wall, invoke this tool from the **Dimension** panel; the **Modify | Place Dimensions** tab will be displayed. In the **Properties** palette, select the dimension type from the **Type Selector** drop-down list. Next, move the cursor near the arc wall and click to specify the radial point. Select the start and end reference points between which the arc length is to be measured. Move the cursor away from the arc wall or beam and click; the arc length dimension will be created, as shown in Figure 7-12.

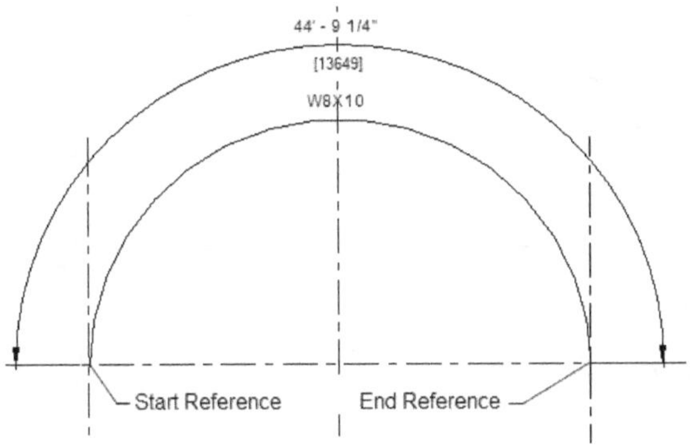

Figure 7-12 *Dimension created using the* ***Arc Length*** *tool*

Spot Elevation Tool

Ribbon:	Annotate > Dimension > Spot Elevation

The **Spot Elevation** tool is used to place a spot elevation at a point in the drawing area. Spot elevations display the elevation level of a point with respect to the base level. They can be placed in a plan, elevation, or 3D view. You can add spot dimensions on non-horizontal and non-planar surfaces. You can use the spot elevation to specify the elevation level of points on a floor, ramp, road, topographical surface, stairs, and other features. To place a spot elevation at a point, invoke the **Spot Elevation** tool from the **Dimension** panel; the **Modify | Place Dimensions** tab along with the **Options Bar** will be displayed. The **Leader** check box is selected by default in the **Options Bar**. As a result, a leader will be displayed along with the spot dimension placed in the drawing. You can clear the check box, if you do not need the leader to be displayed along with the spot dimension in the drawing. In the **Options Bar**, the **Shoulder** check box is also selected by default. As a result, a shoulder line will be displayed along with a leader line in the spot dimension to be placed. You can clear the **Shoulder** check box to remove the shoulder line from the leader line. In the **Relative Base** drop-down list of the **Options Bar**, you can select an option to specify the base level for the spot elevations created using the relative type. By default, the **Relative Base** drop-down list is disabled. This drop-down list will be enabled on selecting a relative type of spot elevation. To add a spot elevation of relative type, select an option from the **Type Selector** drop-down list. The options available in **Type Selector** drop-down list are: **Spot Elevations : Target (Relative)**, **Spot Elevations : Beam (Relative)**, **Spot Elevations : Crosshair (Relative)**, or **Spot Elevations : No Symbol(Relative)**. In the **Options Bar**, select an option from the **Display Elevations** drop-down list to control the display value(s) of the spot elevation to be placed in the project. For example, you can select the **Top & Bottom Elevations** option to display the elevation of the top and bottom faces of an element. In Figure 7-13, you can see the spot elevation of a column. It displays the elevations of top and bottom faces. Next, to place spot elevation in the drawing, move the cursor near any edge of an element; the edge will be highlighted and the desired spot elevation of the highlighted edge will be displayed. Now, you can click on the highlighted edge to place the spot elevation. After clicking once, you may require to click two times in the drawing area to define suitable points for the leader line and the shoulder line.

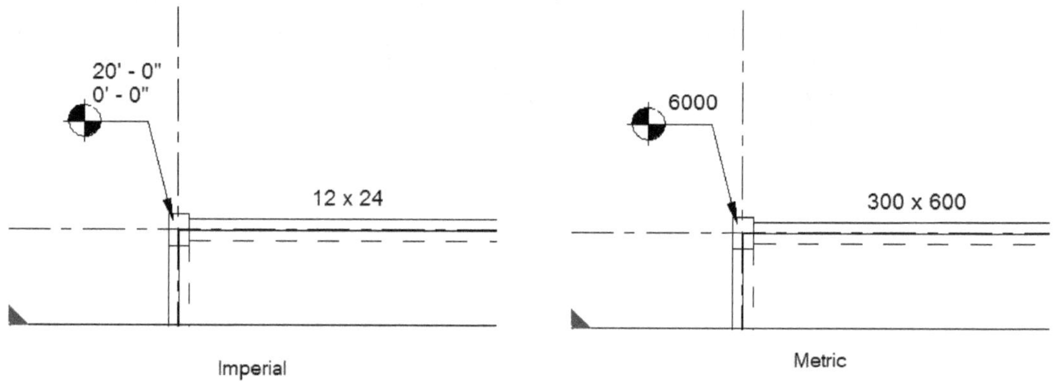

Figure 7-13 *The spot elevation of the column displayed*

Note

The placement of the shoulder line and the leader line in the spot elevation depends on the options that you select from the ***Shoulder*** *and* ***Leader*** *drop-down lists in* ***Options Bar****.*

Spot Coordinate Tool

Ribbon: Annotate > Dimension > Spot Coordinate

The **Spot Coordinate** tool is used to display spot coordinates (Northing and Easting) of a point in a project. You can place spot coordinates on floors, walls, toposurfaces, and boundary lines in a project. Spot coordinates can also be placed on non-horizontal surfaces and non-planar edges.

To place spot coordinates, invoke the **Spot Coordinate** tool from the **Dimension** panel; the **Modify | Place Dimensions** contextual tab will be displayed. The options and the properties for the spot dimension will be displayed in the **Options Bar** and the **Properties** palette, respectively. In the **Options Bar**, you can select the **Leader** and **Shoulder** check boxes to control the display of shoulder and leader along with the spot coordinate dimension. By default, both the check boxes are selected. As a result, leader and shoulder will be displayed along with spot coordinate dimension. To turn off the display of the shoulder and leader, clear the **Shoulder** and **Leader** check boxes. After specifying options from the **Options Bar**, you can select the type of spot coordinate dimension from the **Type Selector** drop-down list in the **Properties** palette. You can select any of these two options: **Horizontal** and **Horizontal (w-Elevation)** from this drop-down list. Select the **Horizontal** option, if it is not selected by default, in the **Type Selector** drop-down list to align the text in the spot coordinate horizontally with respect to the current work plane. Refer to Figure 7-14 to see a spot dimension placed with the **Horizontal** type.

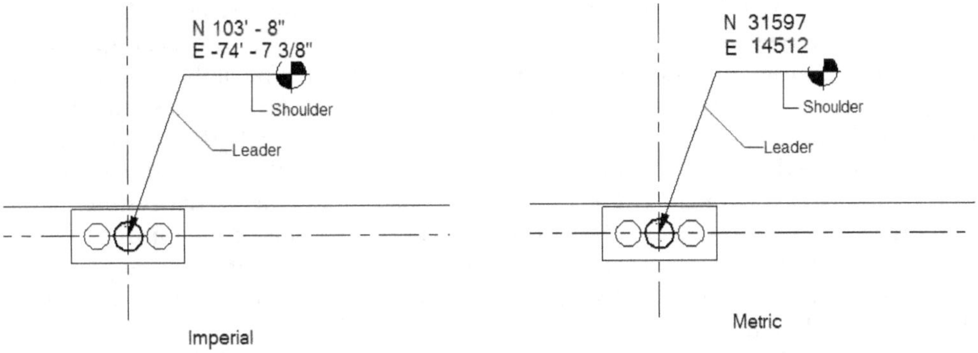

Figure 7-14 *Spot coordinate dimension placed with the **Horizontal** type*

Alternatively, select the **Horizontal (w-Elevation)** option from the **Type Selector** drop-down list to add a spot coordinate dimension along with its elevation. Adding spot coordinates with this type will help you to know the coordinates of a desired point along with its elevation in reference to a predefined base. To add a spot coordinate dimension with the **Horizontal (w-Elevation)** type selected, place the cursor on the point whose spot coordinates you want to display; the point will be highlighted and its spot coordinate dimension will appear. Click on the highlighted point, and then click twice to place the leader and shoulder of the spot coordinate dimension. Figure 7-15 shows a spot coordinate dimension displayed using the **Horizontal (w-Elevation)** option.

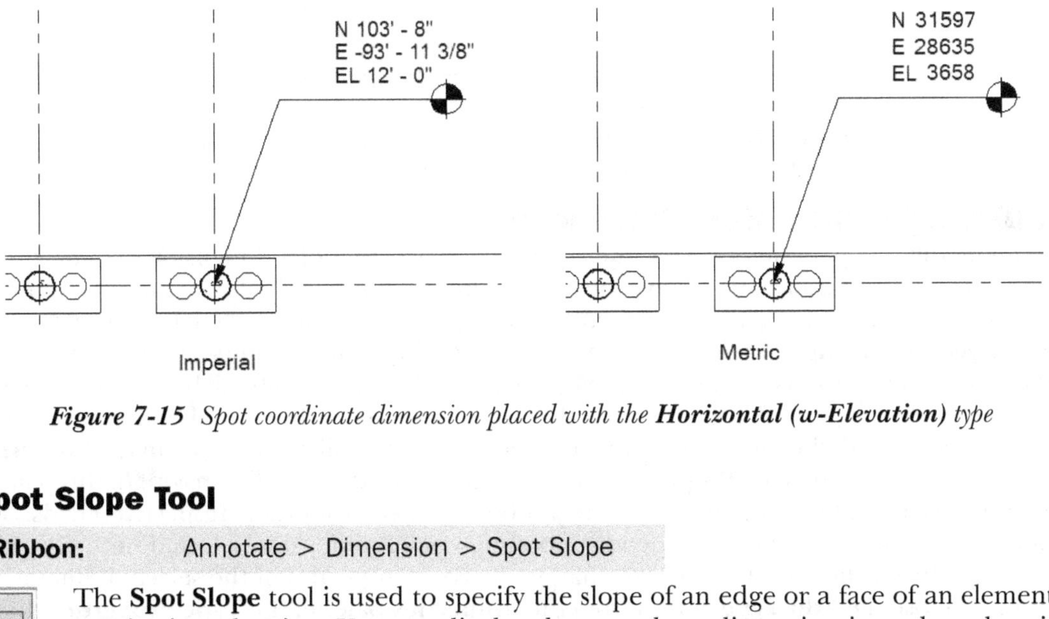

Figure 7-15 *Spot coordinate dimension placed with the **Horizontal (w-Elevation)** type*

Spot Slope Tool

Ribbon:	Annotate > Dimension > Spot Slope

The **Spot Slope** tool is used to specify the slope of an edge or a face of an element at a point in a drawing. You can display the spot slope dimension in a plan, elevation, section, or in a 3d view. To add the spot slope at a point in the drawing, choose the **Spot Slope** tool from the **Dimension** panel; the **Modify | Place Dimension** tab will be displayed. In the **Options Bar**, you can set the representation type for the spot slope dimension that you will place in the drawing. You can set an arrow or a triangle as the representation type for the spot slope dimension. Note that the **Arrow** option is selected by default in the **Slope Representation** drop-down list. You can select the **Triangle** option from the **Slope Representation** drop-down list to set the representation type of the spot slope dimension as a triangle. After setting the desired representation type, you can enter a suitable value in the **Offset from Reference** edit box in the **Options Bar**. This will specify the offset distance of the slope symbol and the dimension text from the point referred for the dimension of the slope. By default, **1/16"** is displayed for the Imperial system or **1.5 mm** for the Metric system in the **Offset from Reference** edit box in **Options Bar**. After setting the options in the **Options Bar**, you can place the cursor on a sloping edge of a model element or a sloping face. As you place the cursor at a point, the slope dimension at that point appears. Click at the desired point to place the slope dimension and then click again to fix the vertical alignment of the slope dimension. Figure 7-16 shows the slope dimensions of an inclined beam and an inclined slab.

Note

*While using the **Spot Slope** tool, if you click on one of the faces of a flat slab or on the straight edge of a beam, the **No Slope** text will be displayed at the point you clicked on.*

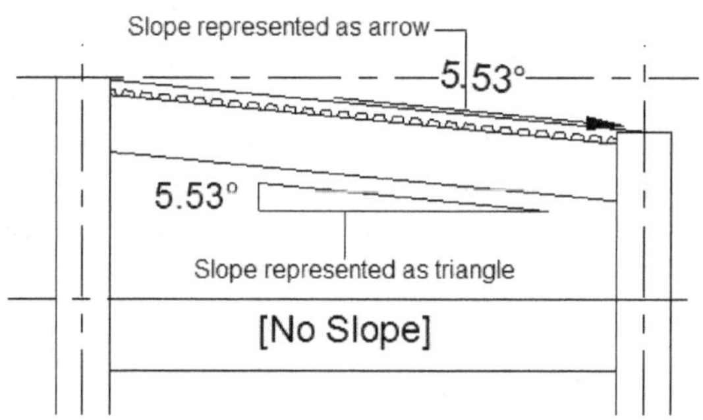

Figure 7-16 *Elevation of a structural model displaying the slope dimensions in different slope representation*

Modifying Dimension Parameters

You can modify the parameters of a dimension such as text font of a dimension, gap of witness line from element, and so on to achieve the desired dimension style. To do so, select the required dimension from the drawing; the contextual tab corresponding to the selected dimension type will be displayed. For example, for linear dimension, the **Modify | Dimensions** tab will be displayed, and for the spot elevation dimension, the **Modify | Spot Elevations** tab will be displayed. In this tab, you can use various editing tools from the concerned panel, if required. While the desired dimension is selected, the properties of the selected dimension will be displayed in the **Properties** palette. Figure 7-17 shows the **Properties** palette of the selected linear dimension. In this palette, there are various instance parameters that can be modified from their respective heads. The appearance of these parameters depends upon the type of dimension selected in the drawing area. In the **Properties** palette, you can change the type properties of the selected dimension. To do so, choose the **Edit Type** button from the **Properties** palette; the **Type Properties** dialog box will be displayed, as shown in Figure 7-18. In the **Type Properties** dialog box, you can also create a new dimension style with the parameters most suitable to the project by using the **Duplicate** button. Various type parameters and their usage are discussed next.

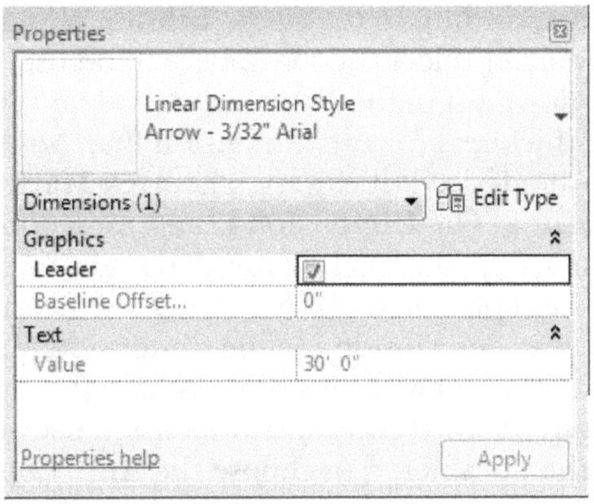

***Figure 7-17** The **Properties** palette displaying various properties of the linear dimension type*

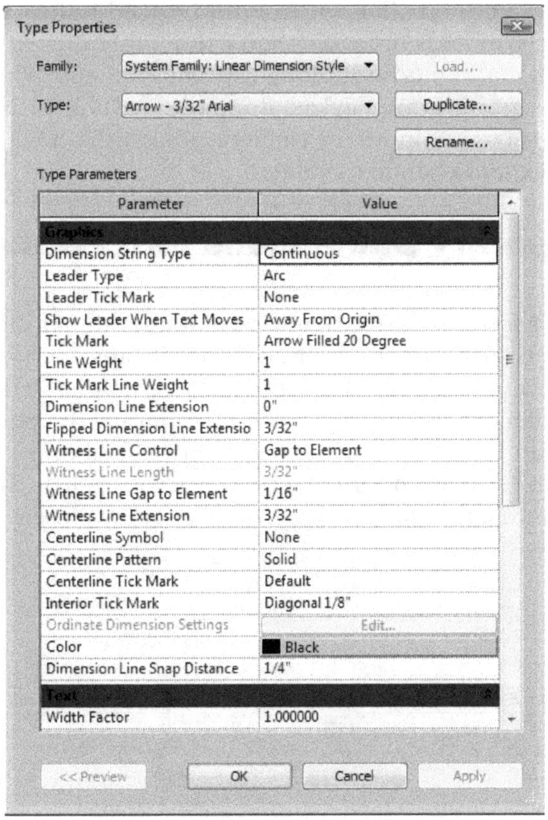

***Figure 7-18** The **Type Properties** dialog box*

Using the **Leader Type** parameter, you can specify the type of line to be drawn for the leader. You can select the **Arc** or **Line** option from the drop-down list displayed on clicking on the **Value** field corresponding to it. You can specify the type of tick mark for the leader by selecting an option from the **Value** column corresponding to the **Leader Tick Mark** parameter. You can specify the condition for the display of the leader. To do so select an option from the drop-down list in the **Value** column corresponding to the **Show Leader When Text Moves** parameter. For the **Tick Mark** parameter, you can select the mark type to be used as the tick mark. Select the desired mark from the drop-down list in the **Value** column. The **Line Weight** parameter is used to specify the line weight or thickness for the dimension line. You can select the value ranging from 1 to 16, depending on the desired thickness. The **Tick Mark Line Weight** parameter sets the thickness of the tick mark line. The **Dimension Line Extension** parameter is used to specify the distance by which the dimension line can be extended to the point where it intersects with the witness line. In the **Equality Text** parameter under the **Other** head, you can change the default label from **EQ** to a different text description for the dimension type. This custom label will be displayed for all equality texts created with this dimension type. Apart from various dimension settings, you can also set various parameters for the dimension text such as **Text Size**, **Text Offset**, and so on. After modifying the parameter(s), choose the **Apply** button to apply the changes.

Locking Permanent Dimensions

When you add a permanent dimension, a lock control symbol is displayed. It also appears when you select a permanent dimension. This symbol can be used to lock or unlock the dimension of an element. When the symbol is unlocked, you can modify the extent of the dimension. In such cases, the dimension is also modified along with the element. When you lock a dimension, the element corresponding to it is also locked along with it. This means you cannot modify its dimension value. You can, however, move the element along with dimension. Once the dimension is locked, you must unlock it to change its value.

Converting Temporary Dimensions into Permanent Dimensions

As discussed in the earlier sections of this chapter, temporary dimensions appear while creating various structural elements such as grids, walls, columns, beams, and other structural elements. They can be used to specify the size and location of an element in the structural model. You can move or resize an element by clicking on the temporary dimension and entering a new value. When you create an element, a conversion control symbol is also displayed along with the temporary dimension, as shown in Figure 7-19. When you click on this symbol, the temporary dimension is converted into a permanent dimension, as shown in Figure 7-20.

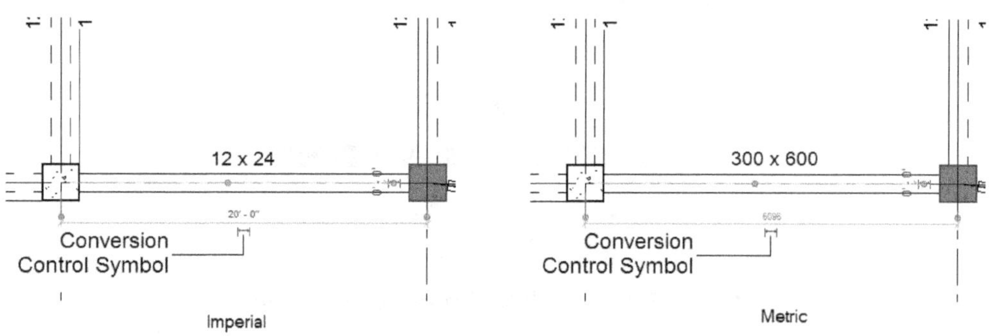

Figure 7-19 *The conversion control symbol displayed in the selected temporary dimension*

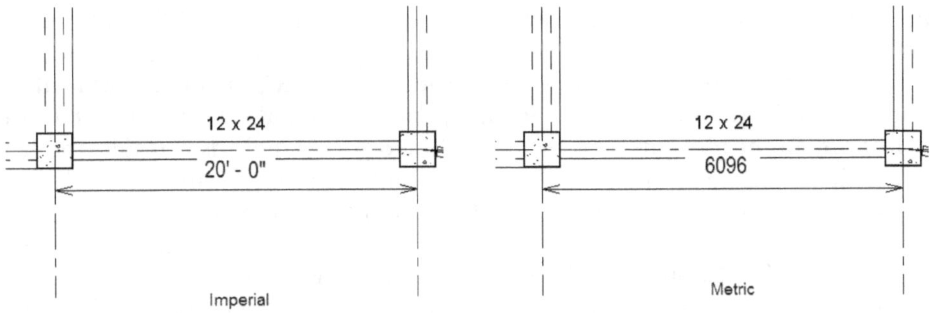

Figure 7-20 Temporary dimension converted into a permanent dimension

TEXT NOTES

Text notes are important parts of a project detail. They not only help in adding the specification of various structural elements but also in conveying the specific design intent. Revit provides a variety of options to add text notes to a structural model detail view by using the **Text** tool. Various options to add and modify a text in a model view are discussed next.

Adding Text Notes

Ribbon:	Annotate > Text > Text

In Revit, text notes are view-specific entities. You can add different text notes to different views. Generally, text notes are added to the plan view, section view, callout view, or in drafting view. The selection of a view in which the text notes are added depends on the project requirement. To add a text note to a desired view, invoke the **Text** tool from the **Text** panel; the **Modify | Place Text** tab will be displayed, as shown in Figure 7-21.

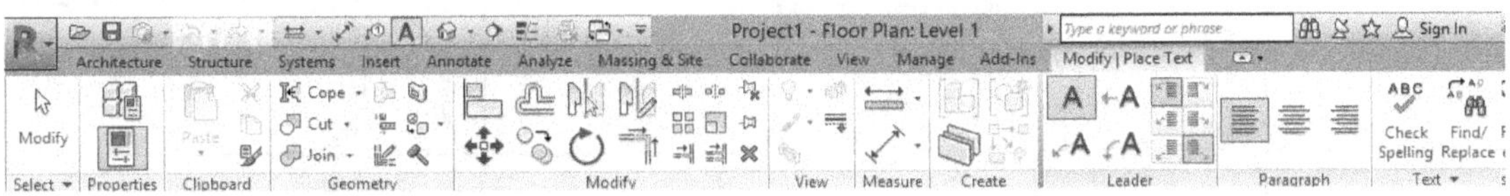

Figure 7-21 Various options in the ***Modify / Place Text*** *tab*

In this tab, you can use the tools in the **Format** panel to control the writing format of the text note to be added in the project view. For instance, using the text note, you can attach a straight leader line that has a single segment or double segments. Alternatively, you can also attach a curved leader line to a text note. If you insert a text note into project view, no leader will be attached to the text note because the **No Leader** tool is chosen by default in the **Format** panel. To attach a straight leader line with one segment, choose the **One Segment** tool from the **Format** panel. You can choose the **Two Segments** tool or the **Curved** tool from the **Format** panel to attach a double segment straight leader line or a curved leader line to the text note. Figure 7-22 illustrates the usage of the leader options discussed earlier. After choosing the desired option for the display of the leader line, you can choose various tools from the **Format** panel to locate the leader line in reference to the text note. For instance, you can place the leader line on the left of a text note at three different places: top left, middle left, or bottom left corner of the text note. To do so, choose the **Leader at Top Left**, **Leader at Middle Left**, or **Leader at Bottom Left** tool respectively, from the **Format** panel. Similarly, you can place the leader line on the right of the text note. To do so, choose the **Leader at Top Right**, **Leader at Middle Right**, or **Leader at Bottom Right** tool from the **Format** panel. By default, the **Leader at Top Left** and **Leader at Bottom Right** tools are chosen from the **Format** panel. As a result, the leader will be attached to the text note at the top left corner or at the bottom right corner, depending on the side the leader is placed on. After setting the options for the placement of leader, you can set the alignment of the text to be inserted. To do so, you can choose any of the three tools, namely **Align Left**, **Align Center**, or **Align Right**, from the **Format** panel. The **Align Left** tool is chosen by default in this panel. As a result, the text will be aligned on the left margin in the text box. You can choose the **Align Center** tool to align the text evenly between the left margin and the right margin of the text box. You can choose the **Align Right** tool to align the text to the right margin of the text box.

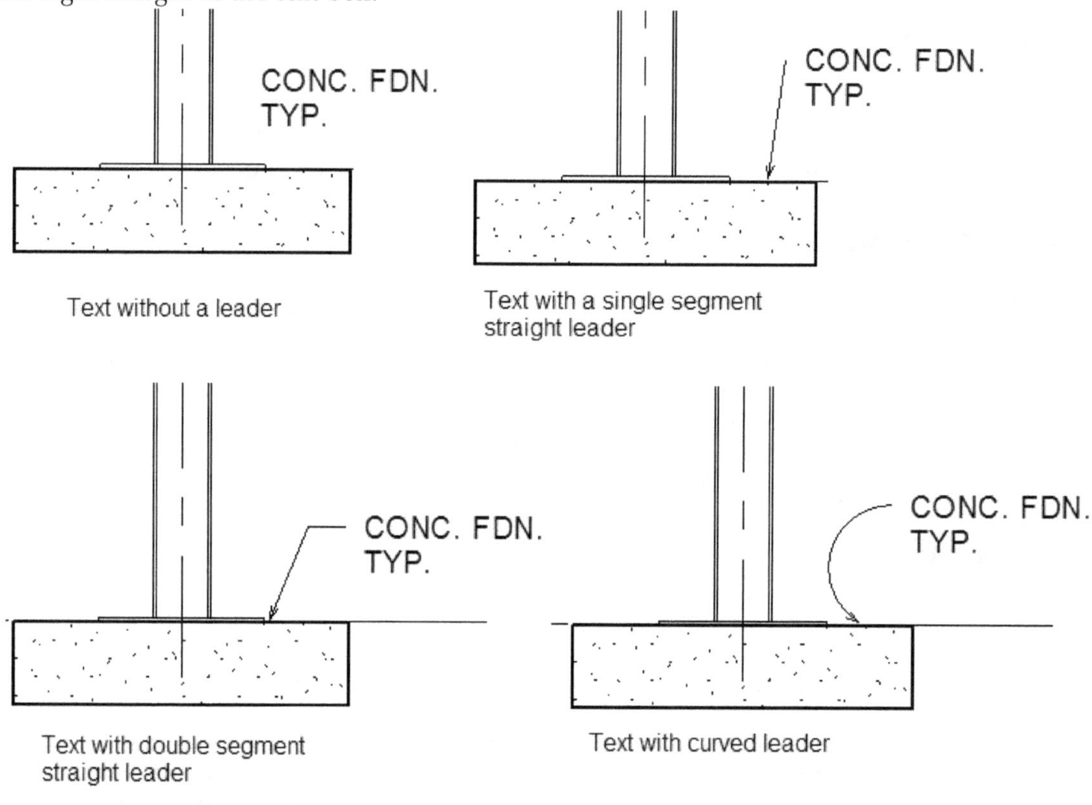

***Figure 7-22** Illustration showing various options of a leader in a text note*

After setting the desired options for placing the leader line and the text alignment, move the cursor near the desired location and click to add the text note. On doing so, a leader will emerge from the location you have clicked on, if the settings for the leader allow the display of leader. You may need to click once or twice depending upon the option chosen for the leader. After specifying the leader, if required, click to display a text box along with a text symbol. After the text box has been displayed, you can add the desired text in it. Note that before adding text in the text box, you can set various text formatting options from the **Format** panel. You can use the **Bold**, **Italics**, or **Underline** tool, or their combinations to make the text as specified. After setting the options for the text formatting, you can write the desired text in the text box.

While entering text in a text box, you may need to formatting options like numbering or bulletting sentences. To format a paragraph, choose any of the five options from the **Paragraph Format** drop-down menu in the **Format** panel. The options displayed in the menu are **None**, **Bullets**, **Numbers**, **Letter-Lowercase**, and **Letter-Uppercase**. After writing text in the text box, click outside the text box in the drawing area; the text note with drag controls will be highlighted. After entering the text in the text box, you can check the spellings as well. To do so, choose the **Check Spelling** tool from the **Tools** panel of the **Modify | Place Text** tab; the **Check Spelling** dialog box will be displayed, as shown in Figure 7-23. In this dialog box, the **Context** area displays the context or sentence in which the word(s) is misspelt. The **Not in Dictionary** edit box displays the word that is misspelt in the context displayed in the **Context** area. The **Change to** area displays a list box containing the words suggested to replace the misspelled word. You can click on the suitable word in the list box and then choose the **Change** button to replace the misspelt word with the suggested word. On doing so, the **Revit** message box will be displayed, informing you that the spell-check operation has completed. Choose the **Close** button and press ESC twice to exit the text editing option.

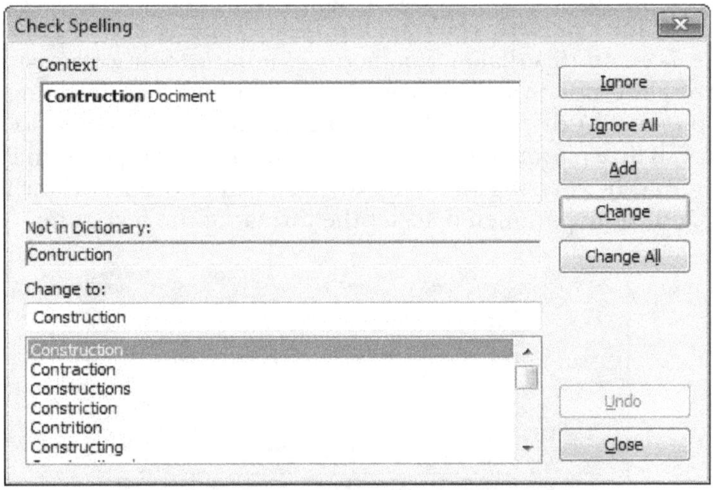

Figure 7-23 *The* ***Check Spelling*** *dialog box*

Note

*In the **Context** area of the **Check Spelling** dialog box, the misspelled word is displayed in boldface.*

Editing Text Notes

You can set the properties of a text note before creating it or edit them later. When you select a text note, its controls are displayed. You can use these controls to edit the text note parameters. The controls consist of one blue dot on each side of the text box that represents the stretch controls that can be used to modify the size of the text box. The rotation controls are also placed at the top right corner of the text box. You can use it to rotate the text box The location of the leader arrow-head remains in its original position and only the text box is rotated. The leader tail automatically adjusts to the rotated text box. You can also drag the location of the leader elbow and leader head by using the drag control dots. The drag control is used to change the location of the text box.

As you select the text note, the **Modify | Text Notes** tab is displayed. In this tab, you can use various editing tools such as **Copy**, **Mirror**, **Rotate**, **Array**, and so on to edit and arrange the text note. The alignment of the text note can be modified by choosing appropriate alignment tool from the **Format** panel. In this panel, you can choose the **Add Left Side Straight Leader** or **Add Right Side Straight Leader** tool to add a leader to the left or right side of the text note. You can select the **Remove Last Leader** option from the **Format** panel to remove the leader that is added last to the text note. To modify a text note, click inside the text box and then select the text note. Next, enter the new note. You can also modify the instance properties of the text note such as text font and text size. To do so, you can use various parameters in the **Properties** palette. In this palette, the value assigned to the **Horizontal Align** parameter is used to specify the horizontal text alignment of the selected text. To change the value assigned to the **Horizontal Align** parameter, click on the value column corresponding to it and select the desired alignment option from the drop-down list displayed. The **Arc Leaders** instance parameter is used to specify whether or not an arc leader should be attached to a text. The value column of this parameter displays a check box, which is cleared by default. You can select this check box to attach an arc leader to a text.

You can also edit the type properties of a text using the **Properties** palette. To do so, choose the **Edit Type** button from the **Properties** palette; the **Type Properties** dialog box will be displayed, as shown in Figure 7-24. In this dialog box, you can edit the font of a text by clicking in the **Value** column corresponding to the **Text Font** parameter and then selecting the required text font from the drop-down list displayed. The **Text Size** parameter is used to specify the size of the text note. The **Tab Size** parameter is used to set the tab spacing used in text notes. The tab spacing can be inserted by pressing the TAB key. You can also select the check boxes for the **Bold**, **Italic**, and **Underline** parameters to set the format of the text note.

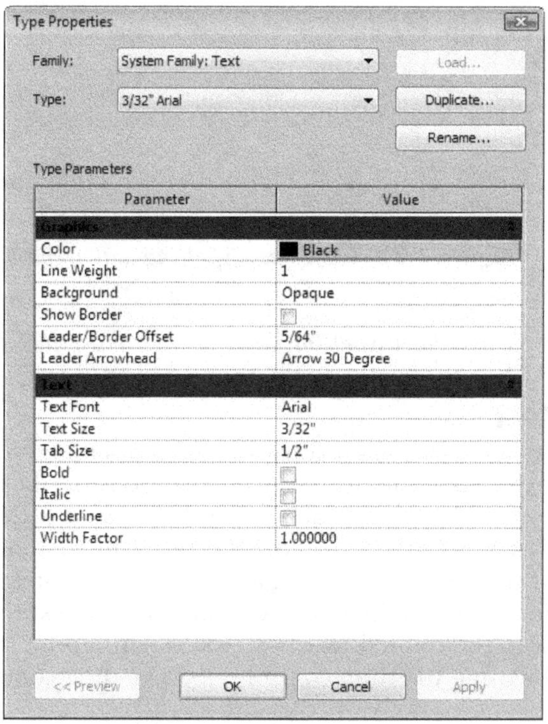

***Figure 7-24** The **Type Properties** dialog box for the selected text*

The **Type Selector** drop-down list in the **Properties** palette of the selected text note displays the available text types. You can select the appropriate type from the available text types before creating text notes. Alternatively, you can select a text note that was created earlier and then select the new text type from the **Type Selector** drop-down list to replace the old text type. You can also use the **Copy to Clipboard** and **Paste Aligned** tools from the **Modify | Text Notes** tab to copy the text notes from one level to multiple levels.

ADDING TAGS

You can easily add tag to various elements in a structural model. A tag is a useful annotation that assists in identifying elements in a drawing. While designing complex structural models using Revit, tags play an important role in arranging various elements in schedules. You can then add necessary description for each tagged element in a tabular form.

Revit provides various tools to add and edit tags. When you open a default template file, the tags of a certain category of elements are pre-loaded. Therefore, when you add the elements such as columns and beams to a model, Revit automatically tags them. For the other elements, you need to load their respective tags from the Revit library. Like other annotations, tags are also view-specific and therefore they appear only in the view they have been created in. You can control the visibility of tags by choosing the **Visibility/Graphics** tool from the **Graphics** panel of the **View** tab. On doing so, the **Visibility/Graphics Overrides for <current view >** dialog box will be displayed. In this dialog box, the **Annotation Categories** tab contains the list of tag categories such as **Analytical BraceTags**, **Analytical BeamTags**, **Analytical Column Tags**, **Analytical Floor Tags, Structural Annotation Tags**, **Structural Column Tags**, **Structural Framing Tags**, **Structural Rebar Tags**, and so on. You can select appropriate check boxes to control the visibility of each category of tags. The method used for tagging an element is discussed next.

Tagging Elements by Category

Ribbon: Annotate > Tag > Tag by Category

To attach a tag to an element based on its category, choose the **Tag by Category** tool from the **Tag** panel; the **Options Bar** displays the parameters related to the placement and orientation of the tag. In the **Options Bar**, you can select the **Horizontal** or **Vertical** option from the drop-down list displayed on the left to specify the direction of the text in the tag. You can use the **Leader** check box to enable or disable the display of a leader with the tag. On selecting this check box, the options in the drop-down list displayed on the right of the **Leader** check box will be enabled. You can select two options from this drop-down list: **Attached End** and **Free End**. If you select the **Attached End** option from the drop-down list, you will not be able to move the end of the leader with the tag away from the element category it is attached to. On selecting the **Attached End** option from the drop-down list, an edit box on its right will be enabled. In this edit box, you can specify the length of the leader or the distance of the tag from the attached element. If you select the **Free End** option from the drop-down list, you can move the end of the leader away from the element it is attached to.

Loading the Tags

By default, Revit loads tags for certain categories of elements such as structural area reinforcement, structural beam systems, structural columns, structural foundation, and so on. To display the tags that are loaded in the project, choose the **Loaded Tags And Symbols** tool from the expanded **Tag** panel of the **Annotate** tab. On doing so, the **Loaded Tags And Symbols** dialog box will be displayed. This dialog box displays the category-wise list of the tags loaded in the project, refer to Figure 7-25.

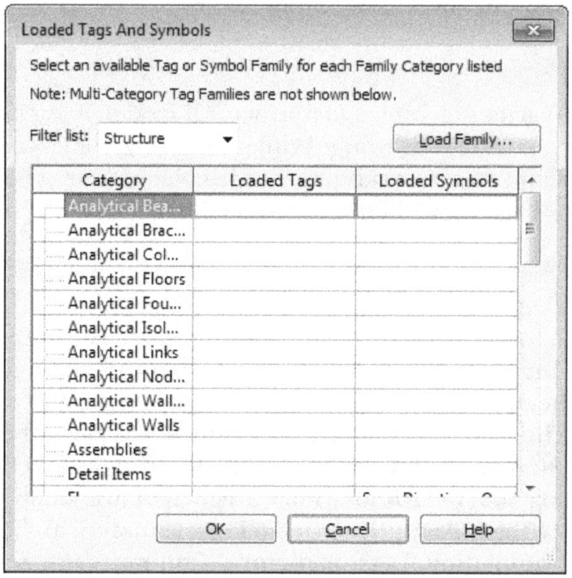

*Figure 7-25 The **Loaded Tags And Symbols** dialog box*

In case the category of element to be tagged does not have its corresponding tag already loaded, choose the **Load Family** button from the **Loaded Tags And Symbols** dialog box; the **Load Family** dialog box will be displayed. You can select various categories of tags from the **US Imperial > Annotations** or **US Metric > Annotations** folder path based on your requirement. The **Preview** area of the **Load Family** dialog box displays the preview image of the selected tag, as shown in Figure 7-26. Select an appropriate tag from the list and choose the **Open** button to load it into the project. On doing so, the desired tag will be added to the list of loaded tags for the corresponding element category in the **Tags** dialog box.

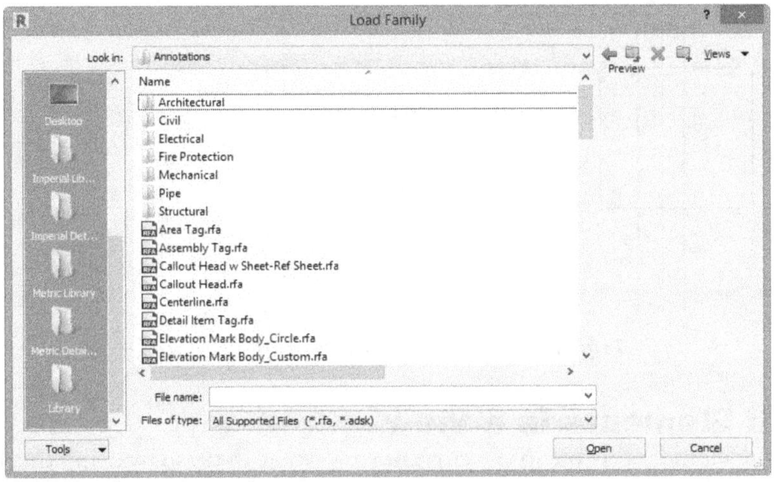

*Figure 7-26 The preview of the selected tag in the **Load Family** dialog box*

Another way of loading the tags is to invoke the **Tag By Category** tool and select the element to be tagged. If you select an element that does not have a tag corresponding to it already loaded, the **No Tag Loaded** message box will be displayed, as shown in Figure 7-27. This message box shows that you have not loaded a tag for the selected object type. Choose the **Yes** button in this message box; the **Load Family** dialog box will be displayed. In this dialog box, select the desired file and choose the **Open** button; the tag will be loaded. After loading the tag if you move the cursor near the element, Revit will display the preview image of the tag. Move the cursor to the desired location and click when the tag appears; the tag will be placed. Then, select different elements of the same category and tag them individually. When you select a tag, its controls will be displayed, as shown in Figure 7-28. The move control displayed with the tag is used to move the tag to the desired location. The leader elbow control is used to adjust the leader. The '?' control is used to enter text for the label of the tag. When you add a label, Revit displays a message box, informing you that you are changing the type parameter of the tag and that it could affect many elements. Choose **Yes** in the message box to continue working in the project; the value entered will appear as the label in the tag.

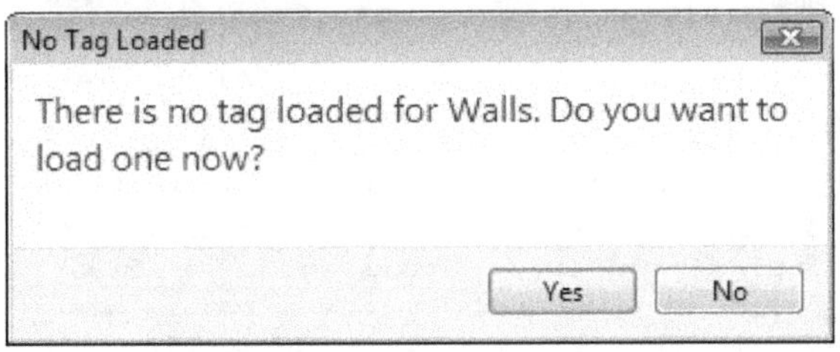

*Figure 7-27 The **No Tag Loaded** message box*

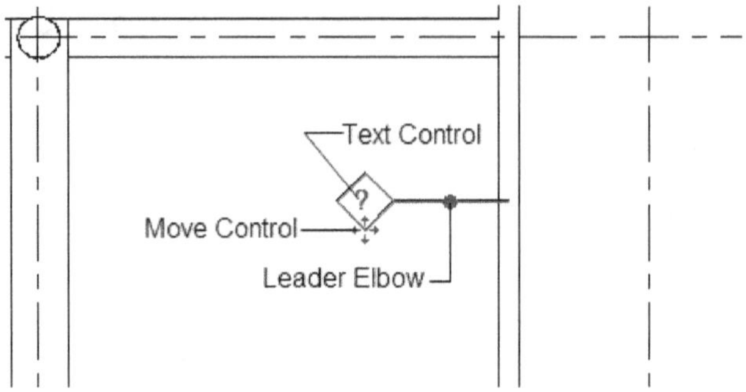

***Figure 7-28** The selected tag with its control*

Tagging All Elements in a View

You can tag all the elements visible in the current view. You can do so by using the **Tag All** tool. So, to tag all elements, choose this tool from the **Tag** panel; the **Tag All Not Tagged** dialog box will be displayed, as shown in Figure 7-29. This dialog box has two radio buttons: **All objects in current view** and **Only selected objects in current view**. By default, the **All objects in current view** radio button is active, and the **Only selected objects in current view** radio button is inactive. If you have selected an object before invoking the **Tag All Not Tagged** dialog box only then the **Only selected objects in current view** radio button gets activated. The selection of the **All objects in current view** radio button ensures that all the objects present in the current view are tagged, provided their categories are selected in the **Category** column of the **Tag All Not Tagged** dialog box. The table displayed below the radio buttons shows the listing of category tags of all elements present in the current view and their corresponding loaded tags.

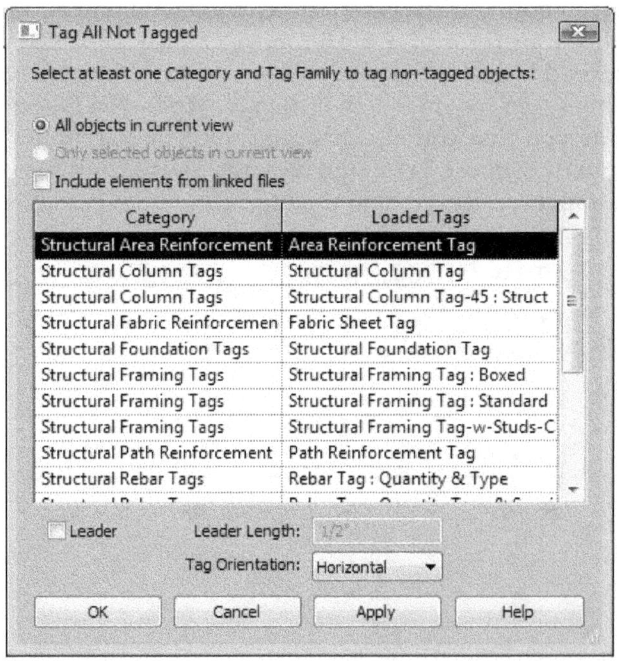

***Figure 7-29** The **Tag All Not Tagged** dialog box*

Tip

*You can load tags for multiple categories of components by holding down the CTRL key and selecting multiple tags from the drop-down list in the **Loaded Tags** column of the **Tags** dialog box.*

Note

*If a category of elements you wish to tag does not appear in this list, you need to load the corresponding tag family and then use the **Tag All** tool.*

From the table displayed in the **Tag All Not Tagged** dialog box, select a desired category. The **Leader** check box can be used to add a leader to them. You can also select the orientation and length of the leader in this dialog box. Choose **Apply** to apply the settings to the current view. For example, you can tag all structural framing members in a layout plan view by using the **Tag All** tool, as shown in Figure 7-30.

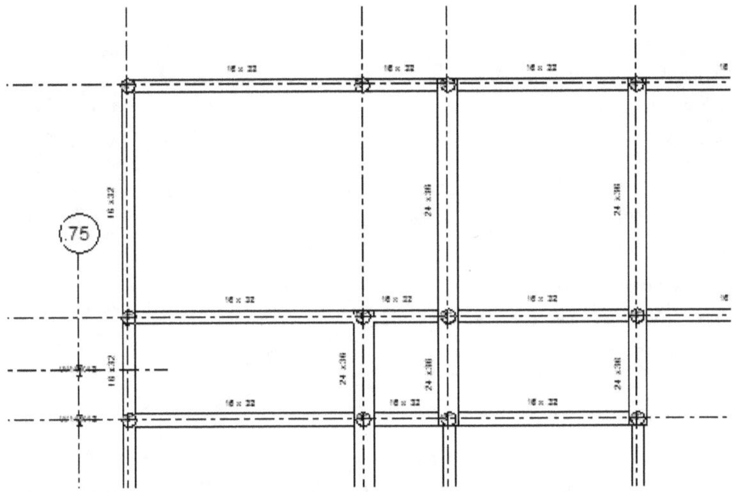

Figure 7-30 *All structural framing members tagged using the **Tag All** tool*

Beam Annotations

Ribbon:	Annotate > Tag > Beam Annotations

In Revit, you can place tags, annotations, and spot elevations on all or selected beams in the current views by using the **Beam Annotations** tool. To do so, invoke this tool from the **Tag** panel; the **Beam Annotations** dialog box will be displayed, as shown in Figure 7-31. This dialog box contains two areas: **Placement** and **Annotation location and type**. The options in these two areas are discussed next.

Placement Area

The options in the **Placement** area are used to specify the placement of beam annotations in the current view. In this area, the **All beams in current plan view** radio button is selected by default. As a result, Revit annotates all beams present in the current view. By default, the **All selected**

beams in current plan view radio button in the **Placement** area remains inactive. You can select this radio button to annotate the beams selected in the current view. To make this radio button active, select the beams that you want to annotate before invoking the **Beam Annotations** tool.

In the **Placement** area of the **Beam Annotations** dialog box, the **Remove existing beam tags and spot elevations** check box is cleared by default. As a result, Revit will not place new annotations on the beams that have already been annotated. You can select the **Remove existing beam tags and spot elevations** check box to replace the existing beam tags and spot elevation in the beam with the new ones. You can choose the **Settings** button to specify the offset distances of annotation tags and spot elevations from the beam. On choosing this button, the **Placement Settings** dialog box will be displayed, as shown in Figure 7-32. In this dialog box, the image displayed illustrates various settings for the offsets of the tag in reference to the beam. In the **Horizontal End Offset** area that is marked **1**, you can enter a suitable value in the edit box displayed inside it to specify the offset distance of tags or spot elevations from

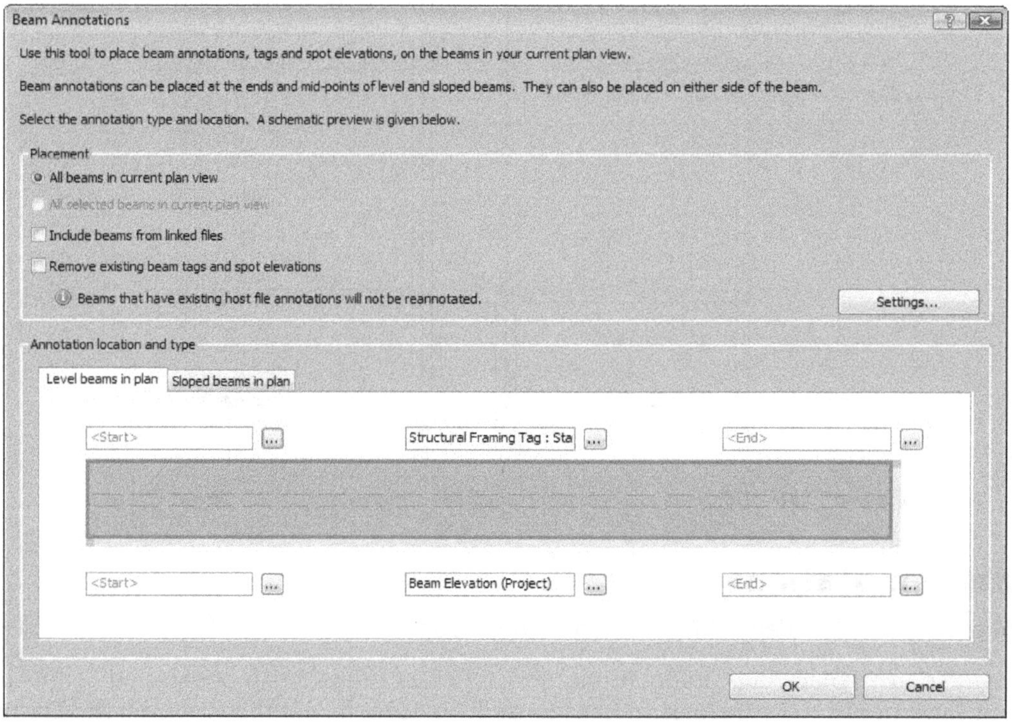

*Figure 7-31 The **Beam Annotations** dialog box*

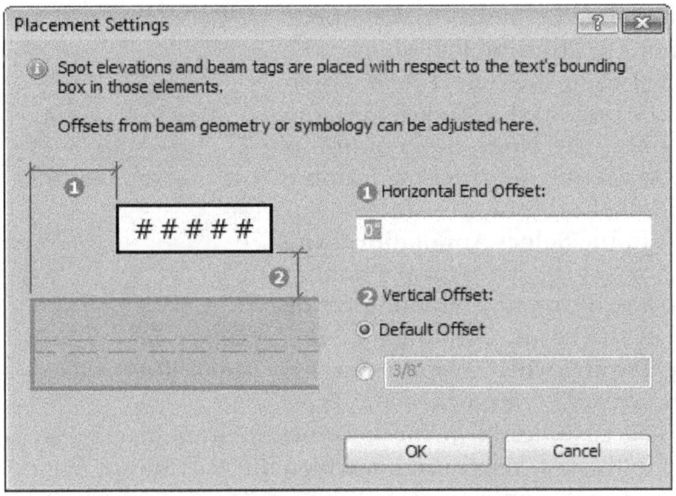

*Figure 7-32 The **Placement Settings** dialog box*

their points of attachment. Similarly, you can specify the settings for the vertical offset distance of the tags or spot elevations from their attachment points. To do so, you can use various settings displayed inside the **Vertical Offset** area, marked **2**, in the **Placement Settings** dialog box. In the **Vertical Offset** area, the **Default Offset** radio button is selected by default. As a result, "3/8" for Imperial system or 9.25mm for Metric system is set as the value for the vertical offset distance. To override the default value for the vertical offset distance, select the radio button below the **Default Offset** radio button in the **Vertical Offset** area; the edit box next to it will become active. You can enter a positive value less than 360000" for Imperial system or 9144000mm for Metric system in the edit box. After specifying the desired offset distances for the beam annotations, choose the **OK** button; the **Placement Settings** dialog box will be closed and the settings will be designated to the beam annotations.

Note

The values entered for the horizontal distance and the vertical distance represents the drawing scale to be used in the project.

Annotation location and type Area

The options in the **Annotation location and type** area are used to specify the annotation type and location for the beam annotations to be placed in a beam in the current view. This area has two tabs: **Level beams in plan** and **Sloped beams in plan**. The **Level beams in plan** tab is chosen by default. This tab displays the image of a level beam in the plan. This image is surrounded by six text boxes: three above it and three below it. The text box displayed at the top left of the image displays the type of tag or spot elevation used to annotate the beam at its start end. This text box will be inactive and displays the **<Start>** text inside it.

To assign a suitable spot elevation or structural framing tag type for the spot elevation or tag, choose the button displayed next to it; the **Select Annotation Type** dialog box will be displayed, as shown in Figure 7-33. In this dialog box, the **None** radio button is selected by default. As a result, no structural framing tag or spot elevation will be displayed at the top left corner of the beam. In the **Select Annotation Type** dialog box, you can select the **Structural Framing Tag** radio button to add a structural framing tag at the top left corner of the beam. On doing so, the **Structural Framing Tag** area will be displayed, refer to Figure 7-34. In this area, you can click on the **Type** drop-down list and select a desired type for the structural framing tag from the list. If the desired type is not available in this list, select the **Load Structural Framing Tag** option from the drop-down list; the **Beam Annotation** window will be displayed. Choose the **Close** button in this window; the **Load Family** dialog box will be displayed. In this dialog box, browse to **US Imperial > Annotations > Structural** or **US Metric > Annotations** folder path based on your requirement and select the **Structural Framing Tag-w-Start Reactions** file. Choose the **Open** button in the **Load Family** dialog box; the selected tag family will be added to the **Type** drop-down list.

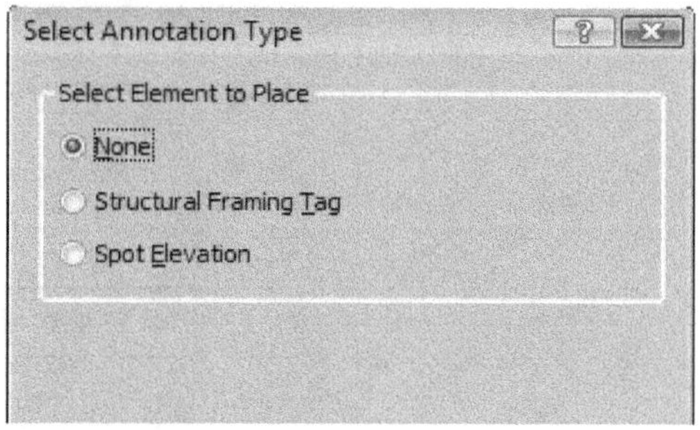

***Figure 7-33** The partial view of the **Select Annotation Type** dialog box*

Alternatively, you can select the **Spot Elevation** radio button in the **Select Element to Place** area to add a spot elevation at the top left corner of the beam (start end of the beam). On selecting the **Spot Elevation** radio button, the **Spot Elevation** area will be displayed in the **Select Annotation Type** dialog box, as shown in Figure 7-35. In the **Spot Elevation** area, you can click on the **Type** drop-down list to select the desired type for the spot elevation.

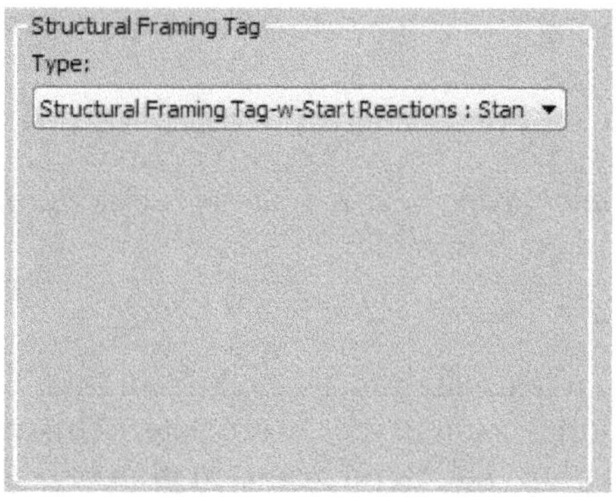

***Figure 7-34** The **Structural Framing Tag** area*

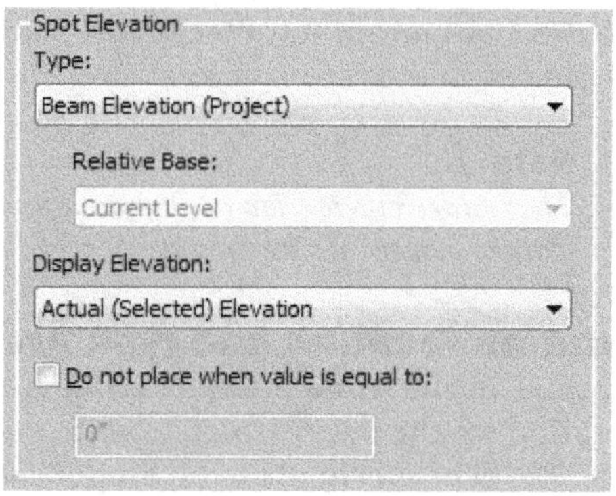

***Figure 7-35** The **Spot Elevation** area*

By default, the **Beam Elevation (Project)** is selected in the drop-down list. As a result, the spot elevation of the beam will be calculated from the lowest level of the project. Alternatively, you can select the **Beam Elevation (Relative)** option from the **Type** drop-down list to display the spot elevation relative to the selected level in the project. On selecting the **Beam Elevation (Relative)** option from the **Type** drop-down list, the **Relative Base** drop-down list will be enabled. In this drop-down list, you can select the desired level to specify the reference level for the spot elevation of the beam. From the **Type** drop-down list in the **Spot Elevation** area, you can also select the **<Create /modify types>** option. On selecting this option, the **Beam Annotation - Loading Spot Elevation Types** message box will be displayed, as shown in Figure 7-36.

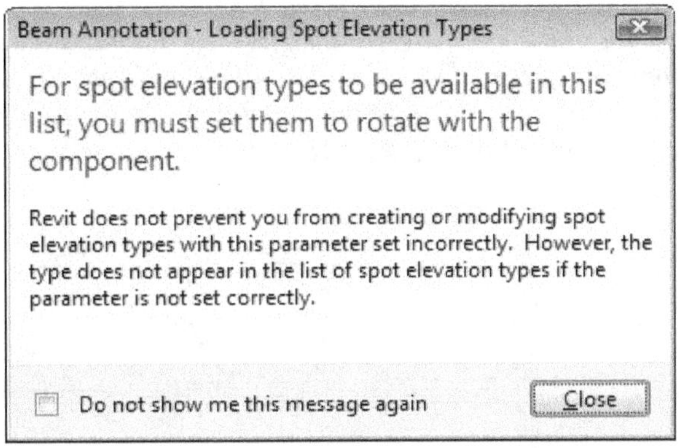

Figure 7-36 *The* ***Beam Annotation-Loading Spot Elevation Types*** *message box*

This message box will inform that a type will be added in the **Type** drop-down list only when spot elevations are allowed to rotate with the components to which they are attached. Next, choose the **Close** button from the message box; the **Type Properties** dialog box will be displayed. In this dialog box, select the check box corresponding to **Rotate with Component parameter**. In the **Type Properties** dialog box, you can set other parameters to change the display of various elements in the spot elevation. After specifying the desired parameters in the **Type Properties** dialog box, choose the **OK** button; the dialog box will be closed. Also, you will notice that a new type of spot elevation has been added to the **Type** drop-down list. In the **Spot Elevation** area, you can select various options from the **Display Elevation** drop-down list to specify the location of the measured elevation on the beam. From the **Display Elevation** drop-down list, you can select any of the four options: **Actual (Selected) Elevation**, **Bottom Elevation**, **Top&Bottom Elevation**, and **Top Elevation**. The **Actual (Selected) Elevation** option will be selected by default from the **Display Elevation** drop-down list. As a result, when the beam is annotated, the spot elevation will display the actual spot elevation of the location line of the beam with respect to the base level. Alternatively, you can select any of the remaining three options available in the **Display Elevation** drop-down list to specify the type of spot elevation to be displayed. In the **Select Annotation Type** dialog box, you can select the **Do not place when value is equal to** check box to prevent the spot elevation from being displayed for certain elevation values. As you select the **Do not place when value is equal to** check box, the edit box below it gets activated. You can enter a value in this edit box to specify elevation for which the spot elevation will not be displayed for the beam. By default, **0"** for metric system or **0.00 mm** for Metric system is displayed in the edit box. After specifying the desired options in the **Select Annotation Type** dialog box, choose the **OK** button; the **Select Annotation Type** dialog box will close and you will return to the **Beam Annotations** dialog box.

In the **Level beams in plan** tab of the **Beam Annotations** dialog box, you can choose other buttons located at the top and bottom of the image of the beam and specify the desired tag using the **Select Annotation Type** dialog box.

In the **Beam Annotations** dialog box, you can choose the **Sloped beams in plan** tab to assign the tags to the sloping beams displayed in the plan view. The options in this tab are similar to that discussed for the **Level beams in plan** tab. After specifying various parameters and options in the

Beam Annotations dialog box, choose the **OK** button; the beam annotations will be displayed in the view at the locations specified in the **Beam Annotations** dialog box; refer to Figure 7-37.

Note

*On selecting the **Remove existing beam tags and spot elevations** check box in the **Beam Annotations** dialog box, the existing beam tags and spot elevations will be removed from the beam locations in the plan view and the annotations in the beams will be placed as per the settings in the **Beam Annotations** dialog box.*

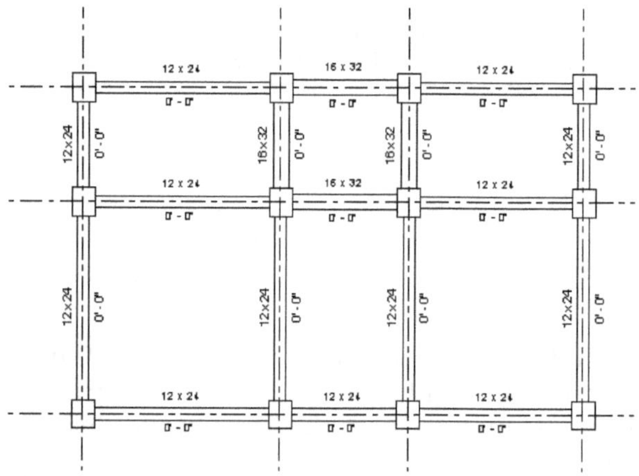

Imperial

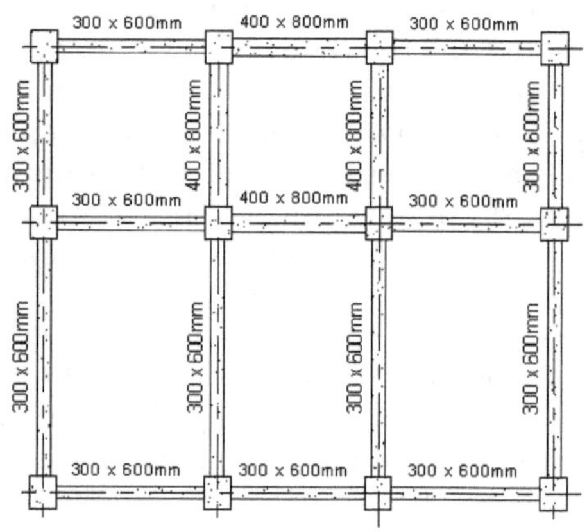

Metric

Figure 7-37 *The beam annotations displayed in a structural plan view*

CREATING FAMILIES

Revit family is defined as a collection of parametric components or elements having similar graphical representation, utility, and common parameters. By nature, all the elements used in Revit are family-based. A Revit family has a wide range of collections starting from a wall to a window and extends up to two-dimensional annotations or symbols required for your project. In Revit, each family inherits the character of a specific type. A type is a member of Revit familythat consists of certain parameters known as **Type Parameters**. These parameters are constant for identical types. For example, the **C-Channel-Column** family can have different types such as **C-Channel-Column C12x30**, **C-Channel-Column C15x40**, and others. You can access these types from the **Type Selector** drop-down list.

In Revit, families are useful tools as they are driven by the Revit parametric change engine. This engine helps to reflect the changes made in a family throughout the project. You can edit the parameters of a family after you have created it in a project using the **Revit Family Editor**. The **Revit Family Editor** is an interface that allows you to create or make changes in a family. At the same time, it helps you add or modify specific parameters of your model while defining it as a family type.

While creating a project, you may need to define a family inside another family. This type of family is called nested family. For example, if you are creating a truss family, you can create another family inside it for the profiles of the members inside it. The use of nested family adds details to the family you have created.

In Autodesk Revit, there are three types of families: System families, In-place families, and Standard component families. These families are differentiated on the basis of their creations and utilities in a project.

The System families include the most common building components that are frequently used in a project such as walls, floors, roofs, and ceilings. You cannot create a new system family of your own, but you can create duplicate families and modify their properties as per your need. The In-place families are project-specific, therefore, you can use them when you need to create a unique component for a specific project. For example, you may need to design a unique table, featuring curved edges and material finishes specific to the flooring of a dining area. In such a case, you can use the In-place families for creating the table. Also, you can save this table for further use. In Revit, the Standard component families are the basic components that are commonly used in a building design. These families are not project-specific. Whenever a need arises, you can download these families from common libraries. You can place the Standard component families in a project template, so that they can be used whenever you start a new project. Revit allows you to load the required family of door, window, and so on from the in-built Revit library. To do so, choose the **Load Family** tool from the **Load from Library** panel in the **Insert** tab; the **Load Family** dialog box will be displayed. Select the required file for the family from the appropriate folder and then choose **Open**; the family will be loaded into the project.

In Revit, you can save a selected family or all families loaded in the current project or template to a location on your system or to a network location. To save a family loaded in the current project, choose **Save As > Library > Family** from the **Application Menu**; the **Save Family** dialog box will be displayed. In this dialog box, browse to a folder location to save the family file. Next, in the **File name** edit box, enter the name of the file and then select an option from the **Family to save** drop-down list to specify a family or all families loaded in the project. Next, choose the **Save** button; the selected family will be saved at the specified location. Note that in-place families and system families, such as walls, and patterns cannot be saved. Revit also enables you to create your own customized families and then use them in a project. There are two methods that you can use to create a family. These methods are discussed next.

Creating In-Place Families

This method is used to create custom families such as the furniture for reception and workstations in a building model. You can create a family while working on a project. The family created is saved within the project and gets updated according to the changes made in the project.

To create an In-Place family, choose the **Model In-Place** tool from **Structure >Model > Component** drop-down; the **Family Category and Parameters** dialog box will be displayed. In the **Family Category** area, select an option from the **Filter list** drop-down list to filter the list of categories that will be displayed in the list box displayed below it. Double-click on the required category from the list box; the **Name** dialog box will be displayed. In this dialog box, enter a name for the family in the **Name** edit box and then choose the **OK** button; the sketch mode will be invoked and the options in the **Create** tab will be displayed. Sketch the reference planes and set the work plane to help you sketch the geometry for the family using the required sketching tools from the **Model** panel and the **Forms** panel. After creating the massing geometry, choose the **Finish Model** button from the **In-Place Editor** panel; a family will be created. After creating the geometry of a family, you can assign materials to the family and create family types with different dimensions.

Creating Families Using Standard Family Templates

In this method, you can create a new family by using certain standard in-built family templates as platform. These templates provide you with some parameters, reference planes, and dimensions in different views that are already saved in templates. You have to create the geometry for a family in the required view. To do so, you need to sketch reference planes, dimension reference planes, label dimensions, and then sketch the profile for the family using the required sketching tools. Once you have created the profile, apply materials to it and create family types by assigning different dimensions to the geometry. At the end, give a name to the template, save it, and load the family into the project to use it in your drawing.

The various family templates files that you can use for creating structural elements are: *Structural Stiffner.rft, Structural Trusses.rft, Structural Column.rft, Structural Foundation.rft*, and more.

TUTORIALS

Tutorial 1 Commercial Complex

In this tutorial, you will open the *Commercial Complex* project created in Tutorial 1 of Chapter 6 and add dimensions to it in different structural plan views. Further, you will annotate beams and columns using various annotation tools. Use the following project parameters:

(Expected time: 25 mins)

1. File name to be assigned:
   For Imperial *c07_Commercial-Complex_tut1.rvt*
   For Metric *M_c07_Commercial-Complex_tut1.rvt*

   The following steps are required to complete this tutorial:

a. Open the project file.
   For Imperial *c06_Commercial-Complex_tut1.rvt*

For Metric *M_c06_Commercial-Complex_tut1.rvt*

b. Dimension grid lines.
c. Add the foundation, column, and slab tags.
d. Add beam annotations.
e. Save the project using the **Save As** tool.
f. Close the project using the **Close** tool.

Opening the Existing Project

In this section, you will open the project file created in the Tutorial 1 of Chapter 6.

1. To open the project file, choose **Open > Project** from the **Application Menu**; the **Open** dialog box is displayed.

2. In this dialog box, browse to *C:/revit_2017_structural_projects/c06* location and choose the *c06_Commercial-Complex_tut1.rvt* for Imperial and *M_c06_Commercial-Complex_tut1.rvt* for Metric project file. You can also download this file from *http://www.cadcim.com*. The path of the file is as follows: *Textbooks > Civil/GIS > Revit Structure > Exploring Autodesk Revit 2017 for Structure*.

3. Choose the **Open** button from the **Open** dialog box; the selected project file opens in the drawing window in the **Second Floor** structural plan view.

Dimensioning the Grid Line

In this section, you will place dimensions to the grid lines displayed in the **Entry** structural plan view.

1. Double-click on the **Entry** node in the **Structural Plans** head in the **Project Browser** to open the corresponding structural floor plan view.

2. Choose the **View** tab and then invoke the **Visibility/Graphics** tool from the **Graphics** panel of this tab; the **Visibility/Graphic Overrides for Structural Plan: Entry** dialog box is displayed.

3. In this dialog box, choose the **Annotation Categories** tab if not selected and then select the **Dimensions** check box displayed under the **Visibility** column.

4. Choose the **Apply** button and then choose the **OK** button; the **Visibility/Graphic Overrides for Structural Plan: Entry** dialog box is closed.

5. Choose the **Annotate** tab and then choose the **Aligned** tool from the **Dimension** panel.

6. In the **Properties** palette, ensure that the **Linear Dimension Style : Arrow - 3/32" Arial** option for Imperial or **Linear Dimension Style : Arrow -2.5mm Arial** option for Metric is selected from the **Type Selector** drop-down list. Next, choose the **Edit Type** button; the **Type Properties** dialog box is displayed.

7. In this dialog box, choose the **Duplicate** button; the **Name** dialog box is displayed.

8. In the **Name** edit box of the dialog box, delete the existing text and enter **Small Dot- 3/32" Arial** for Imperial or enter **Small Dot- 2.5mm Arial** for Metric and then choose the **OK** button; the **Name** dialog box is closed.

9. In the **Type Properties** dialog box, click in the **Value** field of the **Tick Mark** parameter; a drop-down list is displayed. Next, select the desired option from this drop-down list.

For Imperial	**Filled Dot Small**
For Metric	**Filled Dot 2mm**

10. Click in the **Value** field of the **Centerline Symbol** parameter; a drop-down list is displayed. Select the **Centerline** option for imperial and for metric select M_Centerline from the drop-down list.

11. Choose the **OK** button; the **Type Properties** dialog box is closed. Also, notice that in the drawing area, the dimension symbol appears with the cursor.

12. Move the cursor above grid A and place it over grid 1; the grid 1 gets highlighted.

13. Click and then move the cursor to the right and place it on grid 2. Next, click when grid 2 gets highlighted. Notice that a dimension between grid 1 and grid 2 is displayed, as shown in Figure 7-38.

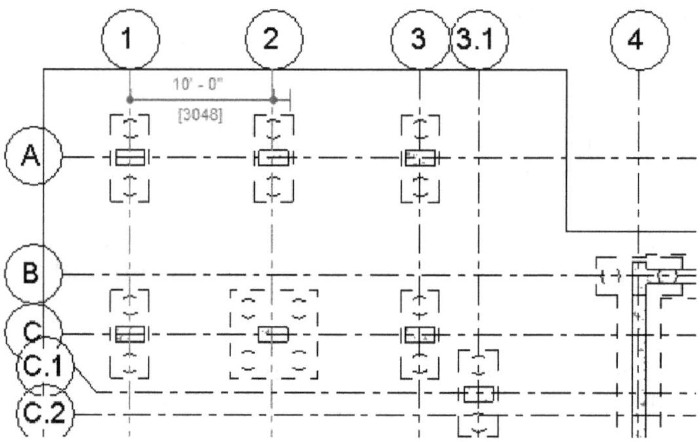

Figure 7-38 *The dimension created between grid 1 and grid 2*

14. Move the cursor over grid 3 and click when it gets highlighted.

15. Using the procedure followed in step 14, provide the dimensions till grid 5.1. After you have clicked over the grid 5.1, move the cursor up and place it above the grid bubbles, as shown in Figure 7-39.

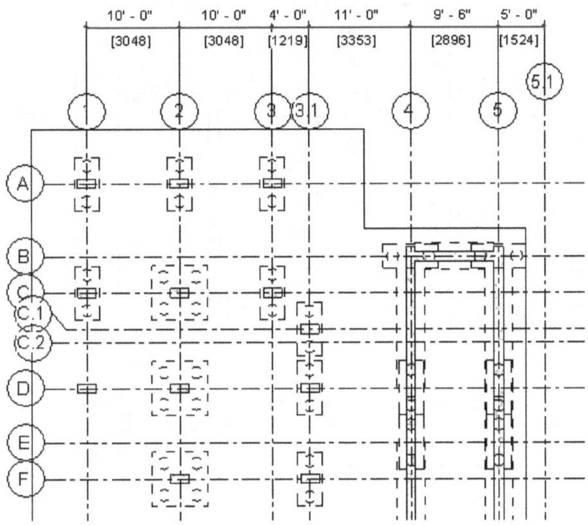

Figure 7-39 Dimensions placed at a location above grid bubbles

16. Now click at the desired location, refer to Figure 7-39; the dimensions are added. Note that you can place the dimensions based on your requirement.

17. Repeat the procedures followed in steps 12 to 16 and add dimensions for all the horizontal grids from A to I. The dimensions are placed on the right of the structural plan view, refer to Figure 7-40.

18. Press ESC twice to exit the tool.

Adding Tags to Foundations, Columns, and the Slab

In this section, you will add tags to foundations, columns, and the slab in the **Entry** structural plan view.

1. Ensure that the **Annotate** tab is chosen and then invoke the **Tag All** tool from the **Tag** panel; the **Tag All Not Tagged** dialog box is displayed.

2. Click in the **Structural Foundations Tag** category in the **Category** column.

3. Next, press and hold the CTRL key and then click in the **Structural Columns Tags** category with the **Structural Column Tag** as the loaded tag; the selected categories are highlighted, as shown in Figure 7-41.

4. Choose the **Apply** button and then the **OK** button; the **Tag All Not Tagged** dialog box is closed and the selected tag categories are displayed in the **Entry** structural plan view.

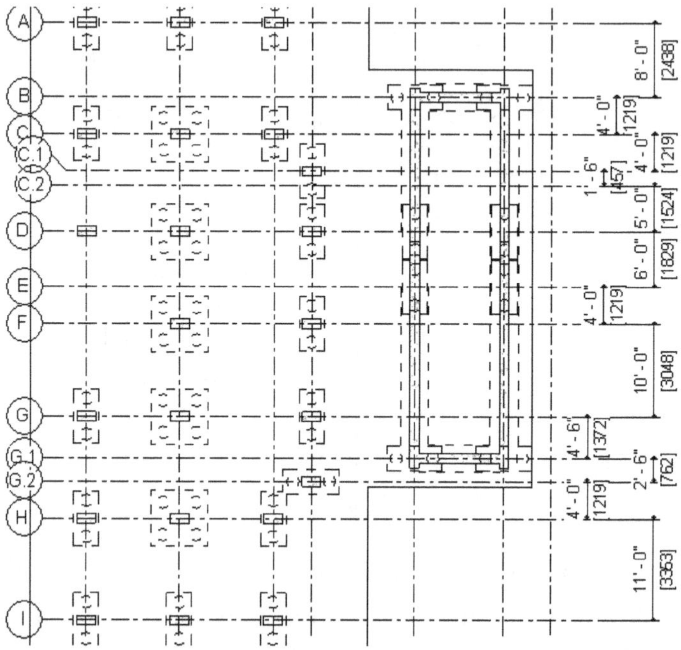

Figure 7-40 The dimensions placed at a location on the right of grid bubbles

Figure 7-41 The ***Tag All Not Tagged*** *dialog box with selected categories highlighted*

5. Next, choose the **Tag by Category** tool from the **Tag** panel and place the cursor over the structural foundation slab; the structural foundation slab is highlighted.

6. Click on the highlighted slab when the preview of the tag appears; the foundation slab tag is added to the foundation slab, as shown in Figure 7-42.

7. Press the ESC key twice to exit the selection mode.

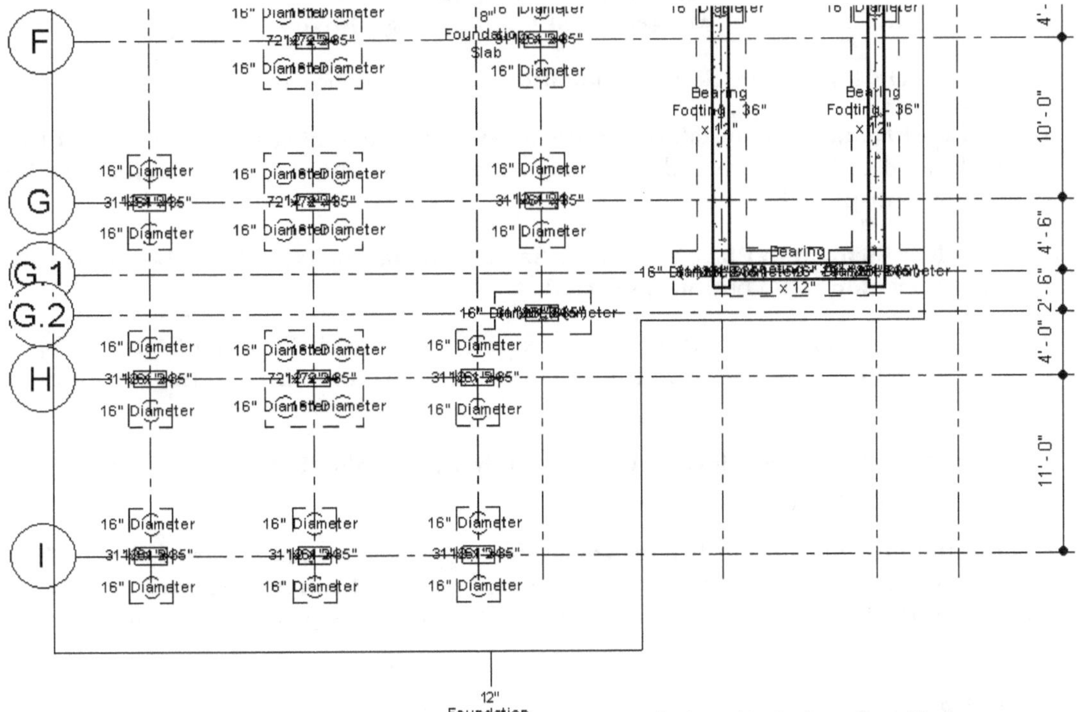

Figure 7-42 The partial view of the structural plan displaying the added foundation slab tag

Adding the Beam Annotations

In this section, you will tag the beams displayed in the **Second Floor** structural floor plan view.

1. Double-click on the **Second Floor** node under the **Structural Plans** head in the **Project Browser** to open the corresponding structural floor plan view.

2. Ensure that the **Annotate** tab is chosen. Choose the **Beam Annotations** tool from the **Tag** panel; the **Beam Annotations** dialog box is displayed.

3. In the **Placement** area of this dialog box, ensure that the **All beams in current plan view** radio button is selected and then select the **Remove existing beam tags and spot elevations** check box.

4. In the **Annotation location and type** area of this dialog box, ensure that the **Level beams in plan** tab is chosen.

5. In this tab, choose the Browse button located at the top middle of the image of the beam; the **Select Annotation Type** dialog box is displayed.

6. In the **Select Element to Place** area of the **Select Annotation Type** dialog box, ensure that the **Structural Framing Tag** radio button is selected. Then, in the **Structural Framing Tag** area, click in the **Type** drop-down list and then select the framing tag option from the list displayed.

   For Imperial **Structural Framing Tag : Boxed**
   For Metric **M_Structural Framing Tag : Boxed**

7. Choose the **OK** button; the **Select Annotation Type** dialog box is closed.

8. In the **Level beams in plan** tab of the **Beam Annotations** dialog box, choose the Browse button located at the bottom middle of the image of the beam; the **Select Annotation Type** dialog box is displayed.

9. In the **Select Element to Place** area of the **Select Annotation Type** dialog box, ensure that the **Spot Elevation** radio button is selected.

10. In the **Spot Elevation** area of this dialog box, select the **Beam Elevation (Relative)** option from the **Type** drop-down list and then select the **Entry** option from the **Relative Base** drop-down list.

11. Select the **Top & Bottom Elevations** option from the **Display Elevation** drop-down list in the **Spot Elevation** area.

12. Choose the **OK** button; the **Select Annotation Type** dialog box is closed. Next, choose the **OK** button in the **Beam Annotations** dialog box; the **Beam Annotations** dialog box is closed and the beam annotations are added to all the beams placed in the view, as shown in Figure 7-43.

Saving the Project

In this section, you will save the project file by using the **Save As** tool.

1. To save the project, choose **Save As > Project** from the **Application Menu**; the **Save As** dialog box is displayed.

2. In this dialog box, browse to *C:/revit_2017_structural_project* and create a folder with the name *c07*. Now, open the *c07* folder and then enter name of file in the **File name** edit box.

   For Imperial **c07_Commercial-Complex_tut1**
   For Metric **M_c07_Commercial-Complex_tut1**

3. Now, choose the **Save** button; the **Save As** dialog box closes and the project file is saved.

Closing the Project

1. To close the project, choose the **Close** option from **Application Menu**.

   The file is closed. This completes Tutorial 1 of Chapter 7.

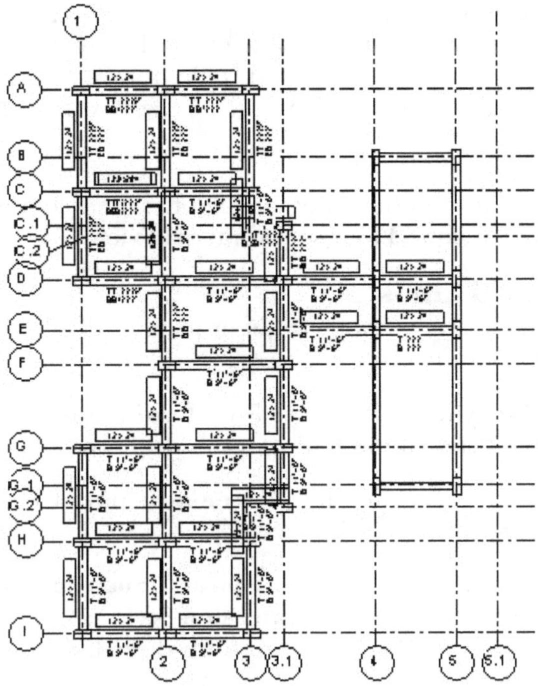

Figure 7-43 *The beam annotation displayed in the* ***Second Floor*** *structural plan view*

Tutorial 2 Industrial Complex

In this tutorial, you will open the *Industrial Complex project* for Imperial or *M_Industrial Complex project* for Metric created in Tutorial 2 of Chapter 6 and then dimension its different structural plan views. Further, you will annotate the beams and columns using the annotation tools. Use the following project parameters:

(Expected time: 25 min)

The following steps are required to complete this tutorial:

a. Open the project file.
   For Imperial *c06_Industrial-Complex_tut2.rvt*
   For Metric *M_c06_Industrial-Complex_tut2.rvt*
b. Dimension the grid lines.
c. Add foundation and column tags.
d. Add beam annotations and floor tags.
e. Save the project by using the **Save As** tool.
f. Close the project by using the **Close** tool.

Opening the Existing Project

In this section, you will open the project file created in Tutorial 2 of Chapter 6.

1. To open the project file, choose **Open > Project** from the **Application Menu**; the **Open** dialog box is displayed.

2. In this dialog box, browse to the location *C:/revit_2017_structural_projects/c06* and then choose the *c06_Industrial-Complex_tut2.rvt* for Imperial *M_c06_Industrial-Complex_tut2.rvt* for Metric project file. You can also download this file from *http://www.cadcim.com*. The path of the file is as follows: *Textbooks > Civil/GIS > Revit Structure > Exploring Autodesk Revit 2017 for Structure*.

3. Choose the **Open** button from the **Open** dialog box; the selected project file opens in the drawing window in the **Second Floor** structural plan view.

Dimensioning the Grid Lines

In this section, you will place dimensions to the grid lines displayed in the **Entry** structural plan view.

1. Double-click on the **Entry** node under the **Structural Plans** head in the **Project Browser**; the structural floor plan view at the **Entry** level is displayed.

2. Choose the **Aligned** tool from the **Dimension** panel of the **Annotate** tab; the **Modify | Place Dimensions** tab is displayed.

3. In the **Properties** palette, ensure that the **Linear Dimension Style : Arrow -3/32" Arial** option for Imperial system or **Linear Dimension Style : Arrow -2.5mm Arial** option for Metric system is selected in the **Type Selector** drop-down list and then choose the **Edit Type** button; the **Type Properties** dialog box is displayed.

4. In the **Type Properties** dialog box, choose the **Duplicate** button; the **Name** dialog box is displayed.

5. In the **Name** dialog box, enter **grid-dim-1/8"** for Imperial or **grid-dim-3mm** for Metric in the **Name** edit box and then choose the **OK** button; the **Name** dialog box closes.

6. In the **Type Properties** dialog box, click in the **Value** field corresponding to the **Tick Mark** parameter; a down arrow is displayed in the clicked field. Click on the down arrow and then select the **Diagonal 1/8"** option for Imperial or **Diagonal 2mm** option for Meric from the drop-down list displayed.

7. Next, click in the **Value** field corresponding to the **Text Size** parameter and enter **1/8"** for imperial and for metric enter 3 mm to replace the existing value.

8. Choose the **OK** button; the **Type Properties** dialog box is closed.

9. In the drawing area, move the cursor below grid A and place it on grid 9; grid 9 is highlighted.

10. Click on the highlighted grid and move the cursor to the right direction and place it on grid 8. Next, click when grid 8 is highlighted.

11. Again, move the cursor in the right direction and place it over grid 7 and click when grid 7 is highlighted. This creates the horizontal dimension of the grids from 7 to 9.

12. Move the cursor down near the grid 8 and then click; the required dimensions are placed, as shown in Figure 7-44.

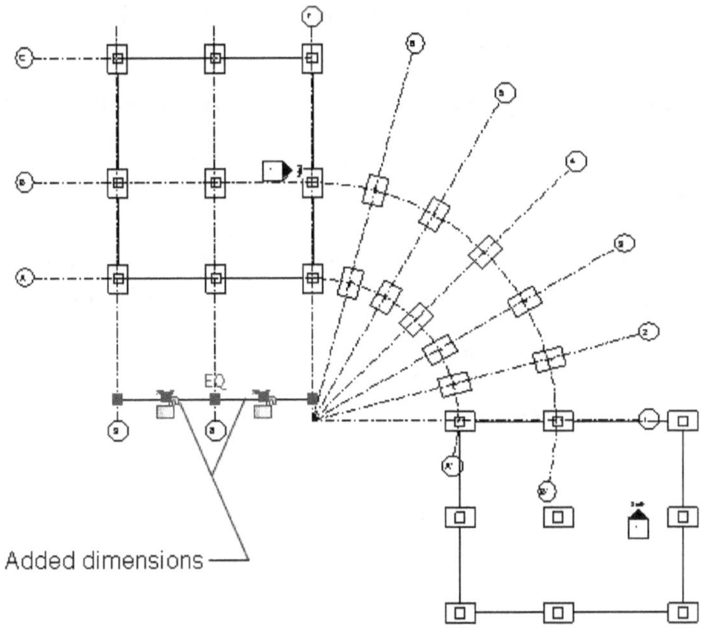

Figure 7-44 *The aligned dimensions added between the grids 7, 8, and 9*

13. Move the cursor and place it near the grid bubble of grid A and then click on the grid A when it is highlighted.

14. Move the cursor in the upward direction and click on grid B and then on grid C. Next, move the cursor and click near the bubble of grid C; the aligned dimensions are created between the horizontal grid lines A, B, and C.

15. Choose the **Modify** button from the **Select** panel; the selection is cleared.

16. To dimension the radial grid lines from 1 to 7, invoke the **Angular** tool from the **Dimension** panel of the **Annotate** tab; the **Modify | Place Dimensions** tab is displayed.

17. In the **Properties** palette, select the **Angular Dimension Style : Diagonal - 3/32"Arial** option for Imperial or **Angular Dimension Style : Diagonal -2.5mm Arial** option for Metric from the **Type Selector** drop-down list and then choose the **Edit Type** button from the **Properties** palette; the **Type Properties** dialog box is displayed.

18. In the **Type Properties** dialog box, choose the **Duplicate** button; the **Name** dialog box is displayed.

19. In the **Name** dialog box, enter **grid-dim-ang-1/8"** for Imperial or **grid-dim-ang-3mm** for Metric in the **Name** edit box and then choose the **OK** button; the **Name** dialog box is closed.
20. In the **Type Properties** dialog box, click in the **Value** field of the **Tick Mark** parameter and then select the **Diagonal 1/8"** option for Imperial or **Diagonal 2mm** option for Metric from the drop-down list displayed in the field you clicked.

21. Next, click in the **Value** field of the **Text Size** parameter and enter **1/8"** for Imperial or **3mm** for Metric to replace the existing value.

22. Choose the **OK** button; the **Type Properties** dialog box is closed.

23. Move the cursor near the grid bubble of grid 7 and then place the cursor on grid 7; the grid 7 gets highlighted.

24. Click on the highlighted grid and then move the cursor in the right direction and click on grid 6; an angular dimension is displayed between grid 7 and grid 6.

25. Move the cursor near the grid bubble 7 and click to place the angular dimension.

    Using the procedure followed in steps 23 to 25 create angular dimensions from grids 6 to 1, as shown in Figure 7-45.

26. Press the ESC key twice to exit the selection mode.

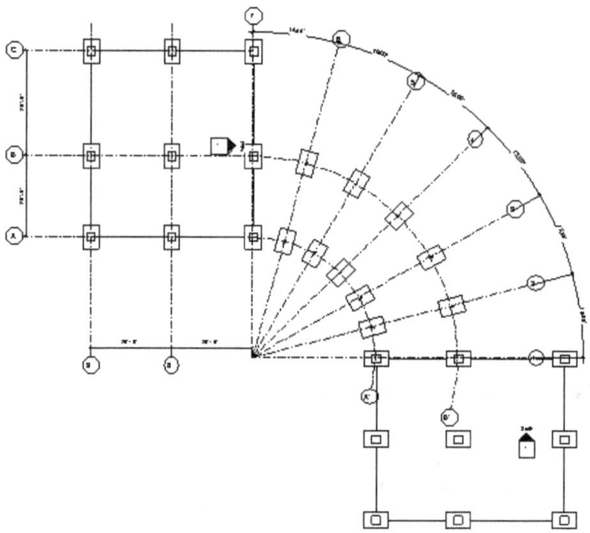

Figure 7-45 *The angular dimensions added from grids 1 to 6*

Adding the Foundation and Column Tags

In this section, you will add tags to the foundations and columns displayed in the **Entry** level view of the structural floor plan

1. Choose the **Tag All** tool from the **Tag** panel of the **Annotate** tab; the **Tag All Not Tagged** dialog box is displayed.

2. Click on the **Structural Column Tags** category and select the **Structural Column Tag -45 : Stuct** from the loaded tag column. Then, press the CTRL key and click on the **Structural Foundation Tags** category.

3. Choose the **Apply** and **OK** buttons; the **Tag All Not Tagged** dialog box is closed and the columns and foundations tags are displayed.

Adding Beam Annotations and Floor Tags

In this section, you will add beam annotations and floor tags to the **Second Floor** structural plan view.

1. Double-click on the **Second Floor** node under the **Structural Plans** head in the **Project Browser**; the structural floor plan view at the **Second Floor** level is displayed.

2. Choose the **Beam Annotations** tool from the **Tag** panel in the **Annotate** tab; the **Beam Annotations** dialog box is displayed.

3. In the **Placement** area of the **Beam Annotations** dialog box, ensure that the **All beams in the current plan view** radio button is selected and then select the **Remove existing beam tags and spot elevations** check box.

4. Choose the **OK** button; the **Beam Annotations** dialog box is closed and the tags are added to all the beams displayed in the view.

5. Next, choose the **Tag By Category** tool from the **Tag** panel; the **Modify | Tag** tab is displayed.

6. In the **Options Bar**, choose the **Tags** button; the **Loaded Tags And Symbols** dialog box is displayed.

7. In this dialog box, choose the **Load Family** button; the **Load Family** dialog box is displayed.

8. In this dialog box, browse to the path **US Imperial > Annotations > Structural** and choose the **Slab Tag**.*rfa* file for Imperial or **US Metric > Annotations > Structural** and choose the **M_Slab Tag**.*rfa* file for Metric unit system. Next, choose the **Open** button; the **Load Family** dialog box is closed.

9. In the **Loaded Tags** dialog box, choose the **OK** button; the **Loaded Tags** dialog box is closed.
10. Place the cursor on the structural floor placed in the left wing of the structural model; the floor gets highlighted and the preview of the tag appears. Click on the highlighted floor; the floor tag is added to the structural floor, refer to Figure 7-46.

11. Next, click on the other two floors one by one and add their corresponding tags, as shown in Figure 7-46.

12. Press the **ESC** key twice to exit the selection mode.

Saving the Project

1. To save the project, choose **Save As > Project** from the **Application Menu**; the **Save As** dialog box is displayed.

2. In this dialog box, browse to *C:/revit_2017_structural_project/c07* and then enter **c07_Industrial-Complex_tut2** for Imperial or **M_c07_Industrial-Complex_tut2** for Metric in the **File name** edit box.

3. Now, choose the **Save** button; the **Save As** dialog box closes and the project file is saved.

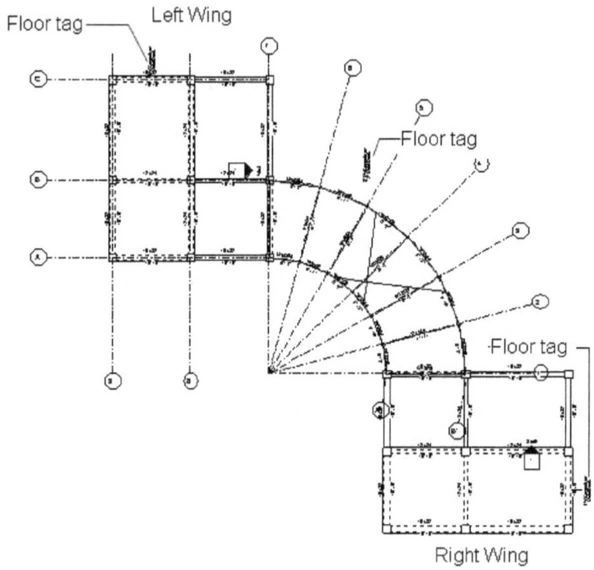

Figure 7-46 *The angular dimensions added from grids 1 to 6*

Closing the Project

1. To close the project, choose the **Close** option from the **Application Menu**.

   The file is closed. This completes Tutorial 2 of Chapter 7.

Self-Evaluation Test

Answer the following questions and then compare them to those given at the end of this chapter:

1. In Revit, there are two types of dimensions: __________ and __________ .

2. __________ dimensions are displayed while placing elements and components in a view.

3. The __________ tool is used to dimension two orthogonal references or points such as the endpoints of a wall.

4. When a dimension is added in a drawing, the __________ are displayed at the intersection of the dimension line with the witness line.

5. The visibility of tags can be controlled by using the __________ tool.

6. The _____________tool is used to dimension the radius of a circular or arc profile.

7. The tags that are loaded in the current view, can be viewed by invoking the __________ tool from the **Annotate** tab.

8. Temporary dimensions can be converted into permanent dimensions. (T/F)

9. Text notes are displayed only in the view in which they are created. (T/F)

10. You can annotate the slope of an edge or a face of an element. (T/F)

Review Questions

Answer the following questions:

1. Which of the following tools is used to write notes in a project?

   (a) **Text Notes** (b) **Text**
   (c) **Notes** (d) **Linear**

2. The __________ tool is used to dimension the overall length of a curved wall.

3. The __________ tool is used to dimension the shortest distance between two points.

4. The __________ tool is used to tag an element based on its category.

5. The __________ tool can be used to dimension the angle between two non-parallel elements in a project.

6. You can display the coordinates (Northing and Easting) of a point in a structural model by using the __________ tool.

7. You cannot correct the spelling of text notes added in a project. (T/F)

8. Tags are displayed only in the view in which they are created. (T/F)

9. You can lock permanent dimensions displayed in the current view for editing. (T/F)

10. You can tag all the untagged elements displayed in the current view of a project using the **Tag All** tool. (T/F)

EXERCISES

Exercise 1 Academic Institution

In this exercise, you will add dimensions to the grid lines and add tags to columns and foundations at the **Entry** structural floor plan view of the structural model created in the Exercise 1 of Chapter 6. Further, you will add Beam Annotations in the **Roof** structural plan view. Refer to Figures 7-47 and 7-48. **(Expected time: 1 hr)**

1. Use the **Entry** structural plan view to dimension the grids.
2. Use the **Roof** structural plan view to add beam annotations.
3. File name to be assigned:

   For Imperial *c07_Academic-Institution_exer1.rvt.*
   For Metric *M_c07_Academic-Institution_exer1.rvt.*

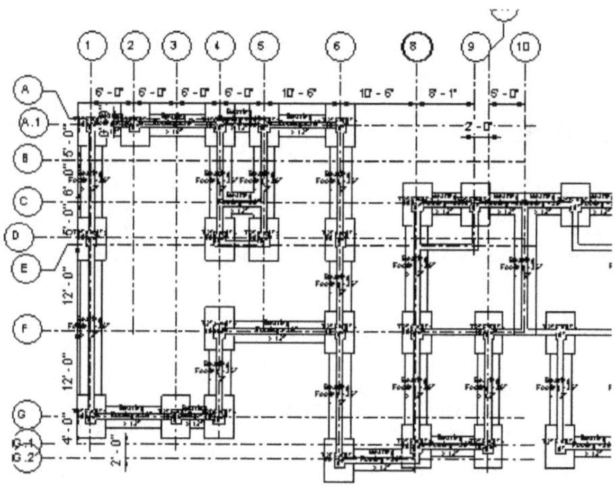

Figure 7-47 *The* ***Entry*** *structural plan view displaying the grid dimensions and the tags added to the foundations and columns*

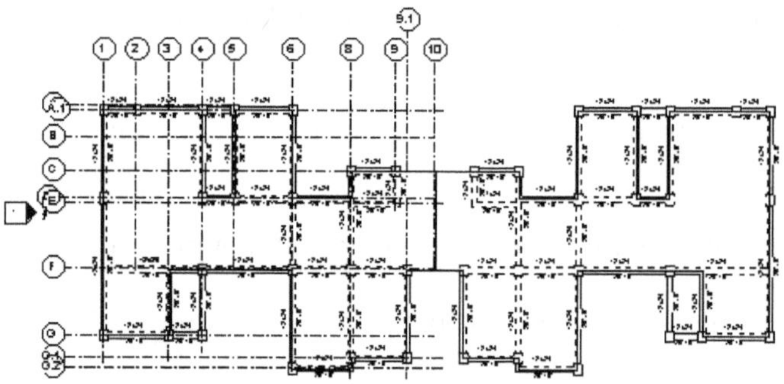

Figure 7-48 *The* ***Roof*** *structural plan view displaying the Beam Annotations*

Exercise 2 Factory Shed

In this exercise, you will add dimensions to the grid lines and add foundation, column, and beam tags at different views to the Exercise 2 of Chapter 6. Refer to Figures 7-49 and 7-50.

(Expected time: 30 min)

1. Use the **Plinth** structural plan view to dimension the grids.
2. Name of the new dimension type-

   For Imperial **dim-grid-1/8"**

For Metric **dim-grid-3mm**

3. Tick Mark-

   For Imperial **Open Dot 1/16"**

   For Metric **Open Dot 2mm**

4. Use the **Structural Column Tags** and **Structural Foundation Tags** categories.
5. Use the **Roof** structu vvral plan view to add beam annotations.
6. File name to be assigned:

   For Imperial *c07_Factory-Shed_exer2.rvt.*

   For Metric *M_c07_Factory-Shed_exer2.rvt.*

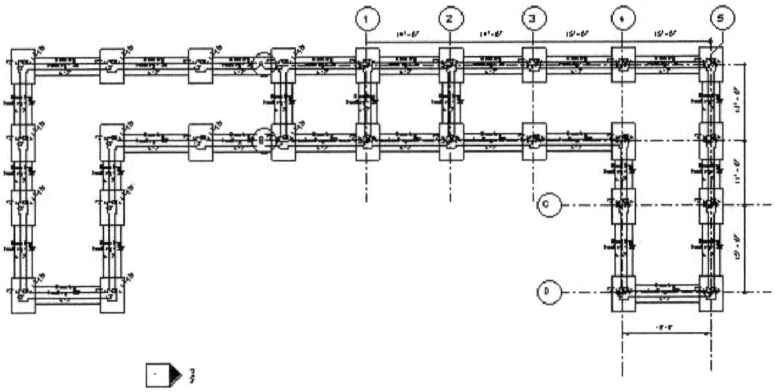

Figure 7-49 *The* ***Plinth*** *structural plan view displaying the dimensions of the grid lines and the tags*

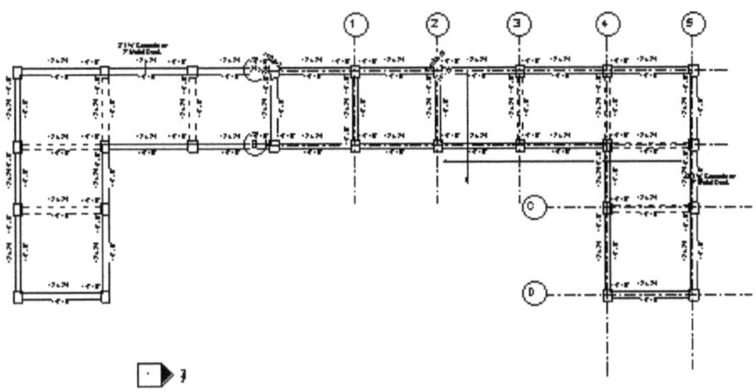

Figure 7-50 *The* ***Second Floor*** *structural plan view displaying the beam annotations*

Answers to Self-Evaluation Test

1. Temporary, Permanent, **2.** Temporary, **3. Linear**, **4.** Dimension Arrows, **5. Visibility/Graphics**, **6. Radial**, **7. Tag All**, **8.** T, **9.** T, **10.** T

Chapter 8

Standard Views, Details, and Schedules

Learning Objectives

After completing this chapter, you will be able to:

- *Create Elevation Views*
- *Create Section Views*
- *Create Drafting Views*
- *Create Duplicate Views*
- *Generate Graphical Column Schedule*
- *Create Schedule*
- *Generate Material Takeoff Schedule*

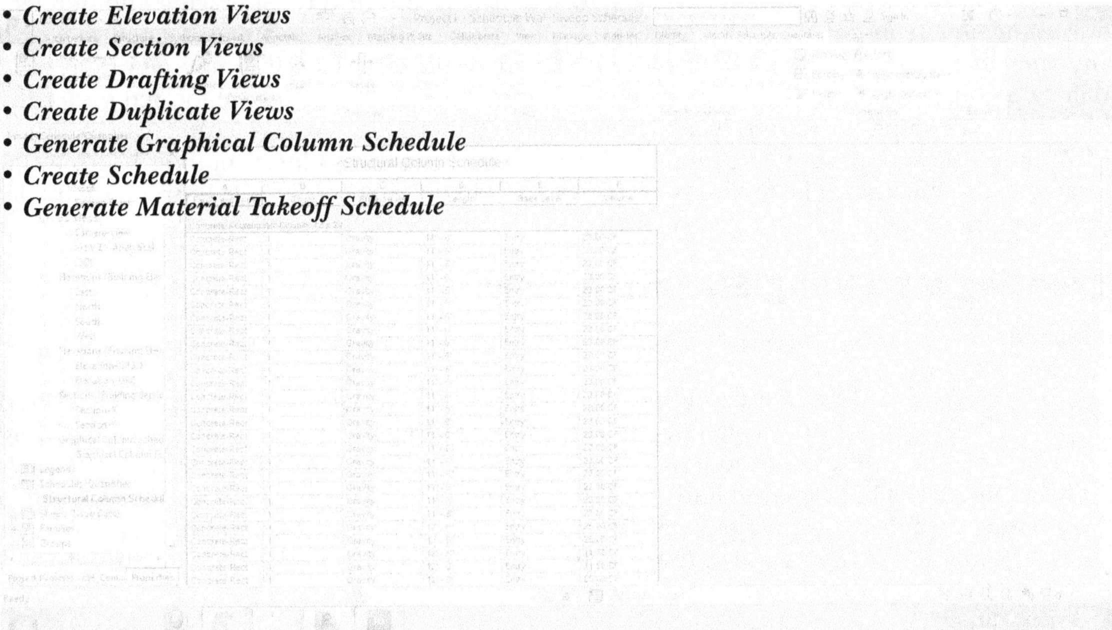

In the previous chapters, you learned to create and document a structural model. In this chapter, you will learn to generate standard views and details as well as schedule reports in a structural model.

In Revit, you can generate sections and elevations very easily. The views generated from sections and elevations will help you prepare the detail drafting of a structural model. Also, you can use these views to add various structural elements to a structural model. For example, you can use section views to add reinforcement details to the structural beam, columns, or floors. Similar to the section and elevation views, you can create drafting and detail views for the structural model.

After creating a structural model and generating the required views, the next important task is to generate the reports related to the structural model. These reports help you in executing the building in a site. These reports are in the form of graphical column schedules, general schedules for estimating the quantities of various structural elements, schedules for Material Takeoffs, and others.

ELEVATION VIEWS

Elevation view is the most commonly used orthographic projection for conveying the appearance of a building model from its exterior or interior. In Revit, elevation views are part of the default template file that you can use to create your project. When you open the default template file to create a project, four elevation views are displayed: North, South, East, and West. However, you can create more exterior and interior elevation views of your structural model using various tools available in Revit.

In Revit, you can create two types of elevation views: building and framing. The building elevation views of a building model are similar to the sectional views of the building model, and are capable of generating four directional views: North, South, East, and West of the building model. Building elevation views in a project are located under the **Elevations (Building Elevation)** head in the **Project Browser**. The Framing elevation views are used to place vertical bracing and moment bracing members in a structural model. The framing elevation views are associated to a grid and therefore, it is necessary to have a grid in your view to create a framing elevation view. When you create framing elevation views in a project, the views created are located under the **Elevations (Framing Elevation)** head in the **Project Browser**. In Revit, you can create building and framing elevation views by using the **Elevation** and **Framing Elevation** tools respectively, located in the **View** tab. The methods of creating building elevation views in a structural model are discussed next.

Creating a Building Elevation View

Ribbon: View > Create > Elevation drop-down > Building Elevation

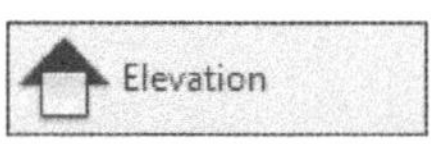

To create a building elevation view of a structural model, open the desired structural plan view from the **Project Browser** and then choose the **Building Elevation** tool from **View > Create > Elevation** drop-down; the **Modify | Elevation** tab will be displayed and an elevation symbol will appear along with the cursor in the drawing area. Next, set various options related to the elevation views in the **Options Bar** before clicking in the drawing area to insert the elevation symbol, as shown in Figure 8-1.

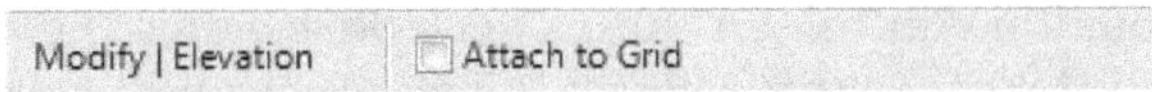

Figure 8-1 *Various option related to the elevation view in the* ***Options Bar***

In the **Options Bar**, the **Attach to Grid** check box is cleared by default. You can select this check box if you want to create a framing elevation view while using the **Building Elevation** tool. While creating the building elevation view of a structural model, ensure that the **Attach to Grid** check box is cleared. In the **Reference** panel, the **Reference Other View** check box is also cleared by default. If you select this check box, a reference elevation will be created in the drawing view. Reference elevation refers to an existing elevation or drafting view in the drawing. The reference elevations are placed in a callout view or plan view. On selecting the **Reference Other View** check box in the **Reference** panel, the drop-down list below it becomes active. In this drop-down list, the **<New Drafting View>** option is selected by default. As a result, on adding the building elevation view, an empty drafting view will be created and added under the **Drafting Views (Detail)** head in the **Project Browser**. The default name given to the drafting view is **<Level name>- Elevation**. You can edit and rename this view in the **Project Browser**. The other options that can be selected from the drop-down list next to the **Reference other view** check box are; **Elevation: North**, **Elevation: East**, **Elevation: West**, **Elevation: South**. Note that this drop-down list can have more options depending on the presence of some other reference views in a project. You can select any of these options from the drop-down list and add elevation to the current structural view plan to reference their corresponding views.

To add a building elevation view to the current structural plan view, make sure that the **Attach to Grid** and **Reference Other View** check boxes are cleared. Next, move the cursor with the elevation symbol in the drawing area. On moving the cursor near a wall in the drawing area, you will notice that the alignment of the arrow head of the elevation symbol changes and it becomes perpendicular to the wall, as shown in Figure 8-2. Next, click when the elevation arrow head symbol points toward the desired direction; the new elevation view will be created and added to the list of elevations in the **Project Browser**.

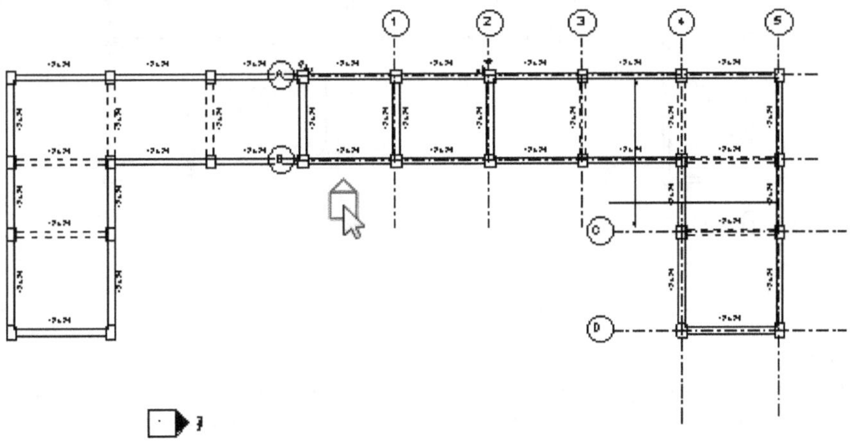

Figure 8-2 *The display of a building elevation symbol in a structural plan view*

In the project view, when you select an elevation symbol, its controls are displayed, as shown in Figure 8-3. The controls of the selected elevation symbol include four check boxes along the four directions and a rotation control. Of these check boxes, only the check box at the placed direction is selected and the other three check boxes are cleared. You can select other check box(es) to create the elevation along that direction. You can use the rotation control to rotate the elevation symbol in the desired direction. To rotate the elevation symbol, move the cursor over the rotation control and then press and drag the left mouse button. On doing so, the elevation symbol will rotate about its center. Release the left mouse button after the rotation of the elevation symbol. Next, set the width of the elevation view using the clip plane control. To view the clip plane control, click on the view arrow of the elevation symbol, refer to Figure 8-3; the clip plane is displayed as a blue line with the drag control dots on its two ends, as shown in Figure 8-4. You can drag the drag control dots to resize the width of the elevation view.

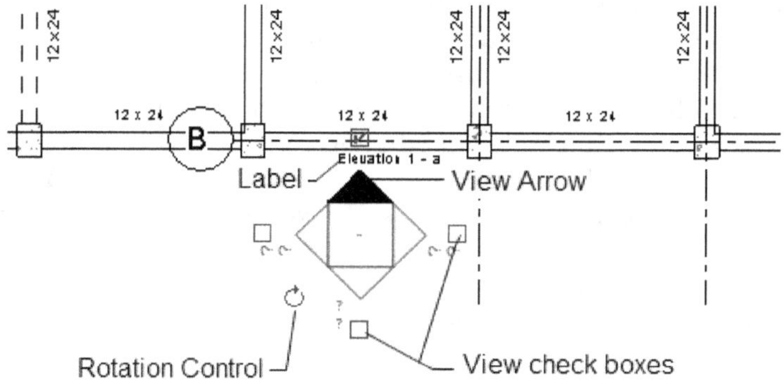

Figure 8-3 *Various controls in the building elevation symbol*

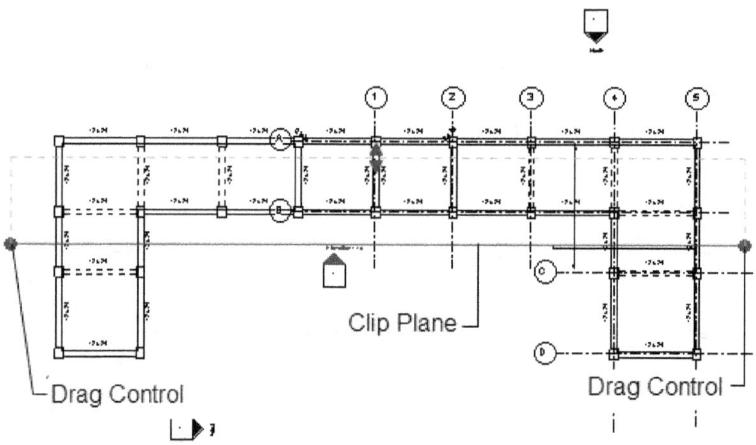

Figure 8-4 *The display of the clip plane*

Note

If the drag control dots are not displayed with the clip plane, select the check box corresponding to the ***Crop View*** *parameter in the* ***Properties*** *palette and then choose* ***Apply****.*

Tip

You can also change the elevation view symbol. To do so, select it from the project view and then choose the ***Type Properties*** *tool from the* ***Properties*** *panel of the* ***Modify / Elevation*** *tab; the* ***Type Properties*** *dialog box will be displayed. Click in the* ***Value*** *field of the* ***Elevation Tag*** *parameter and then choose the Browse button; the Type Properties dialog box will be displayed. In this dialog box, select the required elevation mark from the drop-down list corresponding to the* ***Elevation Mark*** *parameter.*

After placing the elevation symbols and adjusting clipping planes, you can view the elevations in various ways. The first way is to double-click on the elevation name listed under the **Elevations (Building Elevation)** head in the **Project Browser**. You can also double-click on the arrow head of the elevation symbol. The other way to view the elevation is to move the cursor over the elevation symbol and right-click to display the shortcut menu. Choose the **Find Referring Views** option from the shortcut menu; the **Go To View** dialog box will be displayed, as shown in Figure 8-5. Select the desired view name from the list box in the dialog box and choose the **Open View** button or double-click on the view name in the list box to open the view you require.

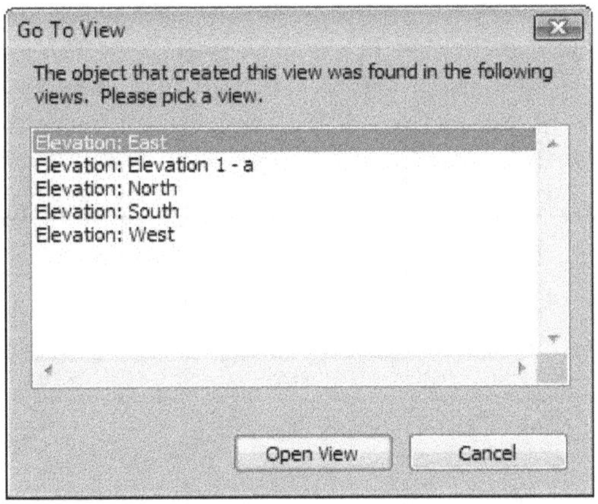

Figure 8-5 *The* ***Go To View*** *dialog box*

Note

While working on a structural floor plan, if the building model created extends beyond the clip planes of the four side elevation views, the corresponding elevations will no longer show the complete exterior views. Instead, an elevation view that is cut through the building model will be displayed. You can drag the clip plane controls symbol beyond the extent of the building profile to retain the view as a complete exterior elevation view.

Creating a Framing Elevation View

Ribbon: View > Create > Elevation drop-down > Framing Elevation

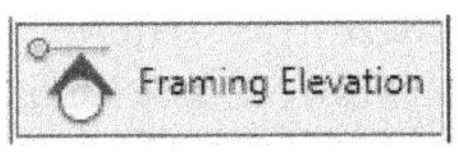

Framing elevation views are used to place the vertical bracing and moment bracing members in a structural model. These views are associated with grid. Framing elevation views in a project are located under the **Elevations (Framing Elevation)** head in the **Project Browser**. You can create a framing elevation view in a structural plan view. To create a structural plan view, invoke the **Framing Elevation** tool from

Elevation drop-down in the **Create** panel of the **View** tab; the **Modify | Framing Elevation** tab will be displayed. In the **Options Bar**, ensure that the **Attach to Grid** check box is selected. The use of the **Reference Other View** check box is the same as in the case of the building elevation view. In the **Properties** palette, you can choose the **Edit Type** button to invoke the **Type Properties** dialog box. The **Type Properties** dialog box can be used to assign various parameters for the tags and label associated with the framing elevation.

After specifying the options and parameters for the framing elevation view, you can add the framing elevation view to your project. To do so, move the cursor and place it over the grid to which you want to associate with the framing elevation view. On doing so, you will notice that a framing elevation symbol appears on the side of the grid where you have moved the cursor, as shown in Figure 8-6. On the display of the framing elevation symbol click on the desired side; the framing elevation symbol will be added at that place. The methods for rotating a symbol and adjusting the clip plane for the framing elevation view are similar to that of the building elevation view. These methods have already been discussed in the **Creating a Building Elevation View** section of this chapter.

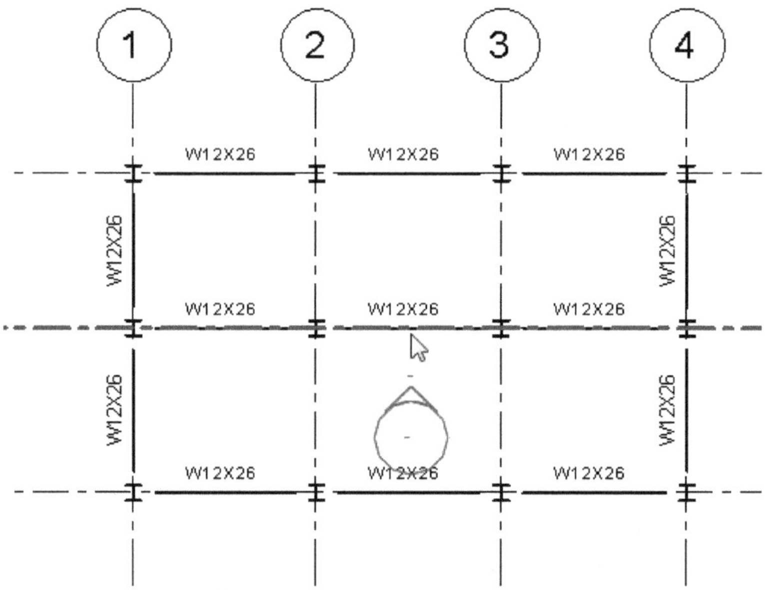

Figure 8-6 *The framing elevation symbol displayed*

SECTION VIEWS

Section views are generated by cutting sections through a building model. These views are created to display the wall elevations, floor heights, cross-sectional details of structural elements, and special vertical features of a project. They are also useful in creating and editing reinforcements for various structural elements in a structural model. For example, to emphasize the structural details of the central atrium of an office building model, you may need to show a section through the central atrium. Revit enables you to create the section with relative ease. You can also modify the sectional view to create a section displaying other details. In the next sections, the method of creating and modifying the section view, controlling the visibility of section lines, and creating different types of sectional views are discussed.

Creating a Section View

Ribbon: View > Create > Section

In Revit, you can create a section view in a plan or elevation view by using the **Section** tool. You can use this tool to create different types of section views such as the building section, wall section, and sections of relevant structural elements like beams, columns, slabs, and so on.

To create a section view, invoke the **Section** tool from the **Create** panel of the **View** tab; the **Modify | Section** tab will be displayed, and various options for the section view will be displayed in the **Options Bar**. The options displayed in the **Options Bar** are the same as those discussed for creating the building elevation view. These options have already been explained in the **Creating a Building Elevation View** section of this chapter. In the **Properties** palette, you can select any of the three types of section views from the **Type Selector** drop-down list. These types are: **Section : Building Section**, **Section : Wall Section**, and **Detail View : Detail**. By default, the **Section : Building Section** option is selected in the **Type Selector** drop-down list.

To create a section view, move the cursor to the viewing area in the project view; the cursor will change into a cross "+" symbol and you will be prompted to draw the section line in the current view. Click at the appropriate location to specify the start point. For creating a section through the entire building, click near the exterior face of the building profile. As you move the cursor, a section line will be displayed with one end fixed at the specified point and the other end attached to the cursor. You can even create a section line at any angle or along the vertical or horizontal axis of the building model, depending upon the project requirement. To create a horizontal section view, move the cursor horizontally across the building model, as shown in Figure 8-7, and click to specify the endpoint. On doing so, the section line along with its controls will be displayed, as shown in Figure 8-8. The section line is represented by a section head and a line. The section head indicates the direction toward which the section will be created. The methods for displaying the created section views are the same as those discussed for the elevation view.

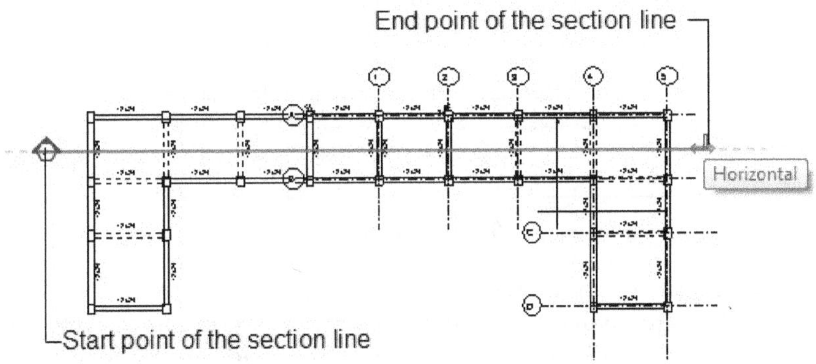

Figure 8-7 *The horizontal section line*

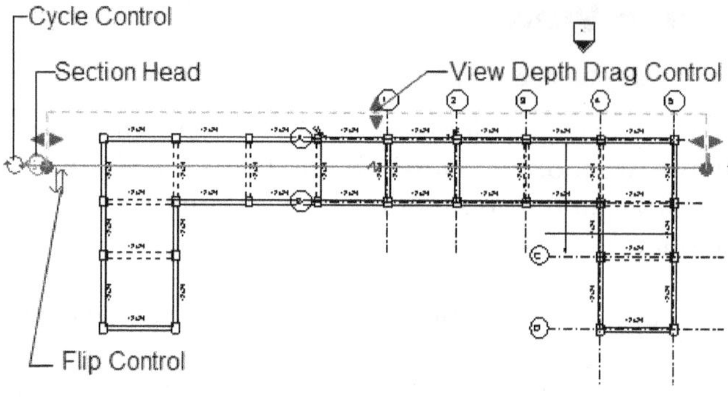

Figure 8-8 *The section line along with its control*

Modifying the Section View

The location of a section line can be modified by dragging it, as shown in Figure 8-9. On dragging the section line, the corresponding section view will get updated immediately, as shown in Figure 8-10.

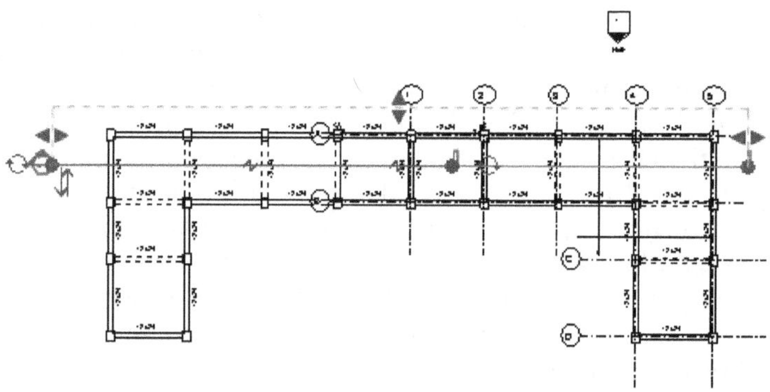

Figure 8-9 *Dragging the section line*

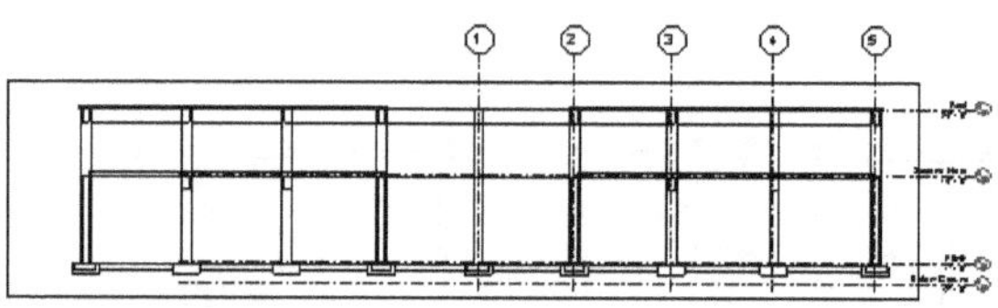

Figure 8-10 *The updated section view*

You can modify the parameters of a section view by using the controls available on the section view line. These controls are displayed when you select the section line. The twin-arrow symbol represents the flip tool that can be used to flip the viewing side of the section view. By default, the section head appears on one side. The cyclic control on both ends of the section line is used to change the visibility of the section head and tail at their respective ends. You can click on the symbol to hide or display the control. You can click on the break line symbol which appears in the middle of the section line, to break it. You can then resize the two section lines to the required extent of the view. To rejoin the section line, click on the break line control again.

When you create a section view, the view depth is created automatically. The view depth is the extent of the view in the current view. It is represented by a dashed line with the blue arrows as the drag controls. To modify the view depth, drag the arrows to the desired location. The section view shows only those elements that are within the view depth. You can modify the instance properties of a section view by selecting its section line from the drawing and specifying various parameters displayed in the **Properties** palette. In the **Properties** palette for the selected section line, you can modify the values such as **View Name**, **View Scale**, **Crop Region Visible**, and so on by clicking in the corresponding value column and selecting the new value from the drop-down list or by entering a new value in that field. In the **Properties** palette, choose the **Edit Type** button to display the **Type Properties** dialog box and modify the type properties of the section view such as type of **Callout Tag**, **Section Tag**, and **Reference Label** in it.

Creating a Segmented Section

In Revit, you can split a section into segments that are orthogonal to the direction of the section view. This enables you to show different parts of a building model in the same section view.

To create a segmented section line, select the section line; the **Modify | Views** tab will be displayed. In this tab, choose the **Split Segment** tool from the **Section** panel; the cursor will change into a knife symbol. Move it over the section line and click at the point from where you want to split the section line; the section line will break from the specified point. Now, you can move the cursor in the desired direction to split the section line along the head or tail side. Click again to specify the location of the split. The segmented section will be created and the section view will be modified immediately, refer to Figure 8-11.

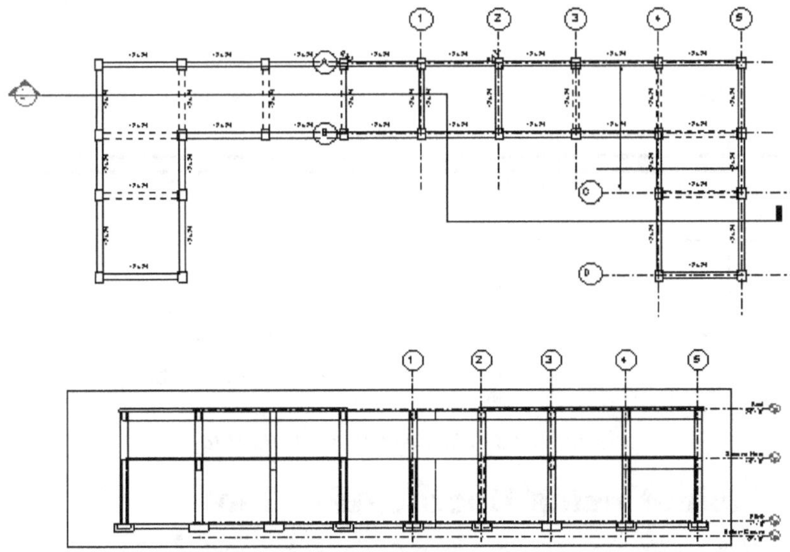

Figure 8-11 *The updated section view after splitting the section line*

Controlling the Visibility of a Section Line

The section line is visible in the plan, section, and elevation views if the view range of the views intersects the crop region of the current view. The section line created in one view is visible and created simultaneously in all other views. You can also control the visibility of the section line in views. To hide a section line in a view, select the section line, right-click, and then choose

Hide in View > Elements from the shortcut menu displayed. The section line will be hidden in the current view.

CALLOUT VIEWS

A callout view is an enlarged view of a part of a building model which requires more detailing. Creating callout views is a common practice among engineers as callout views help them look the project model more precisely and with a higher level of detail. For example, in a structural model, a callout view can be used to show the details of a connection of a beam and column in the plan or an elevation view. In a project, you can create a callout view for a plan view, section view, or elevation view. In these views, the callout tag added to a view will be linked to the callout view. The view in which the callout tag is added is called the parent view of the callout view. If the parent view is deleted, the callout will also be deleted. A callout tag, as shown in Figure 8-12, is an annotation element that represents the location of the callout in the plan, elevation, or section. The callout tag consists of the following parts: Callout bubble, Callout head, Leader line, and Reference labels. The callout bubble is the line drawn around that part of the structural model to which you need to enlarge and view in the callout view. The callout head is a symbol that represents the callout view. When a callout view is placed on a sheet, the callout head displays a detail number and a sheet number by default. The line that connects the callout head to the callout bubble is called the leader line. The reference label in a callout tag is used for the callout views that are referenced.

Various methods for creating callout views, displaying a callout view, modifying callout views properties, and adding detail lines to a callout view are discussed in the next section.

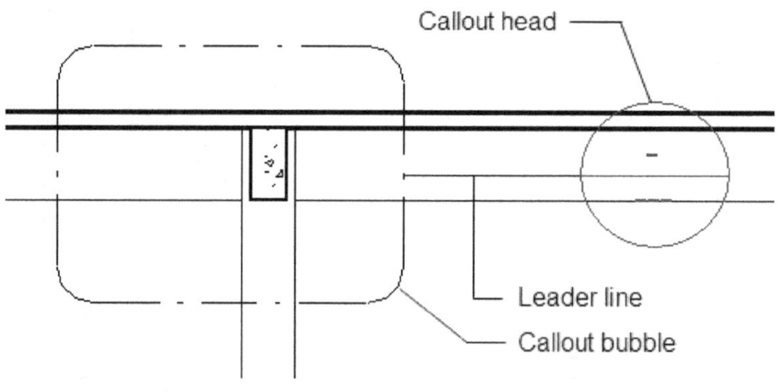

Figure 8-12 *The display of a callout tag*

Creating a Callout using Rectangular Tool

Ribbon:	View > Create > Callout drop-down > Rectangle

To create a Rectangular callout view, choose the **Rectangle** tool from the **Callout** drop-down of the **Create** panel of **View** tab; the **Properties** palette will display the properties related to the callout view to be created. In the **Properties** palette, the **Type Selector** drop-down list displays the type of callout view to be generated. In this drop-down list, you can select any of these two types: **Detail View : Detail** and **Structural Plan : Structural Plan** (for current view). Select the **Detail View : Detail** type if you want to provide detailed information about a specific part in a building model. Select the **Structural Plan : Structural Plan** (for current view) type if you want to provide more information about a part of the current view. The **Scale** drop-down list in the **Status Bar** can be used to set the view scale. In the **Reference** panel, the **Reference other view** check box is cleared, by default. You can select this check box to create a reference callout. As you select this check box, the drop-down list next to it becomes active. In this drop-down list, you can select an option to specify a view that the callout will refer to. After selecting the desired type and option, move the cursor to the top left corner of the area that you want to enlarge and then drag it toward the lower right corner of the area to create a callout bubble, refer to Figure 8-12. Release the left mouse button when the required area is enclosed in it.

Creating a Callout using the Sketch Tool

Ribbon: View > Create > Callout drop-down > Sketch

In Revit you can create a customized callout view by using sketching tools. To do so, invoke the **Sketch** tool from **View > Create > Callout** drop-down; the **Modify| Edit Profile** contextual tab will be displayed. The **Draw** panel of this tab contains various sketching tools such as **Line**, **Rectangle**, **Pick Lines** and so on. Choose any of the desired tool from this panel to sketch a desired callout view.

Displaying a Callout View

When you create a callout in the existing view, a new callout view is added to the **Project Browser** under the parent category. For example, if you create a callout view in the section view, the callout view will be added under the **Sections** heading in the **Project Browser**. In the **Project Browser**, double-click on the name of the callout view to be displayed; the corresponding callout view will be displayed in the drawing window. Alternatively, highlight the callout bubble and right-click to display the shortcut menu. Next, choose the **Go To View** option from the shortcut menu to display the callout view.

Modifying the Properties of a Callout View

You can modify the appearance of a callout bubble in the parent view. It can be modified by using the bubble controls displayed on selecting a callout view. The extents of a callout view can also be modified by using its drag controls. The rotation control can be used to rotate the callout bubble along with its leader and tag. The leader elbow control can be dragged to any desired location.

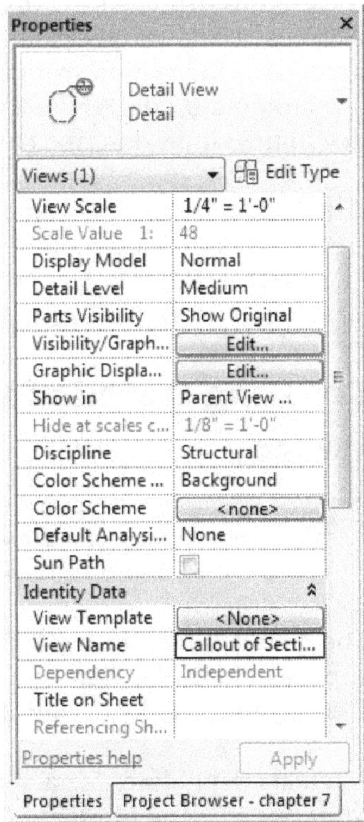

*Figure 8-13 The instance properties of the callout view in the **Properties** palette*

The instance properties of a callout view are different for different types of callouts. To modify the instance properties of a view type callout, select it from the drawing; the instance properties of the selected callout will be displayed in the **Properties** palette, as shown in Figure 8-13. The different instance properties displayed in the **Properties** palette are discussed next.

In the **Properties** palette, you can enter the name to be assigned to a new callout view in the value field of the **View Name** instance parameter under the **Identity Data** head. Similarly, you can enter the title to be given on the sheet in the value field of the **Title on Sheet** parameter.

The **Display Model** parameter under the **Graphics** head is used to set the display type for the view of a building model. By default, this parameter is set to **Normal**. You can set the value of the **Display Model** parameter to **Halftone** to display the model element in the current view as a faded image. You can set the value of this parameter to the **Do not display** option to hide the model element in the current view.

The **Underlay** parameter under the **Graphics** head can be used to set the display of model elements, which are displayed in other levels and are invisible in the current view. By default, the **Underlay** parameter is set to **None**. You can select a level from the drop-down list corresponding to this parameter to make the elements of the selected level visible in the current view. The elements that are displayed as underlay appear dim.

The amount of details to be displayed in a callout view can be controlled by using the **Detail Level** parameter under the **Graphics** head. To display the details, you can select any of the following three options from the drop-down list corresponding to this parameter: **Coarse**, **Medium**, and **Fine**. By default, the **Coarse** option is selected in the drop-down list. As a result, model elements are displayed with fewer details in the current view. You can select the **Fine** option from this drop-down list to display the layers of various materials used in a building model. On selecting the **Fine** option, additional lines are displayed in the callout view. These lines describe composite materials.

The **Edit** button in the value column for the **Visibility /Graphics Overrides** parameter can be used to control the visibility of different models and annotation elements in the callout view. The **View Scale** instance parameter is used to set the scale for a callout view. You can use the drop-down list in the value column to select a scale for the callout view.

Creating Details in a Callout View

Ribbon: Annotate > Detail > Detail Line

Revit provides various tools to create details in a callout view. You can sketch lines using the **Detail Line** tool and also add detail components provided in Revit 's library to details.

The **Detail Line** tool is used to create lines for a detail view. Detail lines are view-specific and appear only in the view in which they are created. You can use the callout view of a building model and trace detail lines over the image using their varying thickness.

To create detail lines for a detail view, choose the **Detail Line** tool from the **Detail** panel of the **Annotate** tab; the **Modify | Place Detail Lines** tab will be displayed. Select the type of detail lines from the **Line Style** drop-down list in the **Line Style** panel of this tab. You can select the appropriate detail line based on its usage. For example, wide lines can be used to show masonry elements, whereas thin lines can be selected to represent lighter materials such as glass and aluminium.

You can draw these lines by using the sketching tools displayed in the **Draw** panel of the **Modify | Place Detail Lines** tab. To add detail lines, you can trace over the underlay elements. When you trace over the underlay elements, you will notice that the cursor snaps at various elements on the underlay lines. The entire detail can be sketched using a variety of line thicknesses to achieve the desired graphical representation. You can also add dimensions and symbols to the details. However, after completing a detail, you can hide the underlay callout view by selecting the callout view from the drawing and then selecting the **Do not display** option from the drop-down list corresponding to the **Display Model** parameter in the **Properties** palette.

Note

After adding the detail lines to the callout view, you can add the filled region to the drafted details by invoking the ***Filled Region*** *tool from the* ***Annotate > Details > Region*** *drop-down.*

DRAFTING DETAILS

Drafted details are created when you want to access the details that are not referenced to the existing project views. These details are not linked to a building model and therefore they do not get updated with it.

To create a drafted detail, first create a drafting view and then use the drafting tools provided in Revit to sketch the detail. You can also import in-built details from Revit 's detail library and use them.

After the drafted detail has been created, this can be used as a reference detail. The method for creating a drafting view and drafting detail are discussed in the next sections.

Creating a Drafting View

Ribbon: View > Create > Drafting View

To create a drafting view, invoke the **Drafting View** tool from the **Create** panel; the **New Drafting View** dialog box will be displayed, as shown in Figure 8-14. In this dialog box, enter a name for the drafting view in the **Name** edit box. Next, select the scale for the detailing from the **Scale** drop-down list. To specify a user-defined scale, select the **Custom** option from the **Scale** drop-down list and then specify a value in the **Scale value 1** edit box. Next, choose the **OK** button; the drafting view will be created and added under the **Drafting Views** subhead in the **Project Browser**.

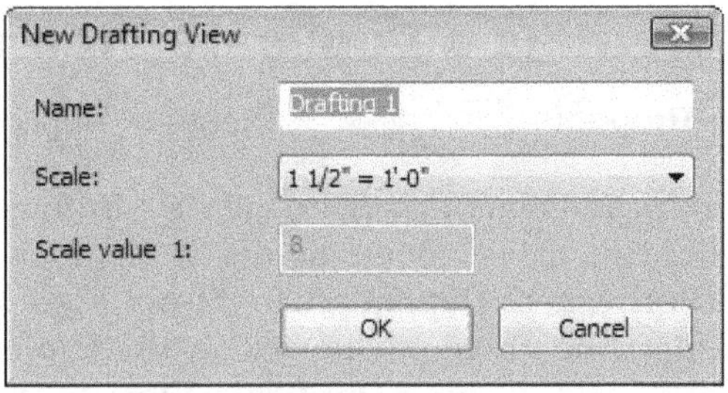

Figure 8-14 *The* ***New Drafting View*** *dialog box*

Note

Drafting views do not use the project views. As a result, when you create drafting views, project views are not displayed in the drawing window.

Drafting a Detail

The method of drafting a detail in a drafting view is similar to the one used in a callout view. In a callout view, lines are traced over an enlarged project view, whereas in a drafting view, the project view is not available and a new detail is drafted as if you are drafting on a blank sheet.

You can draft a detail by using the tools in the **Detail** panel of the **Annotate** tab. The **Detail Lines** tool enables you to draw lines with varying thickness. The **Detail Component** tool is used to add in-built detail components from Revit 's detail library. The **Insulation** tool enables you to add a graphical insulation symbol to the drafted detail. You can create graphical patterns to represent various building materials using the **Region** drop-down. The usage of these tools is similar to the tools in a callout view. You can also add dimensions, text notes, and break lines to complete a drafted detail.

DUPLICATE VIEWS

You can create multiple copies of a project view in Revit. These multiple views are called duplicate views. All duplicate views are dependent on the original view from which they are created. Therefore, these views are updated automatically if any changes are made in the original view. Duplicate views are useful when you want to use or place the same view on more than one sheet. These views are also useful in case of huge projects when the overall project view is too large to fit into a single sheet. In such cases, you can crop the parent view to make small segments by using the crop regions, create dependent views from them, and then place the cropped dependent views on the sheet. In Revit, you can create three types of duplicate views. The tools used for creating these duplicate views are: **Duplicate View**, **Duplicate with Detailing**, and **Duplicate as Dependent**. You can use the **Duplicate View** tool to create a view that contains only the model geometry from the current view. The **Duplicate with Detailing** tool can be used to create a view that contains view-specific elements from the current view. You can use the **Duplicate as Dependent** tool to create a view that is dependent on the current view. The method of creating duplicate view as a dependent view is discussed next.

Creating a Duplicate View as a Dependent View

Ribbon:	View > Create > Duplicate View drop-down > Duplicate as Dependent

You can create a dependent view from a plan, elevation, section, or callout view. To create a dependent view, open the view from which you want to create duplicate views and then choose the **Duplicate as Dependent** tool from **View > Create > Duplicate View** drop-down. Alternatively, select the name of the view in the **Project Browser** and right-click; a shortcut menu will be displayed. Choose **Duplicate View > Duplicate as Dependent** from the shortcut menu; the dependent view will display a crop region boundary and will be added in the **Project Browser** as a dependent view under the primary view.

Tip

*Crop regions help exclude unwanted content from a crop view, thereby reducing the view size. In Revit, there are two types of crop regions that allow you to crop views for the model and annotation categories. You can control the visibility of a crop region by using the **Crop View** or **Do Not Crop View** tool from the **View Control Bar**.*

Next, select the crop region boundary to display drag controls. Using the drag controls, resize the primary view and then crop it to include only the required portion of the view in the dependent view. Next, select the name of the dependent view in the **Project Browser** and right-click; a shortcut menu will be displayed. Choose **Rename** from the shortcut menu; the **Rename View** dialog box will be displayed. In this dialog box, enter a name for the dependent view and choose the **OK** button; the duplicate view will be renamed. Again, open the primary view and choose the **Show Crop Region** button from the **View Control Bar** to display the crop region boundary. Similarly, create another dependent view as explained above. In this way, you can crop the primary view and create multiple dependent views. To navigate to the primary view from the dependent view, select the crop region boundary of the dependent view and right-click; a shortcut menu will be displayed. Choose **Go to Primary View** from the shortcut menu; the primary view will be displayed on the screen. You can also navigate to the dependent views from the primary view. To do so, choose the **Show Crop Region** button from the **View Control Bar** if the crop regions are not displayed in the primary view. On doing so, the crop regions for all dependent views will be displayed. Select the crop region of the required dependent view and then right-click; a shortcut menu will be displayed. Choose **Go to View** from the shortcut menu; the corresponding dependent view will open. You can display the primary or dependent views by double-clicking on their respective view names in the **Project Browser**.

Note

*The procedure to create duplicate views by using the **Duplicate View** and **Duplicate with Detailing** tools is similar to that of creating duplicate as dependent views.*

GRAPHICAL COLUMN SCHEDULES

In a structural project, the Graphical Column Schedule (GCS) is a useful document which contains information related to the size, connection details, and reinforcement details of columns. The Graphical Column Schedule is a tabular representation of the columns in a project. You can use the information in the Graphical Column Schedule to track the structural columns used in the structural model. The columns displayed in the Graphical Column Schedule are linked to grid intersections. For example, if a column is located at the intersection of grid 1 and grid A, the Graphical Column Schedule will display the location of the column as A-1. If a column is not linked to a grid or grid intersection, Revit will link the column to the nearest grid or grid intersection. In the next section, the creation and use of Graphical Column Schedule in a project are discussed.

Creating Graphical Column Schedule

Ribbon: View > Create > Schedules drop-down > Graphical Column Schedule

To create a Graphical Column Schedule, choose the **Graphical Column Schedule** tool from **View > Create > Schedules** drop-down; the Graphical Column Schedule will be displayed in the drawing area in a new view, as shown in Figure 8-15. The name of the new view will appear under the **Graphical Column Schedules** head in the **Project Browser**. At the bottom of the Graphical Column Schedule, refer to Figure 8-15, the **Column Locations** row displays the grid intersections available in the project. Also, on the left and right sides of the schedule, the names of the levels along with their respective elevations are displayed. In this schedule, the columns are displayed in cells and are positioned according to their top and bottom constraints. You can select any of the displayed columns in the schedule and change its type and instance properties. On doing so, the change will be applied to the entire project view and the changed column.

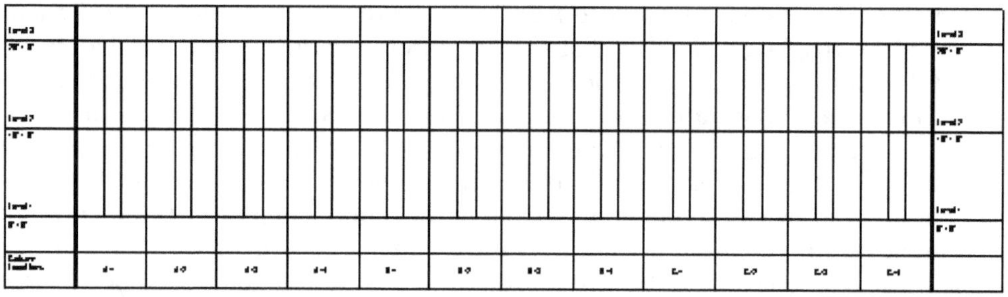

Figure 8-15 The Graphical Column Schedule for a structural model

USING SCHEDULES IN A PROJECT

Schedule is another format for providing project information. In a building project, there may be different items that can be used a number of times at different locations. Schedules are primarily used to provide information regarding these items in a tabular format. For each item, the schedule tables provide information regarding its size, material, cost, finish, level, and so on. These schedules can then be used by associated agencies for various purposes. These schedule tables can assist the fabricating agency to manufacture all similar items. The quantity surveyors and estimators can calculate the quantity and cost of the grouped items. For example, in a building model, a particular column type may be used a number of times in the same layout plan on different floors. With the help of schedules, you can group columns in different ways based on their parameters such as location in a project, size, section, and so on. In a project schedules are created by extracting information from a building model. In the project, all elements are parametric. This means that they have several properties associated with them. When you use these elements, the associated information is automatically added to the building model. This information can then be extracted for creating schedules.

The power of Revit 's parametric change engine enables you to modify elements in a building model at any stage of the project development. Changes are reflected immediately in all project views. As elements are bidirectionally associated, the entire project information is updated immediately and the necessary changes in schedules are reflected automatically. Conversely, when you modify an element in a schedule, all its instances are immediately updated in all project views. Revit also enables you to create schedules based on the common group properties of different categories of items. For example, you can create a schedule of all rooms in a project that has the same floor, ceiling, and wall finish. These are termed as Key Schedules.

Generating a Schedule

Ribbon: View > Create > Schedules drop-down > Schedule/Quantities

You can create a schedule or a quantity report by using the **Schedule/Quantities** tool. To generate a schedule, invoke this tool from the **Create** panel; the **New Schedule** dialog box will be displayed, as shown in Figure 8-16.

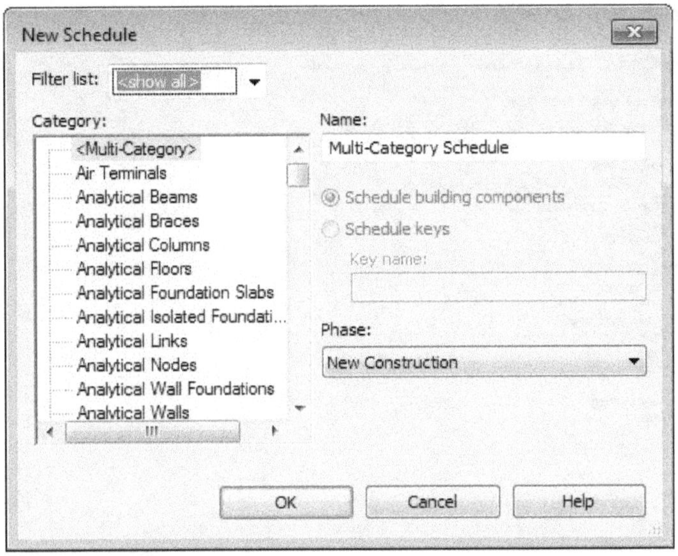

Figure 8-16 *The* ***New Schedule*** *dialog box*

In the **New Schedule** dialog box, various categories of building elements are listed in the **Category** list box. You can select the required category to create a schedule. For example, you can select the **Structural Framing** option from the **Category** list box to generate a schedule for framing members. The category selected for creating a schedule is displayed in the **Name** edit box. The **Schedule building components** radio button is used to create a schedule of building elements. After selecting a category from the **Category** list box, choose the **OK** button; the **Schedule Properties** dialog box for the selected category will be displayed. Figure 8-17 displays the options in the **Schedule Properties** dialog box for the **Structural Framing** category.

Note

To improve the annotation and scheduling and to make the schedule more informative, some other parameters are introduced such as Elevation at Top, Elevation at Top Core, Elevation at Bottom, and Elevation at Bottom Core etc.

The **Schedule Properties** dialog box has options to set various element properties, their formats, and appearance in a schedule. It contains five tabs: **Fields**, **Filter**, **Sorting/Grouping**, **Formatting**, and **Appearance**. The options in the **Fields** tab can be used to select the fields that would appear in a schedule. Select the required fields from the **Available fields** list box and choose the **Add parameter(s)** button to add fields to the **Scheduled fields (in order)** area. Choose the **Remove parameter(s)** button to remove the selected scheduled fields. You can use the **Move parameter up** and **Move parameter down** buttons to move and arrange fields in the desired order. By choosing the **New Parameter** button, the **Parameter Properties** dialog box is displayed from which you can add your own custom field. You can choose the **Add calculated parameter** button; the **Calculated Value** dialog box will be displayed. You can create a field that will be a percentage of another field, or specify a formula to define the field from the **Calculated Value** dialog box. You can choose the **Combined Parameters** button to combine the parameters for the schedule. The **Include elements in link** check box can be selected to include elements from the linked Revit files to the schedules and drawing files of model elements such as columns, walls, beams, and so on.

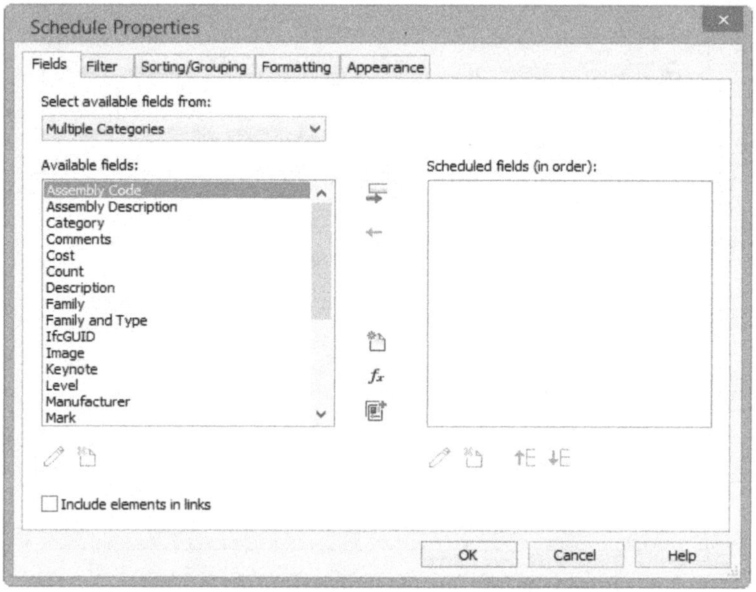

Figure 8-17 *The* ***Schedule Properties*** *dialog box*

The options in the **Filter** tab enable you to restrict the display of elements based on certain conditions. You can use maximum four filters for a schedule based on the parameters and values. For example, in a structural framing schedule, if you want to display the beams constrained to **Level 1**, add the **Level** option available in the **Schedule fields** list box to the **Fields** tab. Next, in the **Filter** tab, select the **Level** option from the **Filter by** drop-down list and then select the **Level 1** option from the drop-down list next to it. This way you can create filters for schedules in a project. This helps in organizing data in schedules.

The options in the **Sorting/Grouping** tab can be used to sort and group rows in a schedule. The parameter for sorting a schedule can be selected from the **Sort by** drop-down list. You can sort it in the ascending or descending order by selecting a radio button. Next, to arrange door types in the ascending order of their width, select the **Width** option from the **Sort by** drop-down list and then select the **Ascending** radio button. The **Grand totals** check box can be selected to display the sum total of all elements.

In the **Formatting** tab, you can set the parameters related to the appearance of headings. You can select a heading from the **Fields** list and enter a new heading name in the **Heading** edit box. The orientation of a heading and its alignment can be specified by selecting desired options from the **Heading orientation** and **Alignment** drop-down list, respectively. The **Field Format** button is used to format the appearance of fields with numerical values. For such fields select the required option from the drop-down list below the **Show conditional format on sheets** check box. By default the **Standard** option is selected from this drop-down list. You can select the **Hidden field** check box to hide a field in a schedule. On doing so, you can sort out or filter a schedule by using the selected field without actually displaying it in the schedule. The options in the **Appearance** tab can be used to change the appearance of a schedule view. You can use various parameters such as **Header text**, **Body text**, and so on to modify the appearance of a schedule.

When you choose the **OK** button from the **Schedule Properties** dialog box, Revit scans the building model for the category of elements and displays the schedule based on the parameters specified. Figure 8-18 shows an example of a structural foundation schedule with the scheduled fields.

Structural Foundation Schedule				
Family and T	Area	Foundation T	Volume	Width
Footing-R	39 SF		36.00 CF	4' - 0"
Footing-R	39 SF		36.00 CF	4' - 0"
Footing-R	39 SF		36.00 CF	4' - 0"
Footing-R	39 SF		36.00 CF	4' - 0"
Footing-R	39 SF		36.00 CF	4' - 0"
Footing-R	39 SF		36.00 CF	4' - 0"
Footing-R	39 SF		36.00 CF	4' - 0"
Footing-R	39 SF		36.00 CF	4' - 0"
Wall Foun		1' - 0"	57.00 CF	3' - 0"
Wall Foun		1' - 0"	18.00 CF	3' - 0"
Wall Foun		1' - 0"	57.00 CF	3' - 0"
Wall Foun		1' - 0"	60.00 CF	3' - 0"
Wall Foun		1' - 0"	33.00 CF	3' - 0"
Wall Foun		1' - 0"	42.00 CF	3' - 0"

Figure 8-18 *Schedule displaying the properties of a structural foundation*

You can modify the properties of a schedule even after creating it. To do so, click on the schedule name in the **Project Browser**. Next, in the **Properties** palette, choose the **Edit** button in the value column corresponding to the **Fields**, **Filter**, **Sorting/Grouping**, **Formatting**, and **Appearance** instance parameters to access their respective tabs in the **Schedule Properties** dialog box.

For example, the schedule shown in Figure 8-18 can be edited to create a simplified schedule in which the foundation with a particular type mark can be grouped together. You can access the options in the **Sorting/Grouping** tab, select the **Grand totals** check box, and clear the **Itemize every instance** check box to summarize the schedule. On doing so, the modified schedule will be displayed, as shown in Figure 8-19.

Structural Foundation Schedule					
Family and T	Area	Mark	Foundation T	Volume	Width
1					
Footing-R	39 SF	1		36.00 CF	4' - 0"
2					
Wall Foun		2	1' - 0"		3' - 0"
14					

Figure 8-19 *Schedule displaying the grouped properties of a structural foundation*

After calculating the required material quantity, you can use the cost per unit to calculate the total estimated cost for that item. For example, to calculate the total cost of walls for a project, first create a wall schedule showing the area of each wall type. Next, enter the value of the cost per square feet for each wall type directly in the created schedule. Revit automatically assigns the cost to all instances of the same wall type. You can then use the **Add Parameter** button in the **Schedule Properties** dialog box to add a new column for **Total Cost** for each type of wall. Use the **Calculated Value** button and assign the formula **Cost*Area/1' ^ 2** to calculate the total cost of each type of wall (the factor 1' ^ 2 is used to suppress the units). In the **Formatting** tab, select the **Calculate total** check box for the **Total Cost** column and generate the total cost of the walls in the building project. This procedure can be adopted for each category of elements such as windows, doors, floor, roof, furniture, and so on and then the gross total cost of all building elements can be calculated.

Editing a Schedule

You can use a schedule to modify elements in it. To do so, select a schedule from the **Project Browser** and then click any of the cells displayed in it; the **Modify Schedule/Quantities** tab will be displayed. You can use the tools displayed in this tab to delete, hide, unhide, highlight, group, and ungroup the properties of the elements displayed in a schedule.

GENERATING THE MATERIAL TAKEOFF SCHEDULE

Ribbon: View > Create > Schedules drop-down > Material Takeoff

Material takeoff schedules display the information of the material used to create structural elements in a model. This schedule helps the quantity surveyor in estimating the materials used to fabricate the structural elements in the site.

The method of generating the material takeoff schedule is the same as that of generating the quantity schedule. To generate a material takeoff schedule, choose the **Material Takeoff** tool from **View > Create > Schedules** drop-down; the **New Material Takeoff** dialog box is displayed as shown in Figure 8-20.

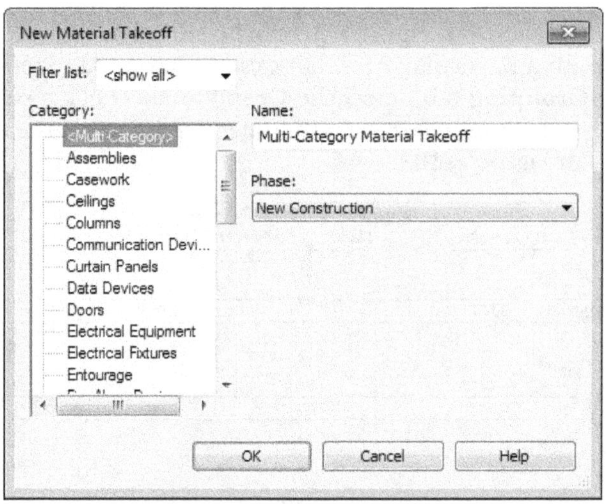

Figure 8-20 *The* ***New Material Takeoff*** *dialog box*

In this dialog box, select a category of structural elements from the **Category** list box for which you want to create the takeoff schedule. On selecting a category from the **Category** list box, you will notice that a default name for the takeoff schedule is displayed in the **Name** edit box. You can enter text in this edit box to specify the name the takeoff schedule to be created. Next, you can select an option from the **Phase** drop-down list to specify the phasing stage of the structural model for which the takeoff schedule will be created. In this drop-down list, you can select the **Existing** option for the working phase or you can select the **New Construction** option for a new phase. In the **New Material Takeoff** dialog box, to include all categories present in the project in the **Category** list box, select the check boxes from the **Filter** drop-down list. After selecting the category from the **Category** list box, choose the **OK** button; the **Material Takeoff Properties** dialog box will be displayed. The options in this dialog box are similar to those in the **Schedule Properties** dialog box as discussed in the previous section. After using various options in this dialog box, choose the **OK** button; the material takeoff schedule will be displayed on the screen. Also, you will also notice that a node is added for the takeoff schedule under the **Schedules/Quantities** head in the **Project Browser**. Figure 8-21 displays a material takeoff schedule for framing members used in a structural model.

Structural Framing Material Takeoff				
Material: Na	Material: Area	Material: Cost	Structural Usage	Material: Vol
Concrete - Cast-in-Place Concrete				
Concrete -	228 SF		Other	95.11 CF
Concrete -	198 SF		Girder	35.56 CF
Concrete -	83 SF		Girder	33.78 CF
Concrete -	87 SF		Girder	35.56 CF
Concrete -	147 SF		Girder	62.22 CF
Concrete -	111 SF		Girder	46.22 CF
Concrete -	82 SF		Girder	26.00 CF
Concrete -	86 SF		Other	27.33 CF
Concrete -	115 SF		Joist	17.67 CF
Concrete -	61 SF		Girder	19.00 CF
Metal - Steel - ASTM A992				
Metal - Steel	27 SF		Other	0.22 CF
Metal - Steel	27 SF		Other	0.23 CF
Metal - Steel	21 SF		Other	0.18 CF
Metal - Steel	28 SF		Other	0.24 CF
Metal - Steel	27 SF		Other	0.23 CF
Metal - Steel	28 SF		Other	0.24 CF
Metal - Steel	28 SF		Other	0.24 CF
Grand total:	1386 SF			400.02 CF

Figure 8-21 *Schedule displaying the grouped properties of a Structural Foundation*

TUTORIALS

Tutorial 1 Commercial Complex

In this tutorial, you will create a section view, an elevation view, and a quantity schedule for the project created in Tutorial 1 of Chapter 7. **(Expected time: 1 hr 15 min)**

The following steps are required to complete this tutorial:

a. Open the *c07_Commercial-Complex_tut1.rvt* project file.
b. Hide annotations and dimensions.
c. Create section views.
d. Create the framing elevation view.
e. Generate Graphical Column Schedule.
f. Create quantity schedule for columns.
g. Save the project.
h. Close the project.

Opening the Project File

In this section, you will open the project file created in Tutorial 1 of Chapter 7.

1. To open the project file, choose **Open > Project** from the **Application Menu**; the **Open** dialog box is displayed.

2. In this dialog box, browse to *C:\rst_2017\c07_rst_2017_tut* folder location and then choose the *c07_Commercial-Complex_tut1.rvt for Imperial and for Metric open M_ c07_Commercial-Complex_tut1.rvt* project file. You can also download this file from *http://www.cadcim.com*. The path of the file is as follows: *Textbooks > Civil/GIS > Revit Structure > Exploring Autodesk Revit 2017 for Structure*.

3. Next, choose the **Open** button from the **Open** dialog box; the selected project file opens in the drawing window.

Hiding Annotations and Dimensions

In this section, you will hide annotations and dimensions in the **Second Floor** structural plan view so that sections can be placed easily.

1. Double-click on the **Second Floor** node under the **Structural Plans** head in the **Project Browser**; the structural floor plan view at the **Second Floor** level is displayed.

2. Choose the **View** tab and then choose the **Visibility/ Graphics** tool from the **Graphics** panel; the **Visibility/Graphic Override for Structural Plan: Second Floor** dialog box is displayed.

3. Choose the **Annotation Categories** tab from this dialog box. In this tab, ensure that the **Structure** option is selected in the **Filter list** drop-down list. Next, in the **Visibility** column, clear the check boxes corresponding to the following categories: **Dimensions**, **Structural Column Tags**, **Structural Foundation Tags**, **Structural Framing Tags**, **Spot Coordinates**, and **Spot Elevations**.

4. Choose the **OK** button; the **Visibility/Graphic Override for Structural Plan: Second Floor** dialog box is closed.

Creating Section Views

In this section, you will create sectional views for the project by using the **Section** tool.

1. Choose the **Section** tool from the **Create** panel of the **View** tab; the **Modify | Section** tab is displayed.

2. In the **Properties** palette, ensure that the **Section : Building Section** option is selected in the **Type Selector** drop-down list.

3. Move the cursor between the grid bubbles of grid D and grid E, and then click to specify the start point of the grid line.

4. Move the cursor horizontally across the building model till you have crossed the vertical grid 5.1. Click at the desired point, as shown in Figure 8-22, to specify the endpoint of the section line. The section view is created and the name **Section 1** is displayed in the **Sections (Building Section)** head in the **Project Browser**.

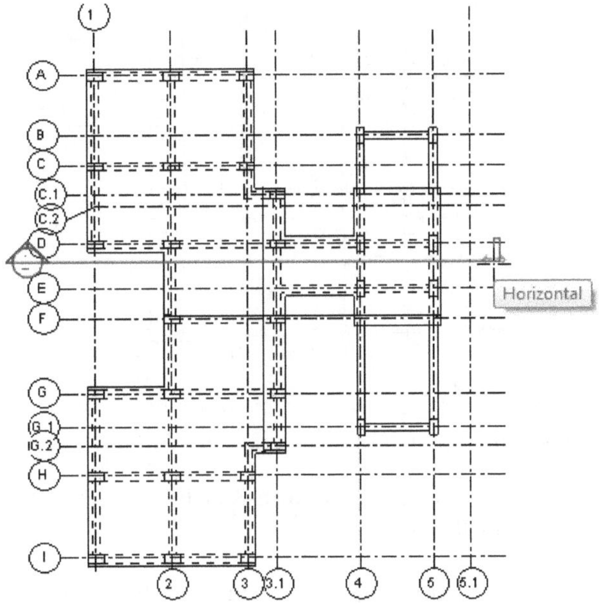

Figure 8-22 *Creating a section line for the section view*

5. In the **Properties** palette, scroll the palette, if required and then click in the value field of the **View Name** parameter under the **Identity Data** area, and enter **Section-X** to replace the existing value.

6. Next, click in the value field corresponding to the **Title on Sheet** parameter under the **Identity Data** area and enter **Sectional View-Longitudinal** in it.

7. In the **Properties** palette, click in the value field corresponding to the **Detail Level** parameter and select the **Fine** option from the drop-down list displayed.

8. Next, click in the value field of the **Far Clip Offset** parameter under the **Extends** area and enter **20'**(**6096** mm) in it. Choose the **Apply** button in the **Properties** palette.

9. Now, choose the **Split Segment** tool from the **Section** panel of the **Modify | Views** tab; a pencil mark appears in place of the cursor.

10. Place the cursor over the section line; the section line gets highlighted. Next, click over the section line at its intersection with grid 4. As you move down the cursor toward the right, the split section line also moves. Place the cursor along with the split cursor line at grid F and click; the section line is displayed, as shown in Figure 8-23.

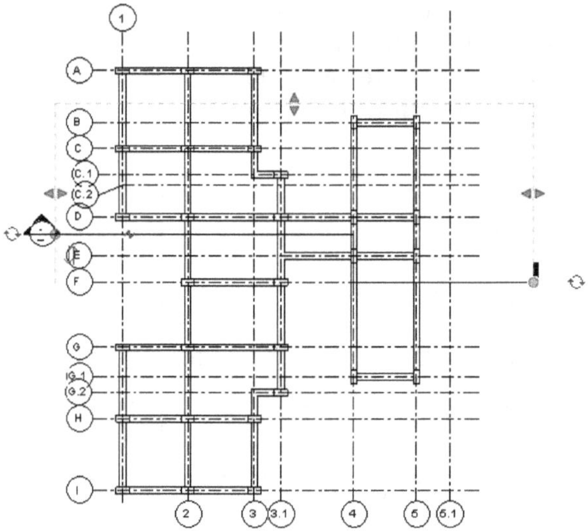

Figure 8-23 *Section line for longitudinal section after splitting the segment*

The section view is created and the name **Section-X** is displayed under the **Sections (Building Section)** head in the **Project Browser**.

11. Repeat steps 1 to 9 to create a transverse section view of the building model. Name the view as **Section-Y** and the title for the sheet as **Sectional View-Transverse** and set the **Far Clip Offset** parameter to **45'**(**13716** mm). For the location of the section line, refer to Figure 8-24.

12. Click on the '+' symbol on the left of the **Sections (Building Section)** head in the **Project Browser** to display the section names created. The **Section X** and **Section Y** is created under the **Sections(Building Section**) head.

13. Press ESC to exit the current selection.

14. Double-click on the name **Section-X** to display the corresponding section view on the drawing window, as shown in Figure 8-25.

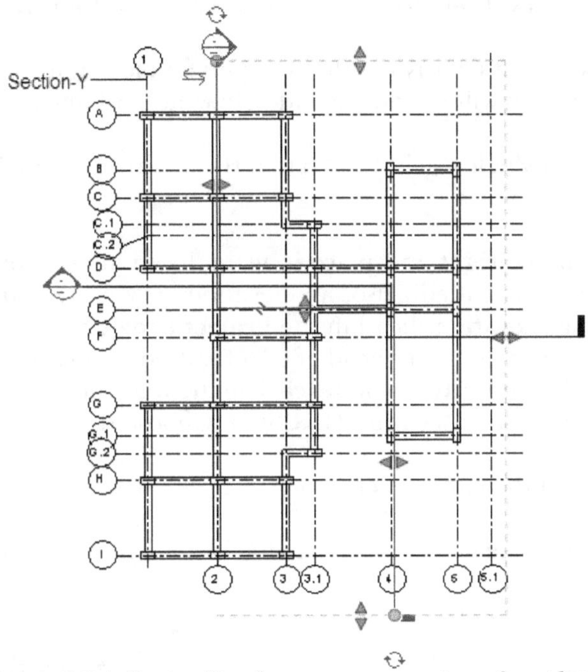

Figure 8-24 *Section line for transverse section after splitting the segment*

15. Similarly, double-click on the name **Section-Y** to display the corresponding section view in the drawing window, as shown in Figure 8-26.

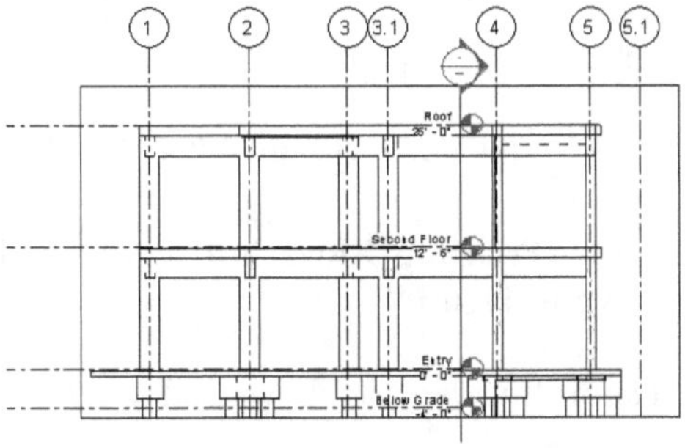

Figure 8-25 *Section view for* ***Section-X***

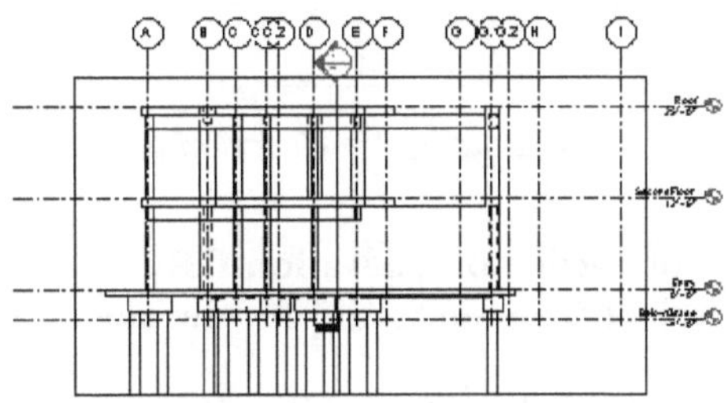

Figure 8-26 *Section view for* ***Section-Y***

Note
You can drag the drag points of the ***Crop Region*** *to customize the view for your section.*

Creating Framing Elevation Views

In this section, you will create the elevation views for the project using the **Framing Elevation** tool and the **Building Elevation** tool.

1. Double-click on the **Second Floor** node under the **Structural Plans** head of the **Project Browser**.

2. Invoke the **Framing Elevation** tool from **View > Create > Elevation** drop-down.

3. Place the cursor over grid C; it is highlighted and an elevation symbol is displayed. Note that the elevation symbol will flip above and below the grid line as you move the cursor.

4. Move the cursor slightly below grid C such that the elevation symbol is displayed below the highlighted grid line.

5. Move the cursor towards right and place it at the location, as shown in Figure 8-27, and click; the elevation tag is added. Also, an elevation view is created and added under the **Elevations (Framing Elevation)** head in the **Project Browser**.
6. Press ESC and then select the arrow head of the elevation tag just added; the **Modify | Views** tab is displayed. In the **Properties** palette, click in the value field corresponding to the **View Name** parameter and enter **Elevation-GRC** in it. Choose the **Apply** button.

7. Press ESC twice to exit the current selection.

8. Repeat steps 2 to 6 and create an elevation at grid 3.1, refer to Figure 8-28. Then, name it as **Elevation GR-3.1**.

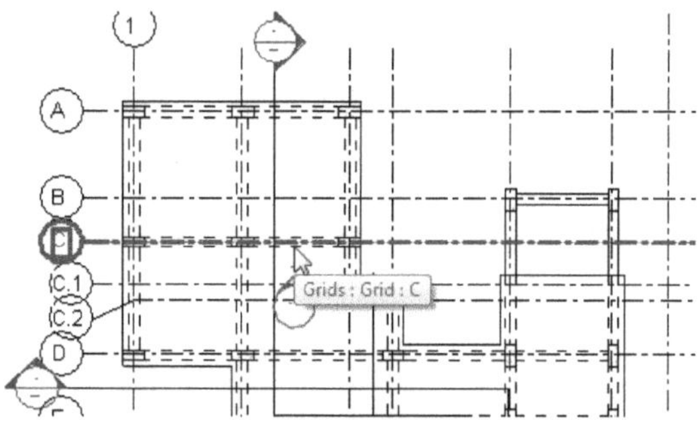

Figure 8-27 *Placing the elevation tag at grid C*

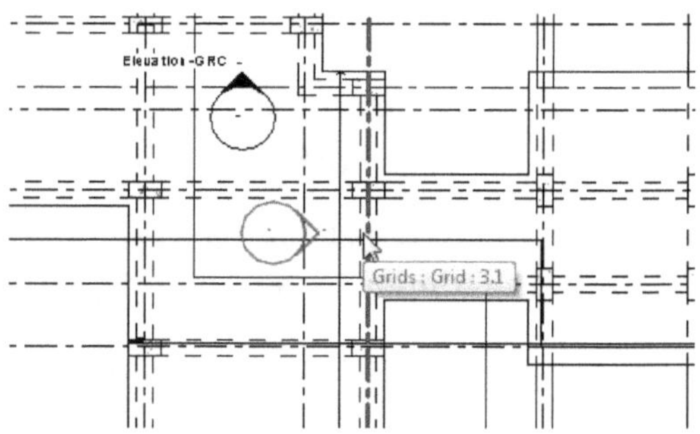

Figure 8-28 *Placing the elevation tag at grid 3.1*

9. Double-click on **Elevation-GRC** under the **Elevations (Framing Elevation)** node in the **Project Browser** to display the elevation view, as shown in Figure 8-29.

10. Double-click on **Elevation-GR3.1** to display its corresponding view, as shown in Figure 8-30.

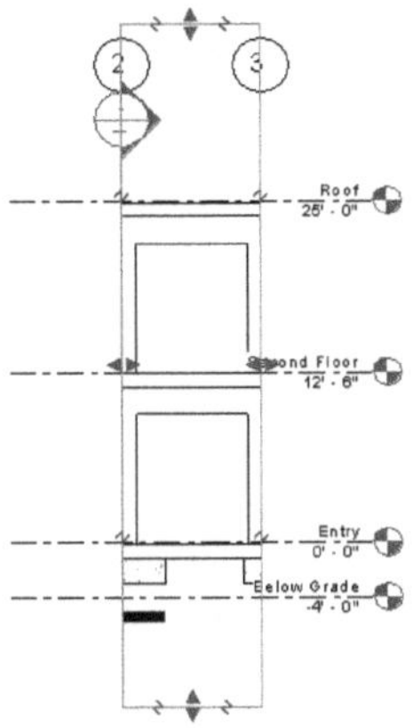

*Figure 8-29 The **Elevation-GRC** sectional view*

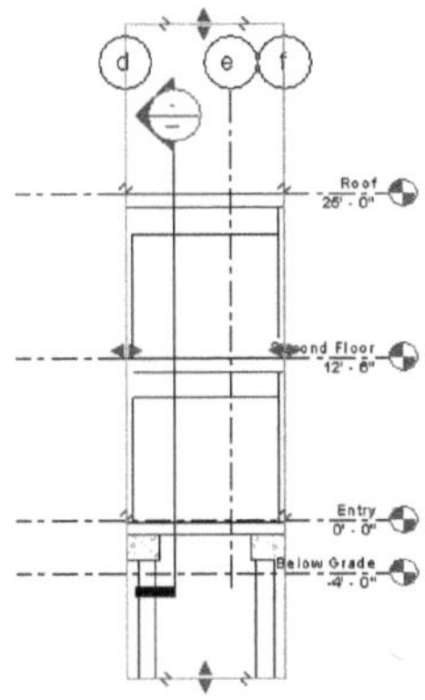

*Figure 8-30 The **Elevation-GR3.1** sectional view*

Generating the Graphical Column Schedule

In this section, you will create the graphical schedule for column and the quantity schedule for columns using the **Graphical Column Schedule** tool.

1. Invoke the **Graphical Column Schedule** tool from **View > Create > Schedule** drop-down; the graphical column schedule is displayed in the drawing window.

2. In the **Properties** palette, select the check box corresponding to the **Group Similar Locations** parameter; the columns at the similar grid line are grouped and displayed in the schedule.

3. Now, click in the **View Scale** parameter and then select the **1/4" = 1'-0" (1:25)** option from the drop-down list displayed. Next, choose the **Apply** button; the viewing scale of the Graphical Column Schedule is changed, as shown in Figure 8-31.

Note

*As you create the schedule, the name of the schedule is displayed as **Graphical Column Schedule 1** under the **Graphical Column Schedule** head in the **Project Browser**.*

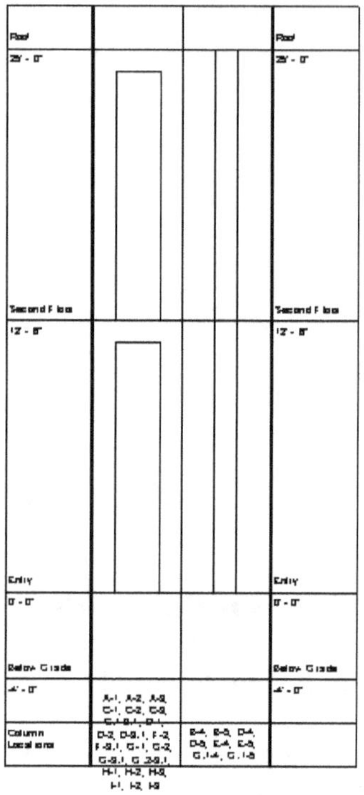

Figure 8-31 The ***Graphical Column Schedule*** *view*

Generating a Quantity Schedule for Columns

In this section, you will create a schedule for displaying the quantity columns and their associated properties used in the project.

1. Invoke the **Schedule/Quantities** tool from **View > Create > Schedules** drop-down; the **New Schedule** dialog box is displayed.

2. In this dialog box, select the **Structural Columns** option from the **Category** list box and then choose the **OK** button; the **Schedule Properties** dialog box is displayed.

3. In the **Available fields** list box, select the **Count** option. Now, press and hold the CTRL key and select the following options from the list box: **Base Level**, **Family and Type**, **Length**, and **Volume**.
4. Next, release the CTRL key and choose the **Add parameter(s)** button; the selected options are added to the **Scheduled fields (in order)** list box.

5. Click on the **Family and Type** option in the **Scheduled fields (in order)** list box and choose the **Move Up** button till the option comes at the top in the list.

6. Repeat step 5 to reorder all the options in the **Scheduled fields (in order)** list box, as shown in Figure 8-32.

7. In the **Schedule Properties** dialog box, choose the **Sorting/Grouping** tab and then select the **Family and Type** option from the **Sort by** drop-down list. Ensure that the **Ascending** radio button is selected.

8. Select the **Header** and **Footer** check boxes and then ensure that the **Title, count, and totals** options are selected in the drop-down lists next to them.

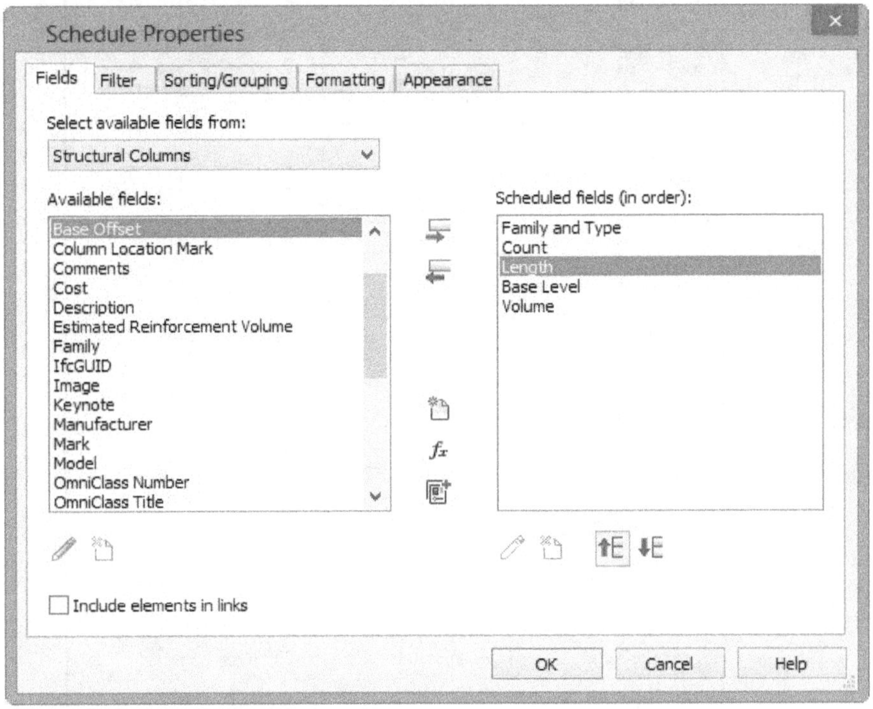

Figure 8-32 *The* ***Schedule Properties*** *dialog box*

9. Select the **Grand totals** check box and ensure that the **Itemize every instance** check box is selected.

10. Choose the **Formatting** tab and then in the **Fields** list box, select **Count**. Next, select the **Calculate totals** option from drop-down list below the **Show conditional format on sheets check box**.

11. Repeat step 10 and select the **Calculate totals** check box for the **Length** and **Volume** options in the **Fields** list box.

12. Choose the **OK** button; the **Structural Column Schedule** is displayed, as shown in Figure 8-33.

<Structural Column Schedule>				
A	B	C	D	E
Family and Type	Count	Length	Base Level	Volume
Concrete-Rectangular-Column: 12 x 24				
Concrete-Recta	1	11' - 6"	Entry	23.00 CF
Concrete-Recta	1	11' - 6"	Entry	23.00 CF
Concrete-Recta	1	11' - 6"	Entry	23.00 CF
Concrete-Recta	1	11' - 6"	Entry	23.00 CF
Concrete-Recta	1	11' - 6"	Entry	23.00 CF
Concrete-Recta	1	11' - 6"	Entry	23.00 CF
Concrete-Recta	1	11' - 6"	Entry	23.00 CF
Concrete-Recta	1	11' - 6"	Entry	23.00 CF
Concrete-Recta	1	11' - 6"	Entry	23.00 CF
Concrete-Recta	1	11' - 6"	Entry	23.00 CF
Concrete-Recta	1	12' - 6"	Entry	11.50 CF
Concrete-Recta	1	12' - 6"	Entry	11.50 CF
Concrete-Recta	1	12' - 6"	Entry	11.50 CF
Concrete-Recta	1	12' - 6"	Entry	11.50 CF
Concrete-Recta	1	11' - 6"	Entry	23.00 CF
Concrete-Recta	1	11' - 6"	Entry	23.00 CF
Concrete-Recta	1	11' - 6"	Entry	23.00 CF
Concrete-Recta	1	11' - 6"	Entry	23.00 CF
Concrete-Recta	1	11' - 6"	Entry	23.00 CF
Concrete-Recta	1	11' - 6"	Entry	23.00 CF
Concrete-Recta	1	11' - 6"	Entry	23.00 CF
Concrete-Recta	1	11' - 6"	Entry	23.00 CF
Concrete-Recta	1	11' - 6"	Entry	23.00 CF
Concrete-Recta	1	11' - 6"	Entry	23.00 CF
Concrete-Recta	1	11' - 6"	Entry	23.00 CF
Concrete-Recta	1	11' - 6"	Entry	23.00 CF
Concrete-Recta	1	12' - 6"	Entry	25.00 CF
Concrete-Recta	1	11' - 6"	Second Floor	23.00 CF
Concrete-Recta	1	11' - 6"	Second Floor	23.00 CF
Concrete-Recta	1	11' - 6"	Second Floor	23.00 CF
Concrete-Recta	1	11' - 6"	Second Floor	23.00 CF
Concrete-Recta	1	11' - 6"	Second Floor	23.00 CF
Concrete-Recta	1	11' - 6"	Second Floor	23.00 CF
Concrete-Recta	1	11' - 6"	Second Floor	23.00 CF
Concrete-Recta	1	11' - 6"	Second Floor	23.00 CF
Concrete-Recta	1	11' - 6"	Second Floor	23.00 CF
Concrete-Recta	1	11' - 6"	Second Floor	23.00 CF
Concrete-Recta	1	12' - 6"	Second Floor	23.00 CF
Concrete-Recta	1	12' - 6"	Second Floor	23.00 CF
Concrete-Recta	1	12' - 6"	Second Floor	23.00 CF
Concrete-Recta	1	12' - 6"	Second Floor	23.00 CF
Concrete-Recta	1	11' - 6"	Second Floor	23.00 CF
Concrete-Recta	1	11' - 6"	Second Floor	23.00 CF
Concrete-Recta	1	11' - 6"	Second Floor	23.00 CF
Concrete-Recta	1	11' - 6"	Second Floor	23.00 CF
Concrete-Recta	1	11' - 6"	Second Floor	23.00 CF
Concrete-Recta	1	11' - 6"	Second Floor	23.00 CF
Concrete-Recta	1	11' - 6"	Second Floor	23.00 CF
Concrete-Recta	1	11' - 6"	Second Floor	23.00 CF
Concrete-Recta	1	11' - 6"	Second Floor	23.00 CF
Concrete-Recta	1	11' - 6"	Second Floor	23.00 CF
Concrete-Recta	1	11' - 6"	Second Floor	23.00 CF
Concrete-Recta	1	11' - 6"	Second Floor	23.00 CF
Concrete-Recta	1	12' - 6"	Second Floor	25.00 CF
Concrete-Recta	1	12' - 6"	Second Floor	25.00 CF
Concrete-Recta	1	12' - 6"	Second Floor	25.00 CF
Concrete-Recta	1	12' - 6"	Second Floor	25.00 CF
Concrete-Rectang	60	706' - 0"		1350.00 CF
Grand total: 60	60	706' - 0"		1350.00 CF

Figure 8-33 *The schedule view for columns*

Note

When you create the schedule, its name is displayed as ***Structural Column Schedule*** *under the* ***Schedule/Quantities*** *head in the* ***Project Browser****.*

Saving the Project

In this section, you will save the project file using the **Save As** tool.

1. To save the project, choose the **Save As > Project** option from the **Application Menu**; the **Save As** dialog box is displayed.

2. In this dialog box, browse to *C:\rst_2017* and create a folder with the name *c08_rst_2017_tut*. Now, open the created folder and then enter **c08_Commercial-Complex_tut1** in the **File name** edit box.

3. Choose the **Save** button; the **Save As** dialog box is closed and the project file is saved.

Closing the Project

1. To close the project, choose the **Close** option from the **Application Menu**.

   The file is closed. This completes Tutorial 1 of Chapter 8.

Tutorial 2 Industrial Complex

In this tutorial, you will create a section view, an elevation view, and a quantity schedule for the project created in Tutorial 2 of Chapter 7. **(Expected time: 1 hr 15 min)**

The following steps are required to complete this tutorial:

a. Open the *c07_Industrial-Complex_tut2.rvt* project file.
b. Generate section view.
c. Generate quantity schedule for beams.
d. Generate material takeoff schedule for foundations.
e. Create Callout View.
f. Save the project.

Opening the Project File

In this section, you will open the project file created in Tutorial 2 of Chapter 7.

1. To open the project file, choose **Open > Project** from the **Application Menu**; the **Open** dialog box is displayed.

2. In this dialog box, browse to the location *C:\rst_2017\c07_rst_2017_tut* and then choose the *c07_Industrial-Complex_tut2.rvt for Imperial and for Metric open M_c07_Industrial-Complex_tut2.rvt* project file. You can also download this file from *http://www.cadcim.com*. The path of the file is as follows: *Textbooks > Civil/GIS > Revit Structure > Exploring Autodesk Revit 2017 for Structure*.

3. Next, choose the **Open** button from the **Open** dialog box; the selected project file opens in the drawing window with the **Entry** structural plan view.

Generating Section Views

In this section, you will generate a sectional view for the project using the **Section** tool.

1. Double-click on the **Foundation Level** node under the **Structural Plans** head in the **Project Browser**; the **Foundation Level** structural plan opens.

2. Choose the **Section** tool from the **Create** panel in the **View** tab; the **Modify | Section** tab is displayed.

3. In the **Properties** palette, ensure that the **Section : Building Section** option is selected in the **Type Selector** drop-down list.

4. Move the cursor between the grid bubbles of grid B and grid C and then click to specify the start point of the section line.

5. Move the cursor horizontally across the building model until it has crossed the building model. Click at the desired point, as shown in Figure 8-34, to specify the endpoint of the section line. The section view is created and the name **Section 1** is displayed in the **Section (Building Section)** head in the **Project Browser**.

6. In the **Properties** palette, scroll the palette, if required, and click in the value field of the **View Name** parameter and enter **Section-X** to replace the existing value.

7 Choose the **Apply** button in the **Properties** palette.

8. Next, choose the **Split Segment** tool from the **Section** panel of the **Modify | Views** tab; a pencil mark replaces the cursor.

9. Place the cursor on the section line; the section line gets highlighted. Next, click on the section line at some suitable point between grid 5 and grid 6. As you move the cursor downward in the right direction, the split section line also moves. Place the cursor along with the split cursor line at some suitable location below grid 1 and then click; the section line appears, as shown in Figure 8-35.

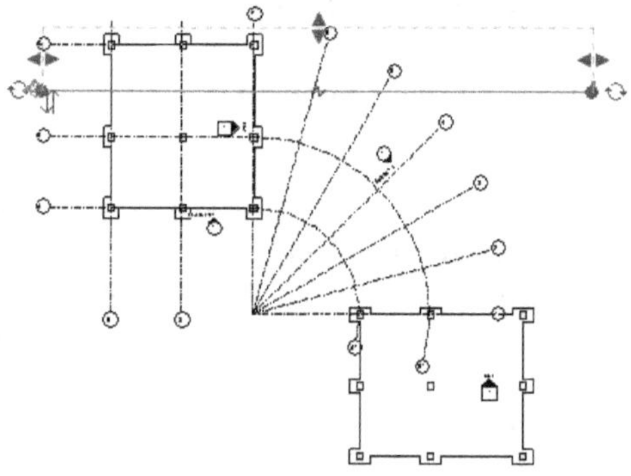

Figure 8-34 *The section line created*

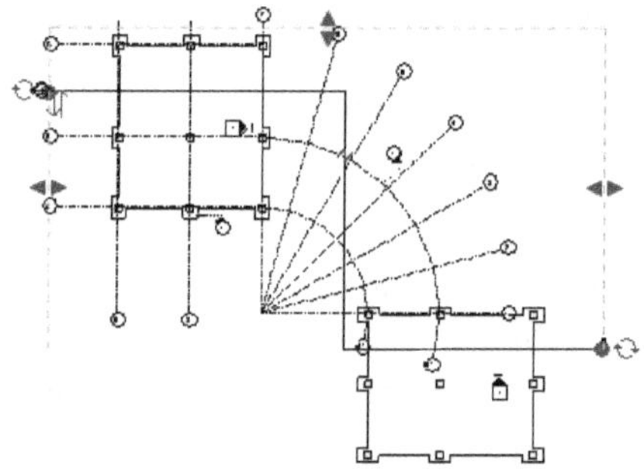

Figure 8-35 *The modified segmented section line*

10. Click on the '+' symbol on the left of the **Sections (Building Section)** head in the **Project Browser** to display the section names created.

11. Double-click on the name **Section-X** to display the corresponding section view in the drawing window, as shown in Figure 8-36.

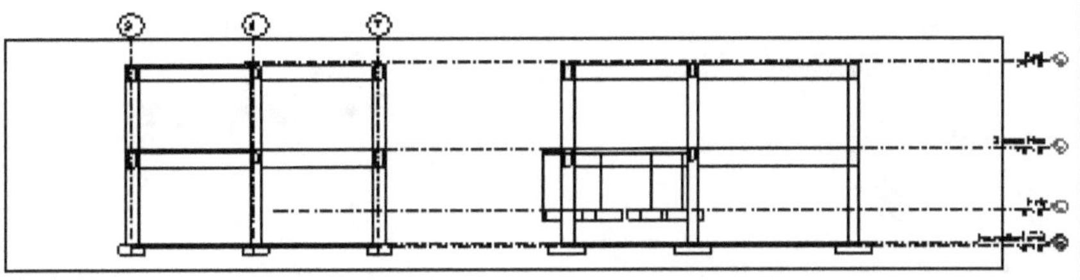

Figure 8-36 The section view displayed for the ***Section-X*** *section line*

Generating the Quantity Schedule for Beams

In this section, you will generate schedule for displaying the quantity of beams and their associated properties used in the project.

1. Invoke the **Schedule/Quantities** tool from **View > Create > Schedules** drop-down; the **New Schedule** dialog box is displayed.

2. In this dialog box, select the **Structural Framing** option from the **Category** list box and then choose the **OK** button; the **Schedule Properties** dialog box is displayed.

3. Select the **Count** option from the **Available fields** list box. Then, press and hold the CTRL key and select the following options from the list box: **Cut Length**, **Family and Type**, **Length**, **Structural Usage**, and **Volume**.

4. Release the CTRL key and choose the **Add parameter(s)** button; the selected options get added to the **Scheduled fields (in order)** list box.

5. Click on the **Family and Type** option in the **Scheduled fields (in order)** list box and click on the **Move Up** button till the option is placed at the top in the list.

6. Reorder all the options in the **Scheduled fields (in order)** list box, as shown in Figure 8-37.

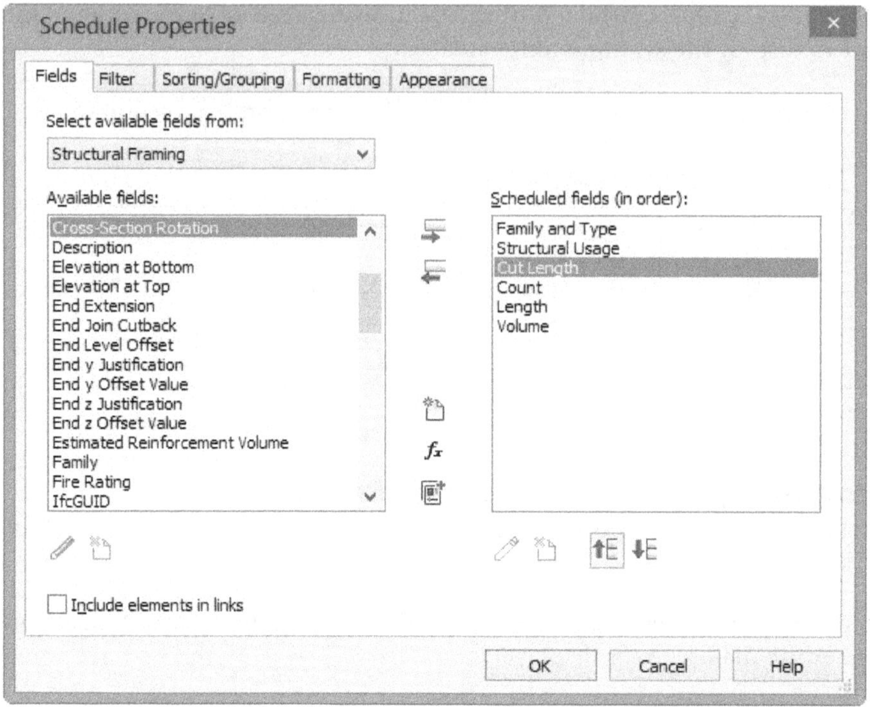

***Figure 8-37** The **Schedule Properties** dialog box for scheduling the beams*

7. Choose the **Sorting/Grouping** tab in the **Schedule Properties** dialog box and then select the **Family and Type** option from the **Sort by** drop-down list. Ensure that the **Ascending** radio button is selected.

8. Select the **Header** check box below the **Sort by** drop-down list.

9. Select the **Grand totals** check box and clear the **Itemize every instance** check box.

10. Choose the **Formatting** tab. In the **Fields** list box, select the **Count** option and then select the **Calculate totals** option from the drop-down list below the **Show conditional format on sheets check box**.

11. Repeat step 10 and select the **Calculate totals** check boxes for the following options in the **Fields** list box: **Cut Length**, **Length**, and **Volume**.

12. Choose the **OK** button; the schedule is displayed, as shown in Figure 8-38.

<Structural Framing Schedule>					
A	B	C	D	E	F
Family and Type	Structural Usage	Cut Length	Count	Length	Volume
Concrete-Rectangular Beam: 12 x 24					
Concrete-Recta	Girder	960' - 0"	48	1056' - 0"	1920.00 CF
W Shapes: W14X30					
W Shapes: W1	Girder	226' - 1 3/4"	18	245' - 3 3/4"	13.76 CF
Grand total: 66		1186' - 1 3/4"	66	1301' - 3 3/4"	1933.76 CF

Figure 8-38 *The quantity schedule for structural framing member*

Generating the Material Takeoff Schedule

In this section, you will generate the Material Takeoff Schedule for the foundation elements used in the project.

1. Invoke the **Material Takeoff** tool from **View > Create > Schedules** drop-down; the **New Material Takeoff** dialog box is displayed.

2. In the **Category** list box, select the **Structural Foundations** option. Next, choose the **OK** button; the **Material Takeoff Properties** dialog box is displayed.

3. In the **Available fields** list box, select the **Count** option. Then, press and hold the CTRL key and select the following options in the list box: **Elevation at Bottom**, **Family and Type**, **Material: Area**, **Material: Name**, and **Material: Volume**.

4. Release the CTRL key and choose the **Add parameter(s)** button; the selected options are added to the **Scheduled fields (in order)** list box.

5. Click on the **Family and Type** option in the **Scheduled fields (in order)** list box and choose the **Move Up** button till the option is placed at the top in the list.

6. Repeat step 5 to reorder all options in the **Scheduled fields (in order)** list box, refer to Figure 8-39.

7. In the **Material Takeoff Properties** dialog box, choose the **Sorting/Grouping** tab and then select the **Family and Type** option from the **Sort by** drop-down list. Ensure that the **Ascending** radio button is selected.

8. Select the **Header** check box below the **Sort by** drop-down list.

9. Select the **Grand totals** check box and ensure that the **Itemize every instance** check box is cleared.

10. Choose the **Formatting** tab, and then in the **Fields** list box, select the **Count** option. Next, select the **Calculate totals** option from the drop-down list below the **Show conditional format on sheets check box** in the **Field formatting** area.

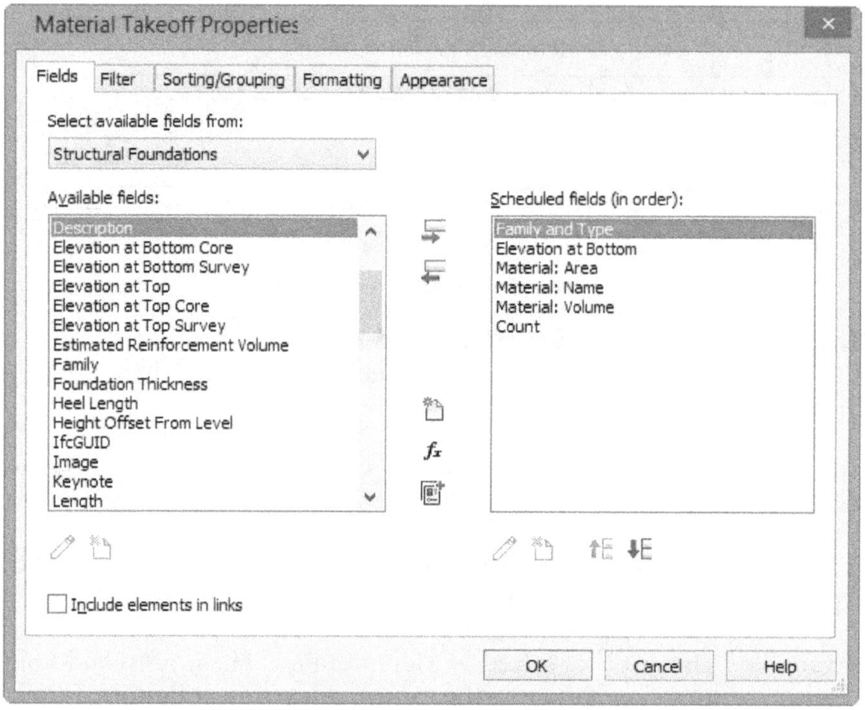

*Figure 8-39 The **Material Takeoff Properties** dialog box*

11. Repeat step 10 and select the **Calculate totals** check box corresponding to following options in the **Fields** list box: **Material: Area** and **Material: Volume**.

12. Choose the **OK** button; the Material Takeoff Schedule is displayed, as shown in Figure 8-40.

<Structural Foundation Material Takeoff>					
A	B	C	D	E	F
Family and Type	Elevation at Bottom	Material: Area	Material: Name	Material: Volume	Count
Footing-Rectangular: 72" x 48" x 18"					
Footing-Rectan	-7' - 0"	2658 SF	Concrete, Cast-in-	878.00 CF	28
Foundation Slab: 6" Foundation Slab					
Foundation Sla	-6' - 6"	4116 SF	Concrete, Cast-in-	2058.00 CF	2
Grand total: 30		6774 SF		2936.00 CF	30

Figure 8-40 The Material Takeoff Schedule for the foundations

Creating a Callout View

In this section, you will create a callout view.

1. Double-click on the **Second Floor** node under the **Structural Plans** head of the **Project Browser**; the **Second Floor** structural plan is displayed.

2. To create a callout view, invoke the **Callout** tool from the **Create** panel of the **View** tab.

3. In the **View Control Bar**, select the **1/2" = 1'-0"** option from the **Scale** drop-down list.

4. Now, draw the callout by clicking at the first and opposite corner, as shown in Figure 8-41 to create a rectangular callout.

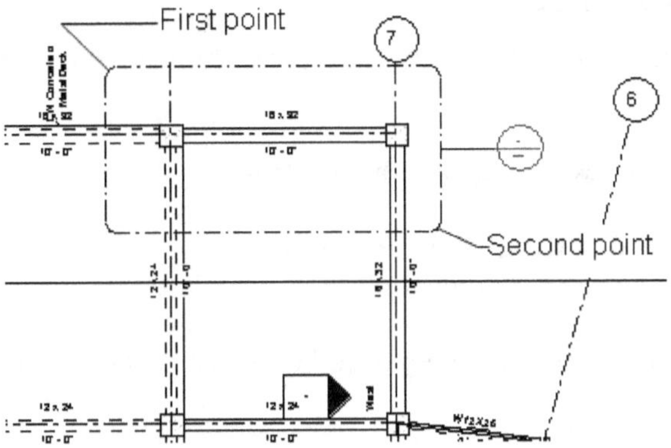

Figure 8-41 *The reference points for creating the callout*

5. After creating the rectangular callout, select it; the **Modify | Views** tab is displayed.

6. In the **Properties** palette of the selected callout, click in the value fields corresponding to the **Display Model** and **Detail Level** parameters and select the **Halftone** and **Fine** options from the drop-down lists, respectively.

7. In the **Properties** palette, click in the value field of the **View Name** parameter and enter the text **Beam -Column Detail** to replace the existing value. Choose the **Apply** button.

8. Double-click on the **Beam-Column Detail** node under the **Structural Plans** head in the **Project Browser** to display the callout view.

Note

You can sketch details in the callout view displayed by using various annotation tools available in the ***Annotate*** *tab.*

Saving the Project

In this section of the tutorial, you will save the project file using the **Save As** tool.

1. To save the project file, choose **Save As > Project** from the **Application Menu**; the **Save As** dialog box is displayed.

2. In this dialog box, browse to *C:\rst_2017\c08_rst_2017_tut* and enter **c08_Industrial-Complex_tut2** in the **File name** edit box.

3. Choose the **Save** button; the **Save As** dialog box closes and the project file is saved.

Closing the Project

1. To close the project, choose the **Close** option from the **Application Menu**.

   The file is closed. This completes Tutorial 2 of Chapter 8.

Self-Evaluation Test

Answer the following questions and then compare them to those given at the end of this chapter:

1. In Revit, you can create two types of elevations: __________ and __________ .

2. When you invoke the **Building Elevation** tool, the __________ check box in the **Options Bar** can be selected to create an elevation view associated to a grid.

3. A __________ view is used to show an enlarged view of a part of a building model that requires more detailing.

4. The __________ tool can be used to display the graphical arrangement of columns in a project.

5. You can create a duplicate dependent view by using the __________ tool.

6. You can use the __________ tool from the **Annotate** tab to add detail components to the drafting views.

7. You can use the __________ tool from the **Annotate** tab to create lines of varying thickness.

8. In Revit, you can create a section view from an elevation view. (T/F)

9. The Framing Elevation views can only be associated to grids. (T/F)

10. In Revit, you can create multiple copies of a project view. (T/F)

Review Questions

Answer the following questions:

1. Which one of the following options can be used to create details in a callout view?

   (a) **Detail Line** (b) **Filled Region**
   (c) **Detail Component** (d) All

2. The instance and type properties of the columns displayed in a Graphical Column Schedule cannot be modified within it. (T/F)

3. In Revit, you can create a segmented section using the __________ tool.

4. A section view of the **Building Section** type is displayed under the _________ head in the **Project Browser**.

5. You can select the _________ type from the **Properties** palette for the **Section** tool to create a detail section view.

6. _________ schedules display the information of the material that are used to create the structural elements in a model.

7. In the **Schedule Properties** dialog box, the _________ check box is selected to include elements from the linked Revit files into schedules.

8. _________ details cannot be referred to the existing project views.

9. You cannot create a dependent view from a callout view. (T/F)

10. In a callout view, you can control the display of elements present in other levels. (T/F)

EXERCISES

Exercise 1 Academic Institution

In this exercise, you will create the section and framing elevation views and a callout view for the *Academic Institution* project created in Exercise 1 of Chapter 7. For the locations of section line, elevation tag, and callout, refer to Figure 8-42. Also, add a callout.

(Expected time: 30 min)

1. Use the **Roof** structural plan view to create the section, elevation, and callout views.
2. File to be referred: *c07_Academic-Institution_exer1.rvt.*
3. File name to be assigned: *c08_Academic-Institution_exer1.rvt.*

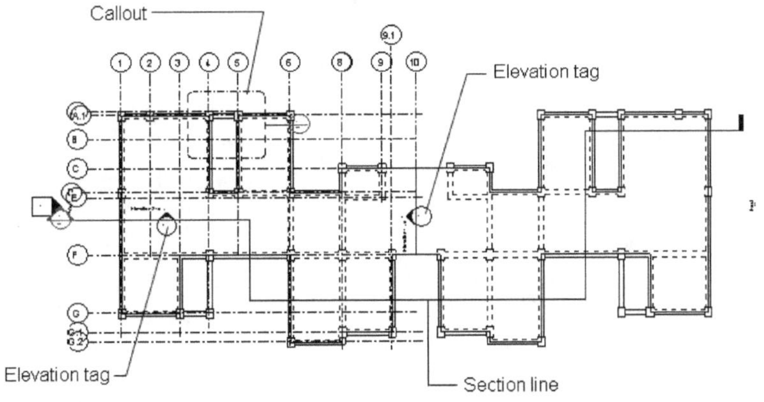

Figure 8-42 *The **Roof** level structural plan view displaying the section line, elevation tag, and callout*

Exercise 2 Factory Shed

In this exercise, you will create Quantity Schedule for beams and foundations to the *Factory Shed* project created in Exercise 2 of Chapter 7. Refer to Figures 8-43 and 8-44.

(Expected time: 30 min)

1. File to be referred: *c07_Factory-Shed_exer2.rvt*.
2. File name to be assigned: *c08_Factory-Shed_exer2.rvt*.

Structural Foundation Schedule								
Family and T	Structural Us	Area	Count	Elevation at	Height Offset	Perimeter	Volume	Width
Footing-Rectangular: 72" x 48" x 18"								
Footing-R		1014 SF	26	-13' - 0"		0' - 0"	936.00 CF	104' - 0"
Wall Foundation: Bearing Footing - 36" x 12"								
Wall Foun	Bearing	0 SF	30	0' - 0"		0' - 0"	798.00 CF	90' - 0"
Grand total: 56		1014 SF	56	-13' - 0"		0' - 0"	1734.00 CF	194' - 0"

Figure 8-43 *The structural foundation schedule*

Structural Framing Schedule				
Family and T	Reference L	Structural Us	Count	Volume
Concrete-Rectangular Beam: 12 x 24				
Concrete-			80	1796.00 CF
Grand total: 80			80	1796.00 CF

Figure 8-44 *The structural framing schedule*

Answers to Self-Evaluation Test

1. Framing, Building, **2. Attached to Grid**, **3.** Callout, **4. Graphical Column Schedule**, **5. Duplicate as Dependent**, **6. Detail Component**, **7. Detail Line**, **8.** T, **9.** T, **10.** T

Chapter 9

3D Views, Sheets, Analysis, Reinforcements, and Massing

Learning Objectives

After completing this chapter, you will be able to:

- *Create 3D views*
- *Create sheets*
- *Use analytical models*
- *Add reinforcements*
- *Create massing geometry*

In this chapter, you will learn how to generate three-dimensional views and 3D section views, add sheets, generate shadows, work with analytical models, and prepare the analytical model for analysis. You will also learn about adding reinforcements to concrete structural elements in a project.

THREE-DIMENSIONAL (3D) VIEWS

In Autodesk Revit, you can create and display two types of three-dimensional (3D) views, Orthographic 3D View and Perspective View.

Orthographic 3D View: An orthographic 3D view is a 3D view where all components are of same size regardless of how far they are from the camera. This is a default 3D view. Once this view has been displayed, you can then use various tools to modify its properties.

Perspective View: A perspective view is a real eye view. It can be created by placing a camera at eye elevation and specifying its target position.

Creating Orthographic 3D Views

Ribbon:	View > Create > 3D View drop-down > Default 3D View

An orthographic 3D view is the view in which all elements are displayed in their actual size, irrespective of their distance from the source. In Autodesk Revit, the default 3D view is an orthographic view. To view the orthographic 3D view of a building model, invoke the **Default 3D View** tool from the **Create** panel; the current view will change into the default 3D view. If you are creating the orthographic 3D view for the first time in a project, the camera is automatically placed at a default position (southeast corner) and the corresponding view is displayed in the drawing window. The default view is named as **{3D}** and is added under the **3D Views** head in the **Project Browser**. Figure 9-1 shows an example of the 3D view of the structural model of a building displayed using the **Default 3D View** tool.

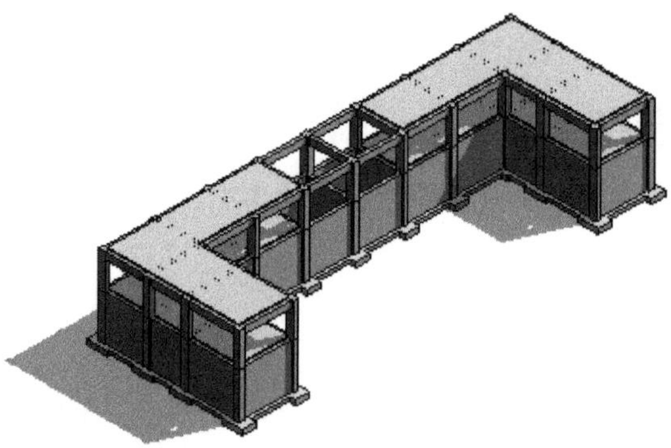

Figure 9-1 *A three-dimensional view of the structural model of a building*

Creating Perspective Views

Ribbon: View > Create > 3D View drop-down > Camera

Perspective views provide a realistic view of the exterior or interior space of a building model. Whenever you see an exterior or an interior view of a building with naked eyes, it appears as the perspective view with the wall edges of the building converging at imaginary points. You can easily create the perspective views of a building model by defining the camera location and the target point. In most building projects, you will find it easier to use a floor plan view to define these points (camera location and camera target). You can then modify the view parameters to achieve the desired result.

To create a perspective view, first open the structural floor plan, elevation, or section view of the building model. Next, in the opened view, invoke the **Camera** tool from the **3D View** drop-down of the **Create** panel; the **Options Bar** will be displayed with various related options for the camera to be placed in the view. In the **Options Bar**, the **Perspective** check box is selected by default. As a result, the camera will create a perspective 3D view. If you clear the **Perspective** check box, the camera will generate an orthographic 3D view. In the **Options Bar**, you can set the elevation of the eye point by selecting a level in the **From** drop-down list and entering a value in the **Offset** edit box. The value that you enter from the **Offset** edit box will determine the distance of the eye point from the level selected from the **From** drop-down list. For example, if the current project plan view is **Level 2** and you want to generate a camera view from the eye point located at **5'0"** for Imperial or **1524mm** for Metric above level 1, then select **Level 1** from the **From** drop-down list and enter the value **5'0"** for Imperial or **1524mm** for Metric in the **Offset** edit box in the **Options Bar**. This process should be followed before selecting the eye point.

Next, click on the desired location in the drawing area to specify the camera location. After specifying the camera location in the drawing area, you need to specify the target point. To do so, move the cursor away from the camera location; you will notice that three rays are generated from the location point, as shown in Figure 9-2.

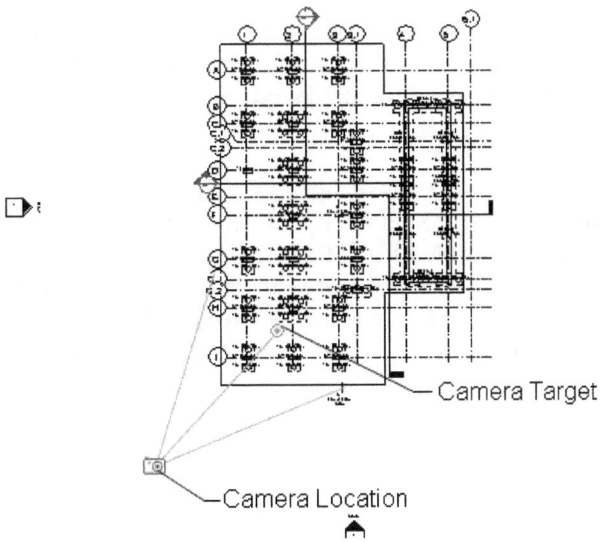

Figure 9-2 *Three rays generated from the camera location*

Note

*You can reset the position of the camera's target point to the center of the crop region. In case the camera target is displayed during modification of the camera field view, you can choose the **Reset Target** tool to center the camera target point.*

These rays depict the field of view for the selected camera point. You can use the **Offset** and **From** options in **Options Bar** to specify the level and the offset elevation of the target point. Next, move the cursor and specify the location of the target point. The target point must be placed inside or beyond the building model for the camera to capture the model. Once you have selected the target point, Autodesk Revit will automatically generate the 3D perspective view of the building model and display it in the drawing window. Next, you can use drag controls to increase the extent of the camera view. After extending the view, press ESC to exit the **Camera** tool. Figure 9-3 shows the camera view of a building model.

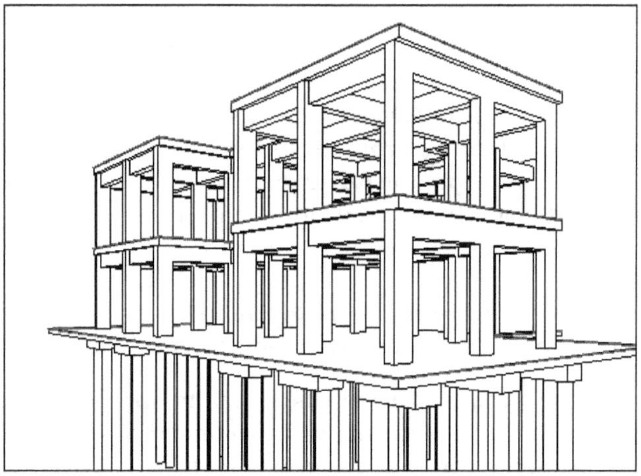

Figure 9-3 *Camera view generated from the camera placed*

The view generated will be added to the **3D Views** node of the **Project Browser** and will be named as **3D View 1**, **3D View 2**, and so on. However, you can rename them by using the **Rename** option from the shortcut menu, which is displayed on right-clicking on the newly created 3D view in the **Project Browser**.

You can also create orthographic views by using the **Camera** tool. To do so, clear the **Perspective** check box in the **Options Bar**; the **Scale** drop-down list will be activated. You can select the scale of the view from this drop-down list. Move the cursor in the drawing window; a camera will be attached to it. Also, you will be prompted to specify the location of the eye point or the point from where you want to view the model. Click to specify its location and then click again to specify the target point, an orthographic 3D view will be generated.

Modifying Perspective View Properties

After creating the perspective view, you can modify its properties to achieve the desired effect. To do so, select the perspective view and modify the instance properties in the **Properties** palette, as shown in Figure 9-4.

The instance properties of the 3D view will be displayed in the **Properties** palette. Some of the commonly used instance properties are discussed next.

In the **Properties** palette, the **View Name** instance parameter under the **Identity Data** head is used to rename a view. The **Detail Level** instance parameter is used to set the level of details required in a view.

The **Edit** button in the value column of the **Visibility/ Graphics Overrides** instance parameter is used to set the visibility of models and annotations. On choosing this button, the **Visibility/Graphics Overrides for 3D View** dialog box will be displayed. In this dialog box, you can set the visibility parameters of the model. Next, to modify the elevation of the eye point, enter the new value in the value column of the **Eye Elevation** instance parameter in the **Properties** palette. Similarly, you can enter a new value for the **Target Elevation** instance parameter in its corresponding value column. As soon as you choose the **Apply** button, the perspective view is modified in the drawing window.

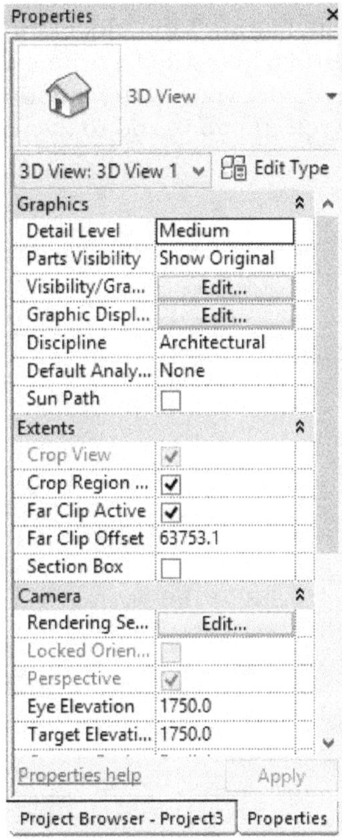

***Figure 9-4** The **Properties** palette for the camera view*

Tip

*To modify the camera and target locations, open the view in which the camera was placed. Right-click on the name of the 3D view in the **Project Browser** and then choose the **Show Camera** option from the shortcut menu displayed. On doing so, the location of the camera and target points will be displayed in the drawing window. Next, drag the camera and target points to a new location; Autodesk Revit will immediately regenerate the 3D view for the new location.*

Creating 3D Section Views

In BIM, sometimes it is difficult to interpret the typical two dimensional views as they do not completely display the design intent of a model. However, the 3D section view gives the better inside view of a model. You can create 3D section views by using the **Selection Box** tool. This tool works in all views and allow you to isolate the element(s) in the 3D view. This is different from isolating an element using Temporary Hide/Isolate mode as in this isolation process, you can isolate the elements as well as create the section views.

To create a 3D section view of a model, open 3D view. In the **Properties** palette choose the

Section Box check box under the **Extends** head, refer to Figure 9-5. The section box will be created around the model. Select the section box, the control points will be activated. These control points will determine the range of section view to be displayed. The selection box helps to enhance your documentation or to view the model. You can use these control points to resize the selection box and to display the different section views of the element, refer to Figure 9-6.

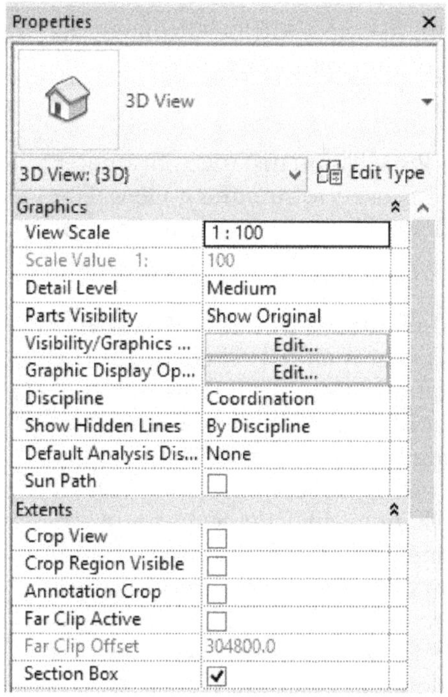

***Figure 9-5** The partial view of the **Properties** palette with **Section Box** selected*

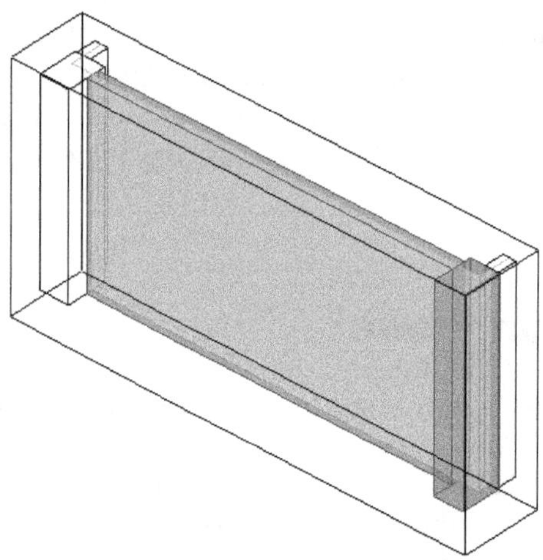

***Figure 9-6** The selection box created around the selected element*

GENERATING SHADOWS

One of the powerful features provided in Autodesk Revit is its ability to generate shadows based on location of the Sun with respect to the building. Shadows can be generated in a plan, elevation, section, or 3D view. To use this feature, open the desired view and click on the inclined arrow symbol (**Graphic Display Options**) in the **Graphic** panel of the **View** tab; the **Graphic Display Options** dialog box will be displayed, as shown in Figure 9-7. In this dialog bo x, you can specify the settings to enhance the visual output of the project view. You can use the **Style** drop-down list to select an option to specify pre-defined visual settings in the project view. In the **Style** drop-down list, if you select the **Shaded** or **Realistic** option, the **Show Edges** check box will be activated. On selecting this check box, the edges of the model will be displayed in the pre-defined visual settings. You can enhance the display of your project by using the **Smooth lines with anti-aliasing** check box. On selecting this check box, the lines and edges in the view will be displayed more smoothly. In the **Model Display** area, you can drag the **Transparency** slider to set the transparency of the entire project view. Also, you can use the **Silhouettes** drop-down list to select an option to create silhouette in the model.

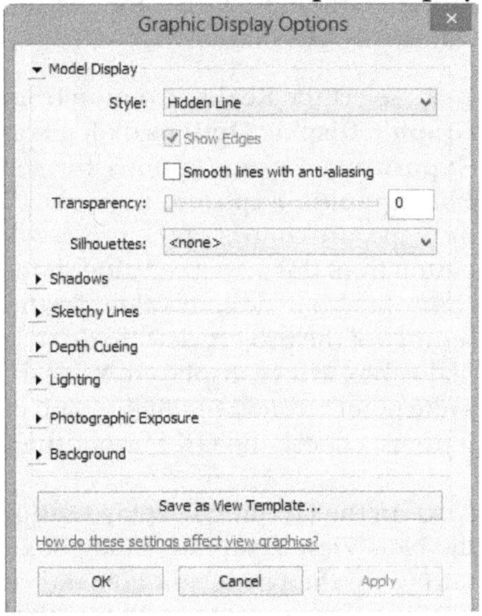

***Figure 9-7** The **Graphic Display Options** dialog box*

The options in the **Shadows** area of the **Graphic Display Options** dialog box are used to control the display of the shadow in the project view. In this area, select the **Cast Shadows** check box to cast shadow of the model. You can select the **Show Ambient Shadows** check box to display the shadow created by ambient light. By default, this check box is cleared.

The options in the **Sketchy Lines** area of the **Graphic Display Options** dialog box are used to control the appearance of sketchy lines in the current view. In this area, select the **Enable Sketchy Lines** check box to turn on sketchy lines for the current view. You can set the degree of the variability of the sketched lines using the **Jitter** slider.

In the **Lighting** area of the **Graphic Display Options** dialog box, choose the button on the right of the **Sun Setting** text; the **Sun Settings** dialog box will be displayed. By default, the **Lighting** radio button is selected in this dialog box. Select the **Still** radio button in the **Solar Study** area and then set the time and date for the solar study in the **Date** and **Time** edit boxes. To specify the location for the solar study, choose the button on the right of the **Location** edit box; the **Location Weather and Site** dialog box will be displayed. In the **Location** tab of the dialog box, select the **Default City List** option from the **Define Location by** drop-down list. On doing so, the options to set the locations will be displayed below this drop-down list. Specify the latitude and longitude in the **Latitude** and **Longitude** edit boxes for a precise solar study or select a city from the **City** drop-down list. Next, choose the **OK** button to close the **Location Weather and Site** dialog box. In the **Sun Settings** dialog box, clear the **Ground Plane at Level** check box to view the shadow pattern on the terrain. If you want the shadows to fall at a particular level or at the entry level, select the **Ground Plane at Level** check box and then select a level from the drop-down list displayed below it. Next, choose the **OK** button to return to the **Graphic Display Options** dialog box.

If you want to create a solar study based on the azimuth and altitude of the Sun from the horizontal surface, select the **Lighting** radio button in the **Sun Settings** dialog box and specify the azimuth and altitude for the solar study. After specifying the settings for the solar study, choose the **OK** button.

Next, select the **Realistic** option from the **Style** drop-down list in the **Model Display** area of the **Graphic Display Options** dialog box. Next, select the **Enabled** check box in the **Photographic Exposure** area; the options for setting the exposure for the view will be enabled. In the **Photographic Exposure** area; the **Automatic** radio button is selected by default. As a result, the exposure control of the project view will be set automatically. If you select the **Manual** radio button from this area; the **Value** slider will be activated. You can use this slider to set the exposure control to your desired value. In the **Photographic Exposure** area, you can specify the color settings of the project view. To do so, choose the **Color Correction** button; the **Color Correction** dialog box will be displayed. You can use this dialog box to specify the intensity of the shadows, white point settings, highlights, and color saturation in a project view. After specifying the desired settings, choose the **OK** button; the **Color Correction** dialog box will be closed.

Next, in the **Graphic Display Options** dialog box, choose the **Save as View Template** button; the **New View Template** dialog box will be displayed. In this dialog box, enter a name for the settings in the **Name** edit box and choose the **OK** button; the **View Templates** dialog box will be displayed. In the **View Properties** area of this dialog box, you can specify options to edit the view properties of the current setting. In this dialog box, choose the **Apply Properties** and the **OK** buttons to close it. You will notice that the shadows are displayed in the model. The shadows displayed will be based on the specified date, time, and location.

Note

*1. In the **Graphic Display Options** dialog box, the **Background** area will be displayed only for 3D views.*

*2. If the shadow does not appear, choose the **Shadow On** button from the **View Control Bar**.*

Solar Study

Solar study helps you to visualize and then analyze the position of the Sun and the solar effects on buildings and sites in the real world at a specified time, date, duration, and location. You can analyze the effects of the Sun on the same building at different locations, such as Boston or Melbourne, by changing the location of the building from Boston to Melbourne. The solar studies help you analyze the amount and impact of the natural Sunlight and shadows on a building and a site at a particular time in a day and at a particular location. It also helps you study the amount of lighting inside a building at different times of a day and year, at a particular location.

To create a still solar study, open the structural floor plan view. In the **Properties** palette of the opened view, set the value of the **Orientation** parameter to **True North** and choose the **Apply** button. Next, choose the **Rotate True North** tool from **Manage > Project Location > Position** drop-down; a rotation symbol will be displayed in the drawing area. In the **Options Bar**, specify the rotation angle in the **Angle from Project to True North** edit box. This angle will be the angle difference of the building from the project **North** direction to the **True North** direction. You can rotate the building to the **East** or **West** of the **True North** direction. In Autodesk Revit, you can choose the **Place** button from the **Options Bar** and change the point of rotation of the view to a different location in the view.

Next, after setting the True North of the project, open the 3D view of the building and adjust the view of the building as required. Choose the **View** tab and then choose the inclined arrow symbol in the **Graphics** panel; the **Graphic Display Options** dialog box will be displayed. In the **Lighting** area of this dialog box, choose the button on the right of the **Sun Setting** text; the **Sun Settings** dialog box will be displayed. In the **Solar Study** area of this dialog box, select the **Still** radio button, and in the **Settings** area, set the time and date for the solar study in the **Date** and **Time** edit boxes. To specify the location for the solar study, choose the button on the right of the **Location** edit box in the **Settings** area; the **Location Weather and Site** dialog box will be displayed. In the **Location** tab of this dialog box, select the **Default City List** option from the **Define Location by** drop-down list; the options to set the locations will be displayed below this drop-down list. Specify the latitude and longitude in the **Latitude** and **Longitude** edit boxes for a precise solar study or select a city from the **City** drop-down list. Next, choose the **OK** button to close the **Location Weather and Site** dialog box. In the **Sun Settings** dialog box, ensure that the **Ground Plane at Level** check box is cleared so that you can view the shadow pattern on the terrain. If you want the shadows to fall at a particular level or at the entry level, select the **Ground Plane at Level** check box and then select a level from the drop-down list displayed below it. Next, choose the **OK** button to return to the **Graphic Display Options** dialog box.

If you want to create a solar study based on the azimuth and altitude of the Sun from the horizontal surface, select the **Lighting** radio button in the **Sun Settings** dialog box and specify the azimuth and altitude for the solar study. After you have specified the settings for the solar study, choose the **OK** button to close the **Sun Settings** dialog box.

Next, in the **Shadows** area of the **Graphic Display Options** dialog box, you can select the **Cast Shadows** and **Show Ambient Shadows** check boxes to enable the ray trace shadows and ambient shadows in the view.

In the **Lighting** area of the **Graphic Display Options** dialog box, you can set the intensities of Sun, Ambient Light, and Shadows by using the **Sun**, **Ambient Light**, and **Shadows** sliders, respectively. You can use the **Shadows** slider to make the shadow light or dark. Next, choose the **OK** button to notice the shadows based on the specified date, time, and location. The shadow patterns generated in a still solar study for a structural model at a specified location at different times during a day are illustrated in Figures 9-8 to 9-11.

Note

Shadows cannot be generated for the following visual styles: ***Wireframe*** *and* ***Consistent Colors****.*

Figure 9-8 *After the sunrise around 6 AM*

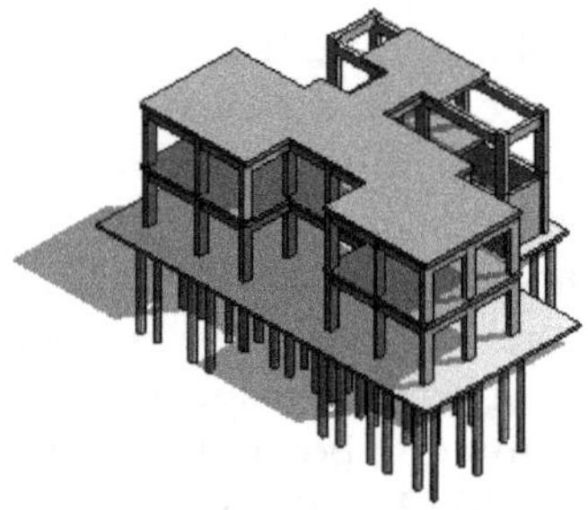

Figure 9-9 *In the morning at 9:30 AM*

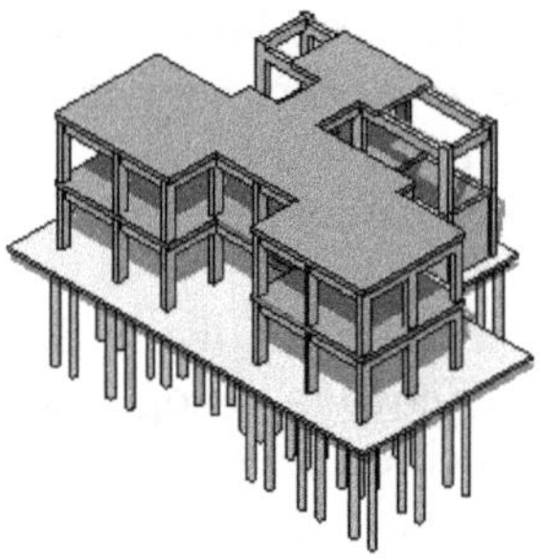

Figure 9-10 *In the afternoon at 1.25 PM*

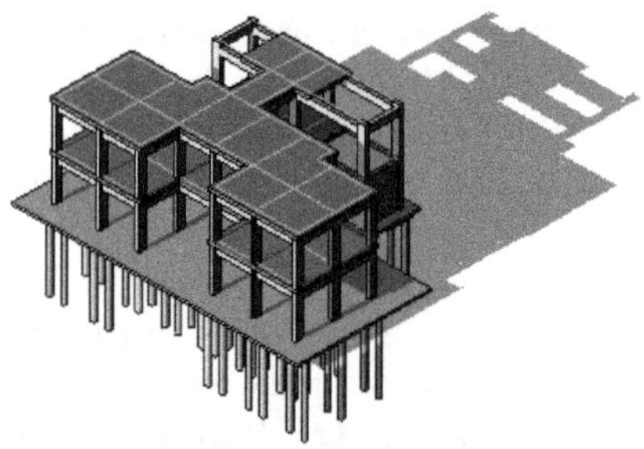

Figure 9-11 *In the evening at 4.30 PM*

In Autodesk Revit, you can see the path of the Sun for creating solar study. The Sun path is a visual representation of the range of the movement of the Sun across the sky at the geographic location specified for a project. To display the Sun path, choose the **Sun Path Off** button from the **View Control Bar** and then select the **Sun Path On** option from the flyout displayed, as shown in Figure 9-12. On doing so, the **Sun Path- Sun Not Displaye**d window will be displayed. In this window, you can choose any of the two options: **Use the specified project location, date, and time instead** and **Continue with current settings**. If you choose the **Use the specified project location, date, and time instead** option from this window, the Sun settings for the Sun path will change to the **<In-session, Still>** setting. If you choose the **Continue with current settings** option from the window, the Sun settings for the Sun path will use the **<In-session, Lighting>** setting. Next, as you choose an option from the **Sun Path- Sun Not Displayed** window, the path of the Sun along with its components are displayed in the drawing area, as shown in Figure 9-13.

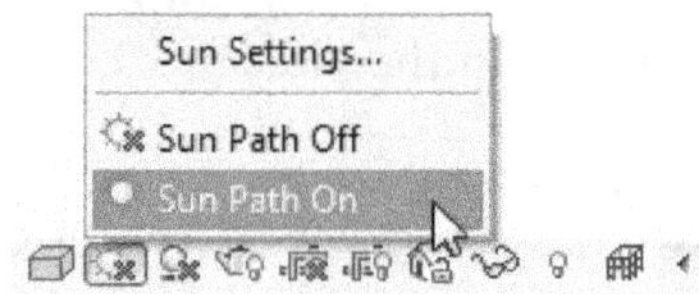

Figure 9-12 *Choosing the* ***Sun Path On*** *option from the flyout*

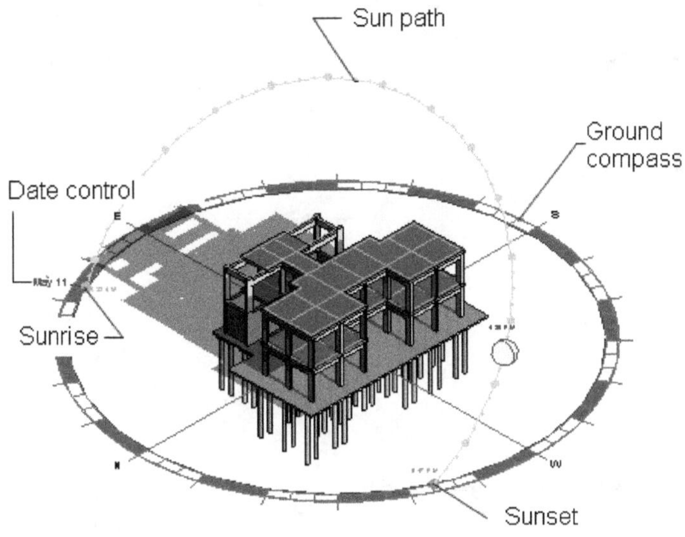

Figure 9-13 *The Sun path along with its components*

SHEETS

Ribbon:	View > Sheet Composition > Sheet

In Autodesk Revit, a sheet or a drawing sheet is a document set that is used for the final working drawings in a site. A drawing sheet contains sheet views that consist of multiple drawing views or schedules added in a project. Sheets are defined by a border and a title block. To create a drawing sheet, you first need to add a sheet view to the project and then add the required project views to the added sheet views.

Adding a Drawing Sheet to a Project

To create a drawing sheet for a project, first you need to decide its title block. In Revit, title blocks are loaded as families. A title block is used to convey project information and drawing sheet title. Project information includes client name, project name, project title, project number, date of issue of drawing, drawing sheet number, scale, and so on. The project information can be added to a project by using the **Project Information** tool, which has been discussed in Chapter 3. The type of title block will decide the size of the sheet. In Revit, there are some predefined title blocks that can be used in a project. To load these title blocks so that they can used in the sheet, choose the **Load Family** tool from the **Load from Library** panel in the **Insert** tab; the **Load Family** dialog box will be displayed, as shown in Figure 9-14. In this dialog box, browse to **US Imperial > Titleblocks** for Imperial system or **US Metric > Titleblocks** for Metric system. In this location, you can select the required title block for the project.

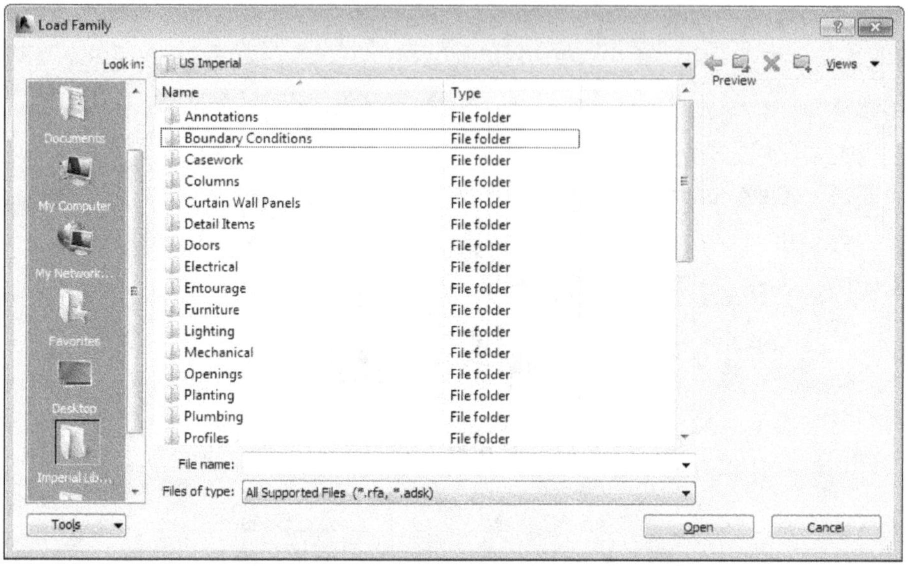

*Figure 9-14 The **Load Family** dialog box*

Note

The sizes of title blocks available in the library are based on ANSI standards for sheet size. These title blocks are available both in vertical (portrait) and horizontal (landscape) alignments.

After selecting a file for the title block from the **Load Family** dialog box, choose the **Open** button; the selected title block(s) will be loaded into the project file. Next, you need to add a drawing sheet to the project. To do so, choose the **Sheet** tool from the **Sheet Composition** panel of the **View** tab; the **New Sheet** dialog box will be displayed. This dialog box has two areas: **Select titleblocks** and **Select placeholder sheets**. Select the required title block for the drawing sheet from the list of title blocks. To load the title block that is not loaded, choose the **Load** button; the selected title block will be loaded into the project and added in the **Select titleblocks** area. On choosing the **OK** button, Autodesk Revit will create a sheet view and display the borders and the title block in the drawing window, as shown in Figure 9-15. As you add sheets to the project, their names and numbers will be displayed in the **Sheets** head in the **Project Browser**.

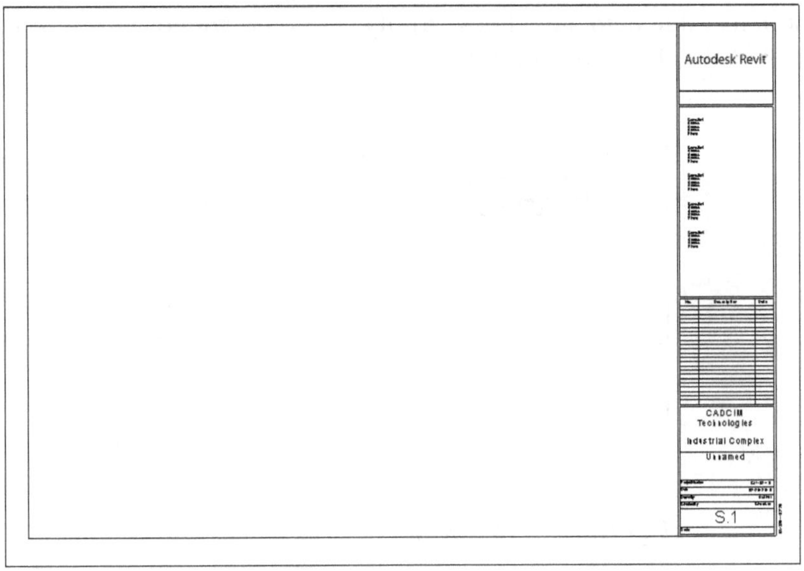

Figure 9-15 Sheet view displayed in the drawing window

Adding Views to a Drawing Sheet

After adding a drawing sheet to a project, you need to add project views to it. To do so, choose the **Place View** tool from the **Sheet Composition** panel of the **View** tab; the **Views** dialog box will be displayed, as shown in Figure 9-16.

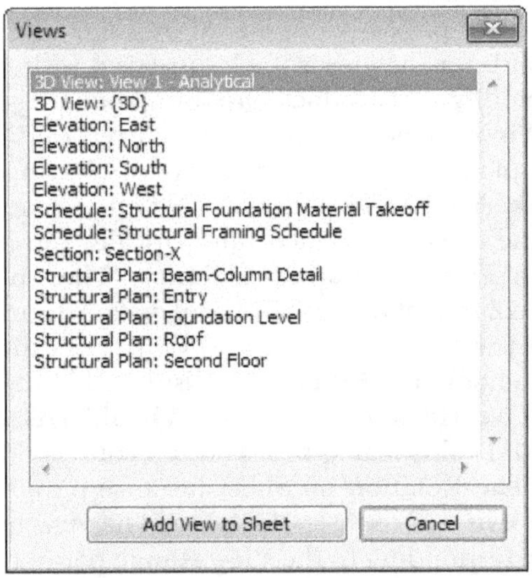

*Figure 9-16 The **Views** dialog box*

This dialog box displays the list of project views available in a project. Select the project view that you want to add to the sheet and choose the **Add View to Sheet** button; the selected view will appear as a viewport represented by a rectangle attached to the cursor. Move the cursor to the drawing area in the sheet and click at the preferred location to add the selected view. Next, in the **Properties** palette, change the **View Scale** parameter by selecting a suitable scale from the drop-down list in its value field and choose **Apply**. The view added will adjust to the specified scale and appear in the sheet, as shown in Figure 9-17.

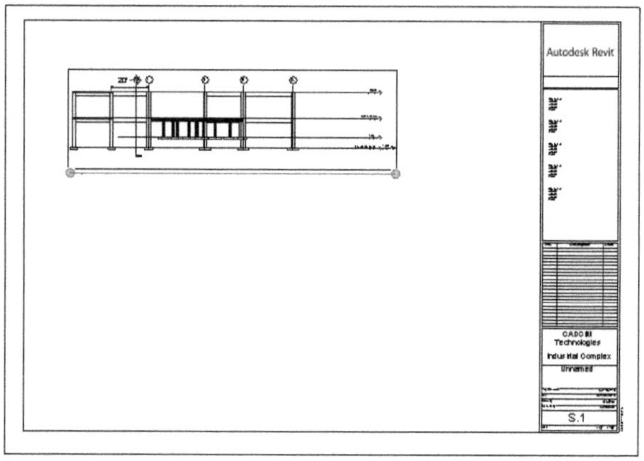

Figure 9-17 *The specified view added to the sheet*

The viewport of the inserted view will display a label line at the bottom left corner. A label line shows the view name, view scale, and view number. You can modify the properties associated with the viewport in the **Properties** palette. You can also activate the project view and then pan it to adjust the view area that you need to display in the viewport. Various options to modify the properties of the views in the viewport and to pan the view areas are discussed in the next section.

Modifying View Properties

To modify the properties of the project views added in a sheet, select the viewport associated with it; the **Properties** palette for the view will be displayed with various instance parameters associated to it. You can modify these values based on your project requirement. When you enter a new value for the **View Name** parameter, the corresponding levels and views are also renamed accordingly. You can modify the scale of a view by selecting a scale from the drop-down list in the value column of the **View Scale** parameter. When you modify the view scale of a view, the corresponding view in the viewport is scaled automatically. You can move the view to another location, if required. You can select the **Coarse**, **Medium**, or **Fine** option from the drop-down list corresponding to the value column of the **Detail Level** instance parameter to define the level of details to be displayed in a view. You can control the visibility of different categories of elements in each viewport. To do so, choose the **Edit** button displayed in the value field corresponding to the **Visibility/Graphics Overrides** parameter; the **Visibility/Graphics Overrides for Sheet** dialog box will be displayed. In this dialog box, you can control the visibility of various model and annotation elements. The **Rotation on Sheet** instance parameter in the **Options Bar** is used to rotate the view clockwise or counter-clockwise. The view range of the view can be set by using the **Edit** button corresponding to the **View Range** parameter. You can also modify the appearance of the label line. To do so, choose the **Edit Type** button in the **Properties** palette; the **Type Properties** dialog box will be displayed, as shown in Figure 9-18. In this dialog box, the visibility of the title and extension lines can be controlled by using their respective type parameters. Some other parameters of the label line such as line weight, color, and line pattern can also be modified by using this dialog box.

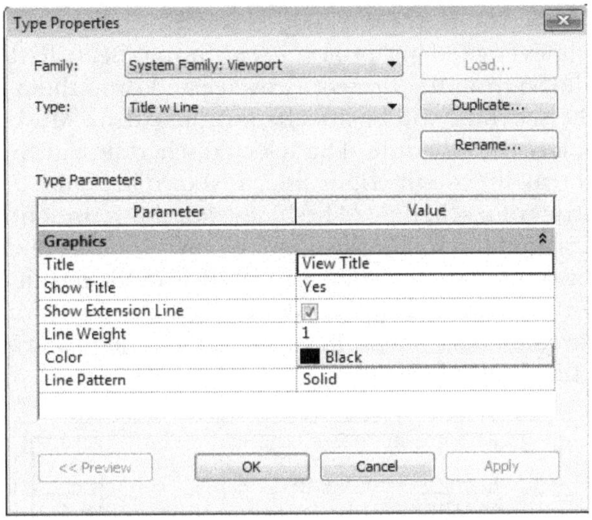

Figure 9-18 The ***Type Properties*** *dialog box*

Panning Views

After adding views and modifying view properties, you may need to pan the viewport in a suitable location in the sheet. To do so, first you need to activate the view in the viewport and then pan or move the views to a desired location in the sheet. To activate the view, select the viewport corresponding to it in the sheet and then right-click to display the shortcut menu. Next, choose the **Activate View** option from the shortcut menu; the view in the selected viewport will be activated. Again, right-click and choose the **Pan Active View** option from the shortcut menu displayed; the cursor will change into a symbol with four arrows. Left-click in the viewport, press and hold the left mouse button, and drag it to move entities in the viewport. Once you have panned and placed the entities at an appropriate location in the sheet, right-click and then choose **Deactivate View** from the shortcut menu to revert to the sheet.

Modifying a Building Model in Sheets

In sheets, sometimes you may need to make quick modifications in the elements of building model. To do so, you need to activate a view and make the desired modifications in it. To activate a view from a drawing sheet, select a viewport. Next, choose the **Activate View** button from the **Viewport** drop-down of the **Sheet Composition** panel in the **View** tab; the view will be activated. Alternatively, you can activate the selected viewport by choosing the **Activate View** button in the **Viewport** panel of the **Modify| Viewports** tab. In the activated view, you can work on the building model as in any other project views. The elements in the building model can be edited using the editing tools. When the building model is edited, the parametric change engine of Autodesk Revit modifies other project views immediately. As a result, all project views update automatically. For example, modifications made to the location of a column in an activated viewport of the drawing sheet are automatically reflected in the corresponding floor plan, sections, and other associated views. After making necessary modifications in the model, ensure that the viewport is selected and then choose the **Deactivate View** tool from **View > Sheet Composition > Viewports** drop-down list to return to the drawing sheet view. Note that new views can also be added to the project when the viewport is activated. You can also create a new plan, elevation, section, callout, detail view, and so on, while working on the activated view. The activated view can then be deactivated. Also, the newly created views can be added to the drawing sheet.

Adding Schedules to a Drawing Sheet

You can easily add schedules created in a project to a drawing sheet. To do so, drag the schedules created (required schedules) from the **Project Browser** and drop them into the drawing sheet; the preview image of the schedule will be attached to the cursor. Move the cursor and click at the desired location to place the schedule. The selected schedule will appear with blue triangles for each column and a break line on the right side, as shown in Figure 9-19. You can modify the appearance and properties of a schedule. The blue triangles representing the controls for the column width can be dragged to modify the width of each column in the schedule. The break line on the right border represents the split control. It is used to split the schedule table into multiple sections. When the split control is used, the schedule splits into two sections. These sections are placed adjacent to each other. In this way, the schedule can be split into a number of sections that fit into the drawing sheet.

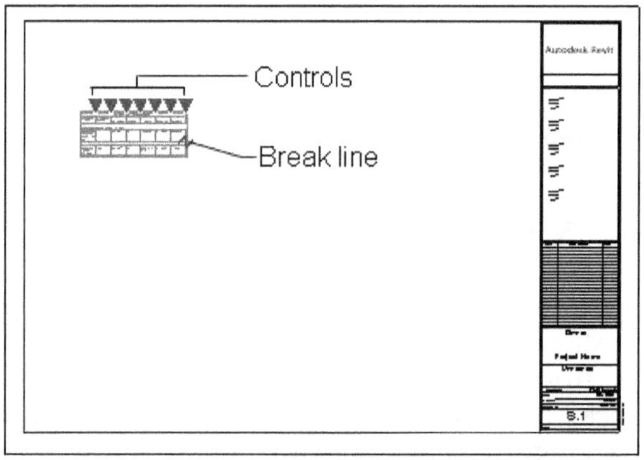

Figure 9-19 *The selected schedule with its controls*

The sections of a schedule can be moved by using the move control available at the center of the schedule section. They can also be resized by dragging the blue dot of the schedule's end. When the schedule section is resized, the additional or reduced rows are automatically adjusted between the sections. However, the last section cannot be resized, as it contains the remaining rows. The split sections of a schedule can be rejoined. As they are sequential, a section can be joined only to its previous or next section. To join a split section, drag a section by using the move control over the previous or the next section; the two schedule sections will merge into a single schedule section.

ANALYTICAL MODELS

Autodesk Revit is a multi-material (steel, precast, cast-in-place concrete, masonry, wood, and so on) modeling software for structural design, analysis, and documentation. A physical model is made up of beams, walls, columns, slabs, and footings. An analytical model is a simplified representation of the physical model in which all members are connected to each other continuously. An analytical model contains loads, load case and load combinations specific to the project being analyzed. An analytical model also includes material properties and release conditions for structural analysis.

A single structural model integrates the physical model (for layout, coordination and documentation) with an independent editable analytical model. This integration enables you to link the structural model bidirectionally to leading analysis applications such as ADAPT-Builder concrete design, RISA, Fastrak Building Designer, Autodesk Robot Structural Analysis, and other application partners. To know more about various software packages that partner with Autodesk Revit, visit *www.autodesk.com.*

Figures 9-20 and 9-21 show the 3D views of a structural model as a physical model and an analytical model, respectively. In Figure 9-21, each element in the project consists of a set of structural member analytical models. In a project, the following structural elements have structural member analytical models: Structural Columns, Structural Framing elements (such as beams and braces), Structural Floors, and Structural Walls.

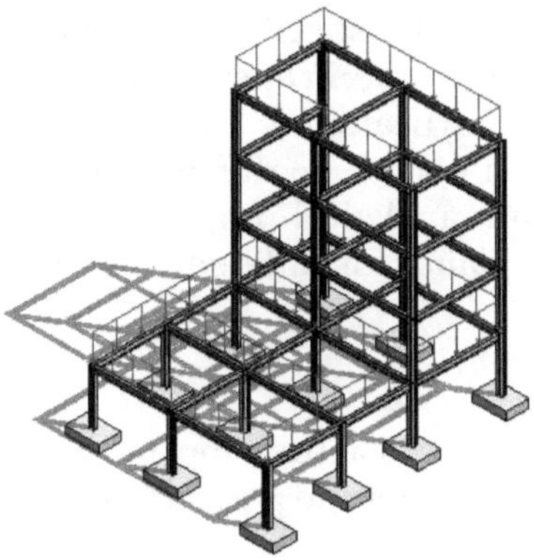

Figure 9-20 *The physical model view of a project*

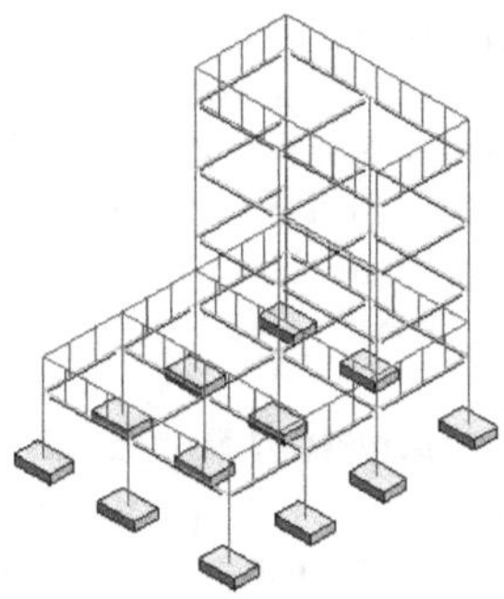

Figure 9-21 *The analytical model view of a project*

In Autodesk Revit, you can control the visibility of the analytical model in a project. You can hide the analytical model in the current view. To display the analytical model in a project view, choose the **View - Analytical** node under the **3D Views** head from the **Project Browser**. If the model is not displayed, you need to choose the **Show Analytical Model** button from the **View Control bar**, as shown in Figure 9-22.

Figure 9-22 *The **Show Analytical Model** button*

In a project, the analytical model contains various instance parameters that can be modified from its corresponding **Properties** palette. Further, you can specify various settings for analytical checks in an analytical model by using the options in the **Analytical Model Settings** tab of the **Structural Settings** dialog box. These options are discussed later in this chapter. The instance parameters and the settings for the analytical model are discussed next.

Note

*To remove the association of an analytical model, select it and then choose the **Disable Analytical Model** button from the **Analytical Model** panel of the contextual tab.*

Instance Parameters for an Analytical Model

In this section, you will learn about the instance parameters of a structural element that are required to be specified for performing an analysis. The instance parameters of structural elements such as beams, columns, floor, and walls have already been discussed in the earlier chapters. The parameters that are required for analysis are called structural analytical instance parameters. These parameters are explained under different heads: Parameters for Analytical projection planes, Parameters for physical material properties, and Parameters for release conditions for beams and columns, and Parameters for rigid links of beams and columns. These headings are discussed next.

Analytical Projection Planes and its Parameters

In Autodesk Revit, major enhancements have been made in the project planes for analytical models. For an analytical model, the projection planes can be specified by using various parameters available under the **Analytical Alignment** head in the **Properties** palette. When you add a structural element in a project, by default, its analytical model will have analytical project planes associated with it. These analytical projection planes are different based on the types of structural element added. For example, the default analytical projection plane for a column differs from that of a slab. The default projection planes for different structural types are described next.

Default Analytical Projection Planes for Columns

The default analytical projection planes for columns are placed at their top and base. These planes define the limits of the structural column analytical model and are at an offset distance from the specified level.

Default Analytical Projection Planes for Structural Walls

The default analytical projection planes for walls are placed at their central plane.

Default Analytical Projection Planes for Beams

The default analytical projection plane for a beam is parallel to its x-axis. This plane is defined at a desired level or relative to the beam section geometry.

Default Analytical Projection Planes for Structural Floors

The analytical projection plane for a structural floor is parallel to the plan level on the first placement. The subsequent structural floor placements reference the last used projection plane.

Default Analytical Projection Planes for Foundation Slabs

The top face of the foundation slab is its default analytical projection plane.

Apart from setting the default location of the project planes for each structural element type, you can set different projection planes for different structural element types. The locations of these projection planes are relative either to the levels of the or to the structural element itself. In the next sections, you will learn about various planes that need to be set for different structural element types.

Analytical Projection Planes for Columns

In Autodesk Revit, a column will have projection planes at its top and base. To modify the projection planes of a column, select its analytical model from the view; the instance properties of the selected column will be displayed in the **Properties** palette. To change the projection plane of the base of the column, click in the **Value** field corresponding to the **Base Alignment Method** parameter under the **Analytical Alignment** head and select the **Projection** option from the drop-down list displayed. Next, to align the projection plane of the column base to x, y, or z axis, select the desired options for the following parameters: **Base x-Direction Projection**, **Base y-Direction Projection**, **Base Extension Method**, and **Base z-Direction Projection**. By default, the **Location line** option will be specified for these parameters. As a result, the projections of these parameters will be based on the location line assigned for the physical model of the concerned analytical model.

To change the projection plane of the top of the column, click in the **Value** field corresponding to the **Top Alignment Method** parameter and select the **Projection** option from the drop-down list displayed. Next, to align the projection plane of the column base to x, y, and z axes, select the desired options for the following parameters: **Top x-Direction Projection**, **Top y-Direction Projection**, **Top Extension Method**, and **Top z-Direction Projection**.

Note

*In the **Properties** palette of the analytical column, the default value specified for the **Base Alignment Method**, **Top Alignment Method**, **Top Extension Method**, and **Base Extension Method** is **Auto-Detect**. On keeping this value intact for these parameters, you will allow the endpoints at the top and base of the column to adjust automatically with its neighboring structural elements in terms of its continuity and consistency of the joints.*

Analytical Projection Planes for Beams

In Autodesk Revit, a beam will have projection planes at its start point and endpoint. To modify the projection planes of a beam, select its analytical model from the view; the instance properties of the selected beam will be displayed in the **Properties** palette.

To change the projection plane of the start point of the beam, click in the value field corresponding to the **Start Alignment Method** parameter under the **Analytical Alignment** head and select the **Projection** option from the drop-down list displayed. Next, to align the projection plane of the start point of the beam to y and z axes, select the desired options for the following parameters: **Start y Projection** and **Start z Projection**. To align the projection plane of the endpoint of the beam to y and z axes, select the desired options for the following parameters: **End y Projection** and **End z Projection**.

Note

*In the **Properties** palette of the analytical beam, the default value specified for the **Start Alignment Method** and **End Alignment Method** is **Auto-detect**. If you retain the default value of these parameters, you will allow the endpoint of the beam to adjust automatically with its neighboring structural elements in terms of its join continuity and consistency.*

Analytical Projection Planes for Structural Floors

Structural floors can have any of the following options as the analytical projection plane: Top of the floor, center of the floor, bottom of the floor, level, and reference plane. To specify the analytical projection plane for the structural floor, you need to use the **Properties** palette. In the **Properties** palette, click in the value field of the **Alignment Method** parameter and select the **Projection** option from the drop-down list displayed. Next, to align the projection plane of the floor, click on the drop-down list corresponding to the **Projection** parameter and select an option from the drop-down list displayed.

Analytical Projection Planes for Structural Wall

Structural walls have projection planes at their top and bottom. The top of a structural wall can have any of the following options as the analytical projection plane: level, top of the wall, and reference plane. The bottom of a wall can have any of the following options as its analytical projection plane: level, bottom of the wall, and reference plane. To specify the analytical projection plane for the top and bottom of the wall, you need to use the **Properties** palette. In the **Properties** palette for the analytical model of the structural wall, click on the drop-down list corresponding to the **Alignment Method** parameter and select the **Projection** option from the drop-down list displayed. Next, to specify the face of the wall that will be used for locating the projection plane, click on the value field corresponding to the **Projection** parameter and select an option from the drop-down list displayed. Next, to align the projection plane of the top and base of the wall, select the desired options for the following parameters: **Top Extension Method**, **Top In-Plane Projection**, **Base Extension Method**, and **Base In-Plane Projection**.

Parameters for Physical Material Properties

All materials have specific structural properties such as density, Poisson's ratio, Young's modulus, and so on. These properties are known as the physical properties of materials. These properties are further used in structural analysis.

To assign a material to an element, you need to invoke the **Material Browser** dialog box. To invoke the **Material Browser** dialog box for a beam or a column, select any of them from the drawing area and then click in the value field corresponding to the **Structural Material** instance parameter. Next, choose the button displayed in the value field corresponding to the **Structural Material** parameter; the **Material Browser** dialog box will be displayed. For structural floors and structural walls, you can invoke the **Material Browser** dialog box from the **Edit Assembly** dialog box that has been discussed in Chapter 4 and Chapter 5.

In the **Project Material: All** area of the **Material Browser** dialog box, double-click on the desired material; the properties of the selected material will be displayed, as shown in Figure 9-23. In the **Material Editor** dialog box, you can edit various properties of the desired material such as **Graphics**, **Appearance**, **Physical**, **Identity** and **Thermal**.

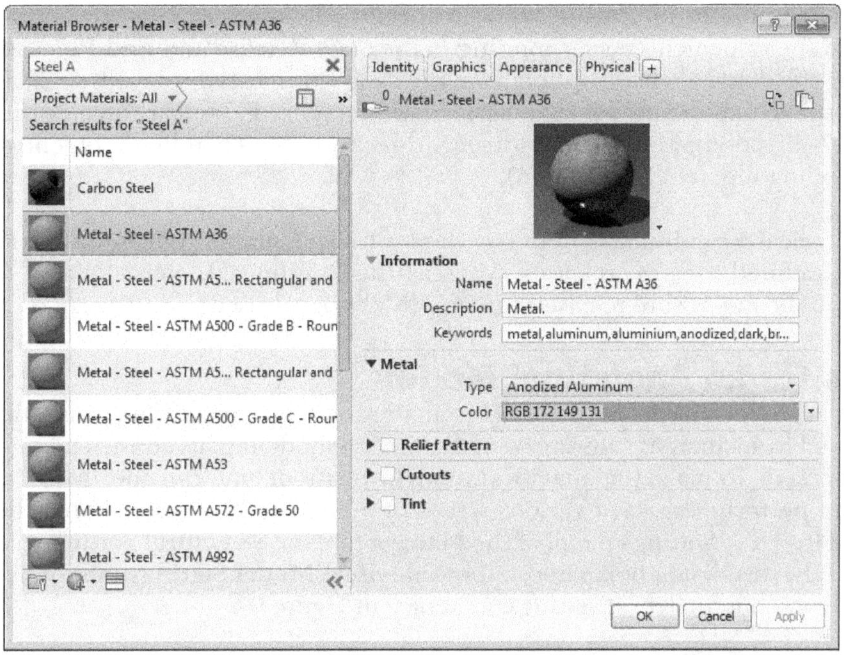

***Figure 9-23** The options in the **Material Browser** dialog box*

The **Graphics** properties of the desired material will be used to assign the shade, surface, and section patterns to it. To edit the **Graphics** properties, choose the **Graphics** tab; the **Graphic** section will be displayed along with various properties. The **Appearance** property of the material is used to assign the texture or pattern to the material for its display after rendering. The **Physical** property of the material is used to assign various intrinsic properties of a material such as its strength, mechanical behavior, and thermal characteristics. In the **Material Browser** dialog box, choose the **Physical** tab to view and edit physical properties of the selected material. The **Thermal** property of the material is used to assign various thermal characteristics of the material such as its thermal conductivity, specific heat, porosity, and so on. You can edit the **Thermal** property of the material. To do so, click on the **Thermal** tab; the **Information** and **Basic Thermal** sections with various editable parameters such as **Name**, **Description**, **Source**, **Specific heat**, **Compressibility** and more will be displayed. You can edit the properties under this section as per your requirement. After editing the properties of the desired material, choose **Apply** and then the **OK** button to save changes and close the dialog box.

Note

*All the physical and thermal properties of materials are a part of the assets defined in the **Asset Browser** dialog box.*

Parameters for Release Conditions

Release conditions are applied for linear elements such as beams and columns. These conditions do not apply to structural walls or structural floors. In Autodesk Revit, the release conditions that can be applied for beams and columns are Fixed, Pinned, Bending Moment, and User-defined. To specify the desired release conditions, select the analytical column and specify the release conditions at its top and bottom by clicking in the value fields corresponding to the **Top Release** and **Bottom Release** parameters and selecting the desired release option from the drop-down lists displayed. Similarly, for beams, you can specify release conditions at their start and end by clicking in the value fields corresponding to the **Start Release** and **End Release** parameters and selecting the desired release option from the drop-down lists displayed. A release condition is defined by six components: Fx - axial force (X direction), Fy - shear force (Y direction), Fz - shear force (Z direction), Mx - torsion force (X direction), My - bending moment (Y direction), and Mz - bending moment (Z direction).

The Pinned release condition is defined when all Moments are released, the Fixed release condition is defined with no releases, whereas the Bending Moment Release Condition is defined when My and Mz are released (torsion (Mx) not released).

Settings for an Analytical Model

The Analytical Model is useful for a project from the analysis point of view. Generally, before you take this model to an analysis software or at the time of modeling, its consistency and continuity has to be checked. To make the analytical model consistent, you can specify tolerance for the elements and perform checks at various stages of modeling. To do so, choose the **Structural Settings** tool from the **Settings** panel of the **Manage** tab; the **Structural Settings** dialog box will be displayed. In this dialog box, choose the **Analytical Model Settings** tab to display options for specifying various analytical settings, as shown in Figure 9-24.

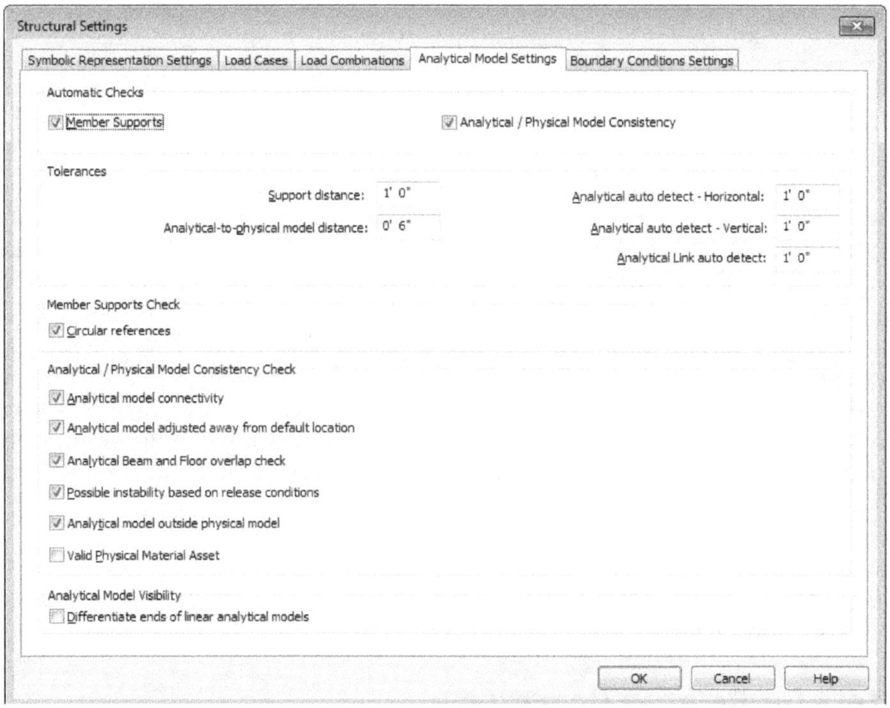

***Figure 9-24** The options in the **Analytical Model Settings** tab of the **Structural Settings** dialog box*

Alternatively, to invoke the **Structural Settings** dialog box with the **Analytical Model Settings** tab chosen, choose the **Analytical Model Settings** button located in the **Analytical Model Tools** panel of the **Analyze** tab, refer to Figure 9-25. The **Analytical Model Settings** tab consists of five areas: **Automatic Checks**, **Tolerances**, **Member Supports Check**, **Analytical/Physical Model Consistency Check**, and **Analytical Model Visibility**. These areas are discussed next.

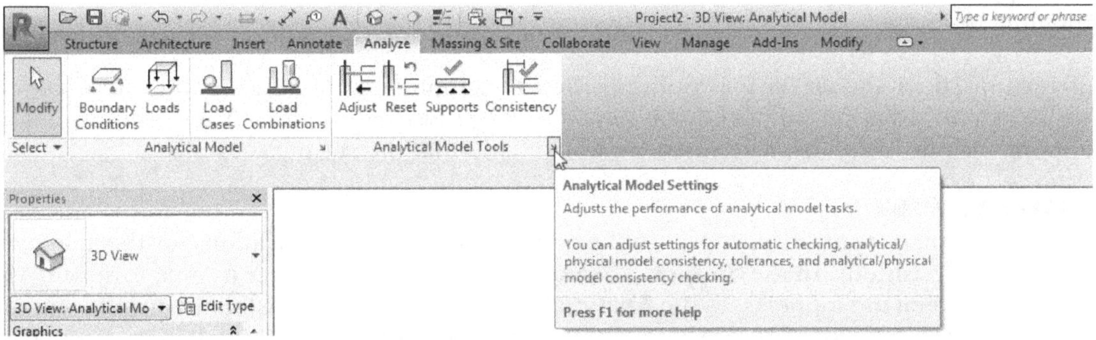

Figure 9-25 *Choosing the* ***Analytical Model Settings*** *button*

Automatic Checks Area

This area consists of two check boxes: **Member Supports** and **Analytical/Physical Model Consistency**. On selecting the **Member Supports** check box, the automatic support checking facility for all structural members will be enabled. Now, if you add a structural member in a project, the **Warning** message box will be displayed.

Note

The selection of the ***Member Supports*** *check box in the* ***Automatic Checks*** *area is useful when most of the modeling tasks for a project are completed and you want to know whether further changes in the elements of the model can affect the stability of the model.*

Next, in the **Automatic Checks** area, you can select the **Analytical/Physical Model Consistency** check box to display warning message at an early stage of the structural modeling if there is any inconsistency in the structural model. This warning message informs about the structural instability of a model and it helps you introspect the design prior to submitting them for complete analysis. Warning message is displayed at the time you add structural members to a project.

However, there are certain rules that must be followed while adding elements to your project. These rules will help you to avoid instability in the structural model so that the model can be easily taken to the analytical software for analysis. The rules that are applicable for structural columns, structural walls, beams, and floors are as follows:

Tip

It is not recommended to select the ***Member Supports*** *check box in the early stage of structural modeling because the number of unsupported elements in the early stage of the project is more and therefore the modeling process may become time consuming.*

Structural column should have a point support at least on one of its ends. The valid support can include any of the following: another column (above or below), isolated or continuous foundations, beams, walls, or structural floors.

Structural wall should be supported by one line support or two-point supports. A valid support can be any of the following: Foundations (continuous or isolated), beams, or structural walls.

Beam should either have two-point supports one each located at the either ends or one point support located at one end with the release condition set to fixed. A valid support to the beam can include intersecting columns, foundations (continuous or isolated) adjoining beams, or walls.

A structural floor should have any of the following supports: at least two-point supports, one line support and a point support that is non-collinear, or two non-collinear line supports. Valid supports include: columns, continuous or isolated foundations, beams, or walls.

Tolerance Area

In this area, you can specify distances in the edit boxes to meet specifications for the analytical model to be consistent. In the **Support distance** edit box, you can enter a value to specify the distances between the supports. In the **Analytical-to-physical model distance** edit box, you can enter a value to specify the distance between an analytical and a physical model. In the **Analytical auto detect - Horizontal** edit box, you can enter a value to specify the distance between the analytical and physical models for horizontal elements. In the **Analytical auto detect - Vertical** edit box, you can enter a value to specify the distance between the analytical and physical models for vertical elements. In the **Analytical Link auto detect** edit box, you can enter a value to specify distance by which an automatic analytical link will be created in 3D space. These analytical links provide the rigidity to the analytical model without adding physical geometry.

Member Supports Check Area

In this area, the **Circular references** check box is selected by default. As a result, circular support conditions in members are checked and identified.

Analytical/Physical Model Consistency Check Area

In this area, all check boxes are selected by default. This enables consistency checks for various specifications related to an analytical model with a physical model.

Analytical Model Visibility

In this area, the **Differentiate ends of linear analytical models** check box remains selected by default. As a result, the color displayed on the ends of the analytical model will be different. If this check box is clear, the color will be uniform throughout for all analytical models.

Applying Loads in Analytical Model

Before you take the analytical model to another software for analysis, it is important to apply various loads to the structural model. These loads can then be classified into various load cases and combined for analysis and design.

In a structural model, you can apply point, line, and area loads. Each of these loads contains instance and type parameters that can be modified before or after placing them in the structural model. You can apply these loads either by sketching or by using host components, such as floors and beams. You can edit load force and moment parameters before or after placing them in the structural model. You can also create load cases and combinations that help you classify loads on the basis of their nature and then combine them based on the design code of practices for analysis.

In the forthcoming sections, you will learn the methods of viewing, adding, and editing load cases, combining load cases, and placing loads in a model.

Load Cases

Ribbon: Analyze > Loads > Load Cases

Load cases are different types of loads that can be applied to a structural model. Load types can be dead, live, wind, seismic, and others, depending on the nature of the load applied to a structural model.

In Autodesk Revit, you need to apply a load case in the analytical model. To view, add, and edit load case in a structural model, choose the **Load Cases** tool from the **Analytical Model** panel of the **Analyze** tab; the **Structural Settings** dialog box chosen will be displayed with the **Load Cases** tab. In this tab, various load cases and options to add load cases and natures will be displayed, as shown in Figure 9-26.

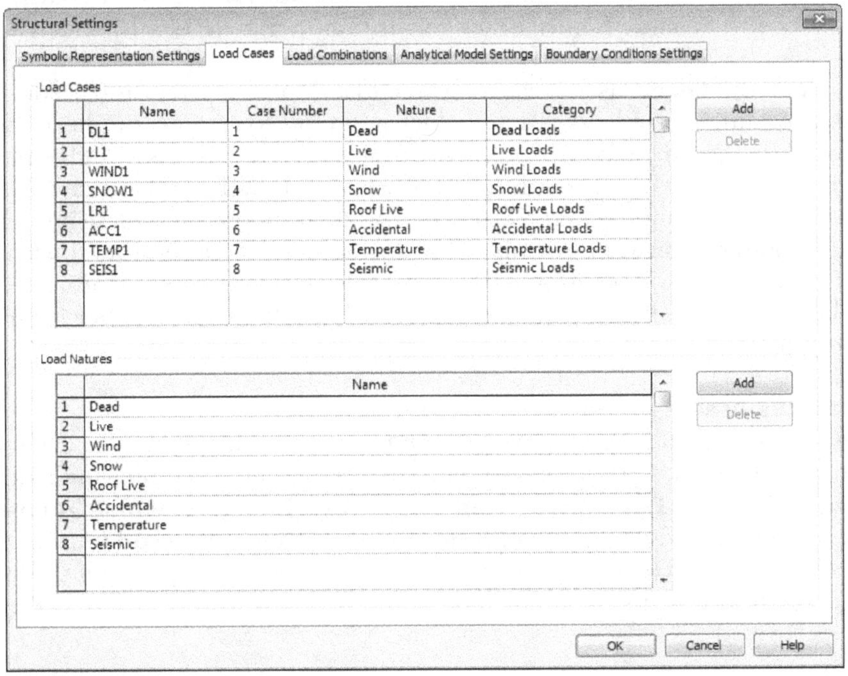

Figure 9-26 *The options in the* ***Load Cases*** *tab of the* ***Structural Settings*** *dialog box*

There are two areas in the **Load Cases** tab: **Load Cases** and **Load Natures**. The options in these areas are discussed next.

Load Cases Area

This area displays a table that contains four columns: **Name**, **Case Number**, **Nature**, and **Category**; and eight rows, each row specifying a load case. For example, row 1 specifies a dead load case, refer to Figure 9-26. In this row, the name of the case in the value field of the **Name** column is specified as **DL1**. You can change the name by clicking in the value field and entering a text. In this table, load case number, nature of load, and load category are specified in the value field of their respective columns. You can change existing values from the value fields by entering a value or selecting them from the drop-down list, wherever required.

The value displayed in the value field of the **Case Number** column for a particular load case is unique and read-only. It represents load case with a numeric value. This value is important for tracing load case at the time of analysis.

The value displayed in the value field of the **Nature** column for a particular load case specifies the nature of the load case. To change the value displayed in the value field, click in the value field and select any of the following options displayed in the drop-down list: **Dead**, **Live**, **Wind**, **Snow**, **Roof Live**, **Accidental**, **Temperature**, and **Seismic**. You can add more options in this drop-down list by choosing the **Add** button in the **Load Natures** area of the **Load Cases** tab. This will be discussed in the next section. In the value field of the **Category** column for a particular load case, you can specify the category of the load case. To do so, click in the value field and select an option from the drop-down list displayed. The options displayed in the drop-down list are: **Accidental Loads**, **Dead Loads**, **Live Loads**, **Roof Live Loads**, **Seismic Loads**, **Snow Loads**, **Temperature Loads**, and **Wind Loads**.

In the table of the **Load Cases** area, you can add a duplicate load case. To do so, select the row corresponding to the load case for which you want to create a duplicate load case and then choose the **Duplicate** button; a copy of the selected load case will be added in a row below the selected row corresponding to a load case. To delete a load case from the table, select the row corresponding to the load case and then choose the **Delete** button; the **Delete Load Case** message box will be displayed. If you choose the **Yes** button, the load case along with its load elements will be deleted from the table as well as from the project. On choosing the **No** button, the load case and its elements will not be deleted either from the table or from the project.

Load Natures Area

This area displays a table that contains the **Name** column and eight rows. Each row displays the nature of the load that can be assigned to a load case. The following load natures are displayed in the table: **Dead**, **Live**, **Wind**, **Snow**, **Roof Live**, **Accidental**, **Temperature**, and **Seismic**. To add a load nature or a row to the table, choose the **Add** button; a row will be added below the last row in the table and it will also have a default name. You can change the name by clicking in its value field and entering a new name.

Placing Loads

In a structural model, you can place any of the three types of load: Point, Line, and Area. You can either place these loads by sketching them in the view or by associating them with a host element such as a floor or a beam.

While placing a load in a structural model, Revit uses any of the following coordinate systems for loads: Project coordinate system, current work plane, and work plane of the host element. The project coordinate system refers to the coordinate system of a project. The current work plane is the plane for placing objects. When the current work plane is used to place loads, loads are placed perpendicular to the current work plane. The host work plane is the plane in which the element hosting a load resides. In the next section, you will learn how the three types of load are placed in the structural model.

Placing Point Load

You can place a point load in a structural model by sketching or by associating it with a host element. These methods are discussed next.

Sketching a Point Load: To sketch a point load in a model, open the structural plan level - analytical view with a framing plan model. Next, choose the **Loads** tool from the **Analytical Model** panel of the **Analyze** tab; the **Modify | Place Loads** tab will be displayed. In this tab, ensure that the **Point Load** tool is chosen. Next, select an option from the **Placement Plane** drop-down list in the **Options Bar** to specify the level or plane in which the load will be placed. If the level selected is different or not the current level, a **Warning** message box will be displayed immediately after placing loads, informing that the load placed will not be displayed in the current view. After selecting an option from the **Options Bar**, you can use the **Properties** palette to specify the instance and type parameters of the point load. In the **Properties** palette, click in the value field corresponding to the **Load Case** parameter and select an option from the drop-down list displayed; the selected option will specify the load case to which this load belongs.

Next, click in the value field corresponding to the **Orient to** parameter and select any of the following options: **Project** and **Workplane**. The selection of any of the options will determine the orientation of the coordinated system. If you select the **Project** option, the coordinate system followed for specifying loads will be oriented on the basis of the project coordinate system. Similarly, if you select the **Workplane** option from the drop-down list, the coordinate system for placing the loads will be oriented based on the coordinate system of the current work plane.

In the **Fx**, **Fy**, **Fz**, **Mx**, **My**, and **Mz** parameters, enter a suitable value in their corresponding value fields to specify the magnitude and directions of the concentrated forces and moments associated with the load definition.

After specifying the instance parameters of the point load, you can edit its type parameters. To do so, choose the **Edit Type** button; the **Type Properties** dialog box will be displayed. In this dialog box, you can use the **Force scale** and **Moment scale** parameters to change the view scale of the forces and moments displayed in the model. To change the scale of these parameters, enter the required values in their respective **Value** fields. In the **Type Properties** dialog box, you can also change the type of arrow heads used for displaying loads (forces and moments) in the model. To do so, you can use the following parameters: **Force arrowhead**, **Moment arrowhead**, and **Moment arrowhead (alternate)**. To change values for these parameters, click in their respective **Value** fields and select an option from the drop-down list displayed. After specifying the type parameters in the **Type Properties** dialog box, choose the **Apply** button and then the **OK** button to apply the changes and close the **Type Properties** dialog box.

You will notice that an arrow is attached to the cursor in the view, as shown in Figure 9-27. This arrow represents the load to be placed. Next, move the arrow and place it at the required position. To add more point loads, keep on clicking on the desired position to place them. Note that while adding point loads you can keep the same parameters or change them by using the **Properties** palette and after placing the loads, press ESC twice.

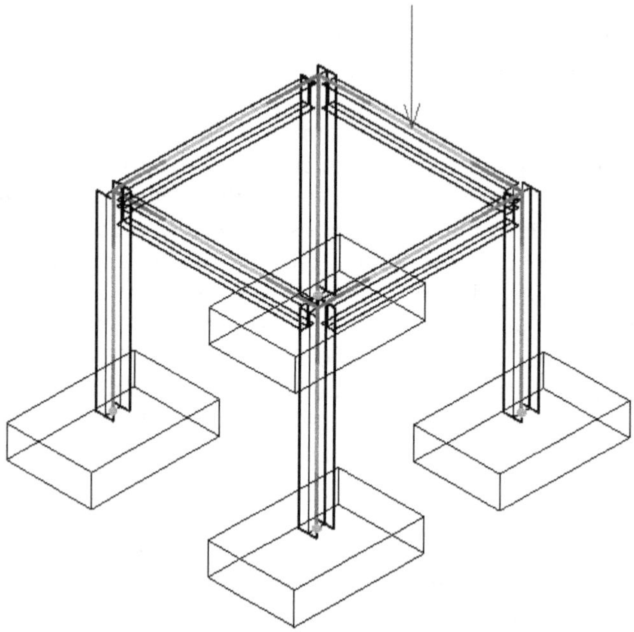

Figure 9-27 The coordinate system symbol and a point load applied in the Y direction

Placing Hosted Point Load: A hosted point load is a concentrated load that is associated with a beam, brace, or column. You can place the hosted point load in an analytical view (structural plan or 3D view). The physical geometry of a model does not accept the load. The hosted point loads are placed at the endpoints of beams, columns, and braces. To place the hosted point load, open an analytical view and then choose the **Loads** tool from the **Loads** panel of the **Analyze** tab. In the **Properties** palette, specify the desired parameters; you will notice that a placing symbol appears with the cursor in the drawing window. Place the cursor on one of the ends of a beam, column, or brace. Click when the desired member is highlighted; the load will be added and associated to the member.

Placing Line Load

Like point loads, you can sketch a line load or associate it with a structural member. The methods for placing the line loads are discussed next.

Sketching Line Load: To sketch a line load, open the structural plan level - analytical view with a framing plan model. Next, choose the **Loads** tool from the **Analytical Model** panel of the **Analyze** tab; the **Modify | Place Loads** tab will be displayed. In this tab, choose the **Line Load** tool from the **Loads** panel. In the **Options Bar**, select an option from the **Placement Plane** drop-down list to specify the level of placement of the line load. In the **Options Bar**, select the **Chain** check box. If this check box is selected, you can add a continuous line load in the model. To add a line load with a break, clear the **Chain** check box.

Next, you can use the **Properties** palette to assign the instance and type parameters to the line load. The use of the **Load Case** and **Orient to** parameters is the same as that used for placing point loads. In the **Properties** palette, the check box displayed in the value field of the **Uniform Load** parameter is selected by default. As a result, the line load will be uniform from start to end. Notice that values for the **Fx 1**, **Mx 1**, **Fy 1**, **My 1**, **Fz 1**, and **Mz 1** parameters can be entered only when the **Uniform Load** check box is selected in the **Properties** palette. The **Fx 2**, **Mx 2**, **Fy 2**, **My 2**, **Fz 2**, and **Mz 2** parameters are disabled for entering values. These parameters will be enabled for entering values when the check box corresponding to the **Uniform Load** will be cleared. Next, you can enter two values of a line load in a particular direction. For example, if the values entered for the **Fz 1** parameter is **-15 kip/ft (-219 kN/m)** and for the **Fz 2** parameter is **-20 kip/ft (-292 kN/m)**, then the line load will start with a load of **-15 kip/ft (-219 kN/m)** and end with a load of **-20 kip/ft (-292 kN/m)**.

You can modify the type properties of the line load. To do so, choose the **Edit Type** button from the **Properties** palette; the **Type Properties** dialog box will be displayed. To change the view scale for the linear force or moment added in the structural model, you can use the **Linear force scale** parameter. To change the distance between the arrows displayed in the loads, you can use the **Distance between arrows** parameter. The usage of other parameters in the **Type Properties** dialog box is the same as that for point loads. After specifying type properties, choose the **OK** button; the **Type Properties** dialog box will be closed. Now, in the project view, move the cursor and click at a desired point to mark the start point of the line load. Move the cursor in the direction of the line load, refer to Figure 9-28; you will notice a magenta-colored line emerging from the start point. At certain points in this line, you will notice symbols of arrows. The number of arrows in this line depends on the value specified for the **Distance between arrows** parameter in the **Type Properties** dialog box. Click again to mark the endpoint of the line load. Next, you can continue adding more line loads or press ESC twice to exit the **Line** load tool.

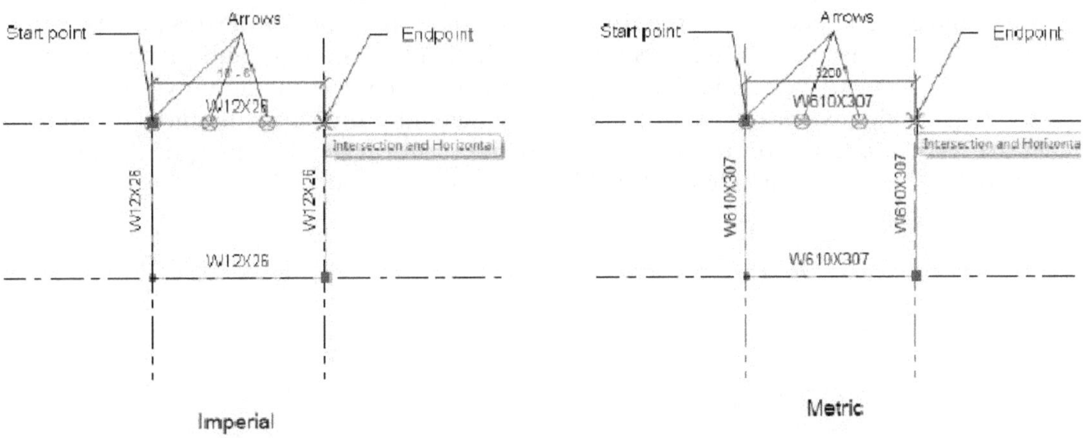

***Figure 9-28** The line load placed in a structural member*

Placing Hosted Line Load: A hosted line load is a linear load that is associated with a beam, brace, or column. You can place the hosted line loads in an analytical view (structural plan or 3D view). The physical geometry of a model does not accept load. The hosted line loads are placed throughout the length of the beams, columns, or braces. To place a hosted line load, open an analytical view and choose the **Loads** tool from the **Analytical Model** panel of the **Analyze** tab. Next, choose the **Hosted Line Load** tool from the **Loads** panel. In the **Properties** palette, enter the desired parameters. In the drawing window, notice that a placing symbol appears with the cursor. Place the cursor on a beam, column, or brace. Click when the desired member is highlighted; the load will be added throughout the length of the highlighted member and will be associated to it.

Placing Area Load

You can sketch an area load or associate it with a structural member. Area loads are applied to a structural model to represent a thrust or pressure force on the members. The methods for placing area loads in a structural model are discussed next.

Sketch an Area Load: To sketch an Area load, open the structural plan level - analytical view with a framing plan model. Next, choose the **Loads** tool from the **Loads** panel of the **Analyze** tab; the **Modify | Place Loads** tab will be displayed. In this tab, choose the **Area Load** tool from the **Loads** panel; the **Modify | Create Area Load Boundary** tab will be displayed. In this tab, you can use various sketching tools from the **Draw** panel. By default, the **Line** tool is chosen. You can use other tools depending on the requirement of the boundary. After choosing an appropriate tool from the **Draw** panel in the **Options Bar**. You can enter a suitable offset distance for the sketched boundary in the **Offset** edit box. In the **Options Bar**, the **Chain** check box is selected for the **Line** tool. As a result, you can draw a continuous chain of the boundary of the area load using the **Line** tool.

After sketching the boundary line in the **Modify | Create Area Load Boundary** tab, you can use the **Reference Point** tool to create sloped area loads by specifying load values for up to 3 vertices of an area load. These vertices refer to the forces F1 (Fx 1, Fy 1, and Fz 1), F2 (Fx 2, Fy 2, and Fz 2), and F3 (Fx 3, Fy 3, and Fz 3) parameters found in the properties of the area load in the **Properties** palette. These parameters will not be enabled until the reference points have been selected with the **Reference Point** tool. In the **Properties** palette, you can assign various instance parameters related to the load magnitudes and plane orientation. In the project view, sketch the closed boundary using the sketching tools in the **Draw** panel to represent the extent of the area load. After sketching the closed boundary, choose the **Finish Edit Mode** button from the **Mode** panel to finish the sketching of the area load. Figure 9-29 shows the area load placed in the structural model.

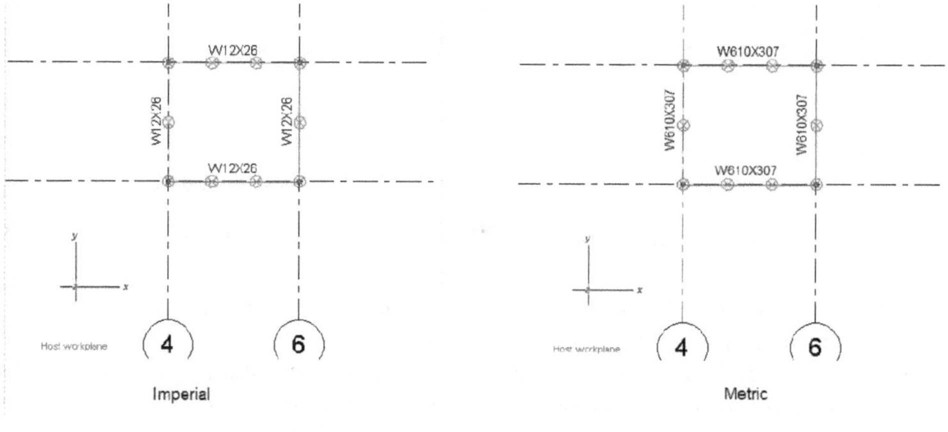

Figure 9-29 *The area load placed in the structural model*

Placing Hosted Area Load: A hosted area load is a pressure load that is associated with a structural floor or a structural wall. You can place an area load in a structural floor or a structural wall. You can place hosted area loads in an analytical view (structural plan or 3D view). The physical geometry of the model does not accept the load. To place an area load, open an analytical view suitable for placing loads and then choose the **Loads** tool from the **Loads** panel; the **Modify | Place Loads** tab will be displayed. In this tab, invoke the **Hosted Area Load** tool from the **Loads** panel. Next, in the **Properties** palette, you can specify the orientation plane, magnitude and direction of forces, and load case. In the **Properties** palette, you can choose the **Edit Type** button to invoke the **Type Properties** dialog box. In this dialog box, you can specify the type parameter for the area load such as the **Area force scale**, **Force arrowhead**, and **Distance between arrows** in their respective **Value** fields. Now, choose the **OK** button; the **Type Properties** dialog box will close. Now, place the cursor at any of the edges of the structural floor or wall and then click when the edge gets highlighted. On doing so, the area load will be added around the perimeter of the structural floor or wall. Figure 9-30 displays a hosted area load placed in a structural floor.

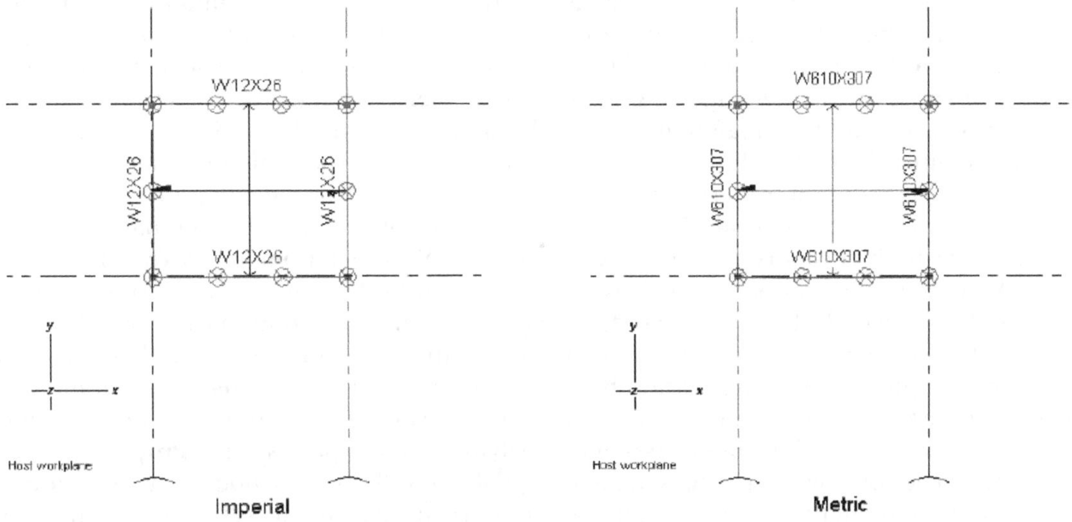

***Figure 9-30** The hosted area load placed in a structural floor*

Note

In Revit 2017, the area hosted load can be placed on the curve walls as well.

Setting Boundary Conditions

Ribbon:	Analyze > Boundary Conditions > Boundary Conditions

In an analytical model, boundary conditions are model elements that define the support conditions of a structural element (beams, columns, or floors). These elements are used to add the assumed support conditions to structural elements so that they can be communicated easily to the analysis software to which a model will be exported for analysis.

To specify a boundary condition for a structural model, open an analytical structural plan view and then choose the **Boundary Conditions** tool from the **Analytical Model** panel of the **Analyze** tab; the **Modify | Place Boundary Conditions** tab will be displayed, as shown in Figure 9-31. The **Boundary Conditions** panel of this tab is used to add different types of boundary conditions. Various tools for setting the boundary conditions are discussed next.

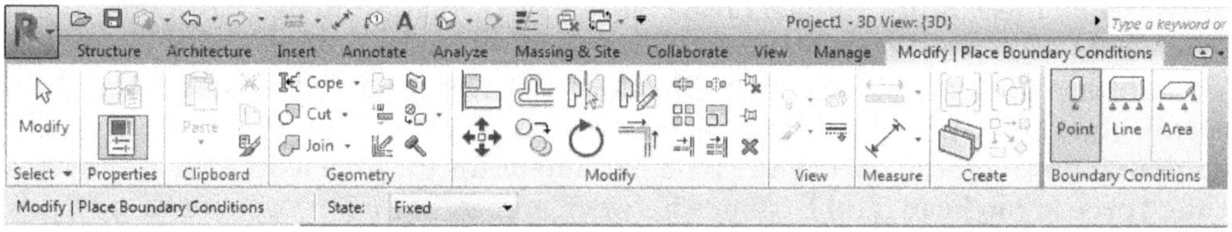

Figure 9-31 Various options in the ***Modify / Place Boundary Condition****s tab*

Area Tool

The **Area** tool is used to add boundary conditions along the perimeter of a structural floor, a foundation slab, or a structural wall. You can invoke the **Area** tool from the **Boundary Conditions** panel. After invoking the **Area** tool, you can select an option from the **State** drop-down list in the **Options Bar** to specify support conditions. The options available in the **State** drop-down list are **Pinned** and **User**. By default, the **Pinned** option is selected in the **State** drop-down list. As a result, when you add the area boundary condition to a structural floor, slab or wall, a pinned support condition is created throughout the perimeter of the structural floor, slab, or wall. You can select the **User** option from the **State** drop-down list, if you want to add a specific boundary condition that is specific to a site. On selecting the **User** option, the following parameters will become active in the **Properties** palette: **X Translation**, **Y Translation**, and **Z Translation**. You can click in the value field of any of these parameters and select any of the following options from the drop-down list displayed: **Fixed**, **Release**, or **Spring**. The option that you select from the drop-down list will specify the support conditions of the structural element in their respective axes. For example, if the **Spring** option has been selected for the **X Translation** parameter, it means that the support in the element along its X axis will be a Spring support. After selecting a state of the support from the **Options Bar** and specifying various options in the **Properties** palette (if required), click on any of the edges of the floor, slab, or wall; the area boundary condition will be applied throughout the perimeter of the selected structural element, as shown in Figure 9-32.

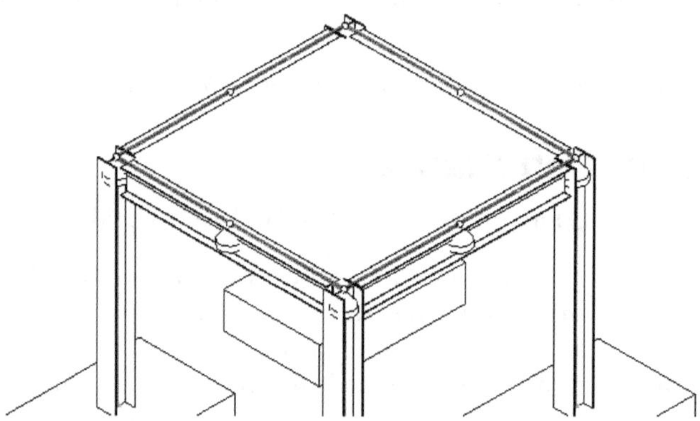

Figure 9-32 *The area boundary condition applied to a structural floor*

Point Tool

Ribbon:	Modify | Place Boundary Conditions > Boundary Conditions > Point

You can use the **Point** tool from the **Boundary Conditions** panel to create boundary condition at the end of a beam, column, or brace. To use the **Point** tool, invoke it from the **Boundary Conditions** panel and then select an option from the **State** drop-down list in the **Options Bar**. The options available in the **State** drop-down list are **Pinned**, **Fixed**, **Roller**, and **User**. By default, the **Fixed** option is selected in the **State** drop-down list. As a result, when you add boundary condition to the end of a beam, column, or brace, a fixed support condition is created at that end. You can select any of the remaining options from the **State** drop-down list based on your design requirement. On selecting the **User** option from the **State** drop-down list in the **Options Bar**, the following parameters in the **Properties** palette will become active: **X Translation**, **Y Translation**, **Z Translation**, **X Rotation**, **Y Rotation**, and **Z Rotation**. You can click in the value field of any of these parameters and select any of the following options from the drop-down list displayed: **Fixed**, **Release**, or **Spring**. The option that you select will specify the support conditions of the structural element in their respective axes. For example, if the **Release** option is selected for the **X Rotation** parameter, it means that the support at the end of the element for its rotational movement along the X axis will be released. After selecting a state of the support from the **Options Bar**, and specifying various options in the **Properties** palette (if required), click on any of the endpoints of a beam, column, or brace; the area boundary condition will be applied at that point of the structural element.

Line Tool

Ribbon:	Modify | Place Boundary Conditions > Boundary Conditions > Line

You can use the **Line** tool from the **Boundary Conditions** panel to create a linear boundary throughout the length of a beam, a column, an edge of a slab, an edge of a wall, or an edge of a floor. To use the **Line** tool, invoke it from the **Boundary Conditions** panel and then select an option from the **State** drop-down list in the **Options Bar**. The options available in the **State** drop-down list are **Pinned**, **Fixed**, and **User**. By default, the **Fixed** option is selected in the **State** drop-down list. As a result, when you add the boundary condition to the end of a beam, column, or brace, a fixed support condition will be created at that end. You can select any of the remaining options from the **State** drop-down list based on your design requirement. On selecting the **User** option from the **State** drop-down list in the **Options Bar**, the following parameters in the **Properties** palette will become active: **X Translation**, **Y Translation**, **Z Translation**, and **X Rotation**. The option that you select will be the support conditions of the structural element in their respective axes.

After selecting a state of the support from the **Options Bar** and specifying various options in the **Properties** palette (if required), click on a beam, column, brace, edge of a wall, edge of a floor, or edge of a wall; the line boundary condition will be applied to the entire length of the selected element or at the edge of the element (wall, slab, or floor).

After specifying the boundary condition, you can set the family symbol for various support conditions and adjust the spacing for each boundary condition representation. Note that you can also apply these settings before adding the boundary conditions to the elements in the project. To set a family symbol and adjust spacing for each boundary condition representation, click on the arrow on the right of the **Boundary Conditions** in the **Analyze** tab, as shown in

Figure 9-33. On doing so, the **Structural Settings** dialog box with the options in the **Boundary Conditions Settings** tab will be displayed, as shown in Figure 9-34. In the **Family Symbol** area of this tab, you can specify different family symbols for different states of support. These family symbols are pre-loaded in structural templates and can also be duplicated and edited to meet the display standards or requirements.

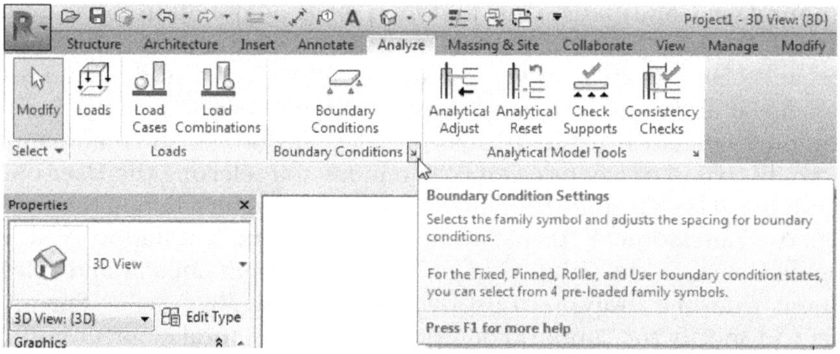

Figure 9-33 *Choosing the* ***Boundary Condition Settings*** *tool from the* ***Boundary Conditions*** *panel*

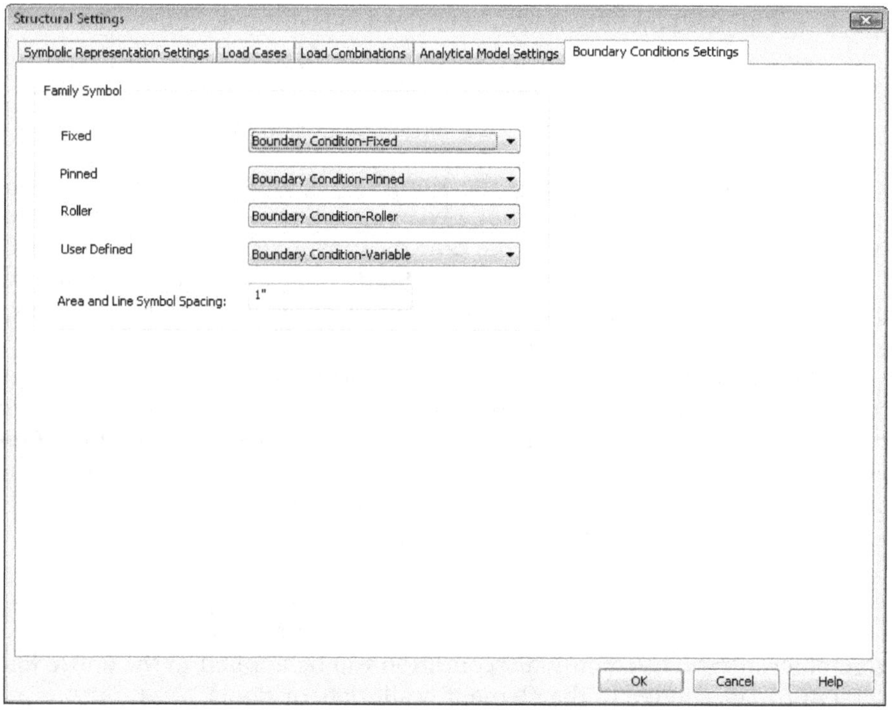

Figure 9-34 *The options in the* ***Boundary Conditions Settings*** *tab of the* ***Structural Settings*** *dialog box*

In the **Family Symbol** area, you can select any of the four drop-down lists, **Fixed**, **Pinned**, **Roller**, and **User Defined** to specify the symbol of the type of support assigned in the boundary condition. The options that you can select for the four drop-down lists are: **Boundary Condition-Fixed**, **Boundary Condition-Pinned**, **Boundary Condition-Roller**, and **Boundary Condition-Variable**. In the **Family Symbol** area, you can enter a distance in the **Area and Line Symbol Spacing** edit box to specify the distance between the symbols that are used for area or line boundary condition. After specifying various options in the **Boundary Conditions** tab, choose the **OK** button; the **Structural Settings** dialog box will close and the settings will be applied to the boundary condition in the structural model.

Analytical Model Tools

The Analytical Model tools are used to refine an analytical model before you export it to another software for analysis and design purpose. The refinement of a structural model includes doing analytical adjustment between members, checking for the consistency of an analytical model, resetting the analytical model to its previous state, and checking the support conditions of the model. Various tools involved in this refinement are available in the **Analytical Model Tools** panel of the **Analyze** tab. The tools available for refinement are **Analytical Adjust**, **Analytical Reset**, **Check Supports**, and **Consistency Checks**. These tools are discussed next.

Adjust Tool

Ribbon:	Analyze > Analytical Model Tools > Adjust

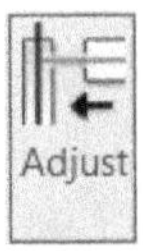

The **Adjust** tool is used to adjust the alignment of an analytical model with reference to the elements it is joined to. The application of this tool has been explained through an example given next. Figure 9-35 shows a 3D analytical view of a structural model.

In this model, you will use the **Adjust** tool to align the analytical model of the beam with that of the structural wall. To do so, choose the **Adjust** tool from the **Analytical Model Tools** panel; the physical model of the structural elements in the project view will be grayed out and various node points of the analytical model will be displayed, as shown in Figure 9-36.

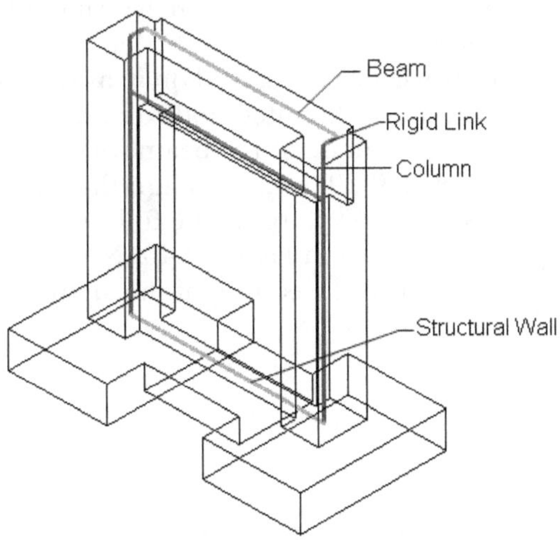

Figure 9-35 *The structural model showing various structural elements*

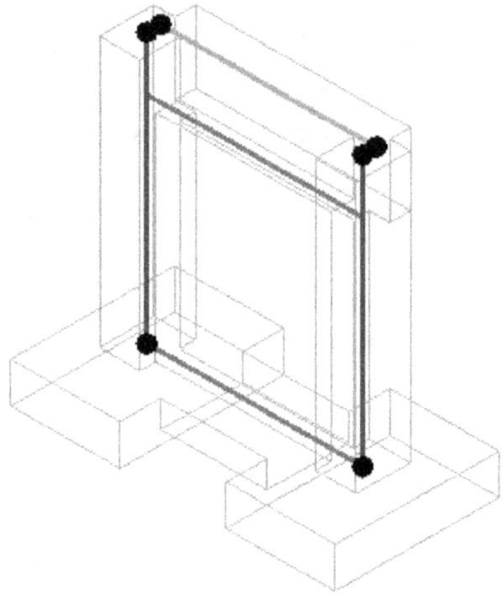

Figure 9-36 *Various node points in the analytical model*

Also, as you invoke the **Adjust** tool, the **Edit Analytical Model** panel will be displayed as a floating panel in the project view, as shown in Figure 9-37.

Figure 9-37 *The options in the* ***Edit Analytical Model*** *panel*

The analytical node points in the analytical model can be directly manipulated at both the ends. As you click on a node, a 3D control to move the node point in the local coordinate system will be displayed. In this example, you need to align the analytical model of the beam with that of the wall. To do so, click on the node point located at the left end of the beam; the node point will get highlighted and a 3D control will be displayed on it, as shown in Figure 9-38. Now, click and hold the left mouse button on the highlighted node and drag the highlighted node to the left node of the structural wall; the left node of the beam will get joined with the left node of the structural wall, as shown in Figure 9-39. Next, to align the right end of the beam with the right end of the structural wall, same procedure can be followed. After both ends are aligned, the structural model will appear, as shown in Figure 9-40. Now, in the **Edit Analytical Model** panel, choose the **Finish** button to finish the editing of the analytical model.

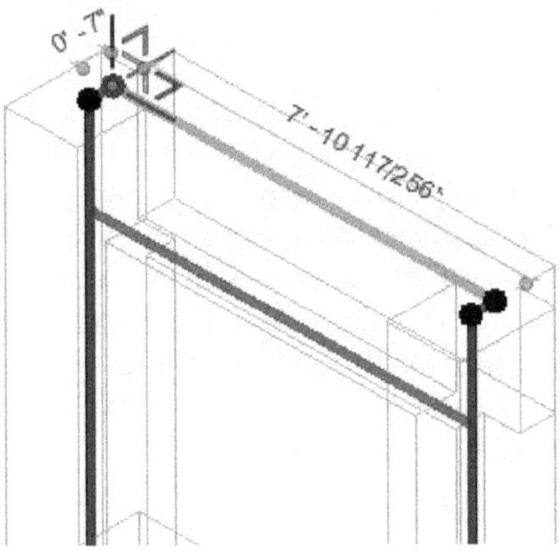

Figure 9-38 *The highlighted node point of the beam*

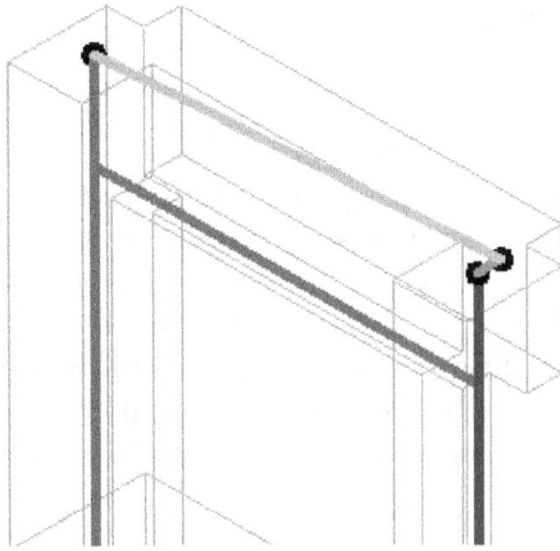

Figure 9-39 *Node of the beam joined with the node of the wall*

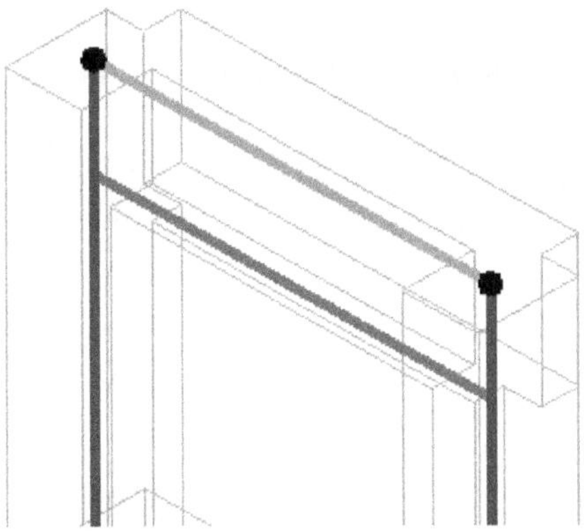

Figure 9-40 *The analytical model of the beam aligned with the analytical model of the structural wall*

Note

From the ***Edit Analytical Model*** *panel, you can choose the* ***Wall Adjustments*** *tool to adjust the analytical wall element in relation to other analytical wall models present in the project.*

Reset Tool

Ribbon: Analyze > Analytical Model Tools > Reset

After adjusting the members of an analytical model, you can bring the structural member analytical model back to its previous state. To do so, choose the **Reset** tool from the **Analytical Model Tools** panel and then select the analytical model member that was selected while using the **Adjust** tool; the selected analytical model member will regain its previous state.

Supports Tool

Ribbon: Analyze > Analytical Model Tools > Supports

This tool is used to check the support stability of an analytical model before it is exported to another software for analysis. To check the support stability, choose the **Supports** tool from the **Analytical Model Tools** panel of the **Analyze** tab; a **Warning** message box informing you about the stability of the supports in the model will be displayed, as shown in Figure 9-41. You can review the warning message by choosing the **Expanding warning dialog** button displayed on the right of the **Warning** message box. On doing so, the **Autodesk Revit 2017** window will open, as shown in Figure 9-42. You can use this window to review the warning displayed in the **Warning** message box.

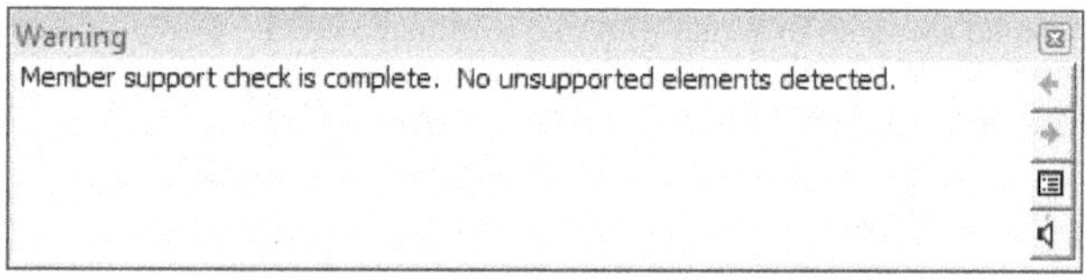

Figure 9-41 *The* ***Warning*** *message box*

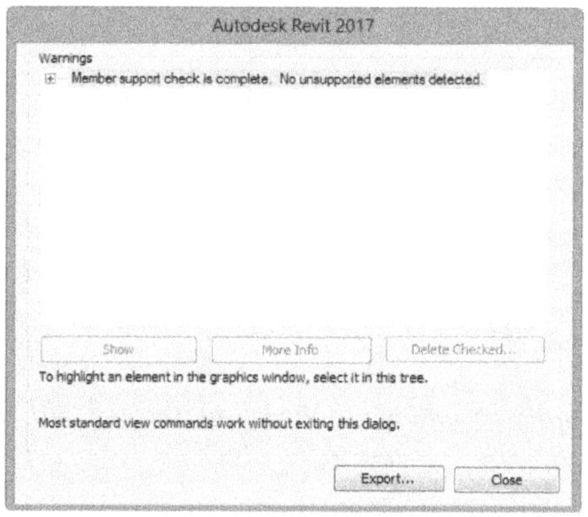

Figure 9-42 *The* ***Autodesk Revit 2017*** *window*

Consistency Checks Tool

Ribbon: Analyze > Analytical Model Tools > Consistency

This tool is used to verify the consistency of the analytical and physical models of a project. To check the consistency of an analytical or physical models, choose the **Consistency** tool from the **Analytical Model Tools** panel of the **Analyze** tab; the **Warning** message box will be displayed. This message box will inform you about any possible inconsistency in the model. You can review the warning message by using the **Autodesk Revit 2017** window which can be invoked by choosing the **Expanding warning dialog** button displayed on the right in the **Warning** message box.

Note

The options in the ***Analytical Model Settings*** *tab of the* ***Structural Settings*** *dialog box can be used to set the criterion for consistency check. These options have already been discussed under the* ***Settings for the Analytical Model*** *section in this chapter.*

ADDING REINFORCEMENTS

Reinforcements are added to concrete structural members based on the design requirement. In Revit, you can add reinforcements to concrete structural members such as floors, beams, columns, and slabs by using various tools. These tools are discussed next.

Place Rebar Parallel to Work Plane Tool

For a concrete column, beam, structural floor, or foundation, you can place rebars parallel to the current work plane. To do so, select a concrete structural member in the current workplane and then choose the **Rebar** tool from the **Reinforcement** panel of the tab; the **Revit** message box will be displayed with the information that the hooks will be included in the Rebar Shape definitions. Now, choose the **OK** button; the **Modify | Place Rebar** tab with the **Rebar Shape Browser** will be displayed. In the **Placement Orientation** panel of this tab, choose the **Parallel to Work Plane** tool; the **Options Bar** displaying various options to place the rebars will be displayed. In the **Options Bar**, you can select a rebar shape from the **Select Rebar Shape type** drop-down list. Alternatively, you can select the rebar shape from the **Rebar Shape Browser** displayed in the drawing area. In case the **Rebar Shape Browser** is not displayed by default, you can choose the **Launch/Close Rebar Shape Browser** button in the **Options Bar**. In the **Placement Plane** panel, you can select an option to specify the workplane used for placing rebars. After selecting the shape of the rebar and specifying the placement plane, you can modify the instance parameters of the rebars from the **Properties** palette. Some of the instance parameters that you can specify are described in the table given next.

Instance Parameter	Description
Schedule Mark	Specifies a mark in the rebars, which will help in identifying them while scheduling.
Style	It is a read-only parameter. This parameter is used to specify the style for controlling the bend radius of the rebars.
Shape	Specifies a number for the rebar shape.
Hook at Start	Specifies the type of rebar hook at the start of the rebar of the selected style. Click in its value field and select a type from the drop-down list displayed.
Hook at End	Specifies the type of rebar hook at the end of the rebar of the selected style. Click in its value field and select a type from the drop-down list displayed.
Reinforcement Volume	It is a read-only parameter and displays the calculated estimate for the reinforcement volume.
Bar Length	It is a read-only parameter and displays the length of individual rebar.

Spacing	This parameter is displayed only after you place the rebars to the hosted elements. This parameter will be active if the **Layout Rule** parameter is set to **Maximum Spacing** layout rule. This parameter specifies the spacing for the rebar placed in the major direction of the hosted element.
Layout Rule	This parameter is displayed only after you place the rebars to the hosted elements. It specifies the type of rebar layout. You can click in its value field and select any of the options: **Maximum Spacing** or **Fixed Number**.
Quantity	This parameter is displayed only after you place the rebars to hosted elements. It specifies the number of rebar instances placed. It is only active if the **Layout Rule** parameter is set to **Fixed Number**.
A, B, C,... (default)	Specifies the variable length of a rebar shape. The quantities of these parameters are defined by the content of the shape.

After specifying the instance parameters of the rebars from the **Properties** palette, you can modify their type properties. To do so, choose the **Edit Type** button from the **Properties** palette; the **Type Properties** dialog box will be displayed. Some of the type parameters in the **Type Properties** dialog box are described in the table next.

Type Parameter	**Description**
Material	Specifies the material assigned to rebars.
Bar Diameter	Specifies the diameter of non-hook bends for a selected type of bar. The change in this parameter does not affect the shape of the rebar.
Stirrup/Tie Bend Diameter	Specifies the diameter of the bend rebar. This bend can be either a standard bend or a stirrup/tie bend.
Hook Lengths	Specifies hook length based on a specific type of rebar. Choose the **Edit** button in its value field to display the **Rebar Hook Lengths** dialog box. You can use this dialog box to specify hook lengths for different hook types.

After specifying the type properties of the rebars from the **Type Properties** dialog box, place the cursor inside the placement area of the selected member; the preview of the selected rebar along with a dotted line representing the cover distance will be displayed, refer to Figure 9-43. Next, click when the preview is displayed; rebars will be placed parallel to the current work plane. Figure 9-44 shows the rebars of the **Rebar Shape:T6** type placed in a concrete rectangular beam by using the **Fixed Number** layout rule.

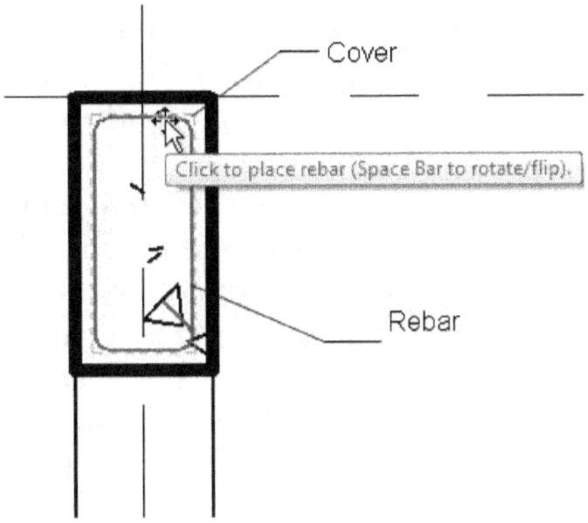

Figure 9-43 *The rebar added to a beam cross-section*

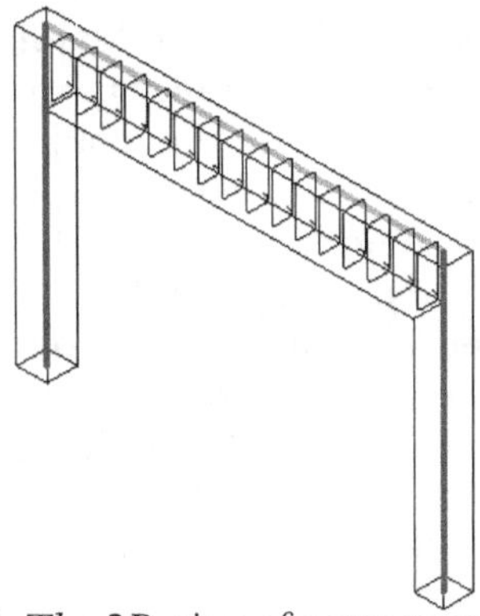

Figure 9-44 *The 3D view of a structural model displaying the added rebars*

Place Rebar Perpendicular to Cover Tool

You can place rebars that are perpendicular to the nearest cover reference in a member. To do so, open the desired view of the member. Next, select it and then choose the **Rebar** tool from the **Reinforcement** panel of the contextual tab; the **Modify | Place Rebar** contextual tab will be displayed. In this tab, choose the **Perpendicular to Cover** tool from the **Placement Orientation** panel and then select a rebar shape from the **Select Rebar Shape type** drop-down list. Now, place the cursor near the edge of the member; the dashed lines will be displayed at an offset from the edges of the member. These dashed lines represent the cover of the member. Also, as you move the cursor, the preview of the rebar will be displayed. Click to add the rebar.

Cover Tool

The **Cover** tool is used to edit the cover settings in the entire length of the host of a rebar. The host of a rebar can be a beam, column, slab, or floor. The cover settings define the offset distance of the bars from the faces of the rebar host. To edit the cover settings, invoke the **Cover** tool from the **Reinforcement** panel of the **Structure** tab; the **Options Bar** displays two tools: **Pick Elements** and **Pick Faces**. You can choose the **Pick Elements** tool to specify the cover distance for the entire length of the host element. Alternatively, you can choose the **Pick Faces** tool to specify the cover distance for the desired face of the host element. After choosing any of the tools and selecting an element or the face of an element, you can select an option from the **Cover Settings** drop-down list to specify the cover distance of the selected element or face. On selecting an option from the **Cover Settings** drop-down list, the cover distance for the placed rebars changes.

Area Tool

Ribbon: Structure > Reinforcement > Area

You can use the **Area** tool to place reinforcement in the entire area of a floor or a slab. To add area reinforcement to a floor or a slab, invoke the **Area** tool from the **Reinforcement** panel of the **Structure** tab; you will be prompted to select a structural floor or a structural wall. Click on the desired structural wall or structural floor; the **Modify | Create Reinforcement Boundary** tab will be displayed. In the **Draw** panel of this tab, ensure that the

Rebar Line tool is chosen, and then choose any of the sketching tools from the list box. After choosing a sketching tool, sketch the boundary of the reinforcement area, wherever required. Next, you can choose the **Major Direction** tool from the **Draw** panel of the **Modify | Create Reinforcement Boundary** tab and define the major direction of the area reinforcement either by sketching a line in the required area or by picking a line from the drawing area. Now, in the **Properties** palette, you can specify various instance parameters for the area reinforcement. Various instance parameters of the area reinforcement that can be specified are the layout rule, diameter of the major and minor rebars to be placed in the top and bottom of the slab or wall, hook type for major and minor directions, hook orientations, and others. You can also specify the type parameters of the area reinforcement. After specifying various instance parameters, choose the **Finish Edit Mode** button in the **Mode** panel, the area reinforcement will be added in the sketched area. Figure 9-45 shows the area reinforcement added to a structural floor. Figure 9-46 shows a section view displaying the distribution of rebars in an area reinforcement.

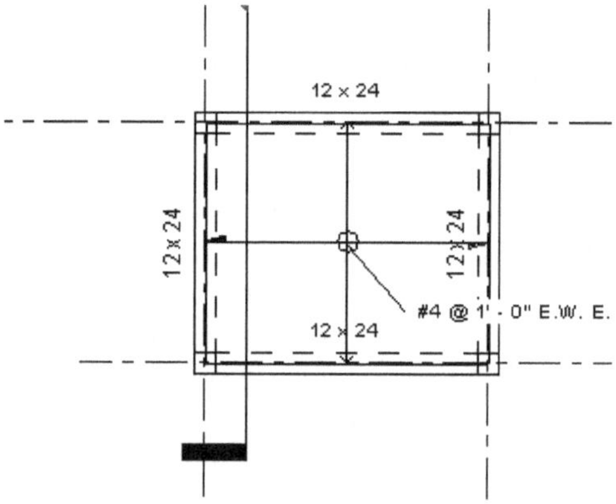

Figure 9-45 The area reinforcement added to a structural floor

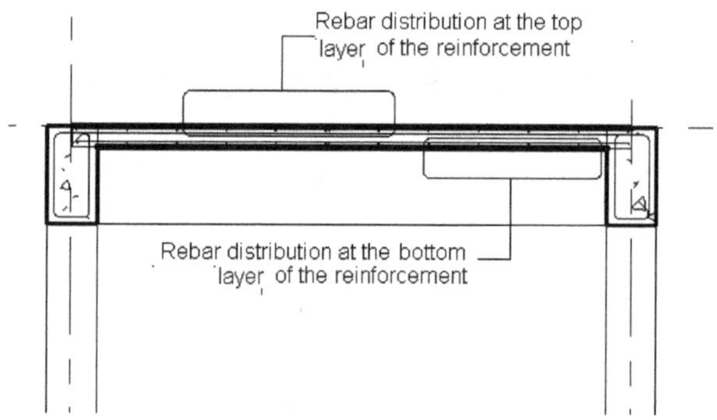

Figure 9-46 Section view showing the placement of the rebars in a structural floor

Path Tool

Ribbon: Structure > Reinforcement > Path

You can use the **Path** tool to sketch lines in a slab or wall. These lines indicate the path of evenly reinforcing bars that are created perpendicular to the sketch lines. To create this path, choose the **Path** tool from the **Reinforcement** panel of the **Structure** tab and then select a wall or floor from the drawing; the **Modify | Create Reinforcement Path** tab will be displayed. You can use the options in this tab to draw the sketch of the reinforcement path.

LINKING BUILDING MODELS AND SHARING COORDINATES

Large projects may require to develop independent building models and later combine these models into a comprehensive project. For example, in a large educational campus project, you need to independently develop the building models of the administrative building, academic complex, faculty residences, student accommodation, and so on, and later add them into a single host project file. You can do so by using the linking and sharing coordinate tools of Autodesk Revit that enable you to link models and share their coordinates within a single host project file. These tools also enable you to work on building models in other project files. When you link projects, you need to share the coordinates of the host model with the linked model so that the linked files retain their positions. To do so, invoke the **Acquire Coordinates** tool from **Manage > Project Locations > Coordinates** drop-down and then select the linked project in your drawing; the origin of the shared coordinates for the linked project becomes the origin for the shared coordinates of the host project. Revit provides flexibility and easy management of the linked models by enhancing the linking of the models and organizing the linked files in the **Project Browser**. You can easily access the linked files, the nested link files, and the Link Manager from the **Project Browser**. The nested Revit links are also listed under the **Revit Links** head with the host link in the **Project Browser**.

Linking or Importing Models

To link or import one project to the other, open the host project file and choose the **Insert** tab; various options in this tab will be displayed, as shown in Figure 9-47. You can choose the options in the **Import** or **Link** panel of the **Insert** tab, depending upon the file format to be imported or linked. Before you import or link a file, make sure that the file contains the data and geometry that is compatible with Revit. To link *.dwg, .dwf, .skp, .dgn,* and other files, choose the **Link CAD** tool from the **Link** panel of the **Insert** tab; the **Link CAD Formats** dialog box will be displayed, as shown in Figure 9-48. In this dialog box, select the file type from the **Files of type** drop-down list. Next, browse to the required location and select the file.

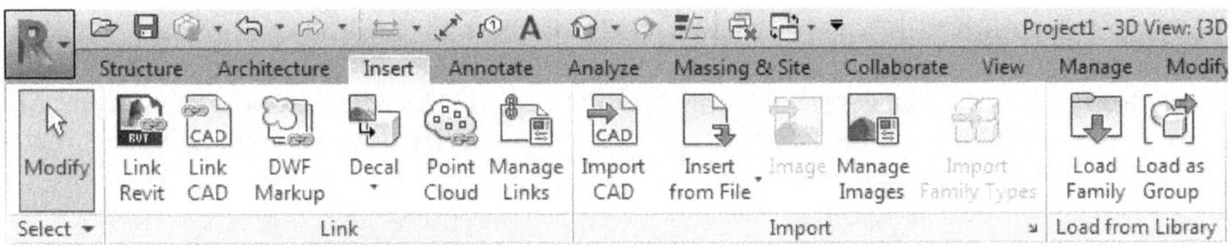

Figure 9-47 *Various options in the **Insert** tab*

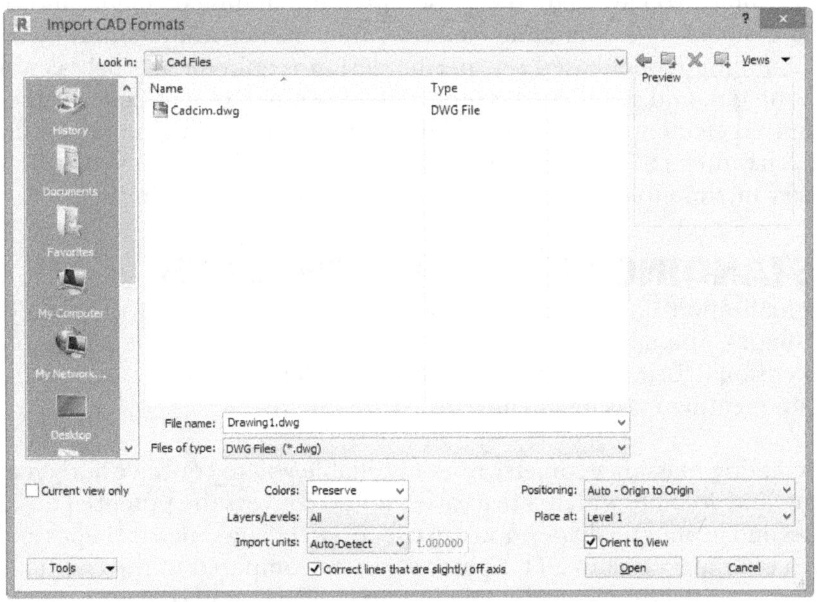

Figure 9-48 *The **Link CAD Formats** dialog box*

Linking or Importing Files

In Revit, you can link or import the CAD files to the current project. If a CAD file is linked to a project, all changes made in it will simultaneously be updated in the project. However, if you import a file in a project, it will remain unchanged, irrespective of the changes made in the file linked to the project.

To link a CAD file to a project, choose the **Link CAD** tool from the **Link** panel of the **Insert** tab of the ribbon; the **Link CAD Formats** dialog box will be displayed. Select the desired CAD file and then choose **Open**; the selected CAD file will be displayed and linked to the Revit project file. Similarly, you can import a CAD file to a project. To do so, choose the **Import CAD** tool from the **Import** panel of the **Insert** tab; the **Import CAD Formats** dialog box will be displayed. Select the desired CAD file and choose **Open**; the selected file will be imported to the Revit file.

While importing or linking a CAD file to your project, you can control its availability either in all views or only in the current view. If you select the **Current view only** check box in the **Import CAD Formats** or **Link CAD Formats** dialog box, you can import or link the drawing in the CAD format only in the active view of the project in Revit. However, if you clear this check box, you can import the drawing in the CAD format in all views of Revit. In this case, the lines and geometries of the drawing will be imported without the text.

The **Layers** drop-down list in the **Import CAD Formats** or **Link CAD Formats** dialog box allows you to specify the layers to be imported from the linked file. On selecting **All** from the **Layers** drop-down list, all layers will be imported or linked. The **Visible** options in this list allow you to import or link the visible layers in the view. The **Specify** option allows you to import the specified layers and levels in the project.

INTRODUCTION TO MASSING

In the earlier chapters, you learned how to use different building elements and components to create a building model. These building elements are parametrically associated and enable you to generate a building model based on specific design requirements such as wall types, door width, window height, and so on. Each of the elements must be assigned specific properties to achieve the desired element parameters, thereby making a building model accurate. Needless to say, this is a fairly time-consuming procedure. Autodesk Revit provides you with an alternative and much easier method to create a building model, known as massing.

UNDERSTANDING MASSING CONCEPTS

At the conceptualization stage of a project, you may want to study it in terms of its building volumes and shapes. You may also want to convey the basic idea of the structure of a building in a three-dimensional form without putting in a lot of detailing. This can be achieved by using various tools for creating massing geometries.

The tools for creating massing geometry not only enable you to conceive and create a variety of building shapes and volumes with relative ease but also conveys the potential design in terms of building masses and geometric shapes. You can create and edit geometric shapes and amalgamate them to form a building structure. This process can be compared to the creation of building a model using foam blocks. You have the freedom of choice to add or cut geometric shapes and join different blocks or masses to form an assembly. Figure 9-49 shows a group of volumes that can be created to represent the building volume and mass.

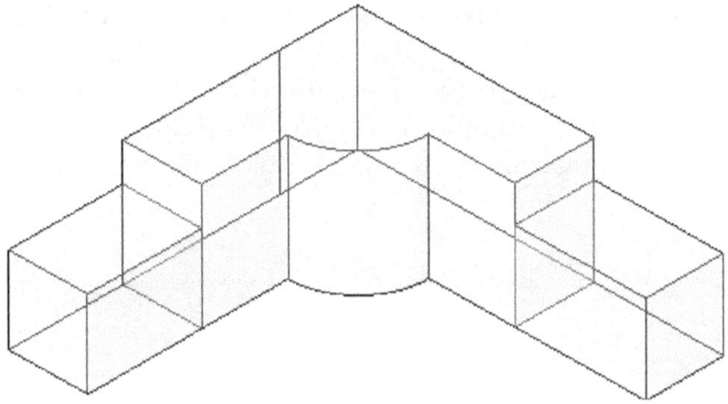

Figure 9-49 *Building blocks created using the massing geometry*

Autodesk Revit also enables you to convert the building mass into building elements. Building blocks can be transformed into an assembly of individual building elements such as walls, roof, and floor. The transformation, carried out by using the tools provided for replacing massing faces with predefined wall, roof, and floor types, allows you to convert the same building mass into a building model. You can then develop it into a detailed building model by incorporating individual building elements with specified parameters.

Autodesk Revit provides a much-needed continuity in the design development of a building project by using the same building model from its conceptualization to completion. It also enables you to control the visibility of geometry between the building volumes (massing) and building elements (shell) during the initial stages of project development. Other project information such as its total area can also be extracted from the building massing.

It is, however, important to understand the limitations of the massing tool. This tool is only meant for conceptual design development using simple geometric shapes. You can place the predefined massing family elements provided in additional libraries. Autodesk Revit attempts to translate the massing geometry into building elements or shell. So, it is recommended not to use massing geometry for the development of a detailed geometry such as columns, cavity walls, footing, and so on.

CREATING THE MASSING GEOMETRY

In Autodesk Revit, you can create the massing geometry in any of these three environments: Family Editor, Conceptual Design, and Project.

To create the massing geometry in the Family Editor environment, choose **New > Family** from **Application Menu**; the **New Family - Select Template File** dialog box will be displayed. In this dialog box, choose the *Generic Model.rft* file (commonly used) from the **English_I** folder for Imperial or choose the *Metric Generic Model Adaptive.rft* file from the **English** folder and then choose the **Open** button; a new file will open in the Family Editor environment. In this new file, you can create the massing geometry using various tools available in the ribbon.

The Conceptual Design environment is a new environment in Revit. This environment is a type of Family Editor that provides the advanced modeling tools and techniques for creating massing families. To start creating a mass in this environment, choose **New > Conceptual Mass** from **Application Menu**; the **New Conceptual Mass - Select Template File** dialog box will be displayed. In this dialog box, select the **Mass** template file from the **Conceptual Mass** folder and then choose **Open**; a new file will open in the Conceptual Design environment. In the new file, you can create the massing geometry using various tools available in the ribbon, as shown in Figure 9-50.

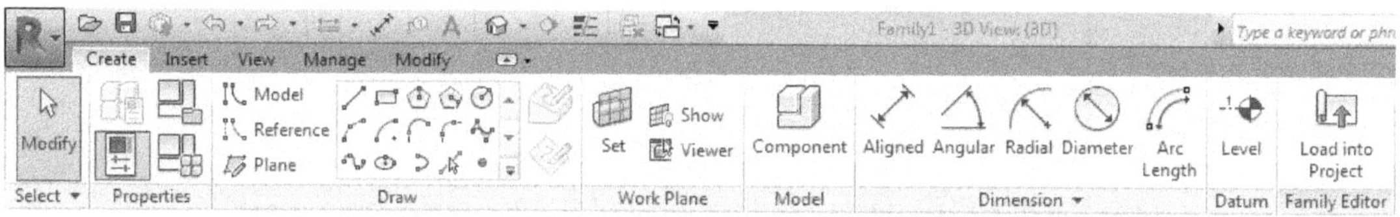

***Figure 9-50** Different tools in the ribbon of the conceptual design environment*

The Project environment is the most common environment used in a project design. To start creating a mass in this environment, open a new file or an existing file by choosing **New > Project** or **Open > Project** from **Application Menu**. After opening a new file or an existing file, choose the **Massing & Site** tab. In this tab, the massing tools can be accessed from the **Conceptual Mass** panel. The **Conceptual Mass** panel contains the following tools for massing: **In-Place Mass**, **Show Mass**, and **Place Mass**.

The options in the **Model by Face** panel are used to convert the conceptual mass created into real building elements like walls, floors, roofs, and curtain systems. As such, these tools are also called the Building Maker tools.

When you create shapes in massing, Autodesk Revit creates its corresponding building elements. It is, therefore, imperative to consider the associativity of the massing and shell elements. The massing elements may need to be transformed into individual building elements simultaneously. Therefore, the massing geometry must be created accordingly.

For example, when you create a complex geometric massing shape and convert faces into building elements, you may find that some planes do not acquire the desired building element characteristics. The inclined planes and the curved surfaces are converted into in-place roofs.

Creating a Massing Geometry in the Family Editor

As discussed earlier, the Family Editor environment provides tools to create massing geometry. To create a new mass, choose **New > Family** from **Application Menu**; the **New Family - Select Template File** dialog box will be displayed. In this dialog box, select the *Generic Model.rft* file (commonly used) from the **English_I** folder for Imperial or select the Metric *Generic Model Adaptive.rft* file from the **English** folder and choose **Open**; a new file will open using the selected template file.

In the Family Editor environment, the **Create** tab contains tools to create massing geometry. These tools can be used to create massing geometries in solid or in void form.

To create a solid form, choose the **Extrusion** / **Blend** / **Revolve** / **Sweep** / **Swept Blend** tool from the **Forms** panel of the **Create** tab. These tools are used to create different solid forms. Similarly, to create a void form, choose the **Void Extrusion** / **Void Blend** / **Void Revolve** / **Void Sweep** / **Void Swept Blend** tool from the **Void Forms** drop-down in the **Forms** panel. You can create a massing geometry using any of these massing tools or a combination of the **Solid** and **Void** tools. The usage of these tools for creating simple massing geometries is described next. However, a designer can select the appropriate tool judiciously depending on the massing geometry to be created. Various tools used to create a solid or a void geometry are explained in the next sections.

Creating an Extrusion

Ribbon: Create > Forms > Extrusion

The **Extrusion** tool is used to create a massing geometry by adding height to a sketched profile. You can invoke this tool from the **Forms** panel in the **Create** tab. On doing so, the **Modify | Create Extrusion** tab will be displayed. You can use different tools from this tab to sketch an extrusion profile and create a massing geometry. Figure 9-51 shows different tools in the **Modify | Create Extrusion** tab.

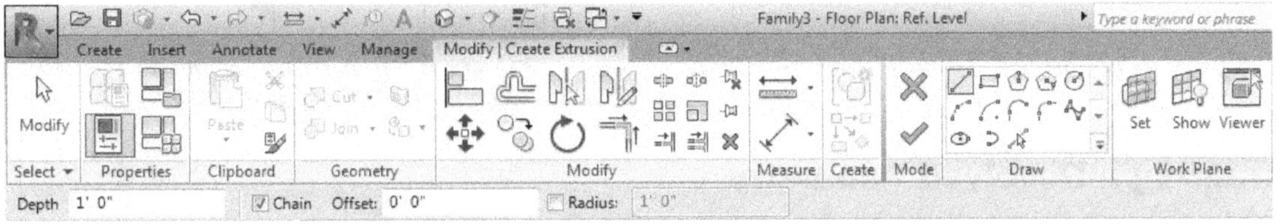

*Figure 9-51 Different tools in the **Modify / Create Extrusion** tab*

To create an extrusion, first you need to define work plane to sketch a profile. To define the work plane, you can invoke the **Set** tool available in the **Work Plane** panel of the **Modify | Create Extrusion** tab. After specifying the work plane, you can invoke the **Reference Plane** tool in the **Datum** panel of the **Create** tab to draw reference planes for locating exact points to sketch a profile. To sketch the 2D profile to be extruded, you can invoke any of the sketching tools available in the **Draw** panel of the **Modify | Create Extrusion** tab. You can invoke the appropriate tool (s) depending on the shape of the profile that must form a closed loop. In the **Properties** panel, choose the **Properties** tool to display the **Properties** palette, if it is not displayed by default. In this palette, you can specify the start and end levels of extrusion and other properties. The **Extrusion Start** instance parameter in the **Properties** palette indicates the start level of extrusion from the base level. The **Extrusion End** parameter indicates the top level of extrusion. The difference between these two parameters is calculated as the depth of the extrusion. You can also enter its value in the **Depth** edit box of the **Options Bar**. Autodesk Revit assumes the depth of extrusion from the base level. After sketching the profile, choose the **Finish Edit Mode** button from the **Mode** panel to finish the sketch of the profile for extrusion.

For example, to create a high-rise building, invoke the **Extrusion** tool from the **Forms** panel. In the **Options Bar**, you can specify the height of extrusion in the **Depth** edit box and also select the **Chain** check box, if it is not selected by default, to sketch a base profile, if required. Next, in the **Draw** panel of this tab, the **Line** tool (invoked by default) can be used to sketch the extrusion profile. You can also use other sketching tools available in the **Draw** panel to sketch the profile. Sketch the base of the specified dimension using temporary dimensions, as shown in Figure 9-52. After the profile is completed, choose the **Finish Edit Mode** button to extrude the sketched profile to the specified depth. The building mass created can be viewed in the 3D view, as shown in Figure 9-53.

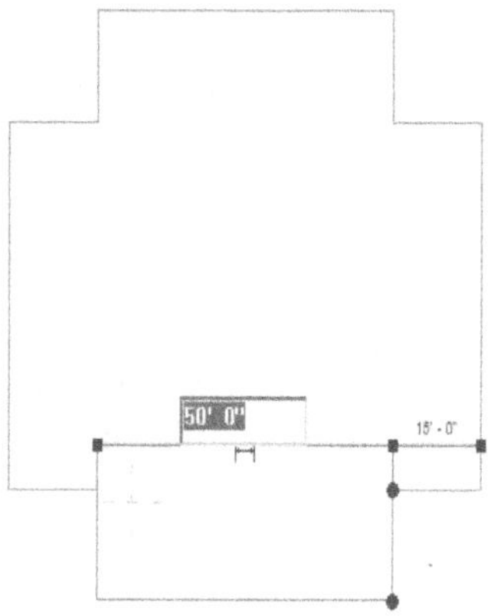

Figure 9-52 Sketch of a building profile to be extruded

*Figure 9-53 Massing geometry created by using the **Extrusion** tool*

Creating a Revolved Geometry

Ribbon: Create > Forms > Revolve

The **Revolve** tool is used to create a solid geometry by revolving a profile about an axis. This tool can also be used to create shapes such as domes, donuts, cylinders, and so on. To create a revolved geometry, invoke this tool from the **Forms** panel; Autodesk Revit enters the sketch mode and the **Modify | Create Revolve** tab will be displayed, as shown in Figure 9-54.

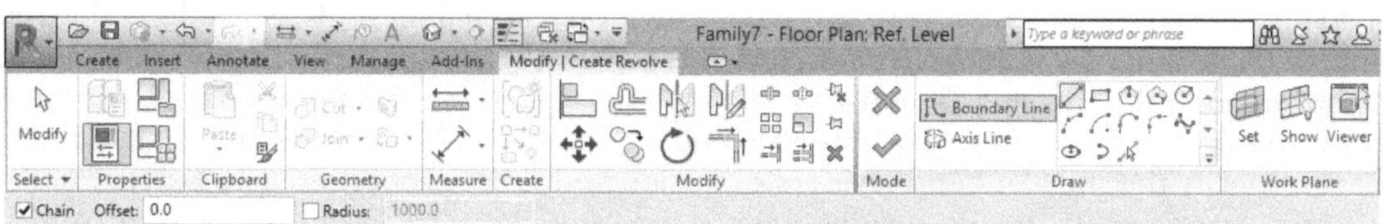

*Figure 9-54 Different options in the **Modify / Create Revolve** tab*

In the **Draw** panel, the **Boundary Lines** tool is chosen by default. As a result, various sketching tools used to sketch the profile of the revolved geometry are displayed in a list box of this panel. Before invoking any of the sketching tools, you need to set the work plane. To do so, choose the **Create** tab, and then choose the **Set** tool from the **Work Plane** panel; the **Work Plane** dialog box will be displayed. In this dialog box, you can use various options to specify the required work plane for sketching the profile of the revolved geometry. After selecting the required plane, you can sketch the profile of the revolved geometry.

To sketch the profile, choose the **Modify | Create Revolve** tab. Next, you can use various sketching tools from the list box in the **Draw** panel. The sketched profile must be a single closed loop or multiple closed loops that do not intersect. After defining the profile, you need to define the axis about which the profile will revolve. To do so, invoke the **Axis Line** tool from the **Draw** panel;

two tools, **Line** and **Pick Lines**, will be displayed in the list. The **Line** tool is used to draw a line that can be used as an axis and the **Pick Lines** tool is used to pick a line or an edge to define the axis of revolution. After completing the sketch and defining the axis of revolution, choose the **Finish Edit Mode** button from the **Modify | Create Revolve** tab; the **Modify | Revolve** tab will be displayed and the revolved geometry will be created.

In the **Modify | Revolve** tab, you can change the instance property of a revolved geometry in the **Properties** palette. In this palette, you can set the start and end angles of revolution. The default values for these two parameters are 0-degree and 360-degree, respectively.

For example, to create a dome, you can sketch the profile and define the axis for the profile, as shown in Figure 9-55. Before you start the sketch, you need to set the work plane that is perpendicular to the horizontal plane. The resultant revolved massing geometry with the axis and its profile is shown in Figure 9-56. Figure 9-58 shows an example of the revolved geometry in which the **End Angle** and **Start Angle** parameters in the **Properties** palette have been assigned the values 360.000° and 0.000°, respectively.

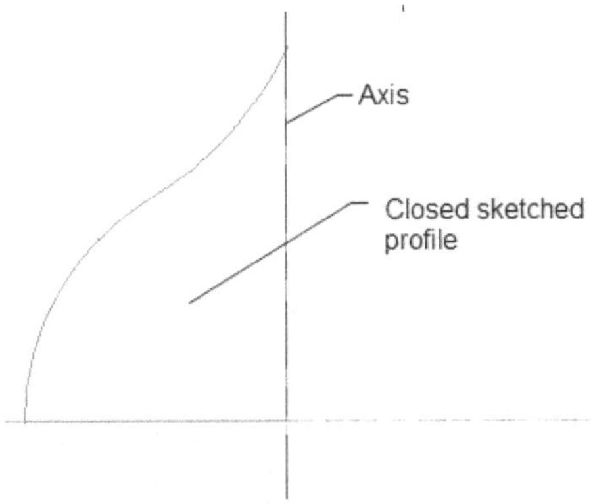

Figure 9-55 *Sketching a close profile for a revolved geometry*

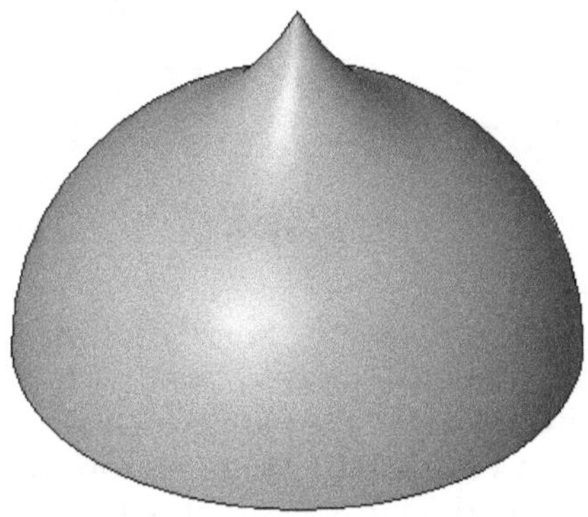

Figure 9-56 *Massing geometry created by using the **Revolve** tool*

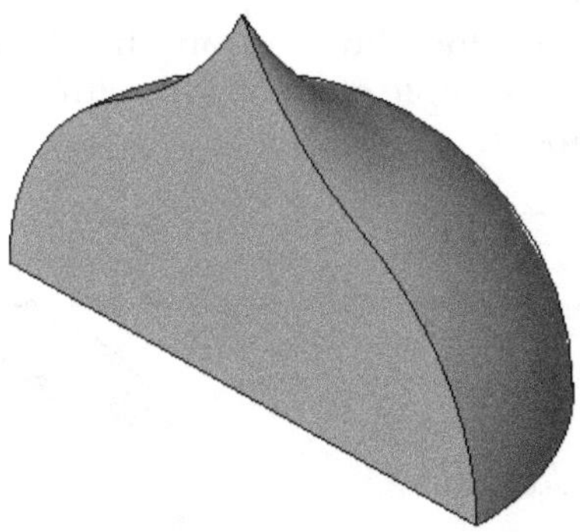

Figure 9-57 *Revolved geometry created by revolving the profile at an angle of 180 degrees*

> **Tip**
> *In Autodesk Revit, you can quickly edit the revolve angle of a revolved geometry by selecting the geometry from the drawing and then changing the temporary angular dimension displayed in the geometry.*

Creating a Sweep

Ribbon: Create > Forms > Sweep

The **Sweep** tool is used to create a massing feature by selecting a profile along a sketched path. To create a massing feature by sweeping a profile, invoke the **Sweep** tool from the **Forms** panel; the sketch mode will be invoked and the **Modify | Sweep** tab with its related tools will be displayed, as shown in Figure 9-58. The **Sketch Path** tool in the **Sweep** panel of the **Modify | Sweep** tab is used to sketch the path to be used for extrusion.

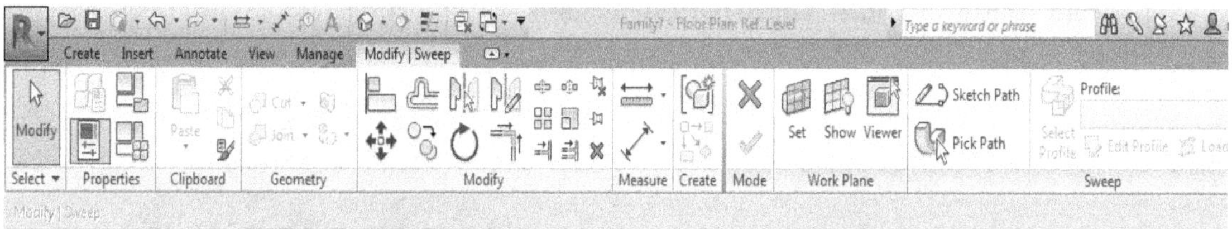

*Figure 9-58 Different options in the **Modify / Sweep** tab*

On invoking this tool, the **Modify | Sweep** tab will be replaced by **Modify | Sweep > Sketch Path** tab, which contains tools to draw path. Using the tools from the **Draw** panel, you can sketch the desired shape of the path, which can be an open or closed profile. After sketching the path, choose the **Finish Edit Mode** button from the **Mode** panel; the **Modify | Sweep** tab will be displayed. In this tab, you can sketch a profile or load a profile for the sweep geometry. To sketch a profile, choose the **Edit Profile** tool from the **Sweep** panel in the **Modify | Sweep** tab; the **Go To View** dialog box will be displayed. In this dialog box, select an appropriate view and sketch the profile using different sketching tools. After sketching the profile, choose the **Finish Edit Mode** button from the **Mode** panel; the **Modify | Sweep > Edit Profile** tab will be displayed. In this tab, choose the **Finish Edit Mode** button from the **Mode** panel; a sweep geometry will be created. Figure 9-59 shows the example of a 2D path with a sketched profile and Figure 9-60 shows the resulting massing shape.

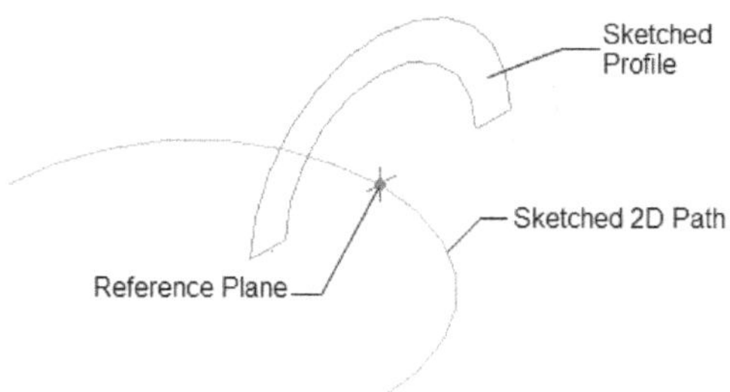

Figure 9-59 Sketching the profile and the 2D path for creating a sweep

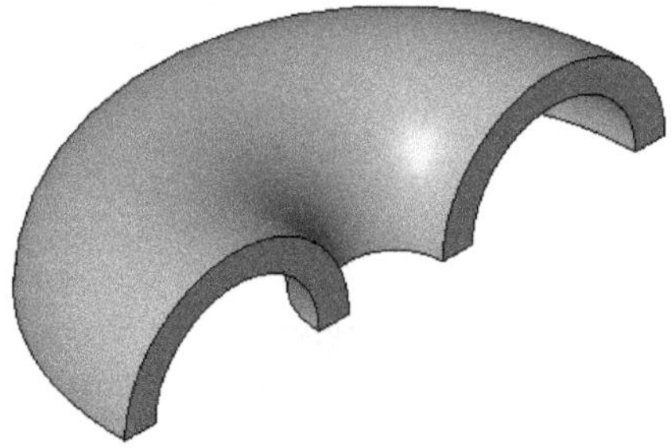

*Figure 9-60 Massing geometry created by using the **Sweep** tool*

Autodesk Revit also provides in-built profiles that can be used in projects. These profiles are available in the **Profile** drop-down list in the **Sweep** panel of the **Modify | Sweep** tab. You can also choose the **Load Profile** button from the **Sweep** panel and access additional profiles in the **Profiles** sub folder in the **US Imperial** folder. You can modify the properties of the sweep profile such as its structural, mechanical, and other properties. To do so, select the mass created by the **Sweep** tool; the **Modify | Sweep** tab will be displayed. In this tab, choose the **Edit Sweep** tool from the **Mode** panel; the properties of the sweep model will be displayed in the **Properties** palette. In this palette, click on the value field corresponding to the **Part Type** parameter under the **Mechanical** head; a drop-down list will be displayed. You can select an option from the displayed drop-down list to specify a part type to the sweep model. In the **Properties** palette, select the check box corresponding to the **Can host rebar** parameter to allow the sweep mass to host rebar for structural detailing.

Creating a Blend

Ribbon: Create > Forms > Blend

Using the **Blend** tool, you can create a massing geometry by blending or linking two profiles. To create a massing geometry using this tool, invoke it from the **Forms** panel; the **Modify | Create Blend Base Boundary** contextual tab will be displayed. The tools in this tab are used to create and edit a blend. Figure 9-61 shows various tools in the **Modify | Create Blend Base Boundary** tab.

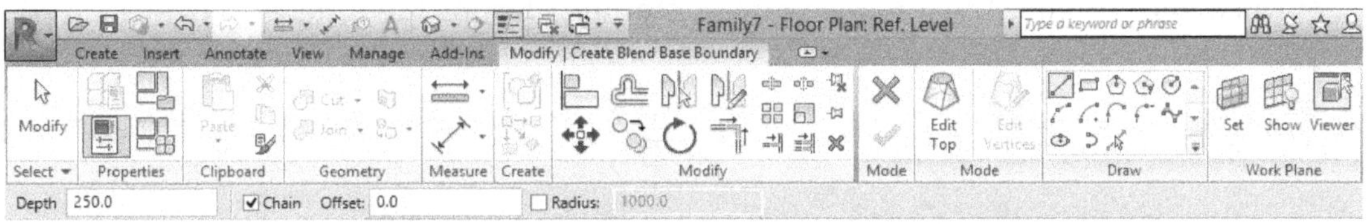

***Figure 9-61** Different tools in the **Modify / Create Blend Base Boundary** tab*

After setting the work plane, you can sketch the base profile using different sketching tools available in the **Draw** panel. In the **Properties** palette, you can specify values in the value fields of the **First End** and **Second End** instance parameters to set the height of the blend geometry. After completing the base profile, you can choose the **Edit Top** option from the **Mode** panel in the **Modify | Create Blend Base Boundary** tab to display the **Modify | Create Blend Top Boundary** tab. In this tab, you can sketch the profile of the top of the blend geometry in any work plane. The depth of the blend geometry can be specified in the **Depth** edit box available in the **Options Bar**.

The **Edit Vertices** tool in the **Mode** panel of the **Modify | Create Blend Top Boundary** tab will be available only after both the base and top profiles are sketched. This tool enables you to specify the connectivity between the vertices of their profiles. After sketching both the profiles, choose the **Finish Edit Mode** button from the **Mode** panel to create the blend geometry. Figure 9-62 shows two circular profiles being sketched as the base and top profiles. The resulting geometry is shown in Figure 9-63.

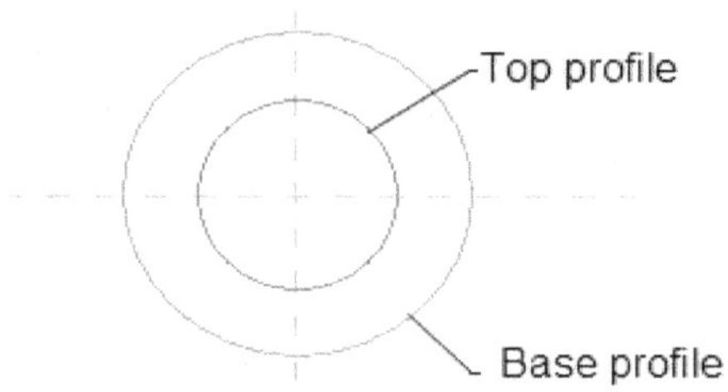

Figure 9-62 *Sketching the base and top profiles to create a blend massing*

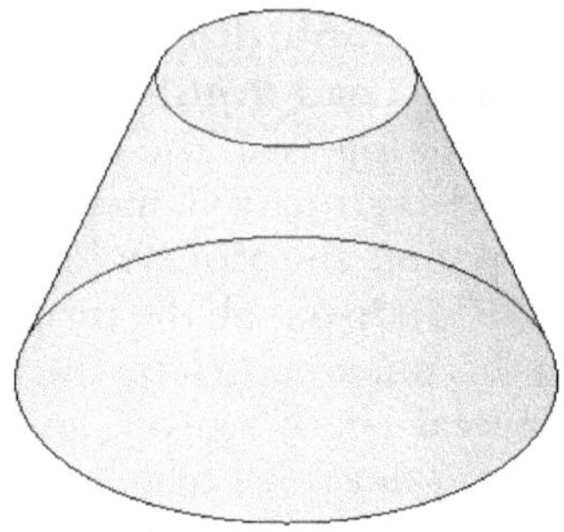

Figure 9-63 *Resulting shape after blending the base and top profiles*

Creating a Swept Blend

Ribbon: Create > Forms > Swept Blend

The solid swept blend geometries are created by using the **Swept Blend** tool. This tool has combined features of the **Sweep** and **Blend** tools. A geometry created using the **Swept Blend** tool consists of a path and two different profiles drawn at the either end of the path.

To create a swept blend geometry, choose the **Swept Blend** tool from the **Forms** panel; the screen will enter into the sketch mode and the **Modify | Swept Blend** tab will be displayed, as shown in Figure 9-64.

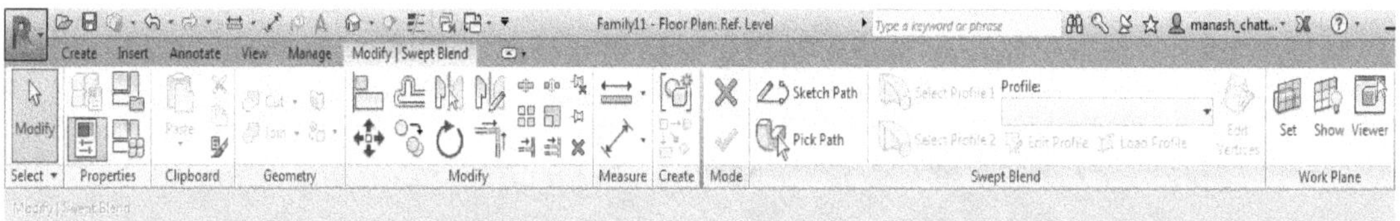

Figure 9-64 *Different options in the **Modify / Swept Blend** tab*

The **Modify | Swept Blend** tab contains various tools to sketch and edit the path and profiles of a solid swept blend geometry. Figure 9-65 shows the graphical representation of the solid Swept Blend geometry.

To create the geometry, first you need to sketch the 2D path. To do so, choose the **Sketch Path** tool from the **Swept Blend** panel; the sketch mode will be invoked. You can use various sketching tools in the **Draw** panel to sketch the 2D path. Next, choose the **Finish Edit Mode** button from the **Mode** panel of the **Modify | Swept Blend > Sketch Path** tab. Alternatively, you can define the 2D path for the solid swept blend geometry by choosing the **Pick Path** tool from the **Swept Blend** panel and then selecting an existing open curve or the edges of elements. The desired 2D path is always an open geometry composed of single segment. Once you have sketched or picked the 2D path, you need to define a profile at either end. Choose the **Select Profile 1** tool from the **Swept Blend** panel and then choose the **Edit Profile** tool from the **Edit** panel; the

Go To View dialog box will be displayed. Choose the required option from this dialog box and then choose the **Open View** button; the view will change based on the option chosen and the **Modify | Swept Blend > Edit Profile** tab will be displayed. Sketch the first profile at the end of the 2D path where you can see the reference plane with a red dot prominently displayed by using the appropriate sketching tools from the **Draw** panel. Next, to finish the sketching of the first profile, choose the **Finish Edit Mode** button from the **Mode** panel of the **Modify | Swept Blend > Edit Profile** tab. Similarly, you can define the second profile. To do so, choose the **Select Profile 2** option from the **Swept Blend** panel and then choose the **Edit Profile** option from the **Mode** panel in the **Modify | Swept Blend** tab; the **Modify | Swept Blend > Edit Profile** tab will be displayed. Sketch the second profile and then choose the **Finish Edit Mode** button from the **Mode** panel; the **Modify | Swept Blend** tab will be displayed. Now, choose the **Finish Edit Mode** button from the **Mode** panel; the solid swept blend geometry will be created, as shown in Figure 9-66.

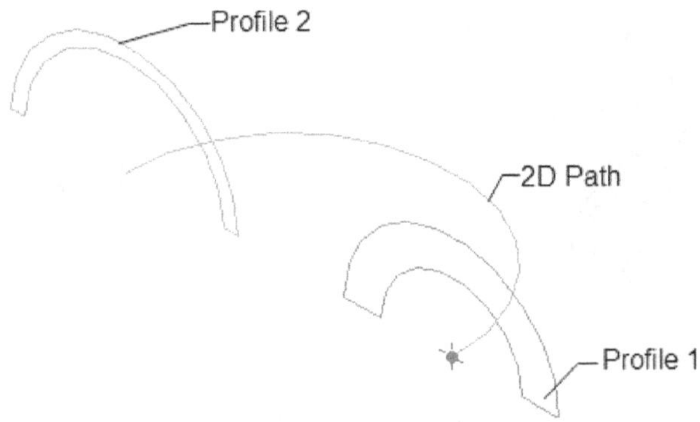

Figure 9-65 *The profiles and the 2D path for creating a Swept Blend geometry*

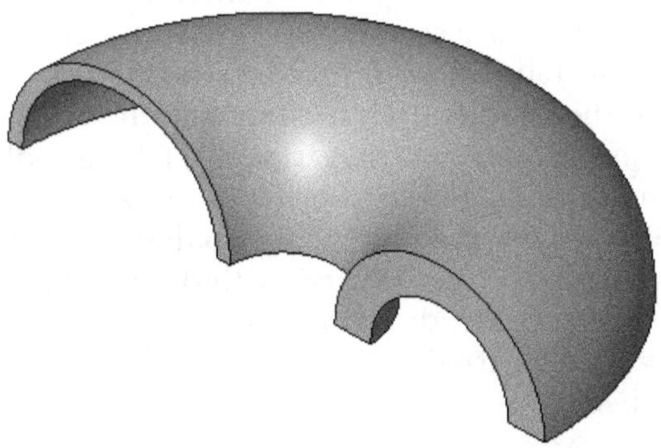

Figure 9-66 *Massing geometry created by using the **Swept Blend** tool*

Editing a Massing Geometry in the Family Editor

In Autodesk Revit, you can easily edit a massing geometry. It can be edited using drag controls or by editing massing parameters.

Resizing a Massing Geometry by Using Drag Controls

On selecting a massing geometry, a number of drag controls will be displayed as arrows. You can use the drag control to drag the desired face. The entire massing geometry is automatically updated, based on the dragged face. For example, Figure 9-67 shows a cuboid with drag controls displayed on all its faces. When you drag a plane of the cuboid using drag controls, Autodesk Revit immediately updates the geometry, as shown in Figure 9-68.

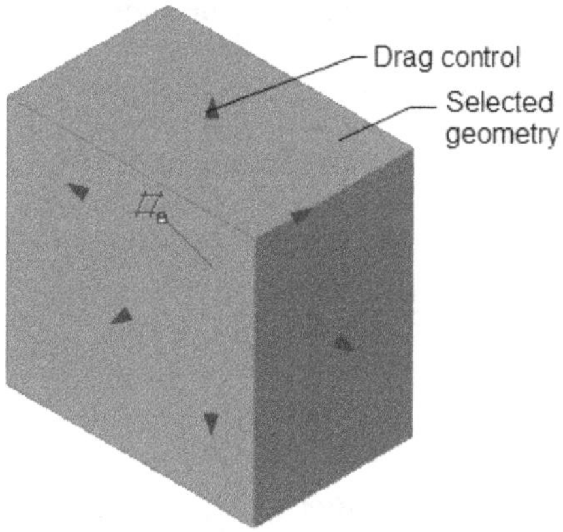

Figure 9-67 *The geometry with the drag controls*

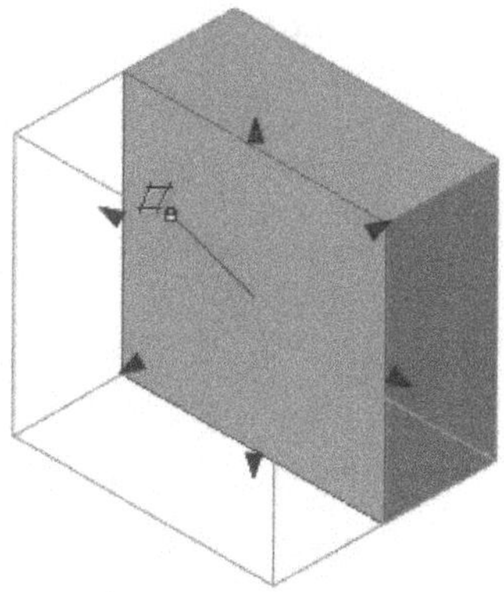

Figure 9-68 *The massing geometry resized using the drag controls*

Editing the Parameters of a Massing Geometry

The parameters of a massing geometry can be modified in the Family Editor, after selecting geometry. To edit the parameters, select a massing geometry created by using the **Extrusion** tool; the **Modify | Extrusion** tab will be displayed. In this tab, choose the **Edit Extrusion** tool from the **Mode** panel; the **Modify | Extrusion > Edit Extrusion** tab will be displayed. From the **Draw** panel of this tab, you can use various sketching tools to alter the sketch of the extrusion profile. You can also change the properties of extrusion such as height, material, and visibility from the **Properties** palette. In this palette, you can change the height of extrusion by changing the values of the **Extrusion End** and **Extrusion Start** parameters. To finish the editing of the selected extruded mass, choose the **Finish Edit Mode** button from the **Modify | Extrusion > Edit Extrusion** tab of the **Mode** panel; the editing of the selected extruded mass will be finished.

Similarly, you can also edit other massing geometries using **Sweep**, **Swept Blend**, **Revolve**, and **Blend** tools.
To a certain extent, massing features can also be modified like other building elements. You can use the editing tools such as **Mirror**, **Copy**, **Group**, **Array**, and so on along with the massing tools.

Creating Cuts in a Massing Geometry Using the Family Editor

You can cut a massing geometry by creating a void form in it. This void form is cut or subtracted from the massing geometry it intersects. You can create void forms by using the tools displayed in the **Void Forms** drop-down in the **Forms** panel of the **Create** tab. The **Void Forms** drop-down displays five tools: **Void Extrusion**, **Void Blend**, **Void Revolve**, **Void Swept Blend**, and **Void Sweep**. You can choose an appropriate tool to generate the shape and volume of the void form. The method of creating a void form using these tools is similar to that of creating a solid form.

When any of the tools from the **Void Forms** drop-down is invoked, a contextual tab is displayed. For example, if you invoke the **Void Extrusion** tool, the **Modify | Create Void Extrusion** tab will be displayed. You can use the options in this tab to sketch the profile for the extruded void geometry. After sketching the profile of the extruded void geometry, choose the **Finish Edit Mode** button from the **Mode** panel of the **Modify | Create Void Extrusion** tab; the **Modify | Void Extrusion** tab will be displayed. From this tab, you can use various editing options to modify the extruded void geometry. Next, click in the drawing area or press ESC; the void form will automatically cut its shape and volume from the intersecting massing geometry.

For example, Figure 9-69 shows the profile of an arcade being sketched using the **Void Extrusion** tool. Now, when you choose the **Finish Edit Mode** button from the **Modify | Create Void Extrusion** tab, the cutting geometry, an arcade in this case, is generated and automatically cut from the larger massing geometry. Figure 9-70 shows the resulting massing geometry.

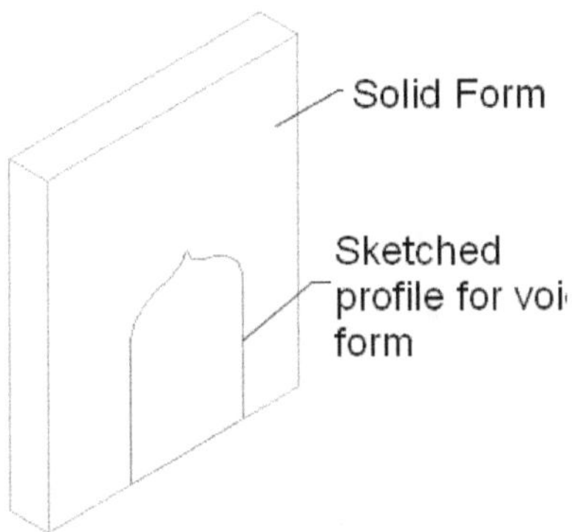

*Figure 9-69 The profile for creating a void form using the **Void Extrusion** tool*

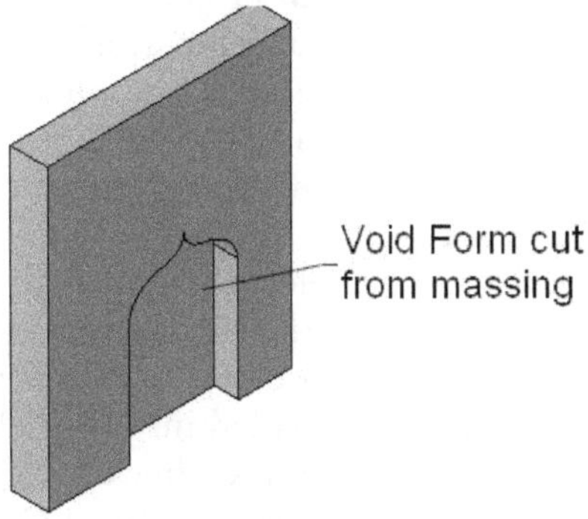

Figure 9-70 The generated void form and the resulting massing geometry

Similarly, you can also use the **Void Revolve**, **Void Sweep**, **Void Blend**, and **Void Swept Blend** tools from the **Void Forms** tool to create the void form. For example, to create a tunnel through a building block, you can use the **Void Sweep** tool to sketch a semicircular profile and specify its path, as shown in Figure 9-71. The resulting void form will be cut from the cuboid to create an arched opening through the building block, as shown in Figure 9-72.

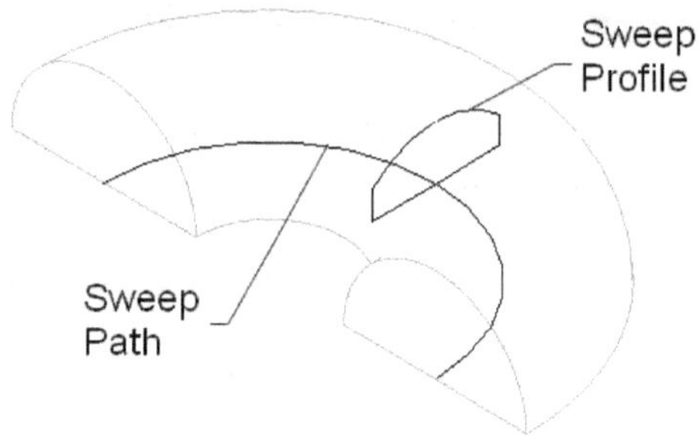

Figure 9-71 Sketching the profile for creating a void form

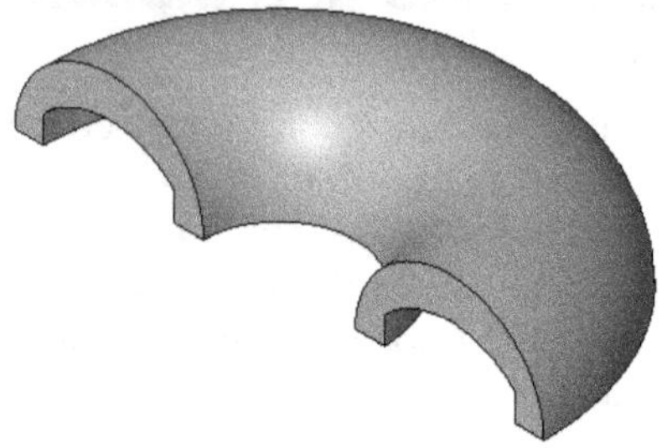

Figure 9-72 The generated void form and the resulting massing geometry

You can also use various editing tools such as **Copy**, **Mirror**, **Array**, and so on to create multiple copies of the void form profile for a single massing geometry. For example, a commercial building block has a semicircular vertical cut as the void form on one of its sides, you can select the 2D profile of the void form and use the **Copy** tool to create its duplicate

Loading Massing Geometry into the Project

After creating massing geometries in the Family Editor, you can load them into the Project environment. To do so, choose the **Load into Project** tool from the **Family Editor** panel of the **Modify** tab; the current project file will appear, if one project file is opened on the screen. Note that if more than one project file is opened in the current session, the **Load into Projects** dialog box will be displayed, as shown in Figure 9-73. In this dialog box, you can select the check box(es) to select the project(s) in which the created mass will be loaded. After selecting the required check box(es), choose **OK**; the mass will be loaded into the selected project file(s) corresponding to the check boxes selected. In this project file, the **Modify | Place Component** tab is chosen and you will notice that the mass created in the Family Editor appears in the drawing area along with the cursor.

*Figure 9-73 The **Load into Projects** dialog box*

Placing the Massing Geometry in a Project

In the Project environment of Autodesk Revit, you have the option to add a predefined massing geometry to the project using the **Place Mass** tool. To add a predefined geometry, invoke this tool from the **Conceptual Mass** panel of the **Massing & Site** tab; the **Massing - Show Mass Enabled** window will be displayed. Choose the **Close** button from the window; the **Revit** message box will be displayed. Choose the **Yes** button from this message box; the **Load Family** dialog box will be displayed. In this dialog box, browse to the **Mass** subfolder in the **US Imperial** folder and then select a massing family. After selecting the massing family, choose the **Open** button; the **Load Family** dialog box will be closed. Also, the **Modify | Place Mass** tab will be displayed and the selected family will be loaded into the project. Now, click in the drawing area to place the loaded massing geometry.

In Revit, you can either align a mass geometry to the selected faces of the components created or place it on the defined work plane based on your requirement. To place a mass geometry on a selected face, choose the **Place on Face** tool from the **Placement** panel of the **Modify | Place Mass** tab and then place the cursor on the face of the component with which you want to align the massing geometry. You will notice that the mass geometry is aligned with the face on which the cursor is placed. Now, you can click to place the mass on the plane. Similarly, while placing a mass in a drawing, you can place it at a desired level. To do so, choose the **Place on Work Plane** tool from the **Placement** panel of the **Modify | Place Mas**s tab; the options for placing the mass will be displayed in the **Options Bar**. In the **Options** Bar, select the required level from the **Placement Plane** drop-down list and click in the drawing area; a mass will be created in the selected level. Before aligning the mass geometry to a desired face or in a work plane, you can also rotate it about an axis. To do so, select the **Rotate after placement** check box from the **Options Bar**. After placing the mass geometry in the drawing, you can edit its geometrical properties such as height, width, radius, and so on, depending upon the type of geometry created. To do so, modify the instance parameters displayed in the **Properties** palette. In the **Dimensions** category of the **Properties** palette, you can select the desired dimension instance parameter and change its value in its respective value field.

Creating the In-Place Mass in a Project

In the Project environment, you have the option to create an in-place massing geometry by using the **In-Place Mass** tool. Invoke this tool from the **Conceptual Mass** panel of the **Massing & Site** tab; the **Name** dialog box will be displayed. In this dialog box, enter the name of the mass in the **Name** edit box and choose **OK**; the **Create** tab will be displayed. You can use the options in this tab to sketch the massing profile and convert it into a solid or void form. To sketch the profile for the mass, choose the **Mode**l tool from the **Draw** panel; the **Modify | Place Lines** tab will be displayed. You can use various sketching tools available in the **Draw** panel of this tab. While sketching, you can also use the **Reference Plane** tool from the **Draw** panel to draw references for the sketch. After sketching the profile, you can use any of the tools displayed in the **Create Form** drop-down in the **Form** panel. The **Create Form** drop-down displays two tools: **Solid Form** and **Void Form**. You can use the **Solid Form** tool to create a solid form and the **Void Form** tool to create a void form. To use any of the tools from the **Create Form** drop-down, you need to select the sketched profile from the drawing and then invoke the **Solid Form** or **Void Form** tool. After invoking any of these tools, the **Modify | Form** tab will be displayed. In this tab, you specify the settings to change the instance property of the mass created, divide the surfaces of the mass, and modify geometrical elements of the mass. You can change the instance properties of the massing geometry like material and visibility from the **Properties** palette. After changing the instance properties of the selected massing geometry, you can change the geometrical property of the surface by adding edges and profiles. The tools for changing the geometrical properties are: **X-Ray**, **Add Edge**, and **Add Profile**. You can invoke these tools from the **From Element** panel of the **Modify | Form** (solid form) tab.

The **X-Ray** tool in the **Element** panel of the **Modify | Form** tab is used to display the skeleton like vertices and edges of the mass. The **Add Edge** tool can be used to add edges to the form of the mass and the **Add Profile** tool is used to add profile to the surface of the mass. Figure 9-74 shows the added edge and profile in the **X-Ray** mode.

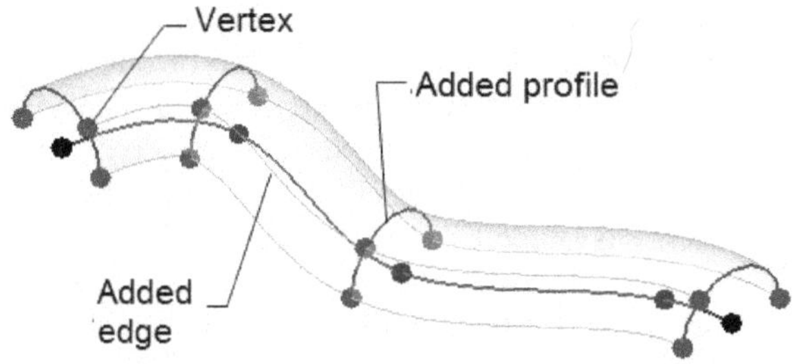

*Figure 9-74 Massing geometry with added edge and profile in the **X-Ray** mode*

Dividing Edges or Paths

You can divide a line or edge of a form to define the nodes on which the components can be placed. On dividing a line or edge, nodes are applied to the division points that denote the position of the points of placement for components. You can divide a line or an edge of a form by using any of the following methods: specifying number of divisions, specifying distance between divisions, or by using the point of intersection between the line or edge with the levels, reference planes, or divided paths. To divide a path on an edge, select a form from the project view; the **Modify | Mass** tab will be displayed. In this tab, choose the **Edit In-Place** tab; the form will be displayed in the editable mode. Select an edge of the form; the **Modify | Form** tab will be displayed. In this tab, choose the **Divide Path** tool from the **Divide** panel; the properties of the path division will be displayed in the **Properties** palette. Also, the selected edges will display 6 nodes in equal division along its edge. You will also notice the number **6** displayed above the edge. You can click on the number and enter a different numerical value in the edit box displayed to increase the number of nodal divisions along the edge.

MASSING IN CONCEPTUAL DESIGN ENVIRONMENT

Conceptual design is the very first phase of a design process. The primary objective of a conceptual design is to create a representation of the idea generated for creating the building mass of a project. For creating a building project, the conceptual design is very important for architects, engineers, and designers as it helps in finding out the final representation of the design intent. It thus enables them to create more specific sets of plans.

In Revit, you can create conceptual designs in any of the following environments: Conceptual Design Environment and Revit Project Environment. The Conceptual Design Environment is used to create massing families as conceptual designs that reside outside the Project Environment. These families can be loaded in the Revit Project Environment that forms the basis to create detail design by applying walls, roofs, floors, and curtain systems to them. In the Revit Project Environment, you can create conceptual designs by using the **In-Place Mass** or **Place Mass** tool from the **Conceptual Mass** panel in the **Massing & Site** tab. In this environment, you cannot create 3D reference planes and view 3D levels.

Interface of the Conceptual Design Environment

To open the interface of the Conceptual Design Environment, choose the **Application** button and then choose **New > Conceptual Mass** from the Application Menu. On doing so, the **New Conceptual Mass-Select Template File** dialog box will be displayed. In this dialog box, select the **Mass.rft** file in the Conceptual Mass folder and then choose the **Open** button; the interface of the Conceptual Design Environment along with the settings of the Mass.rft template file will be displayed, as shown in Figure 9-75. In the interface of the Conceptual Massing Environment, you can use various tools in the ribbon to create the conceptual mass.

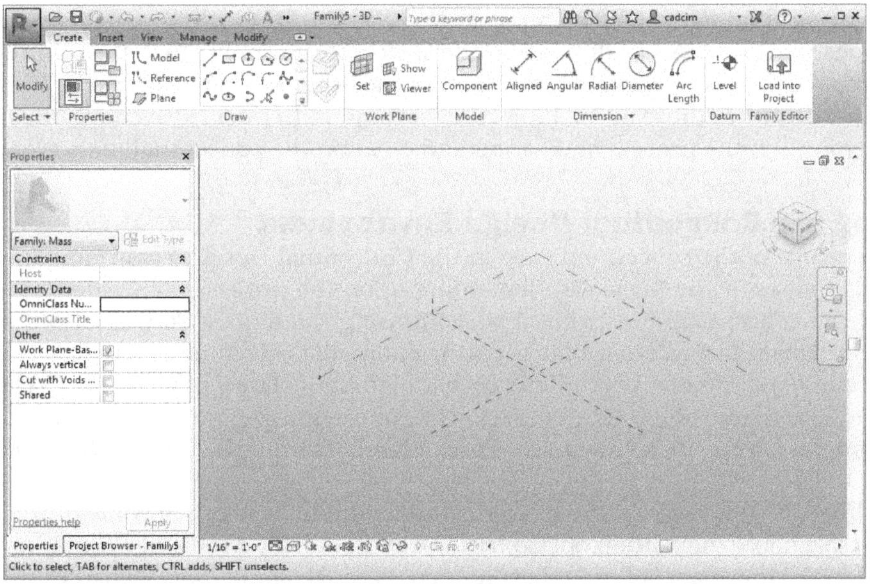

Figure 9-75 The Conceptual Design Environment interface

Creating Masses in the Conceptual Design Environment

In the Conceptual Design Environment, you can create the conceptual mass by using various tools from the ribbon. The ribbon of the Conceptual Design Environment, as shown in Figure 9-76, consists of the following tabs: **Create**, **Insert**, **Views**, **Manage**, and **Modify**.

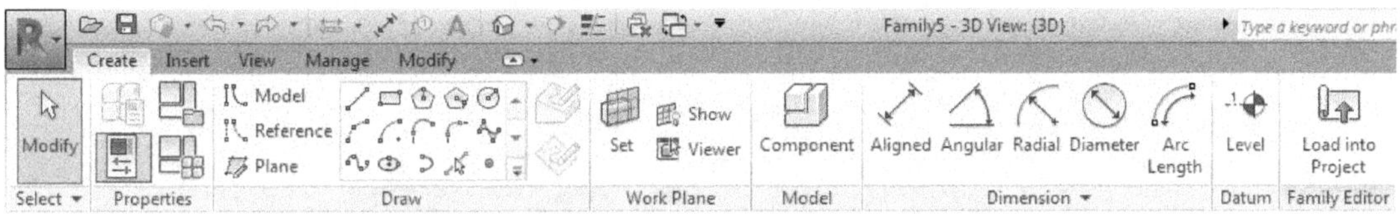

Figure 9-76 The ribbon of the Conceptual Design Environment

The **Create** tab is chosen by default. From the **Create** tab, you can access the tools required to create the Conceptual Mass. From the **Work Plane** panel of the **Create** tab, you can choose any of the tools such as **Set**, **Show**, and **Viewer** that can be used for setting, showing, and viewing workplane to be used to create the conceptual mass. You can choose the **Level** tool in the **Datum** panel to add levels that will assist in the creation of its conceptual mass. You can add levels in 3D and elevation views in the Conceptual Design Environment. The **Draw** panel of the **Create** tab contains tools that can be used to sketch conceptual mass. In this panel, you can use the **Model**, **Reference**, and **Plane** tool to sketch the model lines, reference lines and reference planes to create the conceptual mass. When you choose the **Model** or **Reference** tool from the **Draw** panel, the **Draw on Face** and the **Draw on Work Plane** will be activated. You can use any of these tools to specify the plane in which conceptual mass will be created. Also, on choosing the **Model** or **Reference** tool from the **Draw** panel, a contextual tab will be displayed along with various options in the **Options Bar**. This tab contains options that help in sketching the model or reference line. In the **Family Editor** panel of the **Create** tab, you can choose the **Load into Project** tool to load the conceptual mass into the project file. From the **Dimension** panel of the **Create** tab, you can choose various tools to dimension the sketch of the conceptual mass. In the ribbon of the Conceptual Design Environment, the tools in the other tabs such as **Insert**, **View**, **Manage**, and **Modify** are same as those in the Project Environment. In the next sections, the method and the tools involved in the process of drawing and creating the conceptual mass are discussed.

Sketching the Conceptual Design Environment

To sketch a profile of the conceptual mass in the Conceptual Design Environment, you can use the floor plan views or the 3D views, depending upon the project requirement. You can draw a profile of the conceptual mass with or without using references. You can use the references to create constraints and reference for the conceptual mass. To draw references for the profile conceptual mass, you can invoke the **Reference** tool from the **Draw** panel of the **Create** tab. After invoking the **Reference** tool, you can draw the reference by using various sketching tools such as **Line**, **Rectangle**, **Circle**, **Pick Lines**, and **Point Elements** from the list box displayed next to it.

After sketching the reference, you will sketch the model lines for the profile of the conceptual mass. To do so, choose the **Model** tool from the **Draw** panel; the **Modify | Place Lines** tab will be displayed. In the **Draw** panel of this tab, you can use the sketching tools such as **Line**, **Rectangle**, and others to sketch the conceptual mass. When you invoke a sketching tool, its corresponding options are displayed in the **Options Bar**. These options assist in sketching the profile. After you sketch the conceptual mass, you can also use various editing tools such as **Offset**, **Move**, **Copy**, and others from the **Modify** panel of the **Modify | Place Lines** tab to edit the sketch. Figure 9-77 displays the sketched profile of a conceptual mass.

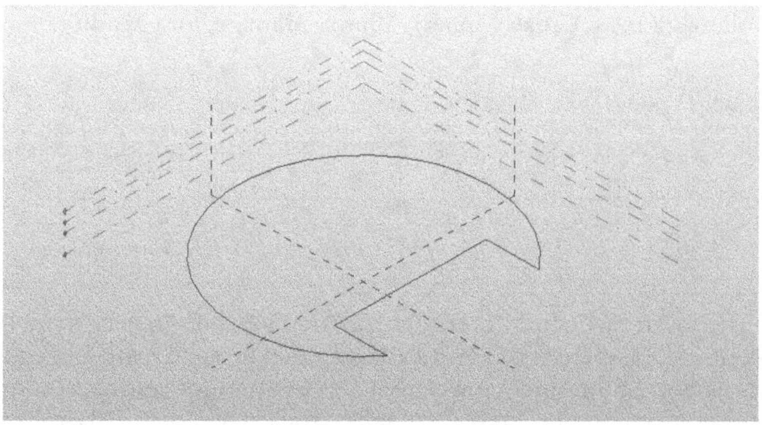

Figure 9-77 *Sketched profile of a conceptual mass*

After sketching the conceptual mass, select it; the **Modify | Line** tab will be displayed. In the **Form** panel of this tab, you can click on the **Create Form** drop-down and choose the **Solid Form** or the **Void Form** tool from the drop-down menu. You can choose the **Solid Form** tool to create a solid from the sketched lines. Alternatively, choose the **Void Form** tool to create a void or cavity in an existing solid. On choosing the **Solid Form** tool from the **Create Form** drop-down, the **Modify | Form** tab will be displayed. Also, the selected sketch line will extrude to a certain height and will appear along with its temporary dimensions. Figure 9-78 displays a solid extruded from the sketched lines.

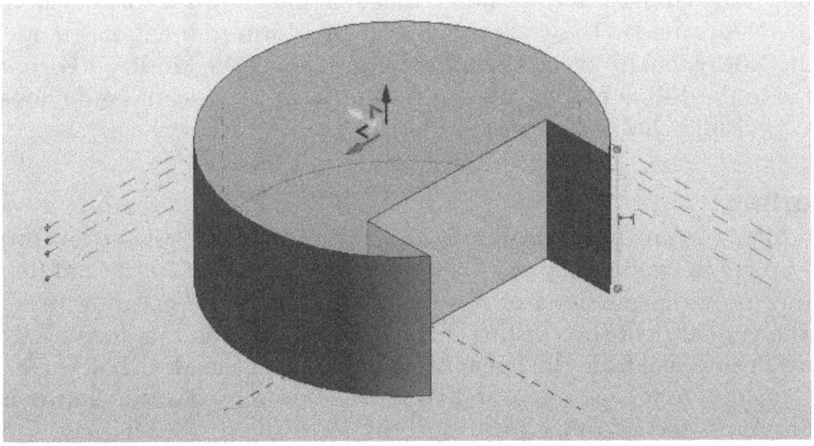

Figure 9-78 *Solid extruded from the sketched lines*

After creating a solid, you can edit its profile. To do so, select the extruded face of the solid profile and then choose the **Edit Profile** tool from the **Mode** panel of the **Modify | Form** tab. On doing so, the profile of the selected face will turn into sketch mode, as shown in Figure 9-79. You can edit the sketch using various tools in the **Modify Form > Edit Profile** tab. After editing the sketched profile, choose the **Finish Edit Mode** button from the **Mode** panel to return to the **Modify | Form** tab.

You can divide the surfaces of a conceptual mass, add profile and edge to its form, display the geometric skeleton of its form, and dissolve the surfaces of its form. The various tools that you can use to perform these functions are **Divide Surface**, **X-Ray**, **Add Profile**, **Add Edge**, and **Dissolve**, which are discussed next.

Figure 9-79 *Profile of the selected face in the editing mode*

Using X-Ray

The **X-Ray** tool can be used to display the underlying geometric skeleton of a selected form. The mode in which the geometric skeleton of the selected form will be displayed is called the X-Ray mode. The advantage of using this mode is that it will allow you to directly interact with the individual elements that comprise the form, as the surfaces become transparent. It is to be noted that at a time only one form in all the models will be visible in the X-Ray mode. To switch the display of a form to the X-Ray mode, select it and then choose the **X-Ray** tool from the **Form Element** panel of the **Modify | Form** tab. Once you switch the form to the X-Ray mode, you can select the element again and choose the **X- Ray** tool to restore its visibility to its previous mode.

Adding Profiles

After you create a conceptual mass, you can add cross-section profiles to it. An edge is a profile that may be a single line, chain of connected lines, or a closed loop, depending upon the geometry of the form. The added profile can be used to generate a form or edit the geometry of the existing form. To add a profile to a form, select it and then choose the **Add Profile** tool from the **Form Element** panel of the **Modify | Form** tab. On doing so, the cursor changes to a cross mark and you will be prompted at the **Status Bar** to insert a new profile at the cross-section of the form element. Move the cursor on the surface of the form, the surface of the form will be highlighted in a blue border and a probable profile will be displayed at the position of the cursor. Click at a desired point on the surface of the form to add the profile. Note that the added profile in the form will only be visible in the X-Ray mode. Figure 9-80 displays a form with added profiles.

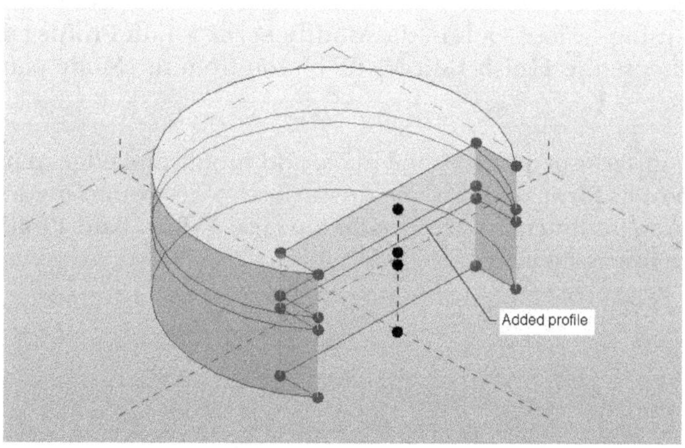

Figure 9-80 *Profiles added in the form*

Adding Edges

After you create a conceptual mass, you can add edges to it. To add an edge, select the form and then choose the **Add Edge** tool from the **Form Element** panel of the **Modify | Form** tab. On doing so, the cursor will change into a cross mark. Place the cursor on the surface of the form; an edge will appear at the point of the cursor. Click at a desired point on the surface and add the edge to the surface of the form, as shown in Figure 9-81.

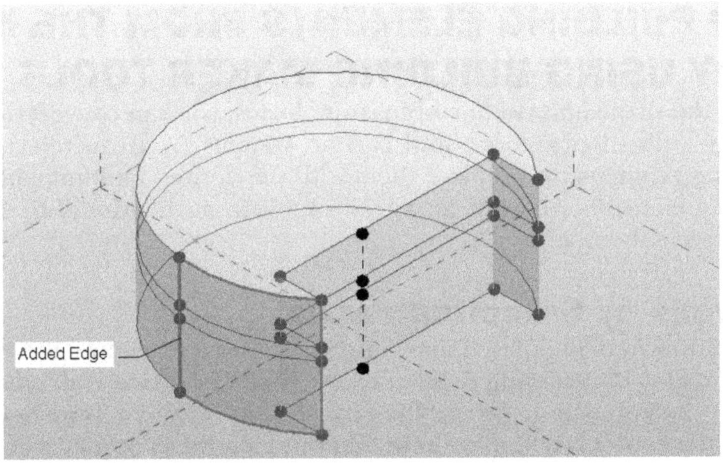

Figure 9-81 *Edge added to the form*

Dissolve

You can delete the surface of the form and retain its profile. You can also edit the retained profiles and recreate a form. To delete the surface of a form, select the form and choose the **Dissolve** tool from the **Form Element** panel of the **Modify | Form** tab. On doing so, the surface(s) will be deleted from the form. Figure 9-82 displays a form with its surfaces removed by using the **Dissolve** tool.

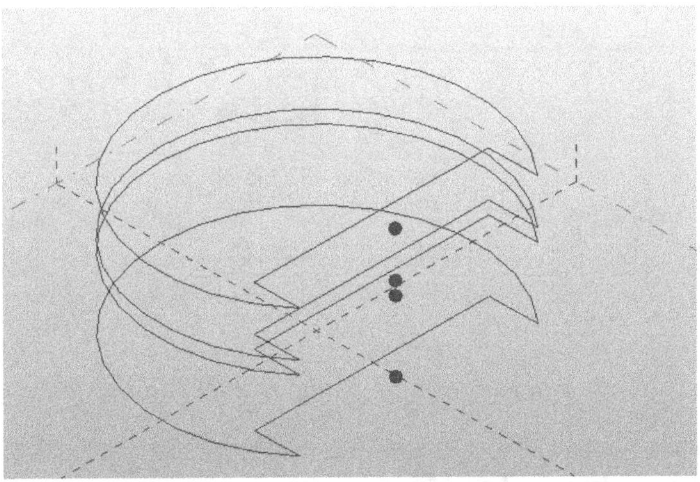

Figure 9-82 *Surfaces removed from the form*

Tip

*You can divide any face of the form. To do so, select the surface of the form and then choose the **Divide Surface** tool from the **Divide** panel of the **Modify / Form** tab. On doing so, the **Modify / Divided Surface** tab will be displayed. You can use various tools in the displayed tab and options in the **Options Bar** to divide the selected surface.*

CREATING BUILDING ELEMENTS FROM THE MASSING GEOMETRY USING BUILDING MAKER TOOLS

After generating the shape and volume of massing forms, you can convert them into building elements such as walls, floors, roofs, and curtain systems. You can select the faces of the generated massing geometry and replace them with the desired building element. To convert faces into building elements, you can use various tools from the **Model by Face** panel of the **Massing & Site** tab in the ribbon.

Creating Walls by Selecting Faces

You can convert the faces of the placed mass into walls. To do so, choose the **Wall** tool from the **Model by Face** panel of the **Massing & Site** tab; the **Modify | Place Wall** tab will be displayed. In this tab, select the wall type to be used for conversion from the **Type Selector** drop-down list. In the **Draw** panel of this tab, make sure that the **Pick Faces** option is chosen. In **Options Bar**, select the appropriate values of the **Level**, **Height**, and **Location Line** from their respective drop-down lists.

Now, to convert the faces into walls, move the cursor over the massing geometry. You will notice that the faces that can be converted are highlighted. Next, click on the desired face to make the conversion. You can use the TAB key to select all the faces of the massing geometry that can be converted into walls. Figure 9-83 shows the selected faces of the massing geometry and Figure 9-84 shows the walls created.

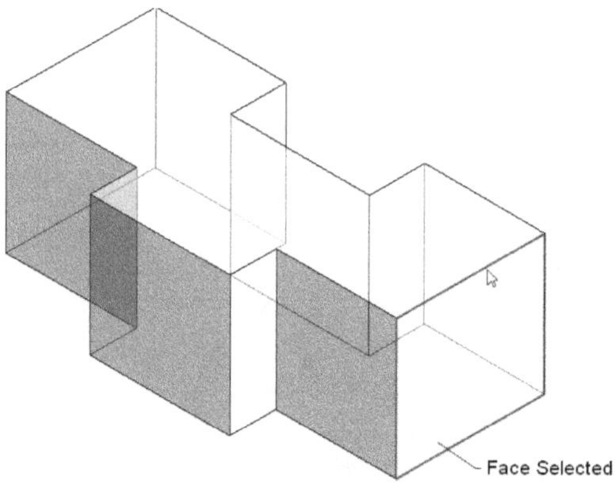

Figure 9-83 *Selected wall face*

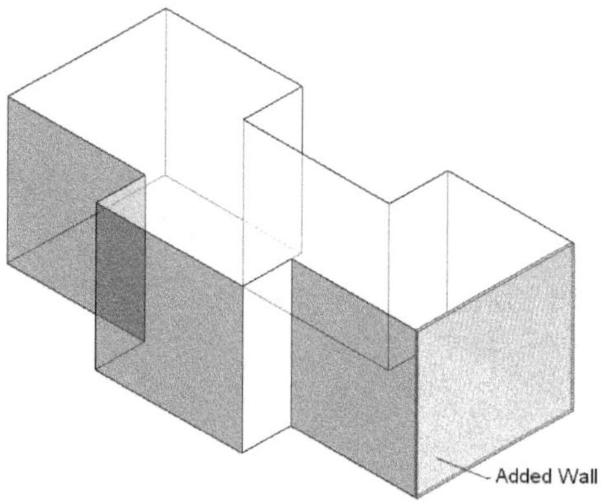

Figure 9-84 *Massing face converted into walls*

Creating Floors by Selecting Faces

To create a floor in a massing geometry, first you need to specify the faces of floor by selecting the massing geometry and then choosing the **Mass Floors** tool from the **Model** panel of the **Modify | Mass** tab. On doing so, the **Mass Floors** dialog box will be displayed. Select the levels at which the floor area faces are required and choose the **OK** button. Autodesk Revit will automatically generate the faces of floor based on the massing geometry and the defined levels.

Next, to add the floor to the faces, you need to invoke the **Floor** tool from the **Model by Face** panel; the **Modify | Place Floor by Face** tab will be displayed. You can select a floor type from the **Type Selector** drop-down list. Next, move the cursor over the massing geometry; the floor area faces will be highlighted. Now, you can choose the **Select Multiple** tool from the **Multiple Selection** panel and use the crossing options to create a selection of multiple floor area faces. Next, choose the **Create Floors** button in the **Multiple Selection** panel of the **Modify | Place Floor by Face** tab to create floors. Autodesk Revit will convert all the selected floor area faces into the selected floor type. Figure 9-85 shows the floor area faces generated for a multistory building. The floors created are shown in Figure 9-86.

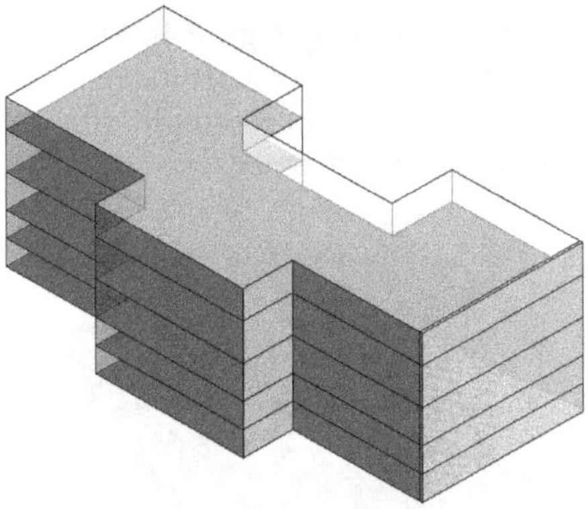

Figure 9-85 Generated floor area faces

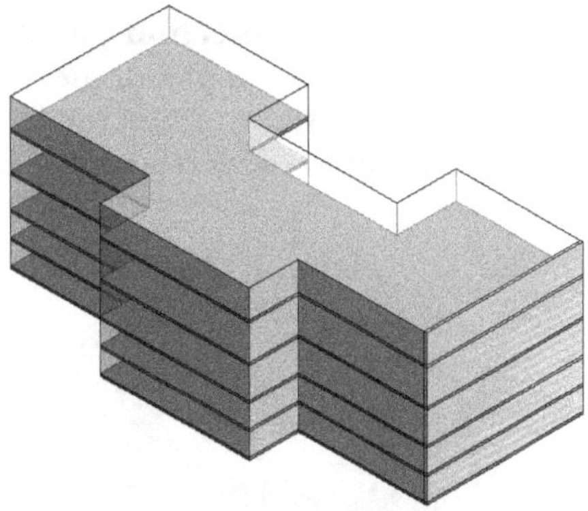

Figure 9-86 Floor area faces converted into floors

Creating Roofs by Selecting Faces

You can use the **Roof** tool in the **Massing & Site** tab to convert the massing faces into roofs. Faces can be horizontal planes of any shape or those created by using the **Extrusion**, **Revolve**, **Blend**, **Sweep**, or **Swept Blend** tool.

Invoke the **Roof** tool from the **Model by Face** panel; the **Modify | Place Roof by Face** tab will be displayed. From this tab, select the roof type to be used from the **Type Selector** drop-down list. Next, move the cursor over the massing geometry and click to select the highlighted face. Next, choose the **Create Roof** button in the **Multiple Selection** panel to convert it into a roof. Figures 9-87 and 9-88 show the top face of a multistory building being converted into a roof.

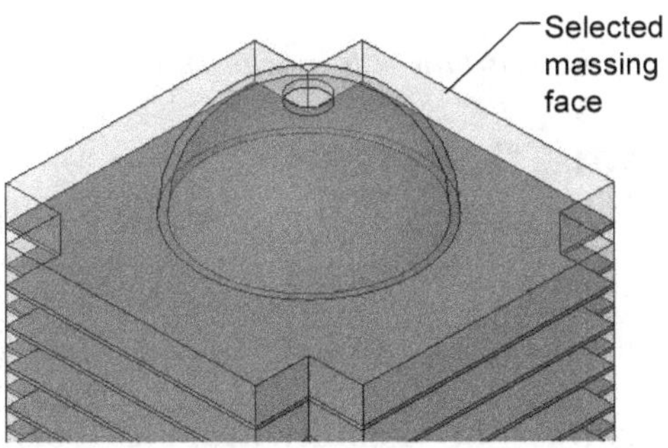

Figure 9-87 The top face selected for conversion into a roof

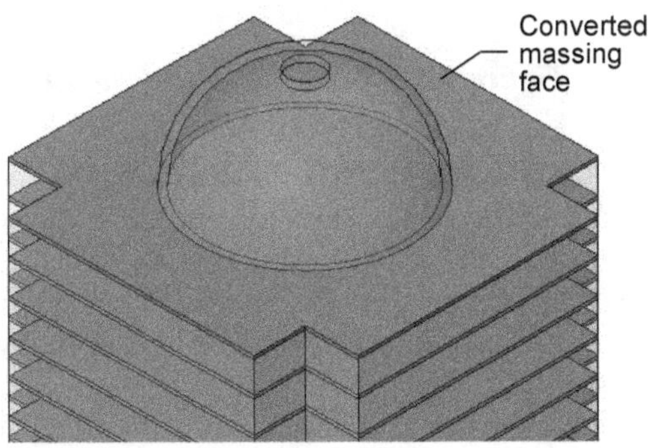

Figure 9-88 The top face of the massing geometry converted into a roof

Creating Curtain Systems by Selecting Faces

Autodesk Revit provides you with the **Curtain System** tool to convert the planar and non-planar faces into curtain systems. Invoke this tool from the **Model by Face** panel; the **Modify | Place Curtain System by Face** tab will be displayed. In this tab, select the curtain system type from the **Type Selector** drop-down list. In the **Properties** palette, you can modify the instance and type properties of the curtain system before creating it. Next, move the cursor over the face that needs to be converted into the curtain system and click to select it. Choose the **Create System** tool from the **Multiple Selection** panel to create the curtain system. Figure 9-89 shows the example of a curtain system of a dome. You can then add mullions to the curtain system using the **Mullions** tool, as shown in Figure 9-90. Figure 9-91 shows another example of using the **Curtain System** tool to convert non-planar massing faces into curtain systems.

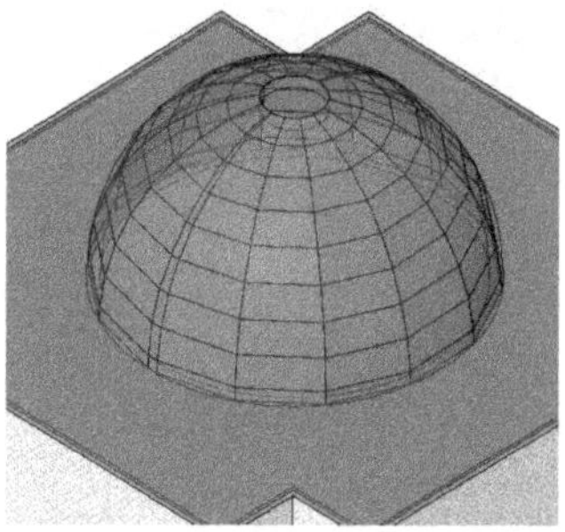

Figure 9-89 *Creating a curtain system using the* ***Curtain System*** *tool*

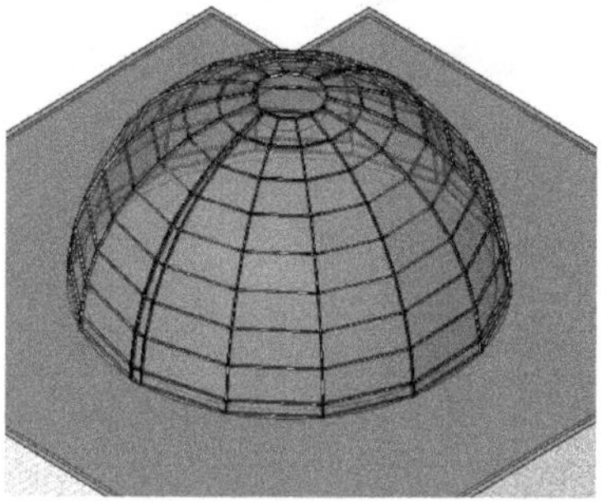

Figure 9-90 *Mullions added to the curtain system using the* ***Mullions*** *tool*

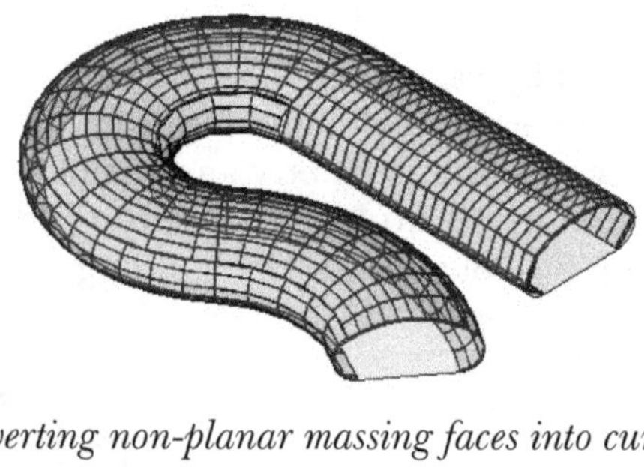

Figure 9-91 *Converting non-planar massing faces into curtain systems with mullions*

Controlling the Visibility of a Massing Geometry

To hide or display a massing geometry, choose the **Show Mass** tool from **Massing & Site > Conceptual Mass**. However, this tool has a temporary effect on the visibility of the massing forms. When a project file is closed, the show mass settings are not saved in it. Therefore, the next time you open the project file, the massing forms will not be visible.

To display massing form permanently in a project file, invoke the **Visibility/Graphics** tool from the **Graphics** panel of the **View** tab; the **Visibility/Graphics Overrides** dialog box for the current view will be displayed. In this dialog box, select the check box for the **Visibility** parameter of the **Mass** model category.

Adding other Building Elements

After converting the massing geometry into basic building elements, you can add other building elements such as doors, windows, roofs, and so on to the basic building elements. You can also use various editing tools to copy, edit, and delete the converted building elements.

Figure 9-92 shows the example of a building model in the plan view with modified wall type and added windows. You can use the editing tools such as **Mirror**, **Array**, and **Paste Aligned** to copy elements and create a building model of desired shape with specified parameters, as shown in Figure 9-93.

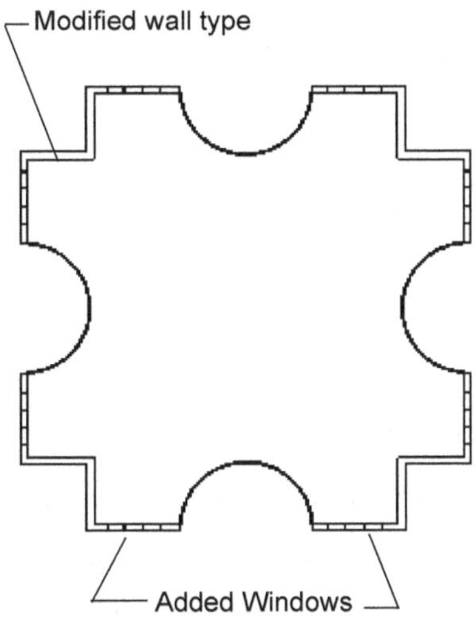

Figure 9-92 *Plan view showing the modified wall type and added windows*

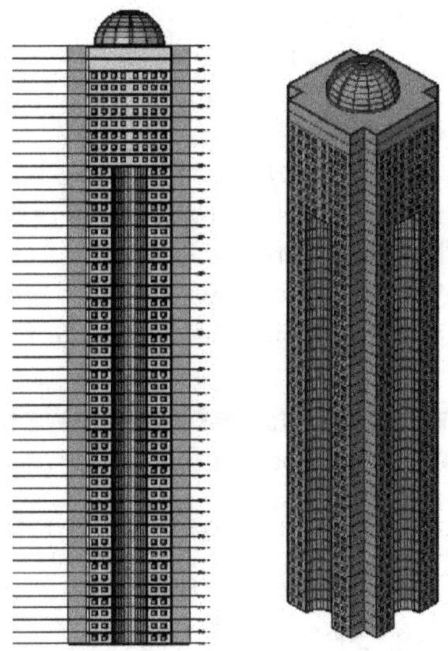

Figure 9-93 *A multistory building with building components*

Like other building models, you can also create the standard building views of the converted massing model. Autodesk Revit also provides the facility to extract the areas and other statistical figures of the created massing geometry.

Tip

In Autodesk Revit, the massing tools enable you to create a variety of shapes and volumes. You can use a combination of these tools on the basis of geometry to be created. The massing geometry thus created can be used to generate the rendered views of conceptual building volumes. The massing tools are very effective in communicating the design intent to the project team before actually implementing the detailed project plans.

TUTORIALS

Tutorial 1 Commercial Complex

In this tutorial, you will create a camera view in the **Entry** structural plan view of the *Commercial Complex* project for Imperial or *M_Commercial Complex* project for Metric created in Tutorial 1 of Chapter 8. Also, you will generate shadows, create a sheet, and add hosted area loads to the structural elements in the project. **(Expected time: 45 min)**

The following steps are required to complete this tutorial:

a. Open the existing project file.
   For Imperial *c08_Commercial-Complex_tut1.rvt*
   For Metric *M_c08_Commercial-Complex_tut1.rvt*
b. Create a camera view.
c. Generate shadows.
d. Create a sheet and add project view to it.
e. Add hosted area loads.
f. Save the project by using the **Save As** tool.
g. Close the project by using the **Close** tool.

Opening the Existing Project

In this section, you will open the project file created in Tutorial 1 of Chapter 8.

1. Choose **Open > Project** from the **Application Menu**; the **Open** dialog box is displayed.

2. In this dialog box, browse to *C:\rst_2017\c08_rst_2017_tut* location and choose the project file *c08_Commercial-Complex_tut1.rvt* for Imperial or *M_c08_Commercial-Complex_tut1.rvt* for Metric. You can also download this file from *http://www.cadcim.com*. The path of the file is as follows: *Textbooks > Civil/GIS > Revit Structure > Exploring Autodesk Revit 2017 for Structure*.

3. Choose the **Open** button from the **Open** dialog box; the selected project file with a schedule opens in the drawing window.

Creating a Camera View

In this section, you will create a camera view for the *Commercial Complex* project. The camera view will be created at a height of **5'6"** for Imperial or **1676** mm for Metric from the **Entry** level of the project.

1. Double-click on the **Entry** node under the **Structural Plans** head in the **Project Browser**.

2. Invoke the **Camera** tool from **View > Create > 3D View** drop-down; the options for inserting the camera are displayed in the **Options Bar**.

3. In the **Options Bar**, ensure that the **Perspective** check box is selected. Enter the value **5'6"** for Imperial or **1676** mm for Metric in the **Offset** edit box and then press ENTER.

4. Move the cursor down and click at a position between the **West** and **South** elevation arrows to specify the camera location. For the location of the camera, refer to Figure 9-94.

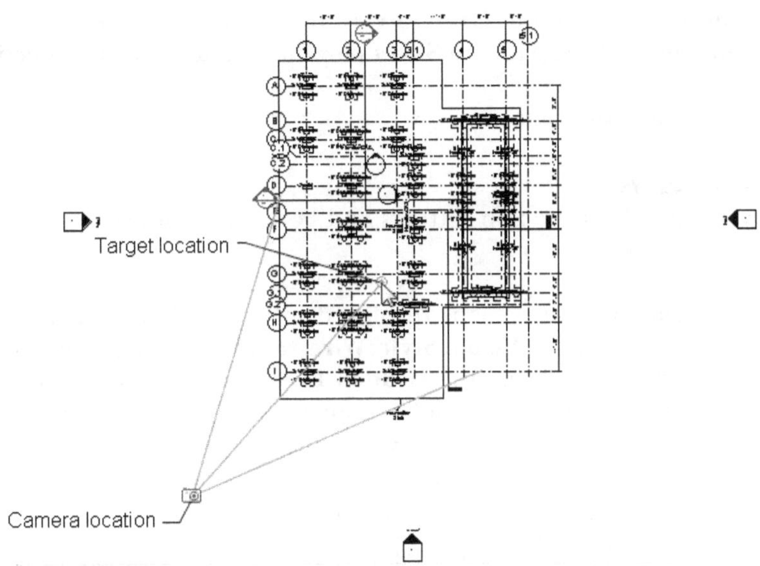

Figure 9-94 *Structural plan view showing the location of the camera and its target*

5. To specify the target location of the camera, move the cursor between the grid G, G.1 and 2, 3 location as shown in Figure 9-94; the project view gets changed and a view similar to the view shown in Figure 9-95 is displayed.

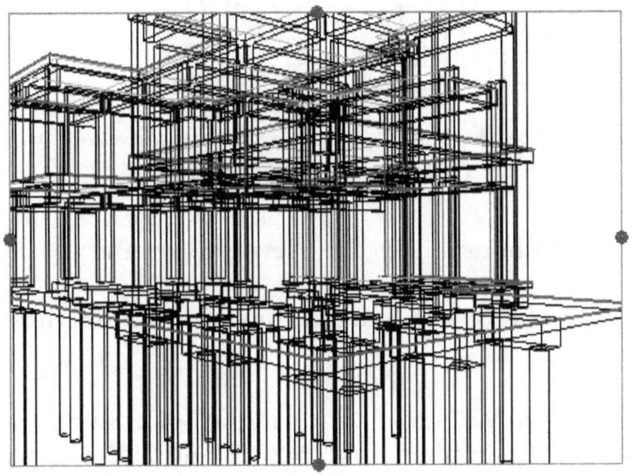

Figure 9-95 *Camera view displaying the structural model in wireframe*

6. In this view, ensure that the crop region of the 3D view is selected and the drag controls (blue dots at the midpoint of the four sides of the rectangle) are displayed.

7. Use drag controls to increase the extent of the view so that the entire structural model is visible inside the crop-region.

8. Ensure that the crop-region is selected. In the **Properties** palette, click in the value field of the

View Name parameter and enter **Camera-view** in it to replace the existing value. Similarly, click in the value field of the **Title on Sheet** parameter and enter **Perspective-View** in it.

9. Press ESC to exit the **Camera** tool.

Generating Shadows

In this section, you will adjust the Sun settings and generate shadows of the structural model displayed in the camera view.

1. Ensure that the **Camera-view** view is displayed in the **3D Views** head from the **Project Browser**. And then, choose the **Visual Style:Wireframe** button from the **View Control Bar**; a flyout is displayed. Next, choose the **Hidden Line** option from the flyout displayed; the hidden line view of the structural model is displayed, as shown in Figure 9-96.

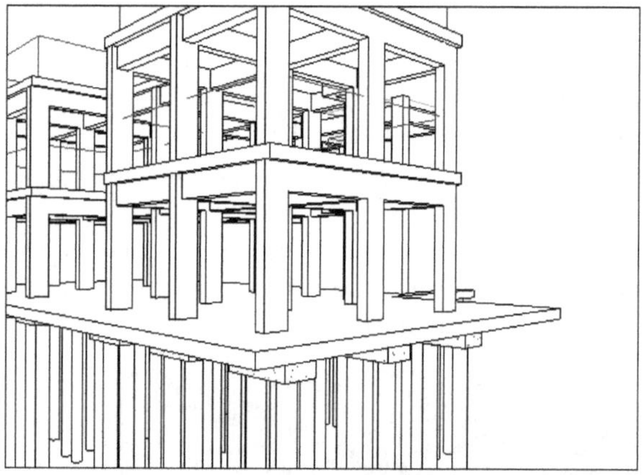

Figure 9-96 *Camera view displaying the structural model in the* ***Hidden Line*** *visual style*

2. Choose the **Sun Path Off** button from the **View Control Bar**; a flyout is displayed.

3. In the displayed flyout, choose the **Sun Settings** options; the **Sun Settings** dialog box is displayed.

4. In this dialog box, select the **Still** radio button in the **Solar Study** area. In the **Settings** area, specify the **Date** and **Time** parameters as **5/5/2016** and **2:45** respectively. Also, ensure that the **Location** parameter is set to **Boston, MA**. Choose the **Disable Updater** button from the **Autodesk Revit 2017** message box if displayed. The **Sun Settings** dialog box is displayed again.

5. Make sure that the **Ground Plane at Level** check box is selected from the **Sun Settings** dialog box. Next, select the **Entry** option from the drop-down list displayed below this check box.

6. Choose the **Apply** and then the **OK** button; the **Sun Settings** dialog box is closed.

7. In the **View Control Bar**, choose the **Shadows Off** button; a shadow is generated in the structural model. Figure 9-97 shows the camera view of the structural model displaying shadows due to the Sun settings.

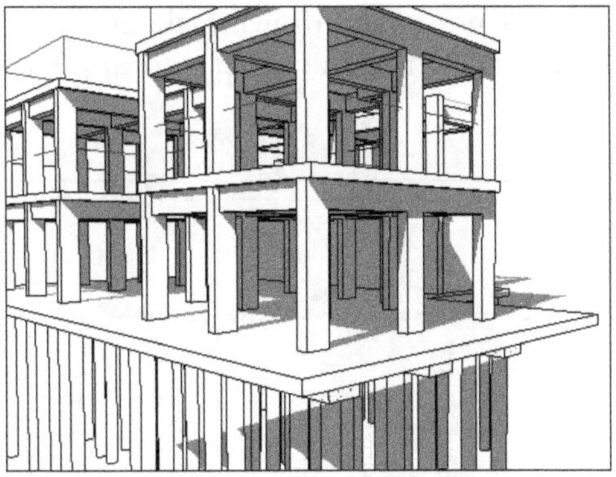

Figure 9-97 *Camera view displaying the structural model with shadows*

Creating a Sheet

In this section, you will create a sheet in the project.

1. Choose the **Sheet** tool from the **Sheet Composition** panel in the **View** tab; the **New Sheet** dialog box is displayed.

2. In this dialog box, choose the **Load** button; the **Load Family** dialog box is displayed.

3. In this dialog box, select the specified title block **C 17 x 22 Horizontal** from the **US Imperial > Titleblocks** folder for Imperial unit system and load **A0 metric** from the **US Metric > Titleblocks** folder for Metric unit system. Choose the **Open** button from the **Load Family** dialog box. On doing so, the specified title block is added to the list in the **Select titleblocks** region of the **New Sheet** dialog box.

4. Choose the **OK** button to create the sheet view by using the loaded title block. The added sheet is now displayed in the drawing window.

Adding Project Views to the Sheet

In this section, you need to add the specified project views to the sheet by dragging their name from the **Project Browser**. Further, based on the sheet layout, you need to place the project views at their designated place in the sheet.

1. In the **Project Browser**, click on the **Entry** node under the **Structural Plans**. Next, in the **Properties** palette, click in the value field corresponding to the **View Scale** parameter and select the **1/16" = 1'-0"** option for Imperial and **1:200** option for Metric from the drop-down list displayed.

2. Now, press and hold the left mouse button on the **Entry** node in the **Project Browser** and then drag the view into the drawing sheet. Release the left mouse button when the project view appears as a rectangle in the sheet.

3. Move the cursor to the lower left area of the title block such that the corner of the rectangle get closed to the lower left corner of the drawing sheet. Next, click to place the view; the **Entry** structural plan view is added to the sheet and appears enclosed in a rectangle, as shown in Figure 9-98.

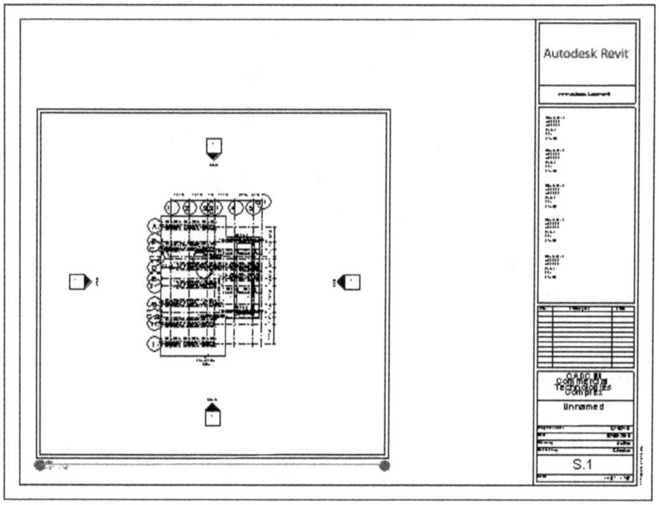

Figure 9-98 *The location of the **Entry** structural plan view in the sheet*

4. Next, ensure that the viewport of the **Entry** structural plan view is selected in the sheet. In the **Properties** palette, ensure that the check box corresponding to the **Crop Region Visible** parameter is cleared.

5. Choose the **Edit Type** button in the **Properties** palette; the **Type Properties** dialog box is displayed. In the **Type Properties** dialog box, select the **Title Only** option from the **Type** drop-down list and ensure that the check box corresponding to the **Show Extension Line** parameter is cleared and then choose the **OK** button.

6. Right-click in the drawing and then choose the **Activate View** option from the shortcut menu displayed.

7. Now, in the project view, click on the **North** elevation arrow head and right-click; a shortcut menu is displayed. Choose **Hide in View > Category** from it.

8. Again, right-click and then choose the **Pan Active View** option from the shortcut menu displayed; a move icon is displayed with the cursor.

9. In the sheet, press and drag the **Entry** structural plan view and then place it at the location, shown in Figure 9-99.

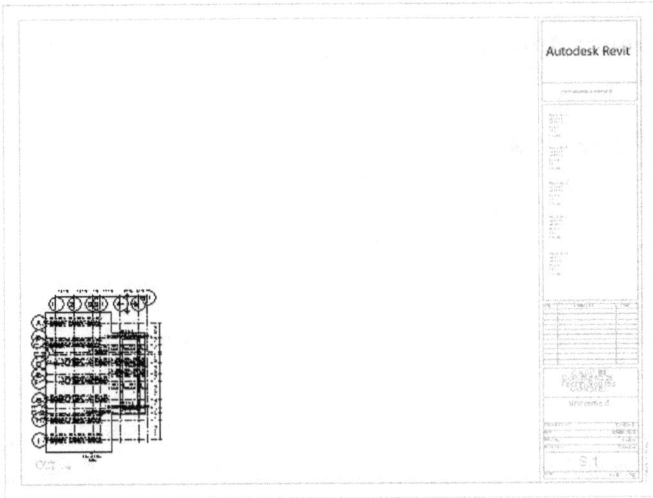

Figure 9-99 Modified location of the ***Entry*** *structural plan view*

10. Next, right-click and choose the **Deactivate View** option from the shortcut menu displayed.

11. Click on the **Camera-view** node under the **3D Views** head in the **Project Browser** and then drag it to the sheet. Next, release the mouse button and click at a location to place the **Camera-view** view, as shown in Figure 9-100.

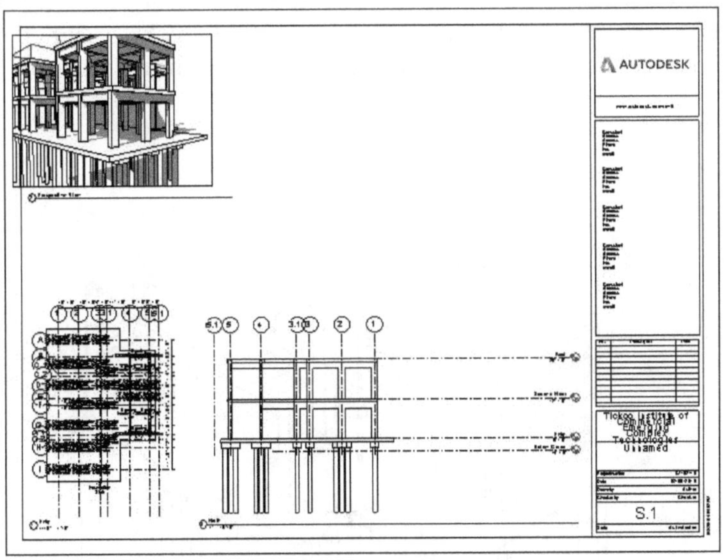

Figure 9-100 Sheet displaying the structural plan, elevation, and camera view

12. Click on the **North** node under the **Elevations** head and then specify the value of the **View Scale** parameter to **1" = 10'-0"** for Imperial and **1:100** for Metric. Next, drag the **North** node from the **Project Browser** and place it at the location, as shown in Figure 9-100. The three views are added to the sheet.

Adding Hosted Area Load

In this section, you will add a hosted area load to the structural floors.

1. Double-click on the **View 1- Analytical** node of the **Project Browser** under the **3D Views** head.

2. Choose the **Visual Style** button in the **View Control Bar** and then choose the **Hidden Line** option from the flyout displayed.

3. Next, choose the **Analyze** tab and then the **Loads** tool from the **Analytical Model** panel; the **Modify | Place Loads** contextual tab is displayed.

4. Choose the **Hosted Area Load** tool from the **Loads** panel. In the **Properties** palette, enter **-.5 ksf** for Imperial and **-23.94 KN/m2** for Metric in the value field of the **Fz 1** parameter and then choose the **Edit Type** button; the **Type Properties** dialog box is displayed.

5. In this dialog box, click on the **Value** field of the **Area force scale** parameter and then enter **10**. Next, choose **Apply** and then **OK**; the **Type Properties** dialog box is closed.

6. Now, in the project view, place the cursor on the edge of the structural floor at the **Roof** level, as shown in Figure 9-101. Next, click on this edge; the assigned load is displayed throughout the area of the floor at the **Roof** level, as shown in Figure 9-102.

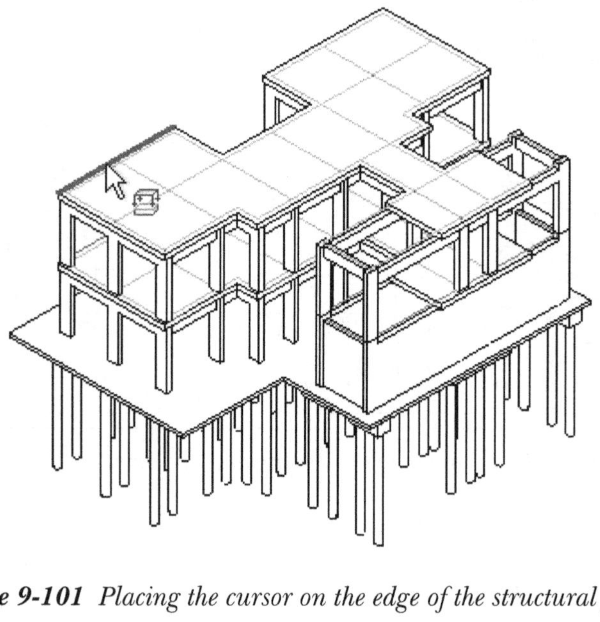

Figure 9-101 *Placing the cursor on the edge of the structural floor*

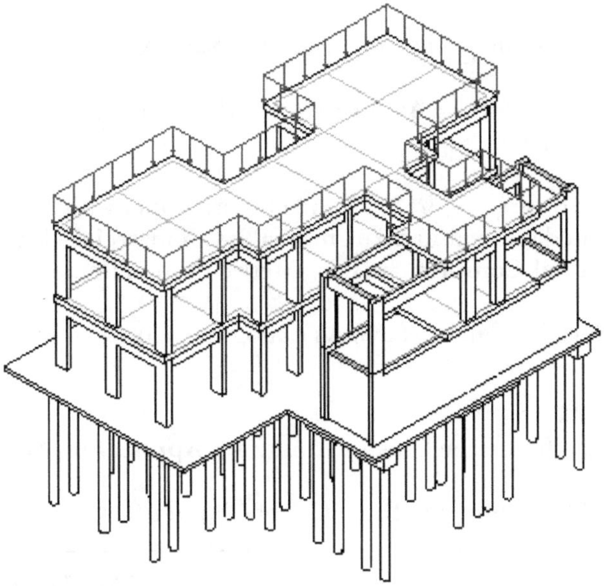

Figure 9-102 *The hosted area load added to the structural floor at the* ***Roof*** *level*

7. Repeat the procedure followed in step 6 to add the area load to the structural floor on the **Second Floor** level. Press ESC to exit the tool. Figure 9-103 shows the structural model after the hosted area load is added to the structural floor at the **Second Floor** level.

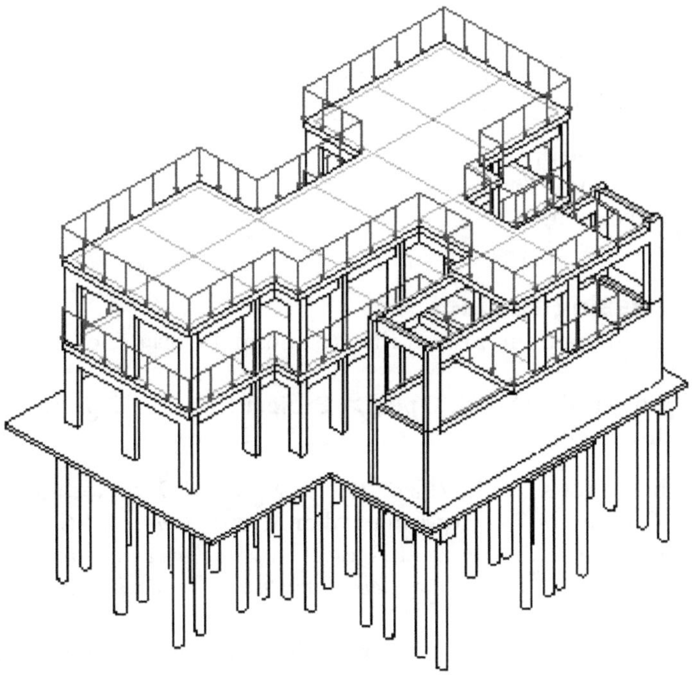

Figure 9-103 *The hosted area load added to structural floor at the* ***Second Floor*** *level*

Saving the Project

In this section, you will save the project file using the **Save As** tool.

1. To save the project, choose **Save As > Project** option from the **Application Menu**; the **Save As** dialog box is displayed.

2. In this dialog box, browse to *C:\rst_2017* and create a folder with the name *c09_rst_2017_tut*. Next, open the created folder and then enter **c09_Commercial-Complex_tut1** for Imperial or **M_c09_Commercial-Complex_tut1** for Metric in the **File name** edit box.

3. Now, choose the **Save** button; the **Save As** dialog box closes and the project file is saved.

Closing the Project

1. To close the project, choose the **Close** option from **Application Menu**.

   The file is closed. This completes Tutorial 1 of Chapter 9.

Tutorial 2 Industrial Complex

In this tutorial, you will set an analytical model and add boundary conditions to the project created in Tutorial 2 of Chapter 8. Also, you will add reinforcements to the structural floor.

(Expected time: 45 min)

The following steps are required to complete this tutorial:

a. Open the project file created in Tutorial 2 of Chapter 8.
   For Imperial *Industrial-Complex*
   For Metric *M_Industrial-Complex*
b. Set the Analytical Model.
c. Add boundary conditions.
d. Add area reinforcement.
e. Save the project.

Opening the Existing Project

In this section, you will open the project file created in Tutorial 2 of Chapter 8.

1. To open the project file, choose **Open > Project** from the **Application Menu**; the **Open** dialog box is displayed.

2. In this dialog box, browse to the *C:\rst_2017\c08_rst_2017_tut* location and then choose the *c08_Industrial-Complex_tut2.rvt* for Imperial or *M_c08_Industrial-Complex_tut2.rvt* for Metric project file. You can also download this file from *http://www.cadcim.com*. The path of the file is as follows: *Textbooks > Civil/GIS > Revit Structure > Exploring Autodesk 2017 for Structure*.

3. Choose the **Open** button from the **Open** dialog box; the selected project file opens in the drawing window.

Setting the Analytical Model

In this section, you will set analytical model to check the consistency of the structure.

1. Double-click on the **{3D}** node under the **3D Views** head in the **Project Browser**; the 3D view of model is displayed.

2. Choose the **Analyze** tab and then choose the **Analytical Model Settings** button in the **Analytical Model Tools** panel; the **Structural Settings** dialog box is displayed with the **Analytical Model Settings** tab chosen.

3. In this dialog box, select the **Member Supports** and **Analytical/Physical Model Consistency** check boxes in the **Automatic Checks** area.

4. Choose the **OK** button; the **Analytical Model Check** window is displayed. Choose the **Yes** button; the **Warning** window is displayed. Close the warning message.

5. Right-click on the View Cube; the shortcut menu is displayed. Select **Orient to a Direction > Southeast Isometric** and the view is displayed, refer to Figure 9-104.

Adding Boundary Conditions

1. In the **Analyze** tab, choose the **Boundary Conditions** tool in the **Analytical Model** panel; the **Modify | Place Boundary Conditions** tab is displayed.

2. Choose the **Area** tool in the **Boundary Conditions** panel, if it is not chosen by default. Move the cursor toward the left wing of the structural model and place it at the edge of the structural foundation slab, as shown in Figure 9-104.

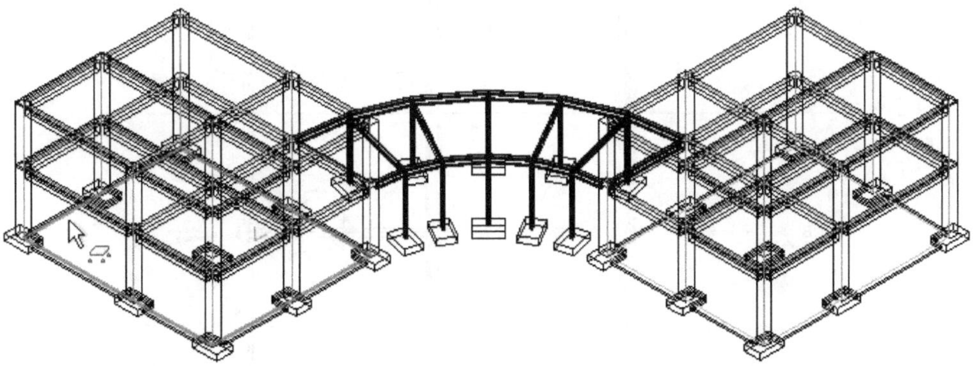

Figure 9-104 *The cursor placed at the edge of the foundation slab*

3. Next, click when the edge is highlighted; the boundary conditions are added to the slab. Similarly, add the boundary conditions to the structural foundation slab of the right wing of the structural model. Figure 9-105 shows the boundary conditions added to the foundation slab of both the wings of the structural model. Press ESC to exit.

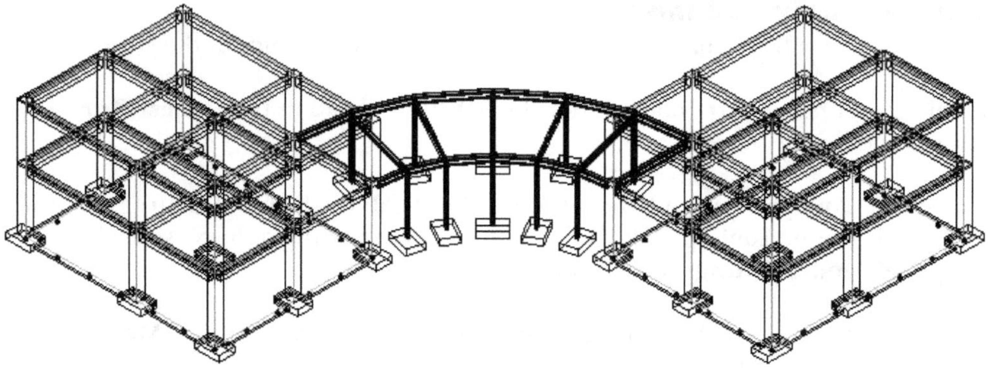

Figure 9-105 *The boundary conditions added to both foundation slabs*

Adding Area Reinforcement

In this section, you will add area reinforcement to the structural floor of the left wing of the structural model.

1. Double-click on the **Second Floor** node under the **Structural Plans** head in the **Project Browser**.

2. Place the cursor at the edge of the structural floor, as shown in Figure 9-106. Click when the edge gets highlighted; the **Modify | Floors** tab is displayed.

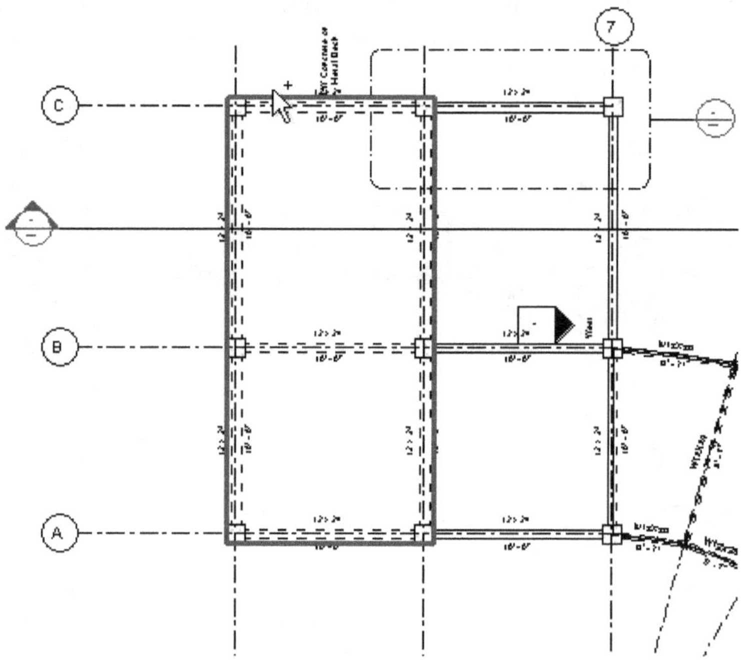

Figure 9-106 *The cursor placed at the edge of the structural floor*

3. In this tab, choose the **Area Reinforcement** tool from the **Reinforcement** panel; the **Revit** message box is displayed.

4. Choose the **OK** button in the **Revit** message box; the **Modify | Create Reinforcement Boundary** tab is displayed.

5. Ensure that the **Rebar Line** tool is chosen in the **Draw** panel. Choose the **Rectangle** tool from the list box in the **Draw** panel and place the cursor at the grid intersection 9 and C and click when the **Endpoint** object snap is displayed.

6. Now, drag the cursor down and place it at the grid intersection A and 8. Click, when the **Endpoint** object snap is displayed; a magenta-colored rectangular boundary is created, as shown in Figure 9-107.

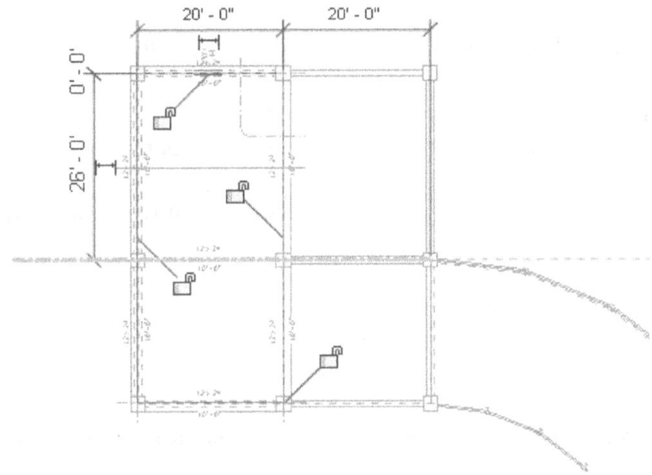

Figure 9-107 *The sketched boundary of the area reinforcement*

7. Choose the **Finish Edit Mode** button in the **Mode** panel of the **Modify | Structural Area Reinforcement** tab.

8. In the **Properties** palette, click in the value field of the **Top Major Bar Type** parameter and select the **#8** value from the drop-down list displayed.

9. Similarly, click in the value field of the **Bottom Major Bar Type** parameter and select the **#6** value from the drop-down list displayed. Choose the **Apply** button.

10. To view the reinforcement added, double-click on the **Section X** node under the **Sections (Building Section)** head in the **Project Browser**; the **Section X** section view is displayed.

11. Zoom in the floor placed in the **Second Floor** level of the left wing; the reinforcements added become visible, as shown in Figure 9-108.

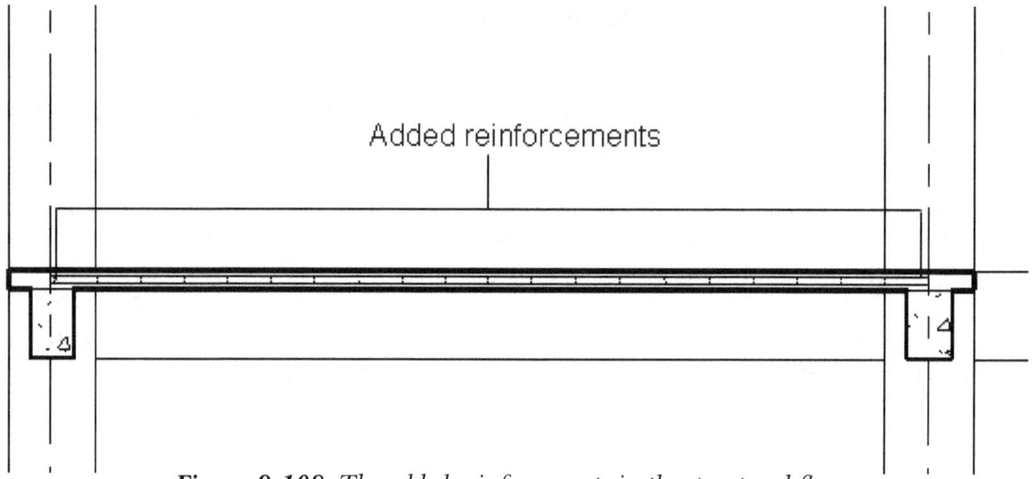

Figure 9-108 *The added reinforcements in the structural floor*

Saving the Project

In this section, you will save the project file using the **Save As** tool.

1. To save the project, choose the **Save As > Project** option from the **Application Menu**; the **Save As** dialog box is displayed.

2. In this dialog box, browse to *C:\revit_2017_structural_project\c09* and then enter **c09_Industrial- Complex_tut2.rvt** for Imperial or **M_c09_Industrial- Complex_tut2.rvt** for Metric in the **File name** edit box.

3. Next, choose the **Save** button; the **Save As** dialog box closes and the project file is saved.

Closing the Project

1. To close the project, choose the **Close** option from the **Application Menu**; the file is closed. This completes Tutorial 2 of Chapter 9.

Self-Evaluation Test

Answer the following questions and then compare them to those given at the end of this chapter:

1. The __________ 3D view is a view of a building model in which all elements are displayed in their actual size, irrespective of their distance from the source.

2. You can add the information related to a project in a sheet by using the __________ tool.

3. __________ study helps you visualize and analyze the position of the Sun and the solar effects on buildings and sites in the real world.

4. Parameters for the projection planes of an analytical model are displayed under the __________ head in the **Properties** palette.

5. You can activate a view from a drawing sheet by choosing the __________ option.

6. To specify the physical properties of a material, choose the __________ tab in the **Materials** dialog box.

7. The __________ tool is used to add reinforcement throughout the area of floor or slab.

8. In Autodesk Revit, you can display the trajectory of the Sun which creates the solar study. (T/F)

9. You can place a load in the physical model view of a structural model. (T/F)

10. You cannot add an existing schedule in a project to a sheet. (T/F)

Review Questions

Answer the following questions:

1. You can invoke the **Camera** tool from the __________ drop-down list in the **View** tab.

2. You can add view and edit load case in a structural model by choosing the __________ tool from the **Analyze** tab.

3. In a concrete column, beam, structural floor, or foundation, you can place rebars parallel to the current work plane by using the __________ tool.

4. The __________ tool is used to check the consistency of the analytical and physical model of a project.

5. The __________ tool is used to adjust the alignment of an analytical structural member in reference to its adjoining analytical member.

6. You can use the __________ tool from the **Boundary Conditions** panel to create a linear boundary throughout the length of a beam.

7. To sketch a point load in a model, you need to invoke the __________ tool.

8. You can check the support stability by using the __________ tool.

9. The default analytical projection planes of columns are placed at their tops and bases. (T/F)

10. After adjusting the members of an analytical model, you can bring the structural member back to its previous state. (T/F)

EXERCISES

Exercise 1 Academic Institution

In this exercise, you will create a sheet and add it to the *Academic-Institution* project for Imperial and *M_Academic-Institution* project for Metric, created in Exercise 1 of Chapter 8. For the location of the views in the sheet, refer to Figure 9-109. Use the following specifications:

(Expected time: 30 min)

1. Insert the following views in the sheet: **3D View: {3D}**, **Elevation: East**, **Elevation 1-a**, **Elevation 2-a**, and **Structural Plan: Entry**.

2 Use the desired sheet type.
   For Imperial **E130x42Horizontal**
   For Metric **A1 Metric**

3. File name to be referred:
   For Imperial *c08_Academic-Institution_exer1.rvt.*
   For Metric *M_c08_Academic-Institution_exer1.rvt.*

4. File name to be assigned:
   For Imperial *c09_Academic-Institution_exer1.rvt.*
   For Metric *M_c09_Academic-Institution_exer1.rvt.*

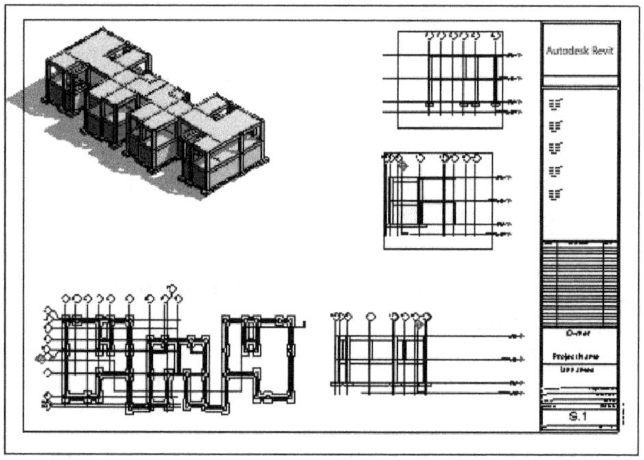

Figure 9-109 *The added views in the sheet*

Exercise 2 Factory Shed

In this exercise, you will add different views to a sheet for the *Factory Shed* project for Imperial or *M_Factory Shed* project for Metric created in Exercise 2 of Chapter 8. For the location of views in the sheet, refer to Figure 9-110. Use the following specifications:

(Expected time: 30 min)

1. Use the desired sheet type.
   For Imperial **E130x42Horizontal**
   For Metric **A1 Metric**

2. Insert the following views into the sheet: **3D View: {3D}**, **Elevation: South**, and **Structural Plan: Second Floor**

3. File name to be referred:
   For Imperial *c08_Factory-Shed_exer2.rvt.*
   For Metric *M_c08_Factory-Shed_exer2.rvt.*

4. File name to be assigned:
   For Imperial *c09_Factory-Shed_exer2.rvt.*
   For Metric *M_c09_Factory-Shed_exer2.rvt.*

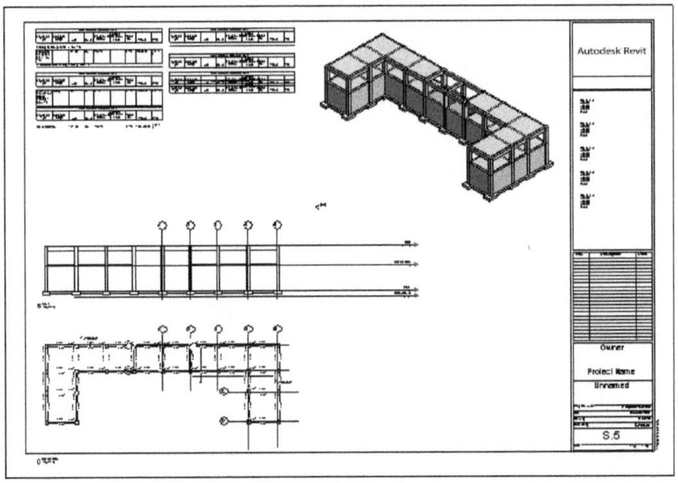

Figure 9-110 *Views added in the sheet*

Answers to Self-Evaluation Test

1. Orthographic, **2. Project Information**, **3.** Solar, **4. Analytical Model**, **5. Activate View**, **6. Physical**, **7. Area**, **8.** T, **9.** F, **10.** F

Chapter 10

Linking Revit Model with Robot Structural Analysis

Learning Objectives

After completing this chapter, you will be able to:

- *Analyze a model using Robot Structural Analysis software*
- *Modify a model in Robot Structural Analysis software*
- *Manage analysis results*
- *Run reinforcement code check*

INSTALLING ROBOT STRUCTURAL ANALYSIS PROFESSIONAL 2017

In Autodesk Revit, after developing a model, annotating it, preparing schedules, defining loads, and checking consistency, the model is ready for analysis and code checking. To analyze the model in Robot Structure Analysis 2017, you have to first install **Robot Structural Analysis Professional 2017** software in your system.

You can start Autodesk Robot Structural Analysis by double-clicking on the **Autodesk Robot Structural Analysis Professional 2017** icon on the desktop. Alternatively, you can choose the **Start** button at the lower left corner of the screen (default position) in the taskbar and then choose **All Programs > Autodesk > Autodesk Robot Structural Analysis Professional 2017 > Autodesk Robot Structural Analysis Professional 2017**; The interface screen of Autodesk Robot Structural Analysis Professional 2017 is shown in Figure 10-1.

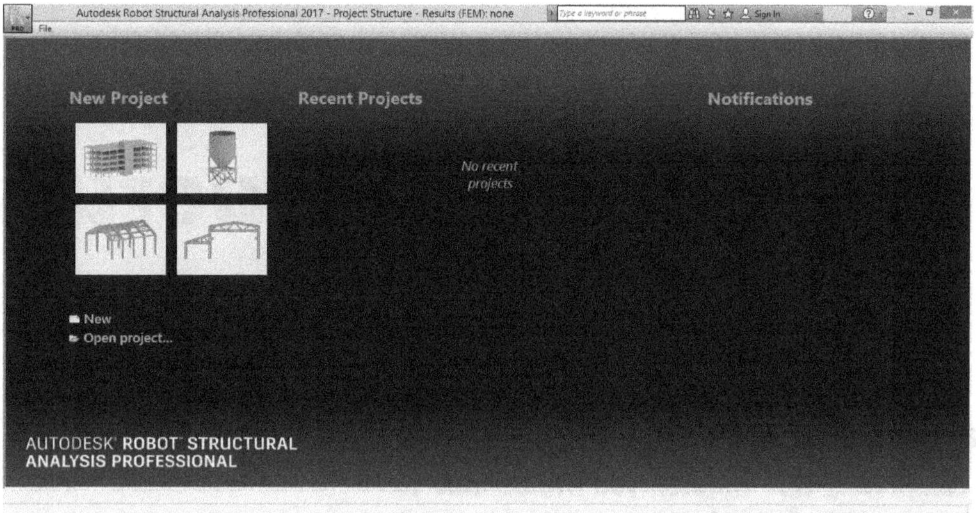

Figure 10-1** The interface screen of **Autodesk Robot Structural Analysis Professional 2017

STRUCTURAL ANALYSIS INTEROPERABILITY

In Autodesk Revit, you can export the analytical model to Autodesk Robot Analysis and perform structural analysis and also check code on it. You can also perform reinforcement design process of a concrete structure directly by linking it to Autodesk Robot Structural Analysis software. The methods of linking the analytical model from Autodesk Revit 2017 to Autodesk Robot Structural Analysis software are discussed next.

Linking the Analytical Model for Analysis and Code Check

To link the analytical model from Autodesk Revit 2017 to Autodesk Robot Structural Analysis software, choose the **Robot Structural Analysis Link** tool from **Analyze > Structural Analysis > Robot Structural Analysis** drop-down; the **Integration with Robot Structural Analysis** dialog box will be displayed, as shown in Figure 10-2.

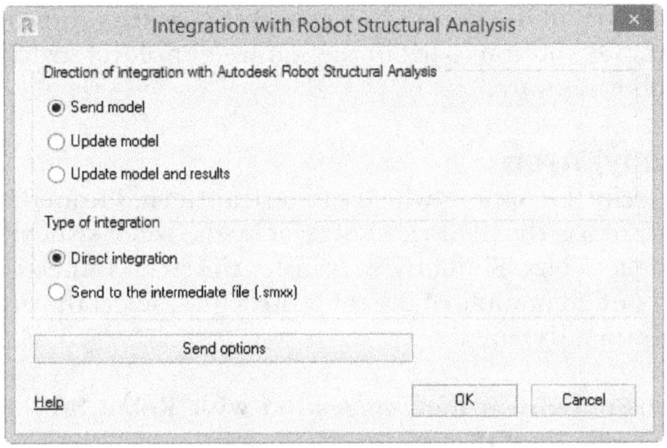

***Figure 10-2** The **Integration with Robot Structural Analysis** dialog box*

In the **Direction of integration with Autodesk Robot Structural Analysis** area, the **Send model** radio button is selected by default. Choose the **Send options** button from this dialog box to send the analytical model to **Robot Structural Analysis** software for analysis; the **Integration with Robot Structural Analysis - Send Options** dialog box will be displayed, as shown in Figure 10-3. In this dialog box, the following areas are displayed: **Scope and correction**, **Specify the case that contains the self-weight**, and **Transfer (optionally)**. These areas are discussed next.

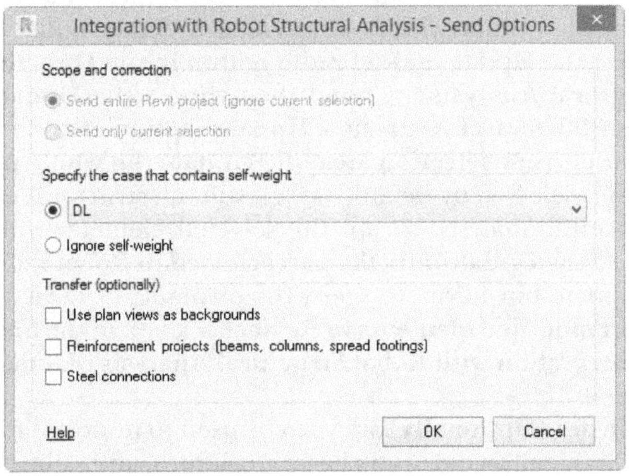

***Figure 10-3** The **Integration with Robot Structural Analysis-Send Options** dialog box*

Scope and correction Area

In this area of the **Integration with Robot Structural Analysis - Send Options** dialog box, the **Send entire Revit project (ignore current selection)** radio button is selected by default. As a result, the entire model, except the currently selected model will be sent to **Robot Structural Analysis** software for analysis. Alternatively, you can select the **Send only current selection** radio button to send the currently selected models to Autodesk Robot Structural Analysis software. Note that this radio button will only be active if a model/element is selected in the project view.

Specify the case that contains the self-weight Area

In this area, the radio button on the left of the drop-down list is selected by default. As a result, you can select the load case that contains the self weight of the structural model from the drop-down list. Alternatively, in the **Specify the case that contain self-weight** area you can select the **Ignore self-weight** radio button to avoid the self-weight of the structure to be sent with the analytical model to Robot software.

Transfer (optionally) Area

In this area, you can specify the options which can be transferred from the revit and can be used by the Robot Structure. To use the plan views of revit by the robot structure, select the **Use plan views as backgrounds** check box. Similarly, to transfer the steel connections defined in the revit to the steel connection design module of the robot structure, select the **Steel connections** check box in the **Transfer (optionally)** area.

After specifying various options in the **Integration with Robot Structural Analysis - Send Options** dialog box, choose the **OK** button; the dialog box will be closed and will return to **Integration with Robot Structural Analysis** dialog box. In this dialog box, choose the **OK** button; the **Send model to Robot Structural Analysis** window will be displayed. This window will display the progress of sending the analytical model to Robot. When the analytical model is completely transferred to the Robot software, the **Send model to Robot Structural Analysis** window will disappear.

After analyzing the structural model in Robot, you need to update the structural model in Revit based on the design changes made in Robot Software. To do so, choose the **Robot Structural Analysis Link** tool in Autodesk Revit from **Analyze > Structural Analysis > Robot Structural Analysis** drop-down; the **Integration with Robot Structural Analysis** dialog box will be displayed. In this dialog box, select the **Update model** radio button in the **Direction of integration with Autodesk Robot Structural Analysis** area and then choose the **Update options** button; the **Integration with Robot Structural Analysis - Update Options** dialog box will be displayed. In the **Scope - consider current selection** area, the **Update the whole project (ignore current selection)** radio button is selected by default. As a result, all structural elements that had been sent to Robot for structural analysis, except the selected element, will be updated in Revit. Alternatively, if you select the **Update only the part selected in Robot** radio button, the elements selected will only be updated in Revit. To select the elements in Revit after they are updated, you can select the **Select modified elements in Revit** check box in the **Scope - consider current selection** area of the **Integration with Robot Structural Analysis - Update Options** dialog box.

The options in the **Transfer (Optionally)** area can be used to import different outputs to Robot such as analysis results and reinforcements in beams, columns, and slabs. After specifying various options in the **Integration with Robot Structural Analysis - Update Options** dialog box, choose the **OK** button; the dialog box will be closed. Now, choose the **OK** button in the **Integration with Robot Structural Analysis** dialog box, the **Update model from Robot Structural Analysis** window will be displayed. This window will display the progress of updating the structural model in Revit. The **Update model from Robot Structural Analysis** window will disappear once the structural model is updated in Revit with the current settings. After that the **Confirmation** message box will appear. In this message box, you can choose the **Yes** button to update the changes and the **No** button to discard any changes.

In Autodesk Revit, you can perform the reinforcement design of RCC beams, columns, and floors elements through Robot Structural Analysis software. To do so, select the elements in the project view for which you want to perform the reinforcement design and then choose the

Reinforcement Code Check tool from **Analyze > Structural Analysis > Robot Structural Analysis** drop-down; the **Reinforcement design** window will be displayed. This window will display the progress of transferring of the selected elements to Robot software. This window will disappear once the selected model is completely transferred to the Robot software. After the **Reinforcement design** window disappears, the Robot software is opened and is ready to perform the reinforcement design for the elements that were selected and sent from Revit.

Structural Analysis in Autodesk Robot Structural Analysis Professional 2017

In the previous chapters of this textbook, you learned about various methods of modeling structural elements, imposing loads, setting boundary conditions, viewing reports, setting analytical models, and importing the structural model to a structural analysis software. Now, you will learn how to link Revit files to Autodesk Robot Structural Analysis Professional 2017, perform analysis, and view analysis results in this software. Moreover, you will also learn to update the structural model in Revit based on the analysis and design process performed in Robot.

Autodesk Robot Structural Analysis software, also known as Robot, is a structural analysis and design software that helps you achieve accurate design results for a project. This software provides a collaborative workflow and interoperability with 3D bidirectional link to Autodesk Revit and other Autodesk companion products. The various features and capabilities of Robot Structural Analysis software are discussed next.

Structural Modeling

Autodesk Robot Structural Analysis software has the capability to model a wide range of structures such as steel, reinforced concrete, and so on. There are various tools in this software along with in-built design codes that enable the engineers to model a variety of structures based on the specification in the design codes.

Mesh Modeling

Autodesk Robot Structural Analysis Software enables the structural engineers to work with complex mesh models. The **Automatic mesh generation** tool in this software can be used for manual manipulation of the mesh, refinement of mesh, and meshing around openings of any shape and size of the structure. There are many meshing tools available in this software that enable the structural engineers to quickly create a high-quality finite element mesh.

Analysis Capabilities

Autodesk Robot Structural Analysis has the capability to perform linear static analysis of building structures. This software can be used by structural engineers to perform non-linear analysis of structures, including nonlinear analysis that includes P-delta analysis, tension/compression members and supports, cables, and plastic hinges, and so on. Autodesk Robot Structural Analysis can also be used to perform dynamic analysis of structures.

Reinforced Concrete and Steel Designing

Autodesk Robot Structural Analysis contains in-built reinforced concrete and steel design codes based on standards relevant to the country. You can use more than 40 international steel codes and 30 reinforced concrete codes to simplify the design process. These codes assist engineers in selecting and evaluating structural elements throughout the world.

Linking Revit Files to Robot

To link Revit files to Robot Structural Analysis software, choose the **Robot Structural Analysis Link** tool from **Analyze > Structural Analysis > Robot Structural Analysis** drop-down; the **Integration with Robot Structural Analysis** dialog box will be displayed. In this dialog box, ensure that the **Send model** radio button is selected and then choose the **OK** button; the **Send model to Robot Structural Analysis** window will be displayed. This window will close when the model is exported to Robot. After the window is closed, the **Integration with Robot Structural Analysis** window will be displayed, informing that the model has been transferred from Revit to Robot. In this window, you can also choose the **Yes** button to view the events report; the **Warning list** window will be displayed. Next, choose the **Close** button to close the window; the Robot Structural Analysis software containing the linked structural model from Revit, will open, as shown in Figure 10-4. In the Robot Structural Analysis software, you can use various tools to analyze and design the linked model from Revit. Before performing analysis and designing, you need to set the job preferences of the structural model linked from Revit. In the next sections, you will learn to set preferences, perform analysis, view analysis results in Robot, and then update the analyzed and designed model in Revit.

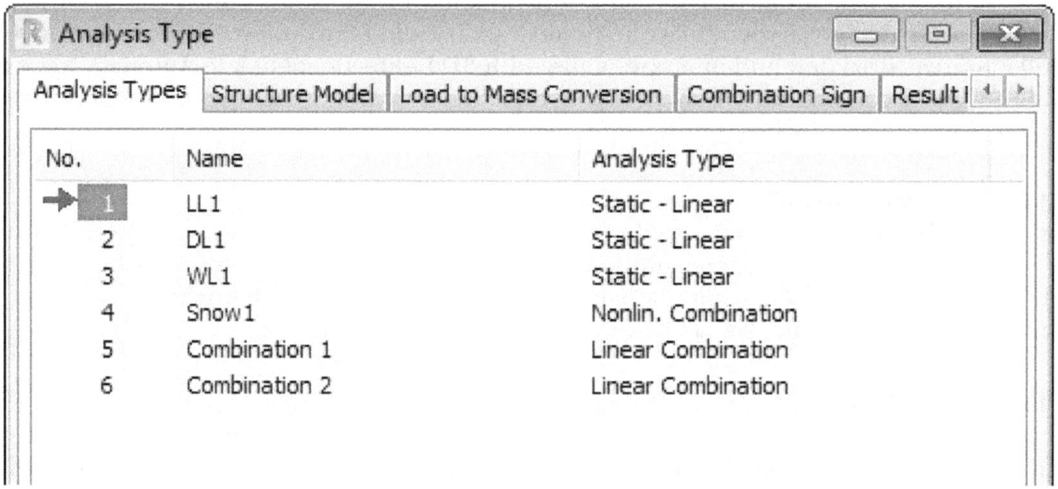

Figure 10-4 *The interface of the Robot Structural Analysis software Showing the linked Revit model*

Setting Job Preferences in Robot

In Robot Structural Analysis software, you can set various job preferences such as units and the codes of practices for materials to be used in the project; databases for various structural entities such as steel and timber sections, vehicle loads, and standard loads applied in the project; design codes to be used for specifying loads such as seismic load, snow or wind load, and load combinations; type of structural analysis to be performed; type of design codes to be used while designing steel or reinforced concrete structure; and various meshing parameters. To set the job preferences, choose the **Job Preferences** tool from the **Tools** menu; the **Job Preferences** dialog box will be displayed, as shown in Figure 10-5. In this dialog box, you can set job preferences.

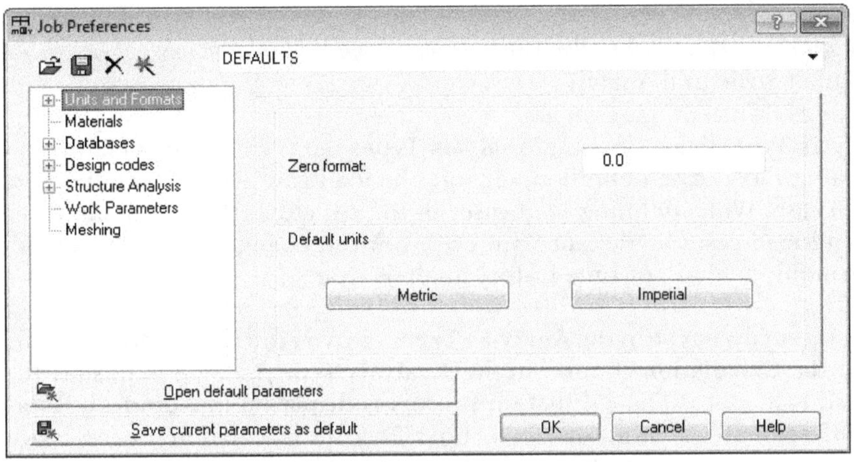

Figure 10-5 *The* ***Job Preferences*** *dialog box*

Analysis in Robot

You can perform the structural analysis of a model and set the analysis type by using the tools available in the **Analysis** menu. To set the analysis type for the structural model, choose the **Analysis Types** tool from the **Analysis** menu; the **Analysis Type** dialog box will be displayed, as shown in Figure 10-6. This dialog box contains six tabs: **Analysis Types**, **Structure Model**, **Load to Mass Conversion**, **Combination Sign**, **Result Filters**, and **Buckling Deformation**. Note that the **Structural Model** tab will be displayed in the **Analysis Type** dialog box, if the **Model generation** check box is selected.

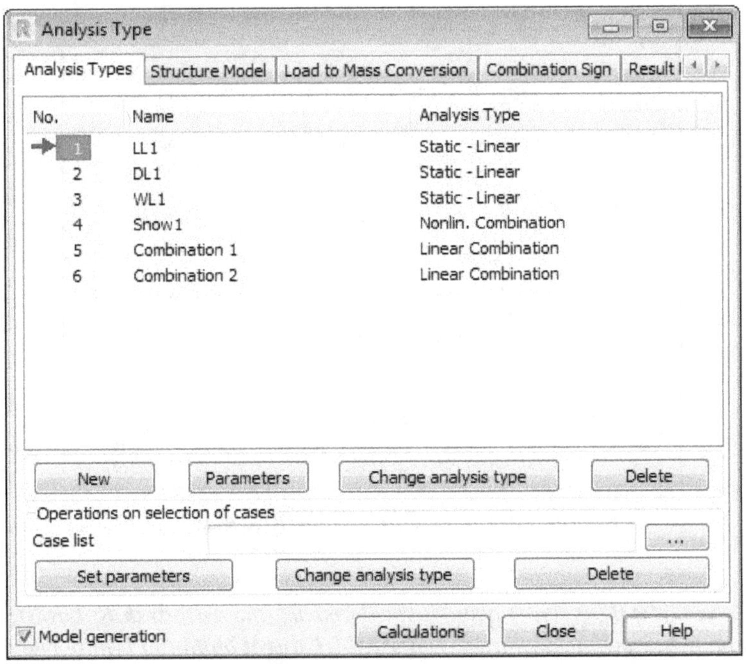

Figure 10-6 *The* ***Analysis Type*** *dialog box*

Note

*The **Analysis Types** tool in the **Analysis** menu will be displayed only when any model is opened in* Robot Structural Analysis.

In the **Analysis Type** dialog box, the **Analysis Types** tab is chosen by default. In this tab, you can specify a new load case definition, change the load case type, or modify parameters of a selected load case. While defining load cases in the structure, the linear static analysis type is assigned to all load cases by default. However, you can change the analysis type to buckling analysis, harmonic analysis, or time history analysis type.

After setting the analysis type in the **Analysis Types** tab, you can specify options in the other tabs. Now, choose the **Calculations** button in the **Analysis Type** dialog box to start the calculation of the desired load cases. The calculation process is displayed in **Autodesk Robot Structural Analysis Professional- Calculations** dialog box. Once the calculation process is over, the dialog box closes automatically. Next, choose the **Calculation Restart** tool from the **Analysis** menu; the **Calculation Restart** dialog box will be displayed. In the **Case list** area of this dialog box, various load cases are displayed. Select the check box(es) corresponding to load case(s) of the structural project you require to perform the calculation. After selecting the desired check box(es), choose the **Restart calculations** button to perform the calculation function. After performing the calculation, you need to prepare the results of the calculation. To do so, choose the **Prepare results** tool from the **Analysis** menu; the **Prepare results** dialog box will be displayed, as shown in Figure 10-7.

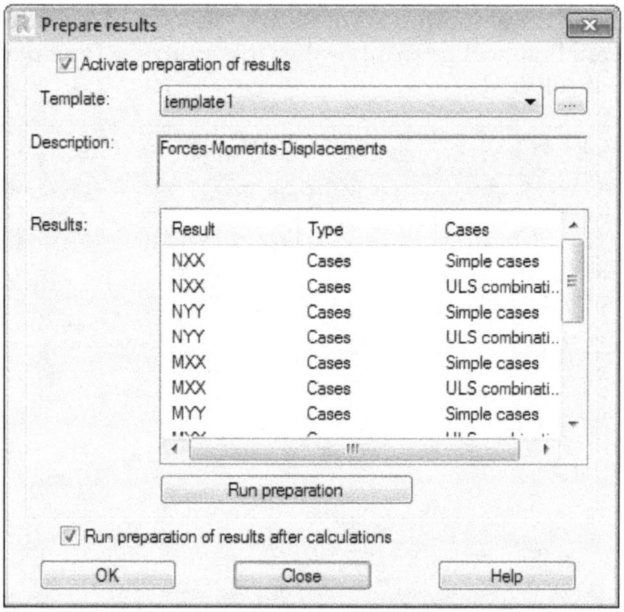

***Figure 10-7** The **Prepare results** dialog box*

Note

*While analyzing, if there is any instability in a model, the **Autodesk Robot Structural Analysis 2017** message box will be displayed. Choose the **Cancel** button to ignore warning. On doing so, the **Calculation Messages** dialog box with the number of warnings will be displayed. Choose the **Close** button.*

In this dialog box, ensure that the **Activate preparation of results** check box is selected. As a result, the options in the **Prepare results** dialog box will be activated. Select an option from the **Template** drop-down list to specify a template file to be used for viewing the results. The **Analytical Model** file is selected by default in this drop-down list. You can edit it as required. To do so, choose the Browse button located next to the **Template** drop-down list; the **Composition of Template for Result Preparation** dialog box will be displayed, as shown in Figure 10-8.

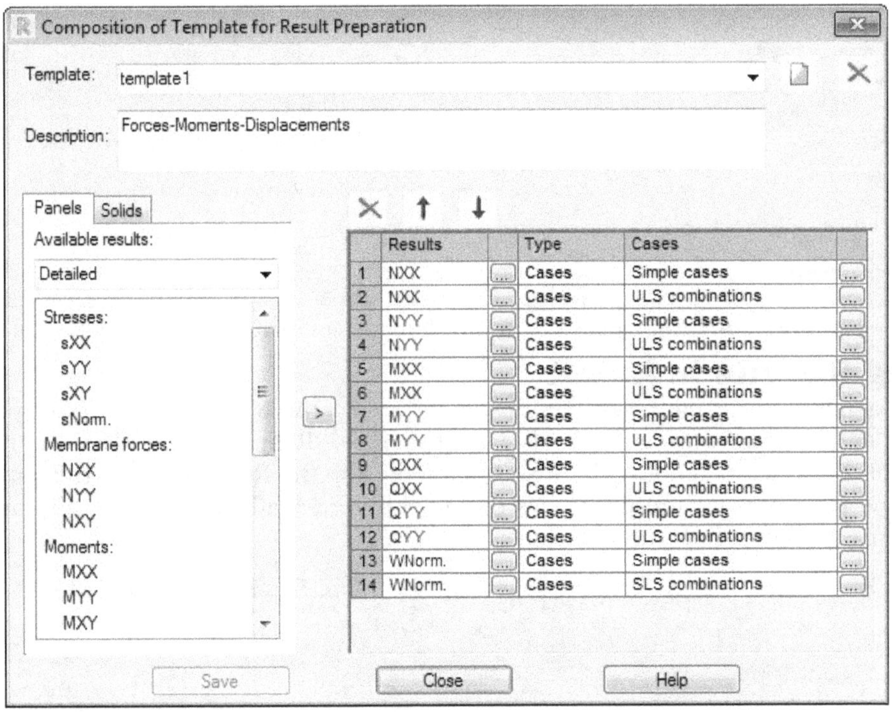

***Figure 10-8** The **Composition of Template for Result Preparation** dialog box*

You can use various options in this dialog box to edit the display of the results of the analysis. Next, choose the **Save** button to save the changes in the template and then choose the **Close** button; the **Composition of Template for Result Preparation** dialog box will be closed and the **Prepare results** dialog box will be displayed again. In this dialog box, the description of the selected template will be displayed in the **Description** text box and it can be edited by specifying a new value in it. In the **Prepare results** dialog box, the **Results** area displays the table of the results contained in three columns, namely, **Result**, **Type**, and **Cases**. The **Result** column shows the list of results along with their type and cases in the **Type** and **Cases** columns, respectively. Next, in the **Prepare** results dialog box, choose the **Run preparation** button and then choose the **OK** button; the preparation of the results will be completed and the **Prepare results** dialog box will be closed. After preparing the results, you can view the calculation report. To do so, choose **Calculation Notes** tool from the **Analysis** menu; a cascading menu will be displayed. In this menu, you can choose any of the two tools: **Simplified Note** and **Full Note**. You can choose the **Simplified Note** tool to view the brief summary of the results in a window. Alternatively, you can choose the **Full Note** tool from the cascading menu to view the detailed summary of the result, as shown in Figure 10-9.

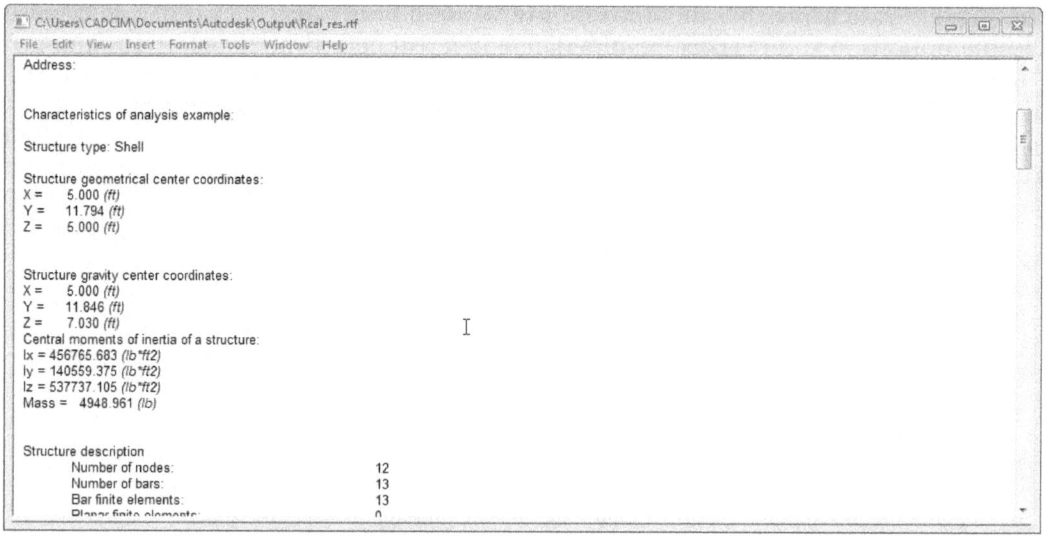

*Figure 10-9 The calculation report displayed on choosing the **Full Note** tool*

Viewing Results in Robot

After preparing and calculating the results, you can view the output of the results in the form of reports and diagrams using the tools available in the **Results** menu. To view the support reactions of the structure, choose the **Reactions** tool from the **Results** menu; the output report showing the reactions at different nodes due to different load cases will be displayed in the drawing area, as shown in Figure 10-10.

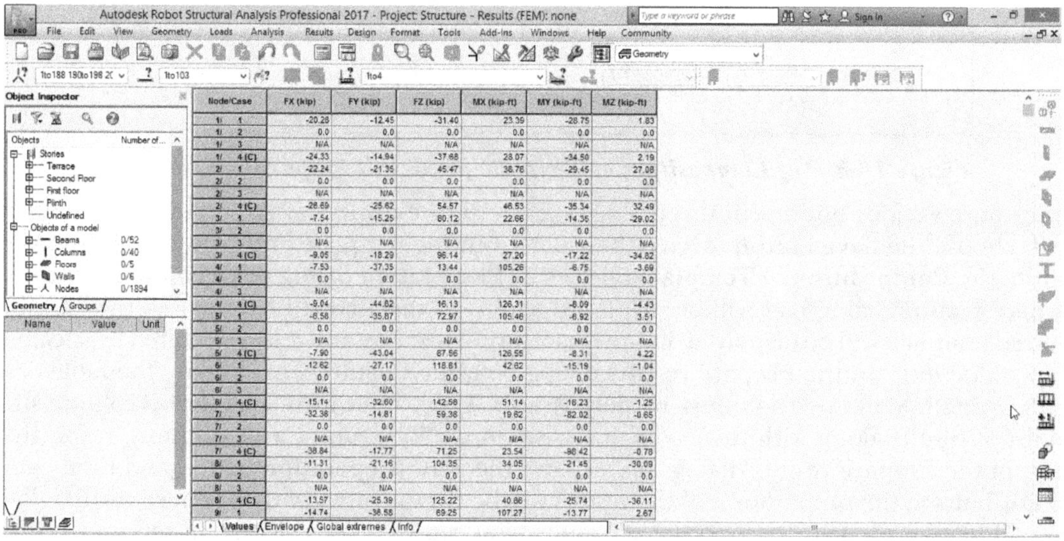

Figure 10-10 The partial view of the report of the support reaction

In the **Status Bar**; the **Values** tab is chosen by default. As a result, all the values related to load cases will be displayed in the report. In the **Status Bar**, you can choose the **Envelope**, **Global extremes**, and the **Info** tab to view the reactions of the nodes for the envelopes, maximum and minimum reactions, and the information of the load cases and the nodes in the reaction, respectively. You can also view the displacement of the nodes, deflection of the members, forces induced in the nodes and members, and stresses induced in the nodes and members by choosing the **Displacements**, **Deflections**, **Forces**, and **Stresses** tools from the **Results** menu, respectively.

To view the diagrams of the forces, moments, and reinforcements in the model, you need to choose the **Detailed Analysis** tool from the **Results** menu. On doing so, the **Detailed Analysis** dialog box will be displayed, as shown in Figure 10-11.

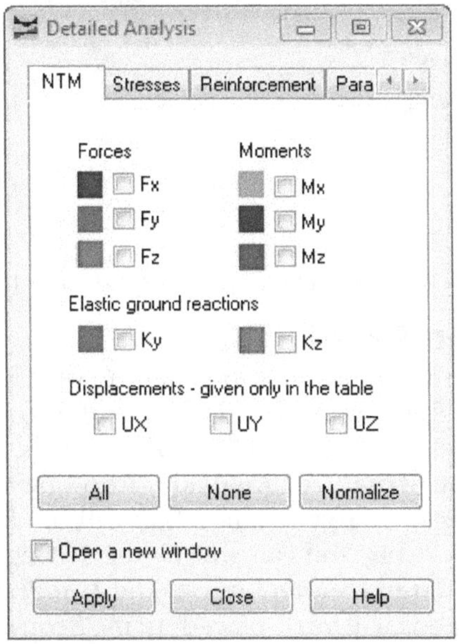

Figure 10-11 *The* ***Detailed Analysis*** *dialog box*

The **Detailed Analysis** dialog box contains the following tabs: **NTM**, **Stresses**, **Reinforcement**, **Parameters**, and **Division points**. In the dialog box, the **NTM** tab will be chosen by default. In this tab, you can select the check boxes related to forces and moments that you want to display graphically in the drawing window. In the **Forces** column of the chosen tab, you can select the **Fy** check box to display the shear forces induced in the members of the project. Similarly, you can select the **Fx** and **Fz** check boxes in the **Forces** column to display the forces induced in the respective directions. In the **Moments** column, you can select the desired check box(es) corresponding to the moments you require to view in the drawing area. In the **Elastic ground** reactions area of the **NTM** tab, you can select the **Ky** and **Kz** check boxes to view the reactions of an elastic ground for bars (the options Ky and Kz) in the drawing window. The Ky and Kz values are obtained as elastic ground reactions (R = K u, where R denotes force of reaction, K is the coefficient of elastic ground reaction, and u denotes displacement) along the member length in the local coordinate system, if elastic ground has been defined for the member. The **Reinforcement** tab of the **Detailed Analysis** dialog box displays the reinforcement in the members of the drawing window. The **Parameters** tab contains options that controls the presentation of the results displayed in drawing window. The **Division points** tab indicates the intermediate points for which the values of selected quantities are to be presented in the table in the drawing window. After specifying these options in the **Detailed Analysis** dialog box, choose the **Apply** button; the diagrams and the table for member 1 will be displayed in the drawing area, as shown in Figure 10-12. Note that the results in the diagram in the drawing window will be based on the load case that is selected from the **Cases** drop-down list in the **Selection** toolbar; refer to Figure 10-12.

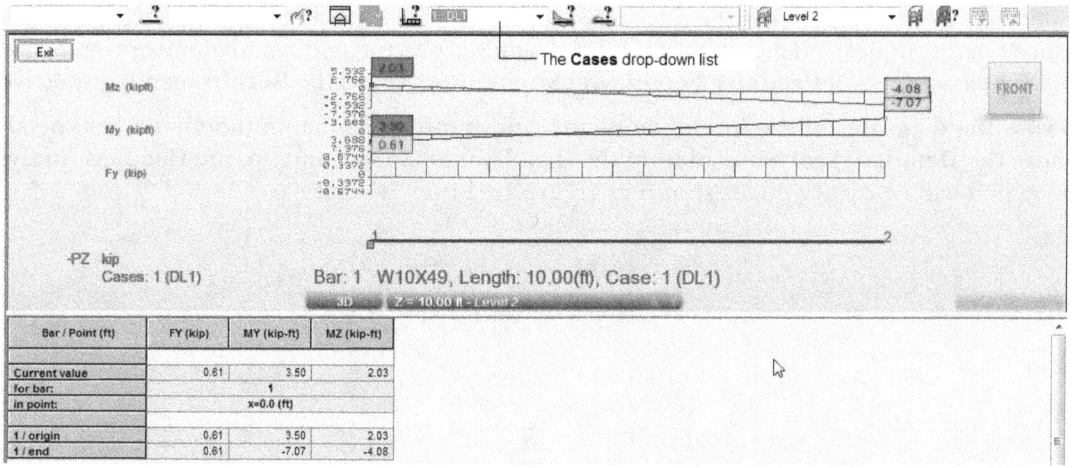

Figure 10-12 The drawing window graphically displaying result of the moments

Updating the Model in Revit

After working in the model, save the project in Robot, before switching back to Revit for updating the BIM model. Next, open the Revit interface containing the structural project which is to be analyzed in Robot. Next, choose the **Robot Structural Analysis Link** tool from the **Analyze > Structural Analysis > Robot Structural Analysis** drop-down in the Revit interface; the **Integration with Robot Structural Analysis** dialog box will be displayed. In this dialog box, select the **Update model** radio button. You can specify the settings for updating the model in Revit. To do so, choose the **Update Options** button; the **Integration with Robot Structural Analysis-Update Options** dialog box will be displayed. In the **Scope - consider current selection** area and **Transfer (optionally)** area of the dialog box, you can specify the scope of the updating the model and also transfer the results from Robot to Revit. Next, choose the **OK** button; the **Integration within Robot Structural Analysis-Update Options** dialog box will close and the **Integration within Robot Structural Analysis** dialog box will be displayed. Choose the **OK** button in this dialog box; the model in Revit will be updated.

TUTORIAL

General instructions for downloading the model file and performing the tutorial:

1. Download the *Analytical Model.rvt* tutorial file from *www.cadcim.com*. The path of the file is as follows: *Textbooks > Civil/GIS > Revit Structure > Exploring Autodesk 2017 for Structure 2017*.

2. Next, browse to *C:\rst_2017* and create a new folder with the name **c10_rst_2017_tut**. Next, save and extract the file in this folder.

3. To perform this tutorial, you are required to download and install the **Autodesk Robot Structural Analysis Professional 2017** software.

Tutorial 1 Building Project

In this tutorial, you will perform the analysis of Revit model in Autodesk Robot Structural Analysis software. **(Expected time: 1 hr 15 min)**

The following steps are required to complete this tutorial:

a. Open the file *Analytical Model.rvt*.
b. Check the model for consistency.
c. Check supports of the model.
d. Send the model to Autodesk Robot Analysis Software.
e. Calculate and prepare results of analysis.
f. Update the Revit model.

Opening the File

In this section, you will open the model.

1. To open the project file, choose **Open > Project** from the **Application Menu**; the **Open** dialog box is displayed.

2. In this dialog box, browse to *C:\rst_2017\c10_rst_2017_tut* and open the file *Analytical Model.rvt* from this folder.

Checking the Model for Consistency and Supports

In this section, you will check the consistency of the model.

1. Choose the **Analyze** tab and then choose the **Analytical Model Settings** button from the **Analytical Model Tools** panel; the **Structural Settings** dialog box is displayed.

2. In this dialog box, the **Analytical Model Settings** tab is chosen by default. Now, in the **Automatic Checks** area under this tab, select both the **Member Supports** and **Analytical/Physical Model Consistency** check boxes, refer to Figure 10-13.

3. Next, choose the **OK** button in this dialog box; the **Analytical Model Check** message box is displayed, prompting you whether to perform analytical model check or not. Choose the **Yes** button in this message box; the message box and the **Structural Settings** dialog box are closed and a warning message box is displayed.

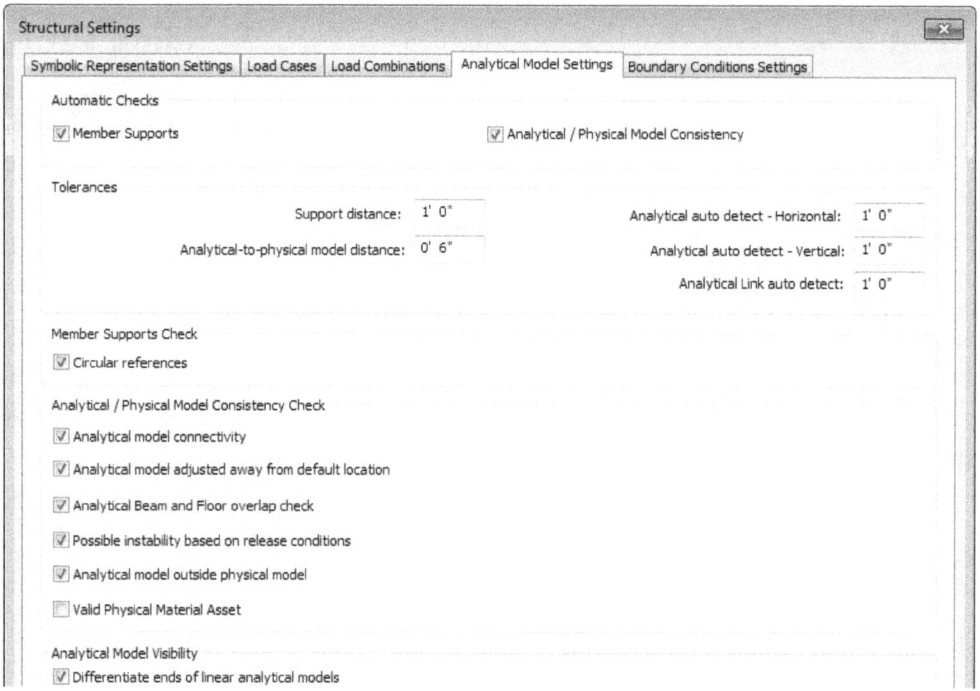

Figure 10-13 *Partial view of the* ***Structural Settings*** *dialog box*

4. In this message box, choose the **Expand warning dialog** button; the **Autodesk Revit 2017** message box is displayed, as shown in Figure 10-14. View the details of the warnings and choose the **Close** button; the message box is closed. 

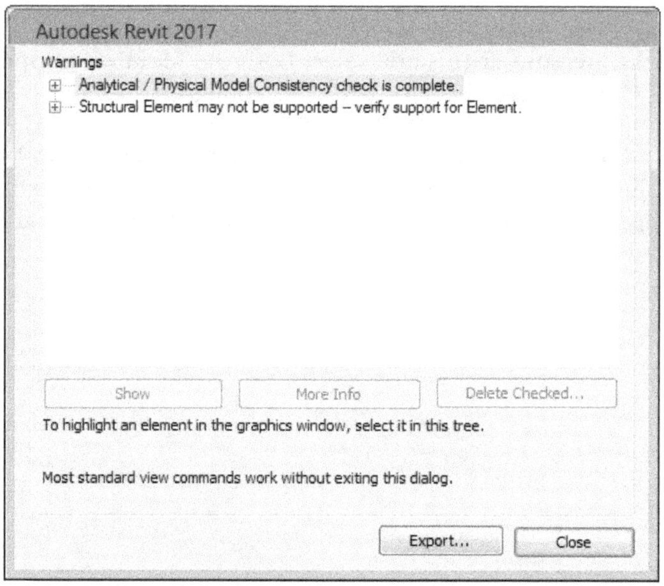

Figure 10-14 *The* ***Autodesk Revit Structure 2017*** *message box*

Sending the Model to Robot Structural Analysis Software

In this section, you will send the model to Autodesk Robot Structural Analysis Professional 2017 software.

1. Choose the **Analytical Model** view under the **3D Views** node in the **Project Browser**, if the current view is different from the **Analytical Model** view.

Note

To view the analytical model, make sure that the ***Show Analytical Model*** *button is chosen from the* ***View Control*** *bar.*

2. Choose the **Robot Structural Analysis Link** tool from **Analyze > Structural Analysis > Robot Structural Analysis** drop-down; the **Integration with Robot Structural Analysis** dialog box is displayed, as shown in Figure 10-15.

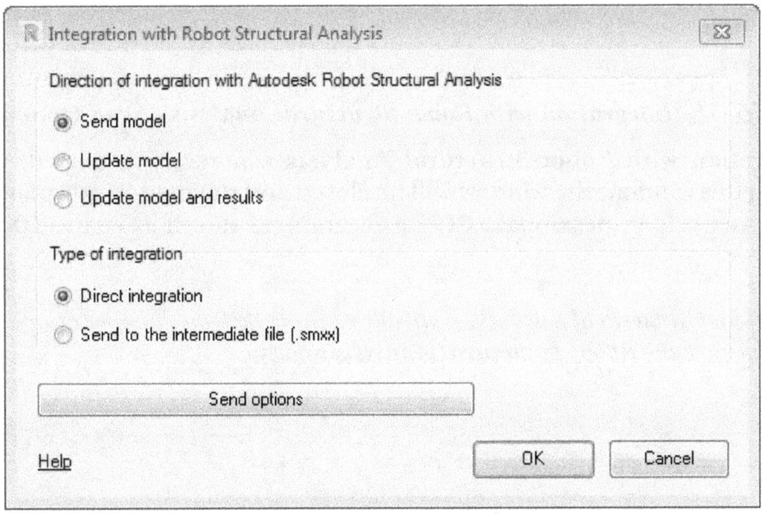

Figure 10-15 The ***Integration with Robot Structural Analysis*** *dialog box*

3. In this dialog box, ensure that the **Send model** and **Direct Integration** radio buttons are selected, and then choose the **Send options** button; the **Integration with Robot Structural Analysis - Send Options** dialog box is displayed, refer to Figure 10-16.

4. In this dialog box, keep the default settings and choose the **OK** button; the **Integration with Robot Structural Analysis - Send Options** dialog box is closed.

5. Choose the **OK** button; the **Integration with Robot Structural Analysis** dialog box is closed and **Send model to Robot Structural Analysis** window is displayed. This window displays the status of file transfer and will automatically close on the completion of the file transfer process.

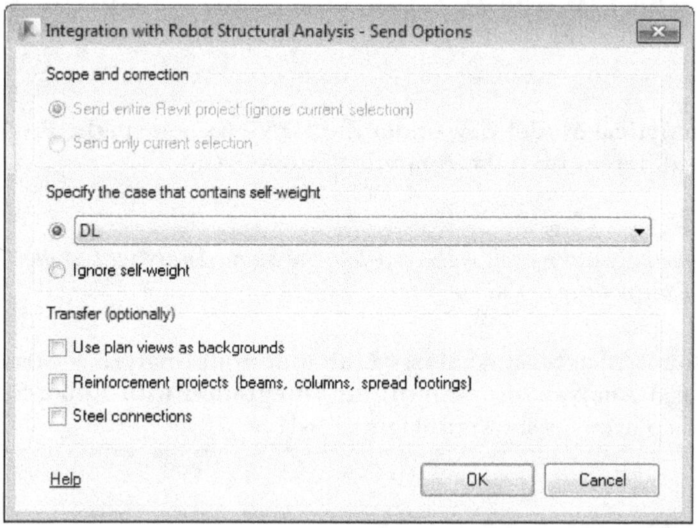

***Figure 10-16** The **Integration with Robot Structural Analysis - Send Options** dialog box*

6. The **Integration with Robot Structural Analysis** window is displayed. Choose the **No** button from this window; the window will be closed and the model is displayed in the **Robot Structural Analysis Professional 2017** application, as shown in Figure 10-17.

Note

*If the **Robot Structural Analysis** is already running then the run time error will occur. So make sure you close the **Robot Structural Analysis** software.*

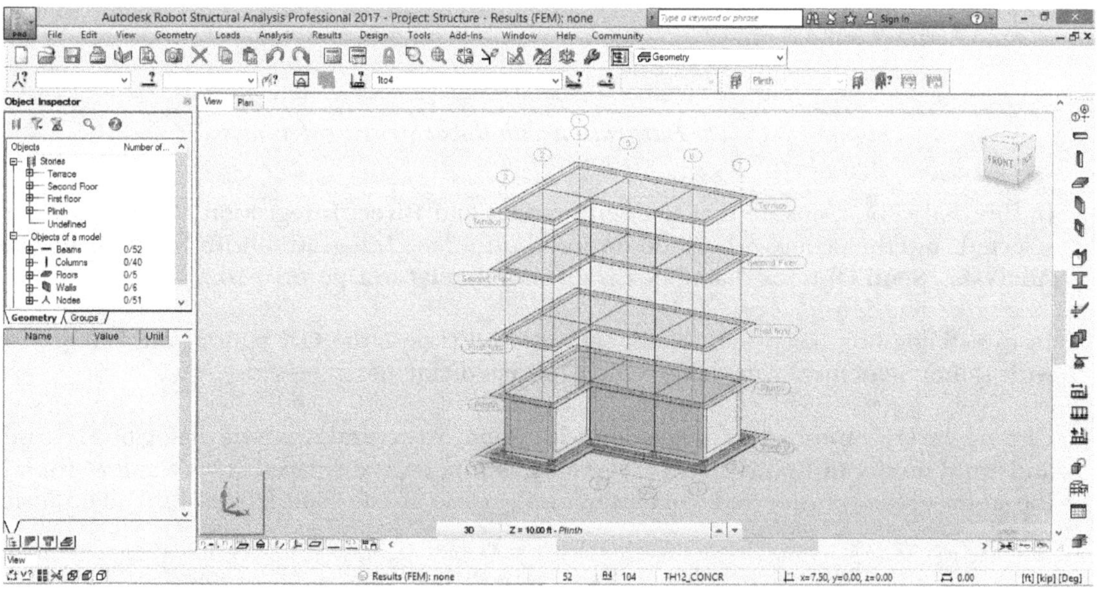

***Figure 10-17** The exported **Revit** model in the **Robot Structural Analysis Professional 2017** window application*

Calculating and Viewing Results

In this section, you will calculate the results in the **Robot Structural Analysis** software.

1. Change the current view of the model to the top view by using the view cube and then select the **Load Types** option from the **Loads** menu; the **Load Types** dialog box with the list of load cases is displayed, refer to Figure 10-18.

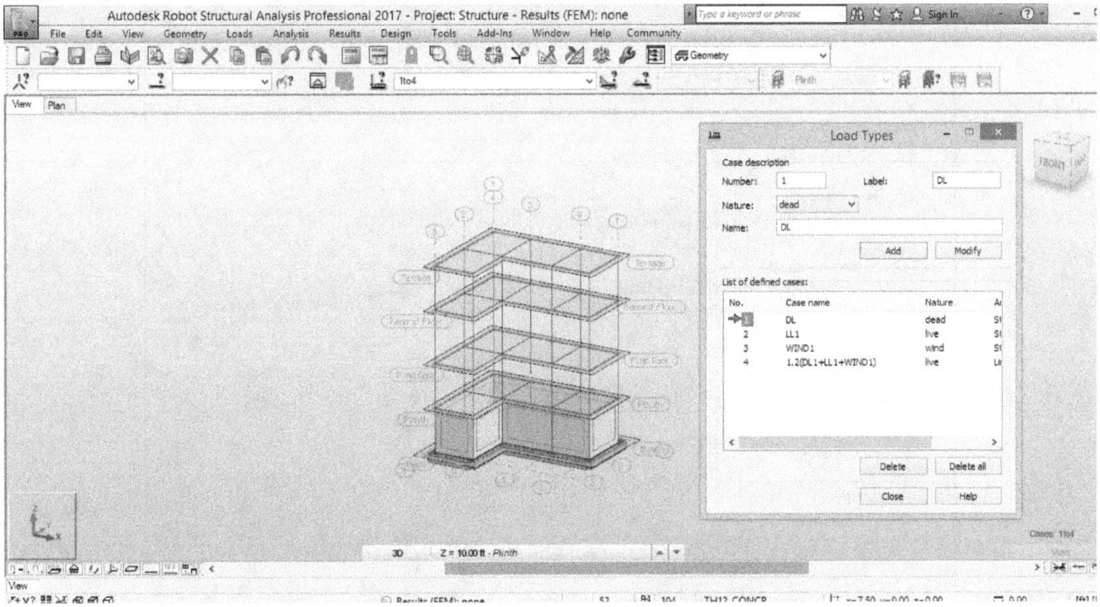

Figure 10-18 *The* ***Load Types*** *dialog box,* ***Load- Cases:1 (DL)*** *window*

2. Choose the **Calculations** tool from the **Analysis** menu, refer to Figure 10-19; the **Autodesk Robot Structural Analysis Professional - Calculations** window is displayed, as shown in Figure 10-20. Note that this window displays the calculation status, and is automatically closed when the calculations are completed.

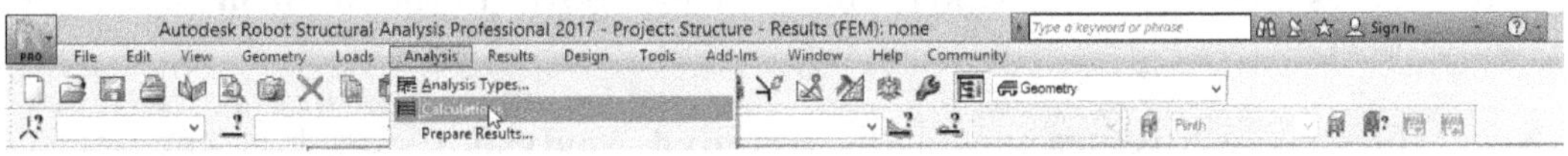

Figure 10-19 *The* ***Calculations*** *tool in the* ***Analysis*** *menu*

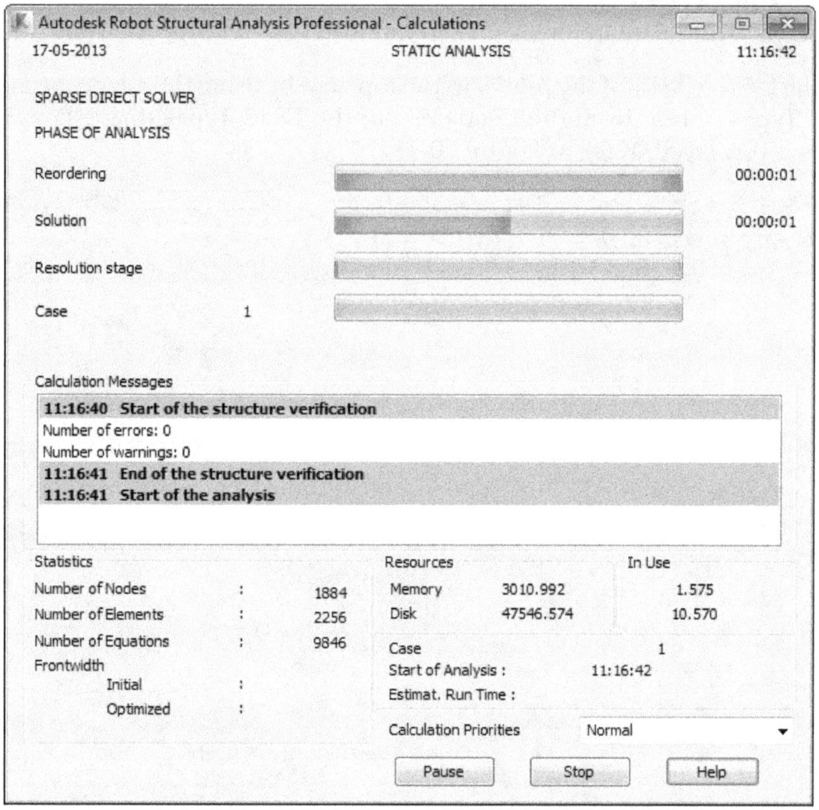

***Figure 10-20** The **Autodesk Robot Structural Analysis Professional - Calculations** window*

3. After performing calculations, choose the **Detailed Analysis** tool from the **Results** menu; the view is changed to **Detailed Analysis** and the **Detailed Analysis** dialog box is displayed.

4. In this dialog box, the **NTM** tab is chosen by default. Select the **Fx**, **Fy**, and **Fz** check boxes in the **Forces** area and the **Mx**, **My**, and **Mz** check boxes in the **Moments** area, refer to Figure 10-21.

5. Next, choose the **Stresses** tab in the **Detailed Analysis** dialog box and select the **maximum- S max** and **minimum - S min** check boxes in the **Normal - maximum** area.

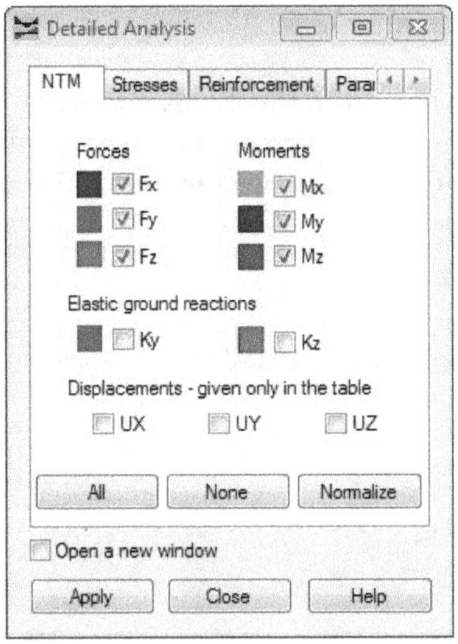

Figure 10-21 The Detailed Analysis dialog box

6. Next, choose the **Stresses** tab in the **Detailed Analysis** dialog box and select the **maximum- S max** and **minimum - S min** check boxes in the **Normal - maximum** area.
7. Choose the **Apply** button; the **Detailed Analysis** window is displayed, showing various graphs for the force and moment analysis, refer to Figure 10-22.

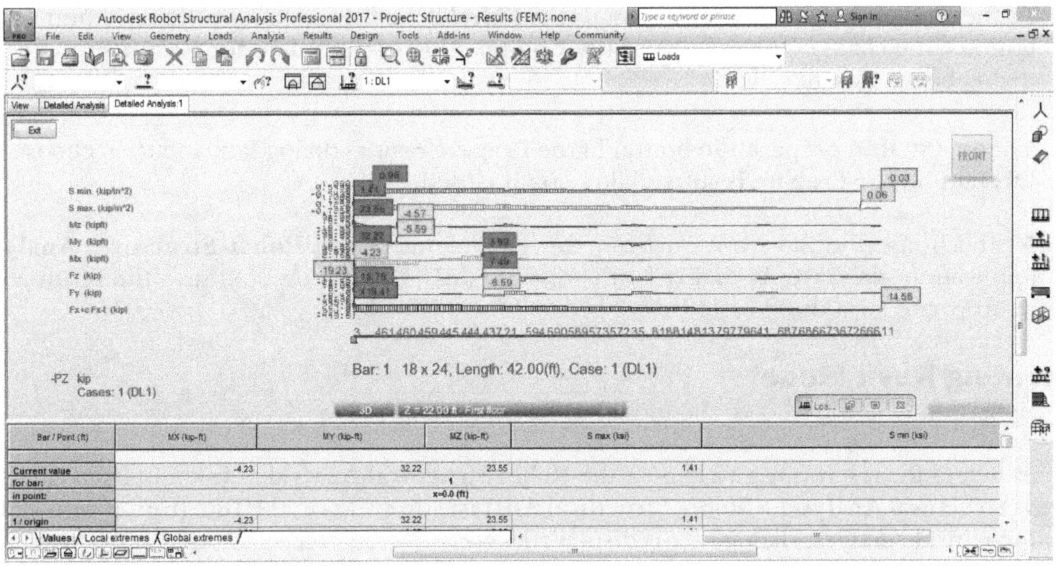

Figure 10-22 Details of the forces and moments in the changed view

8. Next, choose the **Close** button in the **Detailed Analysis** dialog box; the dialog box is closed.

Preparing Results

In this section, you will prepare results of analysis.

1. Choose the **Prepare Results** tool from the **Analysis** menu; the **Prepare results** dialog box is displayed, as shown in Figure 10-23. Choose the Browse button located next to the **Template** drop-down; the **Composition of Template for Result Preparation** dialog box is displayed.

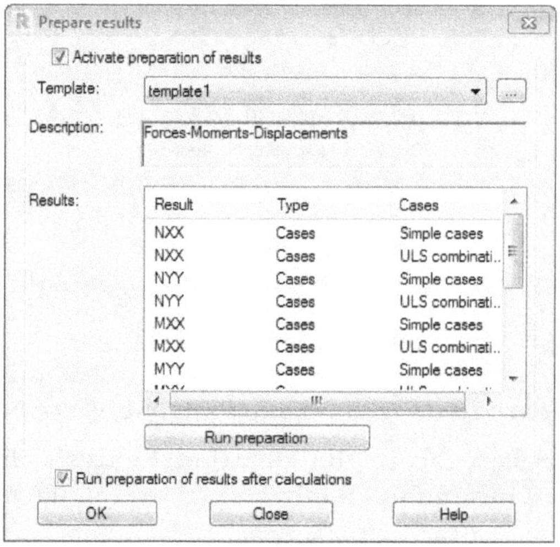

***Figure 10-23** The **Prepare results** dialog box*

2. In the **Template** edit box, enter **Analytical Model** and then choose the **Save** and **Close** buttons; the changes are saved and the **Composition of Template for Result Preparation** dialog box is closed.

3. Choose the **Run preparation** button in the **Prepare results** dialog box and then choose the **OK** button; the **Prepare results** dialog box is closed.

4. Next, choose the **Save** option from the **File** menu of the **Robot Structural Analysis** application; the **Save As** dialog box is displayed. In this dialog box, save the result at an appropriate location.

Updating Revit Model

In this section, you will update the model in Revit.

1. Switch to Revit software and choose the **Robot Structural Analysis Link** tool from **Analyze > Structural Analysis > Robot Structural Analysis** drop-down; the **Integration with Robot Structural Analysis** dialog box is displayed.

2. In this dialog box, select the **Update model and results** and **Direct integration** radio buttons in the **Direction of integration with Autodesk Robot Structural Analysis** and **Type of integration** areas, respectively.

3. Next, choose the **Update options** button in this dialog box; the **Integration with Robot Structural Analysis - Update Options** dialog box is displayed, as shown in Figure 10-24.

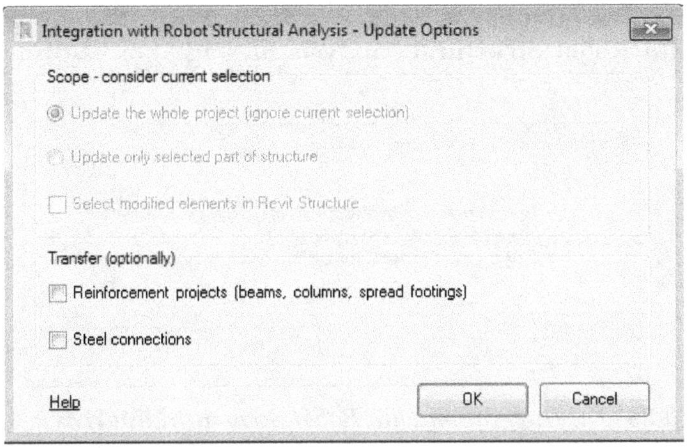

Figure 10-24 *The* ***Integration with Robot Structural Analysis - Update Options*** *dialog box*

4. In this dialog box, select the **Reinforcement projects (beams, columns, spread footings)** and **Steel Connections** check boxes in the **Transfer (optionally)** area. Next, choose the **OK** button; the dialog box is closed.

5. Next, in the **Integration with Robot Structural Analysis** dialog box, choose the **OK** button; the dialog box is closed and the **Send Results to Revit** dialog box is displayed.

6. In this dialog box, click in the edit box next to **Model name** in the **Analysis results package** area and enter **Analytical Model**.

7. Enter **Analytical Model** in the **Model name** edit box in the **Required reinforcement results package** area, refer to Figure 10-25.

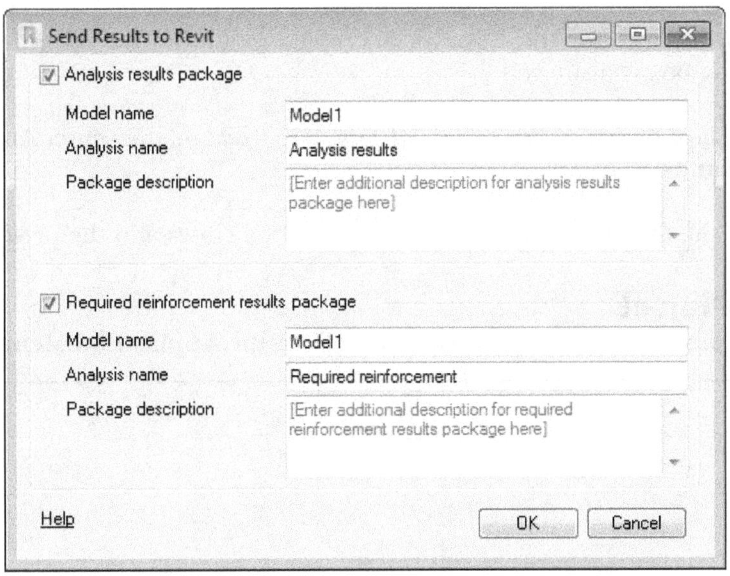

Figure 10-25 *The* ***Send Results to Revit*** *dialog box*

8. Next, choose the **OK** button; the **Send Results to Revit** dialog box is closed and the **Integration with Robot Structural Analysis** message box is displayed, as shown in Figure 10-26.

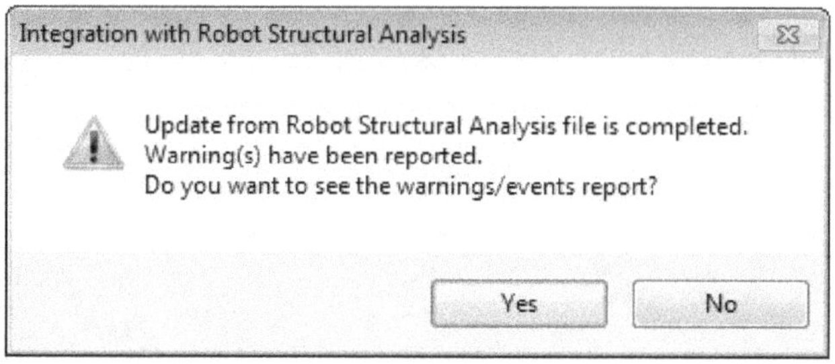

Figure 10-26 *The* ***Integration with Robot Structural Analysis*** *message box*

Note

The process of updating the model will take time and will depend on the processor. The status of the file updation is displayed in the ***Update model from Robot Structural Analysis*** *window.*

9. Choose the **Yes** button from this message box; the **Warning list** message box is displayed, showing update details and warning messages.

10. Choose the **Close** button to close the message box; the **Autodesk Revit 2017** message box is displayed. Choose the **OK** button; the message box is closed and the model is updated and is saved.

Saving the Project

In this section, you will save the project file using the **Save As** tool.

1. To save the project, choose the **Save As > Project** option from the **Application Menu**; the **Save As** dialog box is displayed.

2. In this dialog box, browse to *C:\rst_2017\c10_rst_2017_tut* and then enter **Analytical Model_1** in the **File name** edit box.

3. Now, choose the **Save** button; the **Save As** dialog box closes and the project file is saved.

Closing the Project

1. To close the project, choose the **Close** option from the **Application Menu**.

Self-Evaluation Test

Answer the following questions and then compare them to those given at the end of this chapter:

1. Which of the following check boxes in the **NTM** tab of the **Detailed Analysis** dialog box needs to be selected to view the shear force(s) induced in the structural member(s)?

   (a) **Fx** (b) **My**
   (c) **Mx** (d) **Mz**

2. To invoke the **Job Preferences** dialog box, choose the __________ tool from the **Tools** menu.

3. You can send the current selected model from Revit to Autodesk Robot Structural Analysis software by selecting the __________ radio button in the **Integration with Robot Structural Analysis- Send Options** dialog box.

4. The __________ tool is used to design and analyze reinforcements in a model.

5. You can link the Revit model to Robot Structural Analysis by using the __________ tool.

6. To view the support reactions of the structural model, you need to choose the **Reaction** tool. (T/F)

7. In Robot Structural Analysis, you cannot perform non-linear analysis of a structural model. (T/F)

Review Questions

Answer the following questions:

1. Which of the following buttons from the **Info center** bar is chosen to download and install added application(s) to Revit?

   (a) **Search** (b) **Autodesk 360**
   (c) **Application Exchange** (d) **Communication Center**

2. To view the maximum and minimum reactions in a model, which tool needs to be chosen from the **Status** bar?

   (a) **Envelope** (b) **Global extremes**
   (c) **Info** (d) **Loads**

3. After the analysis is done in Robot Structural Analysis, you can update the model and the results by choosing the __________ radio button from **Integration with Robot Structural Analysis** dialog box.

4. In Robot Structural Analysis, you can set units and code(s) of practice for material used by using the _________ tool(s).

5. To view the graphical representation of moment(s) and stress(es) induced in a model in Revit, you need to choose the _________ tool from the Structural Analysis panel.

6. The **Parameters** tab in the **Detailed Analysis** dialog box controls the presentation of the results displayed in the drawing window. (T/F)

7. Autodesk Robot Structural analysis contains in-built reinforced concrete and steel design codes. (T/F)

EXERCISE

Exercise 1 Industrial Shade

Download the Revit file *c10_Industrial_shade_exer* from *http://www.cadcim.com*. The path of the file is as follows: *Textbooks > Civil/GIS > Revit Structure > Exploring Autodesk Revit 2017 for Structure* . In this exercise, you will analyze the Industrial Shade file in Robot Structural Analysis software. Refer to Figure 10-27 to view the model in Revit window and Figure 10-28 for the graphical representation of the model in Revit. **(Expected time: 1hr 45min)**

1. Project file to be used: *c10_Industrial_shade_exer.*
2. Link the Revit model to Robot Structure Analysis software.
3. Run calculations.
4. Prepare results.
5. Update the model in Revit.
6. Use the **Result Explorer** tool to view the following results graphically in Revit:
   Rotation of the Reactions
   Displacement in members
   Deformation

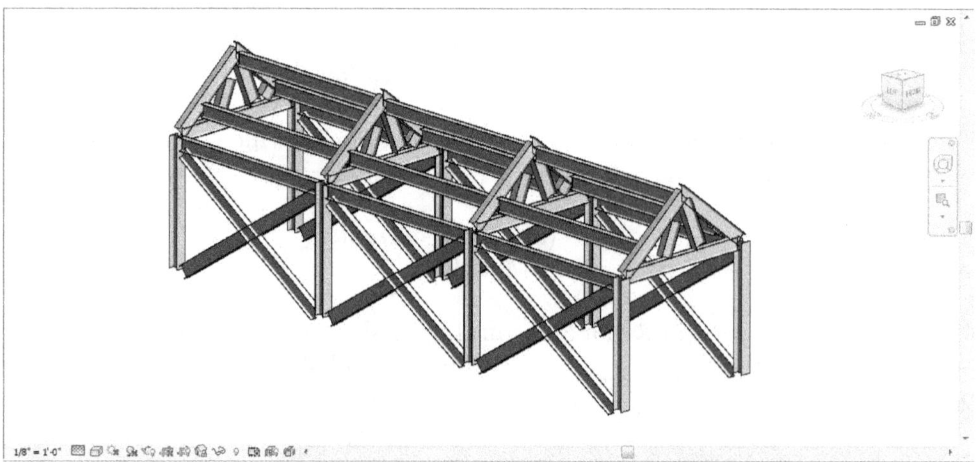

Figure 10-27 *The Steel shade model in Revit window*

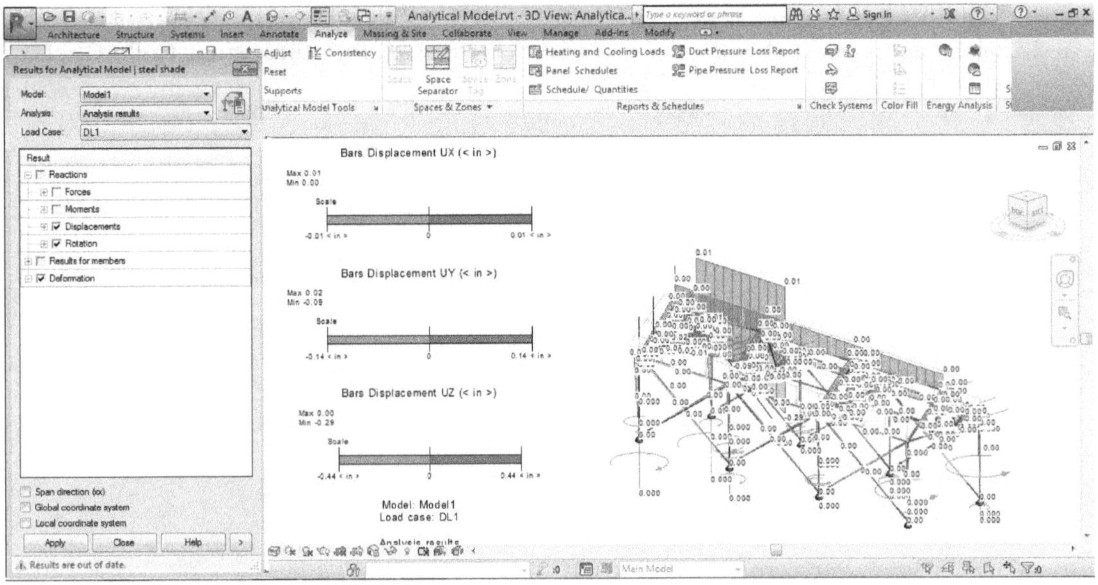

*Figure 10-28 Viewing results by using the **Results Explorer** tool*

Answers to Self-Evaluation Test

1. a, **2.** Job Preferences, **3.** Send only Current Selection, **4.** Reinforcement Code Check, **5.** Robot Structural Analysis Link, **6.** T, **7.** F

Student Project

Student Project — Shopping Complex

In this project you need to create a BIM model of a shopping complex. In this model, you will create a section view, a graphical column schedule, a structural column schedule, a structural foundation takeoff schedule, and a sheet. Refer to Figures 1 through 12 to complete the project. Use the following project parameters and specifications for completing the project:

1. Structural Columns:
   - For Imperial
     - Type: **Concrete-Rectangular-Column**
       Dimension: 24" x 30"
     - Type: **W-Wide Flange-Column**
       Dimension: W10x33
   - For Metric
     - Type: **M_Concrete-Rectangular-Column**
       Dimension: 600 x 750mm
     - Type: **M_W-Wide Flange-Column**
       Dimension: W840x251mm
2. Structural Framing:
   - For Imperial
     - Type: **Concrete-Rectangular**
       Dimension: 24" x 32"
   - For Metric
     - Type: **M_Concrete-Rectangular**
       Dimension: 400x800mm
3. Structural Foundation:
   - For Imperial
     - Type: **Footing-Rectangular**
       Dimension: 72" x 48" x 18"
     - Type: **Foundation Slab**
       Dimension: 6" Foundation Slab
   - For Metric
     - Type: **Footing-Rectangular**
       Dimension: 1800 x 1200 x 450
     - Type: **Foundation Slab**
       Dimension: 150mm Foundation Slab
4. Levels:

For Imperial	For Metric
Name - **Foundation**	Name - **Foundation**
Elevation -10' 0"	Elevation -3000mm
Name - **Ground**	Name - **Ground**
Elevation-0'	Elevation -0
Name - **First**	Name - **First**
Elevation 12 0"	Elevation - 3600mm
Name - **Second Floor**	Name - **Second Floor**
Elevation 21' 0"	Elevation - 6300mm
Name - **Third**	Name - **Third**
Elevation 30' 0"	Elevation - 9000mm
Name - **Fourth**	Name - **Fourth**
Elevation - 39'-0"	Elevation - 11700
Name - **Fifth**	Name - **Fifth**
Elevation -48'	Elevation - 14400

Name - **Terrace**
Elevation - 57'

Name - **Terrace**
Elevation - 17100

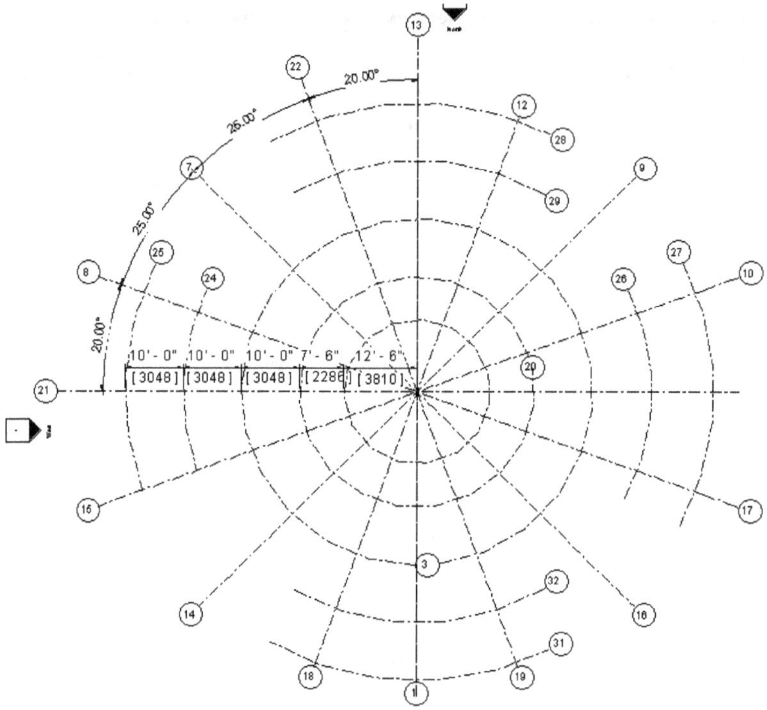

Figure 1 *Grids for the project*

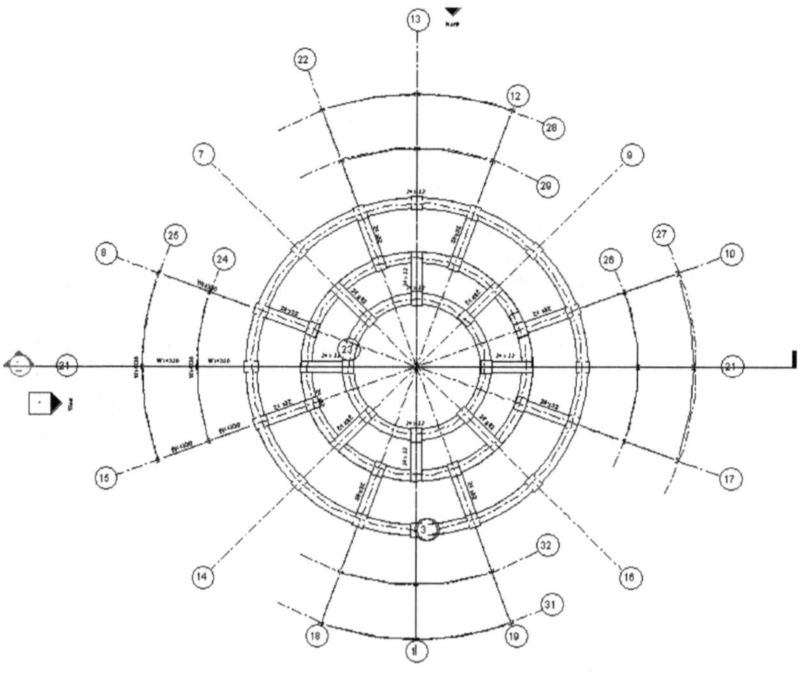

Figure 2 *Structural plan view at* ***First*** *level*

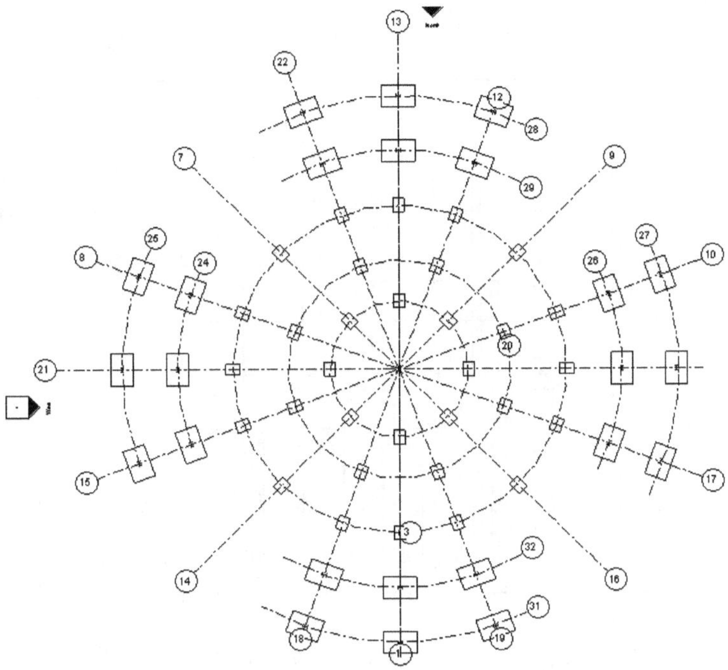

Figure 3 *Structural plan view at* ***Ground*** *level*

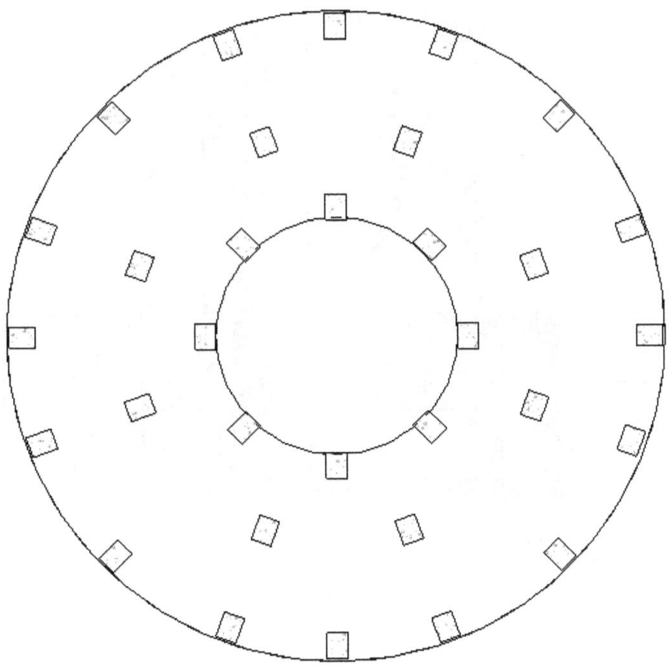

Figure 4 *Structural plan view at* ***Foundation*** *level*

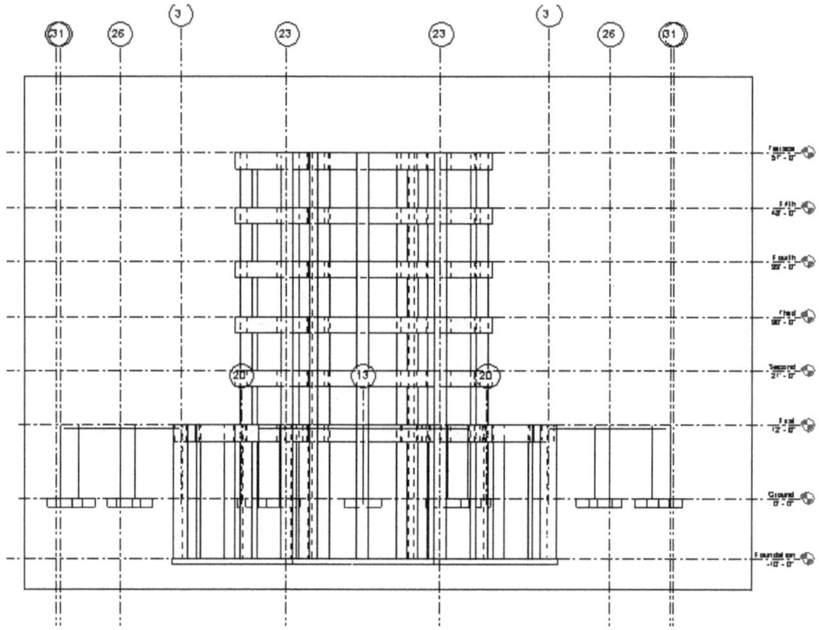

Figure 5 *Sectional view of the project*

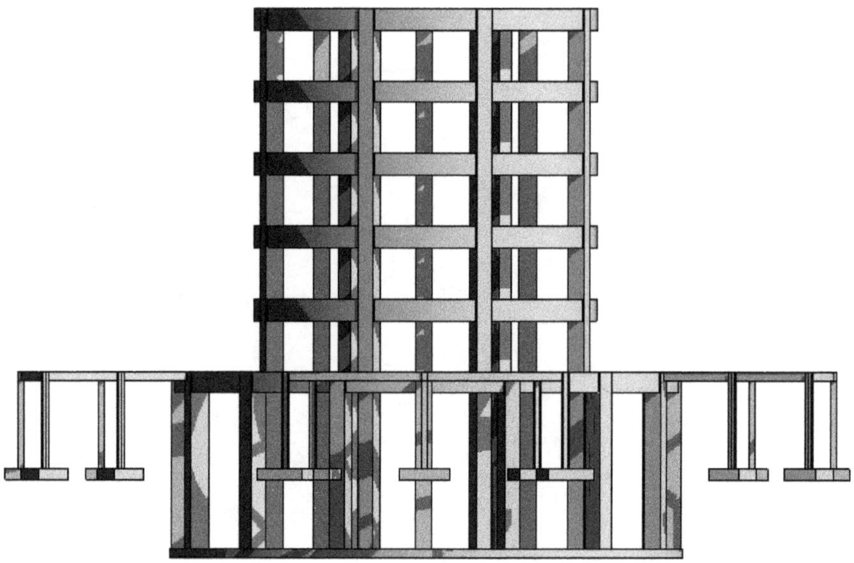

Figure 6 *Front view of the project*

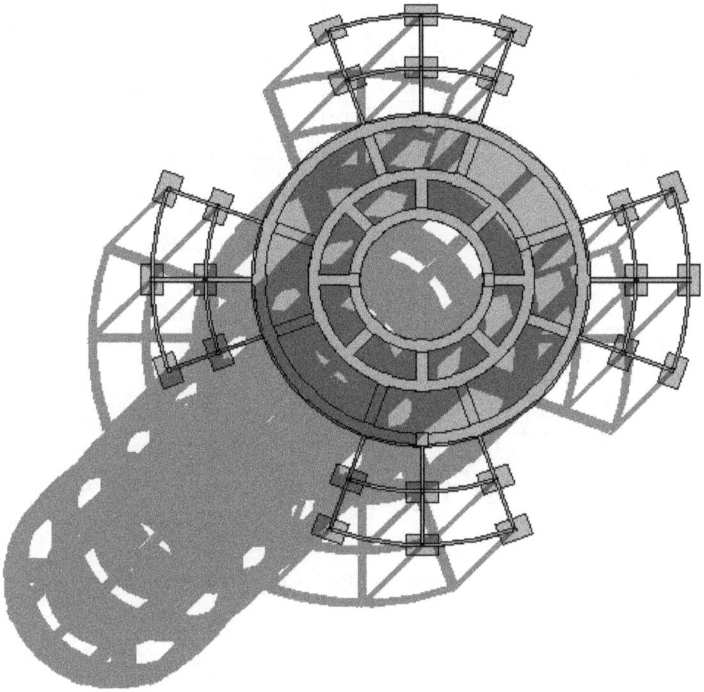

Figure 7 *Top view of the project*

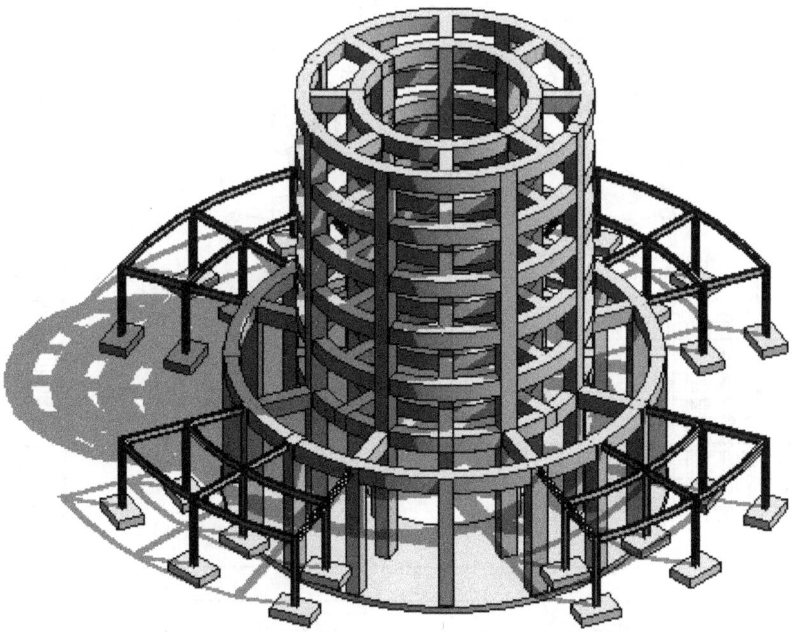

Figure 8 *3D view of the project*

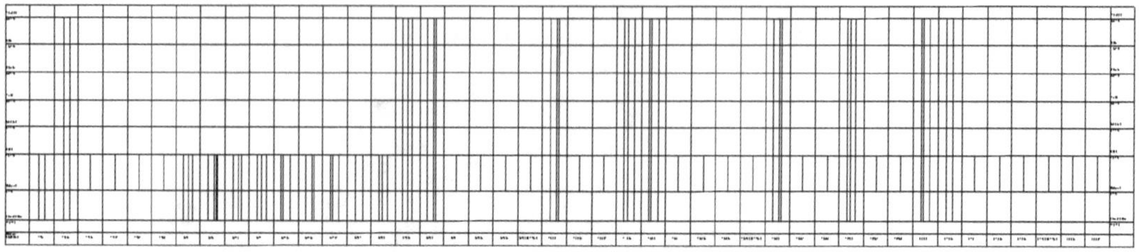

Figure 9 *Graphical Column Schedule of the project*

Structural Column Schedule			
Type	Volume	Count	Length
24 x 30			
24 x 30	7565.00 CF	34	1513' - 0"
W10X33			
W10X33	18.93 CF	24	288' - 0"
Grand total: 5	7583.93 CF		1801' - 0"

Figure 10 *Structural Column Schedule of the project*

Structural Foundation Material Takeoff			
Type	Area	Material: Area	Perimeter
6" Found	2702 SF	2702 SF	270' - 2 1/8"
72" x 48"	39 SF	78 SF	
Grand total: 25			

Figure 11 *Structural Foundation Material Takeoff of the project*

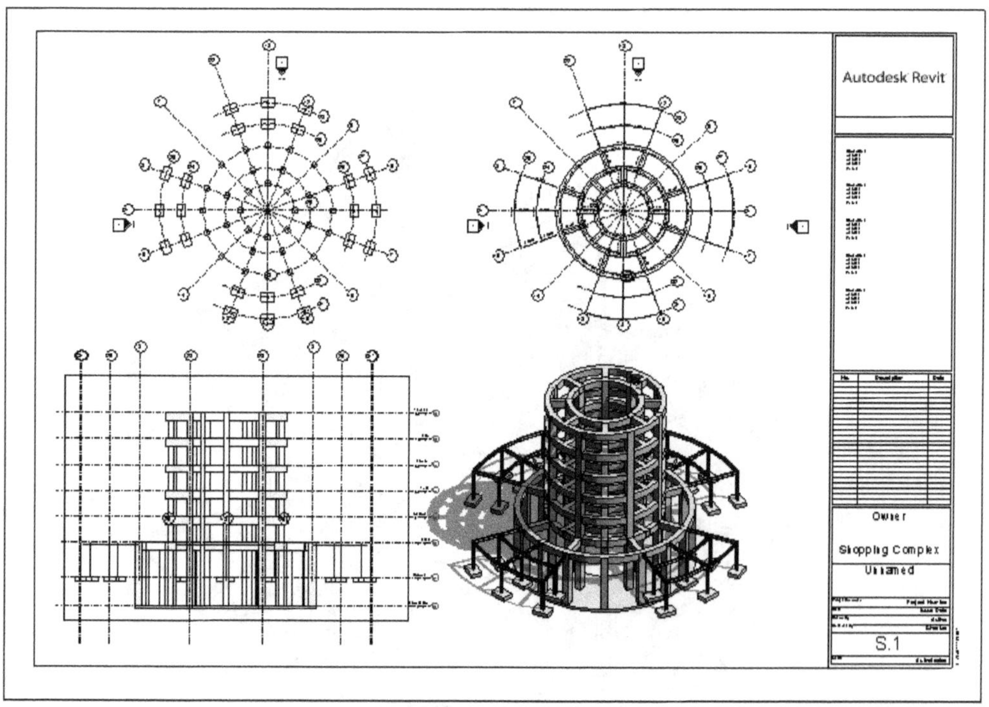

Figure 12 *Sheet view created for the project*

Index

T

U

V

W

X

Other Publications by CADCIM Technologies

The following is the list of some of the publications by CADCIM Technologies. Please visit *www.cadcim.com* for the complete listing.

Autodesk Revit Architecture Textbooks

- Exploring Autodesk Revit 2017 for Architecture, 13th Edition
- Autodesk Revit Architecture 2016 for Architects and Designers, 12th Edition
- Autodesk Revit Architecture 2015 for Architects and Designers, 11th Edition

Autodesk Revit Structure Textbooks

- Exploring Autodesk Revit Structure 2016, 6th Edition
- Exploring Autodesk Revit Structure 2015, 5th Edition

AutoCAD Civil 3D Textbooks

- Exploring AutoCAD Civil 3D 2016, 6th Edition
- Exploring AutoCAD Civil 3D 2015, 5th Edition

AutoCAD Map 3D Textbooks

- Exploring AutoCAD Map 3D 2017, 7th Edition
- Exploring AutoCAD Map 3D 2016, 6th Edition
- Exploring AutoCAD Map 3D 2015, 5th Edition

Autodesk Navisworks Textbooks

- Exploring Autodesk Navisworks 2016, 3rd Edition
- Exploring Autodesk Navisworks 2015

Exploring Oracle Primavera Textbook

Exploring Oracle Primavera P6 v7.0

Exploring AutoCAD Raster Design Textbook

Exploring AutoCAD Raster Design 2016

AutoCAD Textbooks

- AutoCAD 2017: A Problem-Solving Approach, Basic and Intermediate, 23rd Edition
- AutoCAD 2017: A Problem-Solving Approach, 3D and Advanced, 23rd Edition
- AutoCAD 2016: A Problem-Solving Approach, Basic and Intermediate, 22nd Edition
- AutoCAD 2016: A Problem-Solving Approach, 3D and Advanced, 22nd Edition
- AutoCAD 2015: A Problem-Solving Approach, Basic and Intermediate, 21st Edition
- AutoCAD 2015: A Problem-Solving Approach, 3D and Advanced, 21st Edition

Autodesk Inventor Textbooks

- Autodesk Inventor 2017 for Designers, 17th Edition
- Autodesk Inventor 2016 for Designers, 16th Edition
- Autodesk Inventor 2015 for Designers, 15th Edition

AutoCAD MEP Textbooks

- AutoCAD MEP 2016 for Designers, 3rd Edition
- AutoCAD MEP 2015 for Designers

NX Textbooks

- NX 10.0 for Designers, 9th Edition
- NX 9.0 for Designers, 8th Edition

SolidWorks Textbooks

- SOLIDWORKS 2016 for Designers, 14th Edition
- SOLIDWORKS 2015 for Designers, 13th Edition
- SolidWorks 2014: A Tutorial Approach
- Learning SolidWorks 2011: A Project Based Approach

Creo Parametric and Pro/ENGINEER Textbooks

- PTC Creo Parametric 3.0 for Designers, 3rd Edition
- Pro/Engineer Wildfire 5.0 for Designers
- Pro/ENGINEER Wildfire 4.0 for Designers

ANSYS Textbooks

- ANSYS Workbench 14.0: A Tutorial Approach
- ANSYS 11.0 for Designers

Creo Direct Textbook

- Creo Direct 2.0 and Beyond for Designers

Autodesk Alias Textbooks

- Learning Autodesk Alias Design 2016, 5th Edition
- Learning Autodesk Alias Design 2015, 4th Edition

AutoCAD Electrical Textbooks

- AutoCAD Electrical 2017 for Electrical Control Designers, 8th Edition
- AutoCAD Electrical 2016 for Electrical Control Designers, 7th Edition
- AutoCAD Electrical 2015 for Electrical Control Designers, 6th Edition

3ds Max Textbooks

- Autodesk 3ds Max 2017: A Comprehensive Guide, 17th Edition
- Autodesk 3ds Max 2017 for Beginners : A Tutorial Approach
- Autodesk 3ds Max 2016: A Comprehensive Guide, 16th Edition
- Autodesk 3ds Max 2016 for Beginners : A Tutorial Approach
- Autodesk 3ds Max 2016 for Beginners: A Tutorial Approach, 16th Edition
- Autodesk 3ds Max 2015: A Comprehensive Guide, 15th Edition

Autodesk Maya Textbooks

- Autodesk Maya 2016: A Comprehensive Guide, 8th Edition
- Autodesk Maya 2015: A Comprehensive Guide, 7th Edition
- Character Animation: A Tutorial Approach

Fusion Textbooks

- Blackmagic Design Fusion 7 Studio: A Tutorial Approach
- The eyeon Fusion 6.3: A Tutorial Approach

Computer Programming Textbooks

- Introduction to C++ programming
- Learning Oracle 11g
- Learning ASP.NET AJAX
- Learning Java Programming
- Learning Visual Basic.NET 2008
- Introduction to C++ Programming Concepts
- Learning C++ Programming Concepts
- Learning VB.NET Programming Concepts

AutoCAD Textbooks Authored by Prof. Sham Tickoo and Published by Autodesk Press

- AutoCAD: A Problem-Solving Approach: 2013 and Beyond
- AutoCAD 2012: A Problem-Solving Approach
- AutoCAD 2011: A Problem-Solving Approach
- AutoCAD 2010: A Problem-Solving Approach
- Customizing AutoCAD 2010
- AutoCAD 2009: A Problem-Solving Approach

Textbooks Authored by CADCIM Technologies and Published by Other Publishers

3D Studio MAX and VIZ Textbooks

- Learning 3DS Max: A Tutorial Approach, Release 4
  Goodheart-Wilcox Publishers (USA)
- Learning 3D Studio VIZ: A Tutorial Approach
  Goodheart-Wilcox Publishers (USA)

CADCIM Technologies Textbooks Translated in Other Languages

SolidWorks Textbooks

- SolidWorks 2008 for Designers (Serbian Edition)
  Mikro Knjiga Publishing Company, Serbia
- SolidWorks 2006 for Designers (Russian Edition)
  Piter Publishing Press, Russia
- SolidWorks 2006 for Designers (Serbian Edition)
  Mikro Knjiga Publishing Company, Serbia

NX Textbooks

- NX 6 for Designers (Korean Edition)
  Onsolutions, South Korea
- NX 5 for Designers (Korean Edition)
  Onsolutions, South Korea

Pro/ENGINEER Textbooks

- Pro/ENGINEER Wildfire 4.0 for Designers (Korean Edition)
  HongReung Science Publishing Company, South Korea
- Pro/ENGINEER Wildfire 3.0 for Designers (Korean Edition)
  HongReung Science Publishing Company, South Korea

Autodesk 3ds Max Textbook

- 3ds Max 2008: A Comprehensive Guide (Serbian Edition)
  Mikro Knjiga Publishing Company, Serbia

AutoCAD Textbooks

- AutoCAD 2006 (Russian Edition)
  Piter Publishing Press, Russia
- AutoCAD 2005 (Russian Edition)
  Piter Publishing Press, Russia
- AutoCAD 2000 Fondamenti (Italian Edition)

Coming Soon from CADCIM Technologies

- Exploring RISA 3D
- Exploring ETABS
- SOLIDWORKS Simulation 2016 for Designers
- Mold Wizard using NX 10.0

www.ingramcontent.com/pod-product-compliance
Lightning Source LLC
LaVergne TN
LVHW082019200926
848481LV00010B/2890
9781942689355